Triangulated Categories

by

AMNON NEEMAN

PRINCETON UNIVERSITY PRESS

PRINCETON and OXFORD

2001

Published by Princeton University Press, 41 William Street,
Princeton, New Jersey 08540
In the United Kingdom: Princeton University Press,
3 Market Place, Woodstock, Oxfordshire 0X20 1SY

Library of Congress Catalog Card Number 99-068088

ISBN 0-691-08685-0(cloth)
ISBN 0-691-08686-9 (pbk.)

The publisher would like to acknowledge the authors of this
volume for providing the camera-ready copy from which this book was printed

The paper used in this publication meets the minimum requirements of
ANSI/NISO Z39.48-1992 (R 1997) (*Permanence of Paper*)

www.pup.princeton.edu

Printed in the United States of America

1 3 5 7 9 10 8 6 4 2

1 3 5 7 9 10 8 6 4 2
(Pbk.)

Contents

Triangulated Categories

0. Acknowledgements

The author would like to thank Andrew Brooke–Taylor, Daniel Christensen, Pierre Deligne, Jens Franke, Bernhard Keller, Shun–ichi Kimura, Henning Krause, Jack Morava, Saharon Shelah and Vladimir Voevodsky for helpful discussions, and for their valuable contributions to the book.

1. Introduction

Before describing the contents of this book, let me explain its origins. The book began as a joint project between the author and Voevodsky. The idea was to assemble coherently the facts about triangulated categories, that might be relevant in the applications to motives. Since the presumed reader would be interested in applications, Voevodsky suggested that we keep the theory part of the book free of examples. The interested reader should have an example in mind, and read the book to find out what the general theory might have to say about the example. The theory should be presented cleanly, and the examples kept separate.

The division of labor was that I should write the theory, Voevodsky the applications to motives. What then happened was that my part of this joint project mushroomed out of proportion. This book consists just of the formal theory of triangulated categories. In a sequel, we hope to discuss the motivic applications.

The project was initially intended to be purely expository. We meant to cover many topics, but had no new results. This was to be an exposition of the known facts about Brown representability, Bousfield localisation, t–structures and triangulated categories with tensor products. The results should be presented in a unified, clear way, with the exposition accessible to a graduate student wishing to learn the theory. The catch was that the theory should be developed in the generality one would need motivically. The motivic examples, unlike the classical ones, are not compactly generated triangulated categories (whatever this means). The classical literature basically does not treat the situation in the generality required.

My job amounted to modifying the classical arguments, to work in the greater generality. As I started doing this, I quickly came to the conclusion that both the statements and the proofs given classically are very unsatisfactory. The proofs in the literature frequently rely on lifting problems about triangulated categories to problems about more rigid models. Right at the outset I decided that in this book, I will do everything to avoid models. Part of the challenge was to see how much of the theory can be developed without the usual crutch. But there was a far more serious problem. Many of the statements were known only in somewhat special cases, decidedly not including the sort that come up in motives. Thomason once told me that "compact objects are as necessary to this theory as air to breathe". In his words, I was trying to develop the theory in the absence of oxygen.

The book is the result of my work on the subject. It treats a narrower scope of topics than initially planned; we deal basically only with Brown representability and Bousfield localisation. But in some sense we make great progress on the problems. In the process of setting up the theory in the right generality and without lifting to models, we end up with some new and surprising theorems. The book was meant to be an exposition of known results. The way it turned out, it develops a completely new theory. And this theory gives interesting, new applications to very old problems. Now it is time to summarise the mathematical content of the book.

The first two chapters of the book are nothing more than a self–contained exposition of known results. Chapter 1 is the definitions and elementary properties of triangulated categories, while Chapter 2 gives Verdier's construction of the quotient of a triangulated category by a triangulated subcategory. This book was after all intended as a graduate textbook, and therefore assumes little prior knowledge. We assume that the reader is familiar with the language of categories and functors. The reader should know Yoneda's Lemma, the general facts about adjoint functors between categories, units and counits of adjunction, products and coproducts. It is also assumed that the reader has had the equivalent of an elementary course on homological algebra. We assume familiarity with abelian categories, exact sequences, the snake lemma and the 5–lemma. But this is all we assume. In particular, the reader is not assumed to have ever seen the definition of a triangulated category. In practice, since we give no examples, the reader might wish to find one elsewhere, to be able to keep it in mind as an application of the general theory. One place to find a relatively simple, concrete exposition of one example, is the first chapter of Hartshorne's book [**19**]. This first chapter develops the derived category. Note that, since we wish to study mostly triangulated categories closed under all small coproducts, the derived category of most interest is the *unbounded* derived category. In [**19**], this is the derived category that receives probably the least attention. There is another short account of the derived category in Chapter 10, pages 369–415 of Weibel's [**37**]. There are, of course, many other excellent accounts. But they tend to be longer. Anyway, if the reader is willing to forget the examples, begin with the axioms, and see what can be proved using them, then this book is relatively self–contained.

Chapters 1 and 2 are an account of the very classical theory. There are some expository innovations in these two chapters, but otherwise little new. If the reader wants to be able to compare the treatment given here with the older treatments, at the end of each chapter there is a historical summary. In the body of the chapters, I rarely give references to older works. The historical surveys at the end of each chapter contain references to other expositions. They also try to point out what, if anything, distinguishes the exposition given here from older ones.

Starting with Chapter 3, little of what is in the book may be found in the literature. For the reader who has some familiarity with triangulated categories, it seems only fair that the introduction summarise what, if anything, he or she can expect to find in this book which they did not already know. It is inevitable, however, that such an explanation will demand from the reader some prior knowledge of triangulated categories. The graduate student, who has never before met triangulated categories, is advised to skip the remainder of the introduction and proceed to Chapter 1. After reading Chapters 1 and 2, the rest of the introduction will make a lot more sense.

Let me begin with the concrete. Few people have a stomach strong enough for great generalities. Sweeping, general theorems about arbitrary 2–categories tend to leave us cold. We become impressed only when we learn that these theorems teach us something new. Preferably something new about an old, concrete example that we know and love. Before I state the results in the book in great generality, let me tell the reader what we may conclude from them about a special case. Let us look at the special case, where $\mathcal{T}$ is the homotopy category of spectra.

Let $\mathcal{T}$ be the homotopy category of spectra. Let E be a spectrum (ie. an object of $\mathcal{T}$). Following Bousfield, the full subcategory $\mathcal{T}_E \subset \mathcal{T}$, whose objects are called the E–acyclic spectra, is defined by

$$\mathrm{Ob}\left(\mathcal{T}_E\right) \quad = \quad \{x \in \mathrm{Ob}(\mathcal{T}) \mid x \wedge E = 0\}.$$

The full subcategory ${}^{\perp}\mathcal{T}_E \subset \mathcal{T}$, whose objects are called the E–local spectra, is defined by

$$\mathrm{Ob}\left({}^{\perp}\mathcal{T}_E\right) \quad = \quad \{y \in \mathrm{Ob}(\mathcal{T}) \mid \forall x \in \mathrm{Ob}\left(\mathcal{T}_E\right),\, \mathcal{T}(x,y) = 0\}.$$

An old theorem of Bousfield (see [**6**]) asserts that one can localise spectra with respect to any homology theory E. In the notation above, Bousfield's theorem asserts

THEOREM (BOUSFIELD, 1979). *Let E be a spectrum, that is E is an object of $\mathcal{T}$. Let $\mathcal{T}_E \subset \mathcal{T}$ and ${}^{\perp}\mathcal{T}_E \subset \mathcal{T}$ be defined as above. Suppose x is an object of $\mathcal{T}$. Then there is a triangle in $\mathcal{T}$*

$$x_E \longrightarrow x \longrightarrow {}^{\perp}x_E \longrightarrow \Sigma x_E,$$

with $x_E \in \mathcal{T}_E$, and ${}^{\perp}x_E \in {}^{\perp}\mathcal{T}_E$.

Bousfield's theorem has been known for a long time. What this book has to add, are surprising structure theorems about the categories $\mathcal{T}_E$ and ${}^{\perp}\mathcal{T}_E$. We prove the following representability theorems

THEOREM (NEW, THIS BOOK). *Let E be a spectrum. Let $\mathcal{T}_E \subset \mathcal{T}$ and ${}^\perp\mathcal{T}_E \subset \mathcal{T}$ be defined as above. The representable functors*

$$\mathcal{T}_E(-,h) \qquad \text{and} \qquad {}^\perp\mathcal{T}_E(-,h)$$

can be characterised as the homological functors $H : \mathcal{T}_E^{op} \longrightarrow \mathcal{A}b$ (respectively $H : {}^\perp\mathcal{T}_E^{op} \longrightarrow \mathcal{A}b$) taking coproducts in $\mathcal{T}_E$ (respectively ${}^\perp\mathcal{T}_E$) to products in $\mathcal{A}b$. The representable functors

$$\mathcal{T}_E(h,-)$$

can be characterised as the homological functors $H : \mathcal{T}_E \longrightarrow \mathcal{A}b$, taking products to products.

Proof: The characterisation of the functors $\mathcal{T}_E(-,h)$ and ${}^\perp\mathcal{T}_E(-,h)$ may be found in Theorem D.1.12. More precisely, for $\mathcal{T}_E(-,h)$ see D.1.12.1, while the statement for ${}^\perp\mathcal{T}_E(-,h)$ is contained in D.1.12.5. The characterisation of the functors $\mathcal{T}_E(h,-)$ may be found in Lemma D.1.14. □

Representability theorems are central to this subject. What we have achieved here, is to extend Brown's old representability theorem of [**7**]. Brown proved that the functors $\mathcal{T}(-,h)$ can be characterised as the homological functors $\mathcal{T}^{op} \longrightarrow \mathcal{A}b$ taking coproducts to products. We have generalised this to $\mathcal{T}_E$, $\mathcal{T}_E^{op}$ and ${}^\perp\mathcal{T}_E$, but unfortunately not to ${}^\perp\mathcal{T}_E^{op}$. We do not know, whether the the functors ${}^\perp\mathcal{T}_E(h,-)$ can be characterised as the homological functors taking products to products.

Another amusing fact we learn in this book, is that the categories $\mathcal{T}_E$ and ${}^\perp\mathcal{T}_E$ are not equivalent to $\mathcal{T}^{op}$. There are, in fact, many more amusing facts we prove. Let us give one more. We begin with a definition.

DEFINITION. *Let α be a regular cardinal. A morphism $f : x \longrightarrow y$ in $\mathcal{T}$ is called an α–*phantom map *if, for any spectrum s with fewer than α cells, any composite*

$$s \longrightarrow x \xrightarrow{\ f\ } y$$

vanishes.

With this definition, we are ready to state another fun fact that we learn in this book.

THEOREM (NEW, THIS BOOK). *Let $\alpha > \aleph_0$ be a regular cardinal. There is an object $z \in \mathcal{T}$, which admits no maximal α–phantom map $y \longrightarrow z$. That is, given any α–phantom map $y \longrightarrow z$, there is at least one α–phantom map $x \longrightarrow z$ not factoring as*

$$x \longrightarrow y \longrightarrow z.$$

Proof: The proof of this fact follows from Proposition D.2.5, coupled with Lemma 8.5.20. □

REMARK 1.1. It should be noted that the above is surprising. If $\alpha = \aleph_0$, the α–phantom maps are the maps vanishing on all finite spectra. These are very classical, and have been extensively studied in the literature. Usually, they go by the name *phantom maps;* the reference to $\alpha = \aleph_0$ is new to this book, where we study the natural large–cardinal generalisation. From the work of Christensen and Strickland [**9**], we know that every object $z \in \mathcal{T}$ admits a maximal $\aleph_0$–phantom map $y \longrightarrow z$. There is an $\aleph_0$–phantom map $y \longrightarrow z$, so that all other $\aleph_0$–phantom maps $x \longrightarrow z$ factor as

$$x \longrightarrow y \longrightarrow z.$$

What is quite surprising is that this is very special to $\alpha = \aleph_0$.

So far, we have given the reader a sampling of facts about the homotopy category of spectra, which follow from the more general results of this book. I could give more; but it is perhaps more instructive to indicate the broad approach.

The idea of this book is to study a certain class of triangulated categories, the *well–generated triangulated categories.* And the thrust is to prove great facts about them. We will show, among many other things

THEOREM 1.2. *The following facts are true:*

1.2.1. *Let $\mathcal{T}$ be the homotopy category of spectra. Let E be an object of $\mathcal{T}$. Then both the category $\mathcal{T}_E$ and the category ${}^{\perp}\mathcal{T}_E$ are well–generated triangulated categories.*

1.2.2. *Suppose $\mathcal{T}$ is a well–generated triangulated category. The representable functors $\mathcal{T}(-,h)$ can be characterised as the homological functors $H : \mathcal{T}^{op} \longrightarrow \mathcal{A}b$, taking coproducts in $\mathcal{T}$ to products in $\mathcal{A}b$.*

In other words, we will prove a vast generalisation of Brown's representability theorem. Not only does it generalise to $\mathcal{T}_E$ and ${}^{\perp}\mathcal{T}_E$, but to very many other categories as well. The categories that typically come up in the study of motives are examples. And now it is probably time to tell the reader what a well–generated triangulated category is. It turns out to be quite a deep fact that this structure even makes sense.

Let $\mathcal{T}$ be a triangulated category. We remind the reader: a *homological functor* $\mathcal{T} \longrightarrow \mathcal{A}$ is a functor from $\mathcal{T}$ to an abelian category $\mathcal{A}$, taking triangles to long exact sequences. We can consider the collection of all homological functors $\mathcal{T} \longrightarrow \mathcal{A}$. An old theorem of Freyd's (see [**13**]) asserts that

THEOREM (FREYD, 1966). *Among all the homological functors $\mathcal{T} \longrightarrow \mathcal{A}$ there is a universal one. There is an abelian category $A(\mathcal{T})$ and a homological functor $\mathcal{T} \longrightarrow A(\mathcal{T})$, so that any other homological functor $\mathcal{T} \longrightarrow \mathcal{A}$ factors as*

$$\mathcal{T} \longrightarrow A(\mathcal{T}) \xrightarrow{\exists!} \mathcal{A}$$

where the exact functor $A(\mathcal{T}) \longrightarrow \mathcal{A}$ is unique up to canonical equivalence. Any natural tranformation of homological functors $\mathcal{T} \longrightarrow \mathcal{A}$ factors uniquely through a natural transformation of the (unique) exact functors $A(\mathcal{T}) \longrightarrow \mathcal{A}$.

This theorem tells us that, associated naturally to every triangulated category $\mathcal{T}$, there is an abelian category $A(\mathcal{T})$. The association is easily seen to be functorial. It takes the 2–category of triangulated categories and triangulated functors to the 2–category of abelian categories and exact functors, and is a lax functor.

One can wonder about the homological algebra of the abelian category $A(\mathcal{T})$. Freyd proves also

PROPOSITION (FREYD, 1966). *Let $\mathcal{T}$ be a triangulated category. The abelian category $A(\mathcal{T})$ of the previous theorem has enough projectives and enough injectives. In fact, the projectives and injectives in $A(\mathcal{T})$ are the same. An object $a \in A(\mathcal{T})$ is projective (equivalently, injective) if and only if there exists an object $b \in A(\mathcal{T})$, so that*

$$a \oplus b \in \mathcal{T} \subset A(\mathcal{T}).$$

That is, a is a direct summand of an object $a \oplus b$, and $a \oplus b$ is in the image of the universal homological functor $\mathcal{T} \longrightarrow A(\mathcal{T})$. This universal homological functor happens to be a fully faithful embedding; hence I allow myself to write $\mathcal{T} \subset A(\mathcal{T})$.

It turns out to be easy to deduce the following corollary:

COROLLARY 1.3. *Let $F : \mathcal{S} \longrightarrow \mathcal{T}$ be a triangulated functor. If F has a right adjoint $G : \mathcal{T} \longrightarrow \mathcal{S}$, then G is also triangulated, and $A(G) : A(\mathcal{T}) \longrightarrow A(\mathcal{S})$ is right adjoint to $A(F) : A(\mathcal{S}) \longrightarrow A(\mathcal{T})$. But more interesting is the following. If every idempotent in $\mathcal{S}$ splits, then $F : \mathcal{S} \longrightarrow \mathcal{T}$ has a right adjoint if and only if $A(F) : A(\mathcal{S}) \longrightarrow A(\mathcal{T})$ does. That is, if $A(F) : A(\mathcal{S}) \longrightarrow A(\mathcal{T})$ has a right adjoint $\tilde{G} : A(\mathcal{T}) \longrightarrow A(\mathcal{S})$, then $F : \mathcal{S} \longrightarrow \mathcal{T}$ has a right adjoint $G : \mathcal{T} \longrightarrow \mathcal{S}$, and of course $A(G)$ is naturally isomorphic to $\tilde{G}$.*

Proof: Lemma 5.3.6 shows that the adjoint of a triangulated functor is triangulated, Lemma 5.3.8 proves that if G is right adjoint to F then $A(G)$ is right adjoint to $A(F)$, while Proposition 5.3.9 establishes that if $A(F)$ has a right adjoint $\tilde{G}$, then F has a right adjoint G. □

REMARK 1.4. It turns out that many of the deepest and most interesting questions about triangulated categories involve the existence of adjoints. This suggests that Corollary 1.3 should be great. It tells us that finding adjoints to triangulated functors between triangulated categories, a difficult problem, is equivalent to finding adjoints to exact functors between abelian categories. We feel much more comfortable with abelian categories, so the Corollary should make us very happy.

The problem is that the abelian categories that arise are terrible. For example, let $\mathcal{T}$ be the category $D(\mathbb{Z})$, the derived category of the category of all abelian groups. Then the abelian group $\mathbb{Z}$ can be viewed as an object of $D(\mathbb{Z})$; it is the complex which is the group $\mathbb{Z}$ in dimension 0, and zero elsewhere. The universal homological functor

$$D(\mathbb{Z}) \longrightarrow A\big(D(\mathbb{Z})\big)$$

takes $\mathbb{Z}$ to an object of $A\big(D(\mathbb{Z})\big)$. I assert that this object, in the abelian category $A\big(D(\mathbb{Z})\big)$, has a proper class of subobjects. The collection of subobjects of $\mathbb{Z} \in A\big(D(\mathbb{Z})\big)$ is not a set; it is genuinely only a class. The proof may be found in Appendix C.

In the light of Remark 1.4, it is natural to look for approximations to the abelian category $A(\mathcal{T})$. It seems reasonable to try to find other abelian categories $\mathcal{A}$, together with exact functors $A(\mathcal{T}) \longrightarrow \mathcal{A}$, which are "reasonable" approximations. It is natural to want the objects of $\mathcal{A}$ to only have sets (not classes) of subobjects. But otherwise it would be nice if $\mathcal{A}$ is as close as possible to the universal abelian category $A(\mathcal{T})$.

The universal property of $A(\mathcal{T})$ asserts that exact functors $A(\mathcal{T}) \longrightarrow \mathcal{A}$ are in 1–1 correspondence with homological functors $\mathcal{T} \longrightarrow \mathcal{A}$. We therefore want to find reasonable homological functors $\mathcal{T} \longrightarrow \mathcal{A}$, for suitable $\mathcal{A}$.

Let $\mathcal{T}$ be a triangulated category. It is said to satisfy [TR5] if the coproduct of any small set of objects in $\mathcal{T}$ exists in $\mathcal{T}$. If the dual category $\mathcal{T}^{op}$ satisfies [TR5], then $\mathcal{T}$ is said to satisfy [TR5*]. Let α be an infinite cardinal. Let $\mathcal{T}$ be a triangulated category satisfying [TR5] and $\mathcal{S}$ a triangulated subcategory. We say that $\mathcal{S}$ is *α–localising* if any coproduct of fewer than α objects of $\mathcal{S}$ lies in $\mathcal{S}$. We call $\mathcal{S} \subset \mathcal{T}$ *localising* if it is α–localising for every infinite cardinal α.

Given a triangulated category $\mathcal{T}$ satisfying [TR5], and an α–localising subcategory $\mathcal{S} \subset \mathcal{T}$, there is a God–given abelian category one can construct out of $\mathcal{S}$, and a homological functor

$$\mathcal{T} \longrightarrow \mathcal{E}x\big(\mathcal{S}^{op}, \mathcal{A}b\big).$$

Now it is time to define these.

The category $\mathcal{E}x\big(\mathcal{S}^{op}, \mathcal{A}b\big)$ is the abelian category of all functors $\mathcal{S}^{op} \longrightarrow \mathcal{A}b$ which preserve products of fewer than α objects. Recall that $\mathcal{S}$ is α–localising. Given fewer than α objects in $\mathcal{S}$, their coproduct exists in $\mathcal{T}$

because $\mathcal{T}$ satisfies [TR5], and is contained in $\mathcal{S}$ because $\mathcal{S}$ is α–localising. That is, the product exists in the dual $\mathcal{S}^{op}$, and we look at functors to abelian groups

$$\mathcal{S}^{op} \longrightarrow \mathcal{A}b$$

preserving all such products. The reader can easily check (see Lemma 6.1.4 for details) that $\mathcal{E}x\big(\mathcal{S}^{op}, \mathcal{A}b\big)$ is an abelian category. Furthermore, there is a homological functor

$$\mathcal{T} \longrightarrow \mathcal{E}x\big(\mathcal{S}^{op}, \mathcal{A}b\big).$$

It is the functor that takes an object $t \in \mathcal{T}$ to the representable functor $\mathcal{T}(-, t)$, restricted to $\mathcal{S} \subset \mathcal{T}$. We denote this restriction

$$\mathcal{T}(-, t)|_{\mathcal{S}}.$$

This construction depends on the choice of an infinite cardinal α, and an α–localising subcategory $\mathcal{S} \subset \mathcal{T}$. In what follows, it is convenient to assume that the cardinal α is regular. That is, α is not the sum of fewer than α cardinals, all smaller than α.

Starting with any regular cardinal α and any α–localising subcategory $\mathcal{S} \subset \mathcal{T}$, we have constructed a homological functor

$$\mathcal{T} \longrightarrow \mathcal{E}x\big(\mathcal{S}^{op}, \mathcal{A}b\big).$$

It factors uniquely through the universal homological functor, to give an exact functor

$$A(\mathcal{T}) \xrightarrow{\ \pi\ } \mathcal{E}x\big(\mathcal{S}^{op}, \mathcal{A}b\big).$$

It is very easy to show that the functor π has a left adjoint; we denote the left adjoint $F : \mathcal{E}x\big(\mathcal{S}^{op}, \mathcal{A}b\big) \longrightarrow A(\mathcal{T})$. We deduce a unit of adjunction

$$\eta : 1 \longrightarrow \pi F.$$

We prove

PROPOSITION 1.5. *Suppose $\eta : 1 \longrightarrow \pi F$ is the unit of adjunction above. If the functor $\pi : A(\mathcal{T}) \longrightarrow \mathcal{E}x\big(\mathcal{S}^{op}, \mathcal{A}b\big)$ preserves coproducts, then η is an isomorphism.*

Proof: See Poposition 6.5.3. □

By Gabriel's theory of localisations of abelian categories (see Appendix A), the unit of adjunction $\eta : 1 \longrightarrow \pi F$ is an isomorphism if and only if $\mathcal{E}x\big(\mathcal{S}^{op}, \mathcal{A}b\big)$ is a quotient of $A(\mathcal{T})$, with π being the quotient map. It is therefore natural to want to study the $\mathcal{S} \subset \mathcal{T}$ for which this happens. By Proposition 1.5, a particularly interesting case is when the map

$$A(\mathcal{T}) \xrightarrow{\ \pi\ } \mathcal{E}x\big(\mathcal{S}^{op}, \mathcal{A}b\big)$$

preserves coproducts. It turns out that, for a given regular cardinal α, this depends on a choice of the α–localising subcategory $\mathcal{S} \subset \mathcal{T}$. We proceed now to describe the complete answer.

DEFINITION 1.6. *Let α be a regular cardinal. Let $\mathcal{T}$ be a triangulated category satisfying [TR5]. An object $t \in \mathcal{T}$ is called* α–small *if any morphism from t to a coproduct*

$$t \longrightarrow \coprod_{\lambda \in \Lambda} X_\lambda$$

factors through a coproduct of fewer than α objects. There is a subset $\Lambda' \subset \Lambda$, Λ' of cardinality $< \alpha$, and a factorisation

$$t \longrightarrow \coprod_{\lambda \in \Lambda'} X_\lambda \quad \subset \quad \coprod_{\lambda \in \Lambda} X_\lambda .$$

The full subcategory of all α–small objects in $\mathcal{T}$ is denoted $\mathcal{T}^{(\alpha)}$. Next we need

DEFINITION 1.7. *Let α be a regular cardinal. Let $\mathcal{T}$ be a triangulated category satisfying [TR5]. A class T, containing 0, of objects in $\mathcal{T}$ is called* α–perfect *if, for any collection $\{X_\lambda, \lambda \in \Lambda\}$ of fewer than α objects of $\mathcal{T}$, any object $t \in T$, and any map*

$$t \longrightarrow \coprod_{\lambda \in \Lambda} X_\lambda$$

there is a factorisation

$$t \longrightarrow \coprod_{\lambda \in \Lambda} t_\lambda \xrightarrow{\coprod_{\lambda \in \Lambda} f_\lambda} \coprod_{\lambda \in \Lambda} X_\lambda$$

with t_λ in T. Furthermore, if the composite

$$t \longrightarrow \coprod_{\lambda \in \Lambda} t_\lambda \xrightarrow{\coprod_{\lambda \in \Lambda} f_\lambda} \coprod_{\lambda \in \Lambda} X_\lambda$$

vanishes, then each of the maps

$$t_\lambda \xrightarrow{f_\lambda} X_\lambda$$

factors as

$$t_\lambda \longrightarrow u_\lambda \longrightarrow X_\lambda$$

with $u_\lambda \in T$, so that the composite

$$t \longrightarrow \coprod_{\lambda \in \Lambda} t_\lambda \longrightarrow \coprod_{\lambda \in \Lambda} u_\lambda$$

already vanishes.

With these two definitions, we have a theorem

THEOREM 1.8. *Let α be a regular cardinal, $\mathcal{T}$ a triangulated category satisfying [TR5]. Let $\mathcal{S} \subset \mathcal{T}$ be an α–localising subcategory. The natural functor*

$$A(\mathcal{T}) \xrightarrow{\ \pi\ } \mathcal{E}x\big(\mathcal{S}^{op}, \mathcal{A}b\big)$$

preserves coproducts if and only if

1.8.1. *The objects of $\mathcal{S}$ are all α–small; that is $\mathcal{S} \subset \mathcal{T}^{(\alpha)}$.*

1.8.2. *The class of all objects in $\mathcal{S}$ is α–perfect, as in Definition 1.7.*

Proof: Lemma 6.2.5. In the statement of Lemma 6.2.5, we only assert the sufficiency; if $\mathcal{S}$ satisfies 1.8.1 and 1.8.2, then π preserves coproducts. But the proof immediately also gives the converse. □

It is therefore of interest to study α–localising subcategories $\mathcal{S} \subset \mathcal{T}^{(\alpha)}$, whose collection of objects form an α–perfect class. The remarkable fact is that there is a biggest one. For any regular cardinal α, we can define a canonical, God–given α–localising subcategory $\mathcal{T}^\alpha$. It is given by

DEFINITION 1.9. *The full subcategory $\mathcal{T}^\alpha \subset \mathcal{T}$, of all α–compact objects in $\mathcal{T}$, is defined as follows. The class of objects in $Ob(\mathcal{T}^\alpha)$ is the unique maximal α–perfect class in $\mathcal{T}^{(\alpha)}$. It of course needs to be shown that such a maximal α–perfect class exists. It is also relevant to know that $\mathcal{T}^\alpha \subset \mathcal{T}$ is an α–localising triangulated subcategory. All this is proved in Chapters 3 and 4.*

The categories $\mathcal{T}^\alpha$ are, in a certain sense, the optimal choices for $\mathcal{S}$. For each α, we have an exact functor

$$A(\mathcal{T}) \xrightarrow{\ \pi\ } \mathcal{E}x\Big(\{\mathcal{T}^\alpha\}^{op}, \mathcal{A}b\Big)$$

preserving coproducts and products, having a left adjoint F, and so that $\mathcal{E}x\Big(\{\mathcal{T}^\alpha\}^{op}, \mathcal{A}b\Big)$ is the Gabriel quotient of $A(\mathcal{T})$ by the Serre subcategory of all objects on which the functor π vanishes. It is natural to study the subcategories $\mathcal{T}^\alpha \subset \mathcal{T}$.

EXAMPLE 1.10. The previous paragraphs may be a little confusing. But the upshot is the following. Suppose we are given a triangulated category $\mathcal{T}$ closed under coproducts, and a regular cardinal α. There is some mysterious, canonical way to define an α–localising triangulated subcategory, denoted $\mathcal{T}^\alpha \subset \mathcal{T}$. The reader might naturally be curious to know what $\mathcal{T}^\alpha$ is, in some simple examples.

If $\mathcal{T}$ is the homotopy category of spectra, then $\mathcal{T}^\alpha$ is the full subcategory of spectra with fewer than α cells. If $\mathcal{T}$ is the derived category of an associative ring R, then the objects of $\mathcal{T}^\alpha$ turn out to be chain complexes of projective R–modules, whose total rank(=rank of the sum of all the modules) is $< \alpha$. If $\alpha = \aleph_0$, then $\mathcal{T}^\alpha = \mathcal{T}^{\aleph_0}$ is the subcategory of compact objects in $\mathcal{T}$, and its study is very classical. But even for $\alpha > \aleph_0$, we are dealing with a fairly natural subcategory.

Now we return from the examples to the general theory. The first, trivial property of $\mathcal{T}^\alpha \subset \mathcal{T}$ is

LEMMA 1.11. *If $\alpha < \beta$ are regular cardinals, then $\mathcal{T}^\alpha \subset \mathcal{T}^\beta$.*

Proof: Lemma 4.2.3. □

To get further, we need the notion of *generation.*

DEFINITION 1.12. *Let α be a regular cardinal. Let S be a class of objects of $\mathcal{T}$. Then $\langle S\rangle^\alpha$ stands for the smallest α–localising subcategory of $\mathcal{T}$ containing S. The symbol $\langle S\rangle$ stands for the smallest localising subcategory containing S. That is,*

$$\langle S\rangle \quad = \quad \bigcup_\alpha \langle S\rangle^\alpha.$$

We say that S *generates* $\mathcal{T}$ if $\mathcal{T} = \langle S\rangle$. With this definition, we are ready for Thomason's localisation theorem. The theorem is essentially the statement that the subcategories $\mathcal{T}^\alpha$ behave well with respect to quotient maps. We begin with a statement involving only one triangulated category $\mathcal{T}$.

LEMMA 1.13. *Suppose α is a regular cardinal, and $\mathcal{T}$ is a triangulated category satisfying [TR5]. Suppose further that $\mathcal{T}^\alpha$ generates $\mathcal{T}$; that is,*

$$\langle \mathcal{T}^\alpha\rangle \quad = \quad \mathcal{T}.$$

Then for any regular cardinal $\beta > \alpha$,

$$\langle \mathcal{T}^\alpha\rangle^\beta \quad = \quad \mathcal{T}^\beta.$$

Note that the conclusion of the theorem might be confusing. Since we are assuming $\langle \mathcal{T}^\alpha\rangle = \mathcal{T}$, then surely it should be a tautology that

$$\langle \mathcal{T}^\alpha\rangle^\beta \quad = \quad \mathcal{T}^\beta.$$

Just replace the $\langle \mathcal{T}^\alpha\rangle$ by $\mathcal{T}$. But this misses the point that there are two ways to read the symbol $\langle \mathcal{T}^\alpha\rangle^\beta$. One is to note that, for any triangulated category $\mathcal{S}$ closed under coproducts, there is a canonical way to define a subcategory $\mathcal{S}^\beta$; and then we apply the construction to $\mathcal{S} = \langle \mathcal{T}^\alpha\rangle$. But given a collection of objects $S \in \mathcal{T}$, there is also a canonical way to define a subcategory $\langle S\rangle^\beta$, the β–localising subcategory generated by S. And we could do this construction to $S = \mathcal{T}^\alpha$. The assertion of the Lemma is that under reasonable conditions, these two agree, and the notation leads to no confusion.

Proof: Lemma 4.4.5. □

This lemma is actually of great practical use. It says that once we have computed $\mathcal{T}^\alpha$, then as long as $\mathcal{T}^\alpha$ generates, we know all $\mathcal{T}^\beta$ for $\beta > \alpha$. They are just the closure of $\mathcal{T}^\alpha$ with respect to coproducts of fewer than β objects, and triangles. Call this statement zero of Thomason's localisation theorem. The rest of the theorem concerns the situation of a Verdier quotient.

THEOREM 1.14. *Let $\mathcal{S}$ be a triangulated category satisfying [TR5], $\mathcal{R} \subset \mathcal{S}$ a localising subcategory. Write $\mathcal{T}$ for the Verdier quotient $\mathcal{S}/\mathcal{R}$.*

Suppose there is a regular cardinal α, a class of objects $S \subset \mathcal{S}^\alpha$ and another class of objects $R \subset \mathcal{R} \cap \mathcal{S}^\alpha$, so that

$$\mathcal{R} = \langle R \rangle \qquad \text{and} \qquad \mathcal{S} = \langle S \rangle.$$

Then for any regular $\beta \geq \alpha$,

$$\langle R \rangle^\beta = \mathcal{R}^\beta = \mathcal{R} \cap \mathcal{S}^\beta,$$

$$\langle S \rangle^\beta = \mathcal{S}^\beta.$$

The natural map

$$\mathcal{S}^\beta/\mathcal{R}^\beta \longrightarrow \mathcal{T}$$

factors as

$$\mathcal{S}^\beta/\mathcal{R}^\beta \longrightarrow \mathcal{T}^\beta \subset \mathcal{T},$$

and the functor

$$\mathcal{S}^\beta/\mathcal{R}^\beta \longrightarrow \mathcal{T}^\beta$$

is fully faithful. If $\beta > \aleph_0$, the functor

$$\mathcal{S}^\beta/\mathcal{R}^\beta \longrightarrow \mathcal{T}^\beta$$

is an equivalence of categories. If $\beta = \aleph_0$, then every object of $\mathcal{T}^\beta$ is a direct summand of an object in $\mathcal{S}^\beta/\mathcal{R}^\beta$.

Proof: Theorem 4.4.9. □

Since I have been telling the reader that this theory is quite new, the reader may well wonder why Thomason's name is attached to it. Thomason proved the special case where $\alpha = \beta = \aleph_0$, and $\mathcal{S}$ is the derived category of the category of quasi–coherent sheaves on a quasi–compact, separated scheme. (Actually, he studies the slightly more general situation of a semi–separated scheme. A scheme is semi–separated if it has an open cover by affine open subsets with affine intersections.) For details, the reader is referred to Thomason's [**34**]. In all fairness to Thomason, his wonderful observation was that this fact had great applications in K–theory. In any case, what is really new here is the generalisation to arbitrary regular cardinals α.

This raises, of course, the question of why one cares. Thomason proved the theorem where $\alpha = \beta = \aleph_0$, and $\mathcal{S}$ is the derived category of the category of quasi–coherent sheaves on a quasi–compact, separated scheme. The author gave a simpler proof, which also generalised the result to all $\mathcal{S}$, as long as $\alpha = \beta = \aleph_0$. This may be found in [**23**]. The obvious question is: who cares about the case of large regular cardinals?

The short answer is that everybody should. First of all, it has already been mentioned that in the applications to motives, the case $\alpha = \beta = \aleph_0$ does not apply. Only rarely is there a set of objects $T \subset \mathcal{T}^{\aleph_0}$, with $\langle T \rangle = \mathcal{T}$. But even the people with both their feet firmly on the ground, the ones who could not care less about motives, should be interested in the case of large cardinals.

The reason is the following. If $\mathcal{T}$ is the homotopy category of spectra, it has been known for a long time that $\mathcal{T}^{\aleph_0}$ generates $\mathcal{T}$. But now let E be a homology theory. Following Bousfield, let $\mathcal{T}_E \subset \mathcal{T}$ be the subcategory of E–acyclic spectra, and let ${}^{\perp}\mathcal{T}_E$ be the subcategory of E–local spectra. In general, $\{\mathcal{T}_E\}^{\aleph_0}$ and $\{{}^{\perp}\mathcal{T}_E\}^{\aleph_0}$ are small and very uninteresting. It is only for sufficiently large α that the categories $\{\mathcal{T}_E\}^{\alpha}$ and $\{{}^{\perp}\mathcal{T}_E\}^{\alpha}$ start generating. See Remark D.1.15, for an estimate on how large α must be. The moral is very simple. Suppose the main object of interest is a triangulated category to which Thomason's theorem, or my old generalisation of it, apply. That is, the main object of study is a category for which the case $\alpha = \beta = \aleph_0$ is non–trivial. As soon as we Bousfield localise it, we get a category for which we are naturally forced into the large cardinal α generalisation.

So far we have seen that, for each regular cardinal α, it is possible to attach to $\mathcal{T}$ a canonically defined α–localising subcategory $\mathcal{T}^{\alpha}$. We have also seen Thomason's localisation theorem, which says that the subcategories $\mathcal{T}^{\alpha}$ behave well with respect to Verdier quotients. But to convince the reader that the exercise is worthwhile, I must use the subcategories $\mathcal{T}^{\alpha} \subset \mathcal{T}$ to prove a statement not directly involving them. First we need a key definition.

DEFINITION 1.15. *Let α be a regular cardinal. Let $\mathcal{T}$ be a triangulated category with small* Hom*–sets, satisfying [TR5]. If the subcategory $\mathcal{T}^{\alpha}$ is essentially small, and if $\langle \mathcal{T}^{\alpha} \rangle = \mathcal{T}$, we say that $\mathcal{T}$ is* α–compactly generated. *It turns out that if $\mathcal{T}$ is α–compactly generated, then it is also β–compactly generated for any $\beta > \alpha$. A triangulated category $\mathcal{T}$ is said to be* well generated *if*

1.15.1. *$\mathcal{T}$ has small* Hom*–sets.*

1.15.2. *$\mathcal{T}$ satisfies [TR5].*

1.15.3. *For some regular α, $\mathcal{T}$ is α–compactly generated.*

Of course, a well generated triangulated category is in fact β–compactly generated for all sufficiently large β.

REMARK 1.16. It might be worth restating Thomason's localisation theorem (Theorem 1.14) in the above terms. Let $\mathcal{S}$ be a triangulated category satisfying [TR5], let $\mathcal{R} \subset \mathcal{S}$ be a localising subcategory, and let $\mathcal{T} = \mathcal{S}/\mathcal{R}$ be the Verdier quotient. Thomason's localisation theorem asserts that if $\mathcal{S}$ is well generated then, under mild hypotheses, so are $\mathcal{R}$ and $\mathcal{T}$. The hypotheses are that $\mathcal{R}$ be generated by a set of its objects. The precise statement keeps track of the cardinals β for which these categories are β–compactly generated. The reader is refered to Theorem 1.14 for the more refined statement.

Now we begin the main theorems of the book.

THEOREM 1.17. (BROWN REPRESENTABILITY). *Let $\mathcal{T}$ be any well–generated triangulated category. Let H be a contravariant functor $H : \mathcal{T}^{op} \longrightarrow \mathcal{A}b$, where $\mathcal{A}b$ is the category of abelian groups. The functor H is representable if and only if it is homological, and takes coproducts in $\mathcal{T}$ to products of abelian groups. In other words, the representable functors $\mathcal{T}(-,t)$ can be characterised as the cohomological functors taking coproducts in $\mathcal{T}$ to products in $\mathcal{A}b$.*

Proof: Theorem 8.3.3. □

This theorem has several immediate corollaries. One of them is

COROLLARY 1.18. *Let $\mathcal{T}$ be any well–generated triangulated category. Then $\mathcal{T}$ satisfies [TR5*]; all small products exist in $\mathcal{T}$.*

Proof: Given a set $\{X_\lambda, \lambda \in \Lambda\}$ of objects in $\mathcal{T}$, the following functor $\mathcal{T}^{op} \longrightarrow \mathcal{A}b$

$$H(-) \quad = \quad \prod_{\lambda\in\Lambda} \mathcal{T}(-, X_\lambda)$$

is homological, and takes coproducts in $\mathcal{T}$ to products of abelian groups. By Theorem 1.17 it is representable. The representing object is then the product, in $\mathcal{T}$, of $\{X_\lambda, \lambda \in \Lambda\}$. See also Proposition 8.4.6. □

In other words, we learn that a well–generated triangulated category is closed under products. This leads naturally to the question of whether the dual of Brown's representability theorem holds. We prove

THEOREM 1.19. (BROWN REPRESENTABILITY FOR THE DUAL). *Let $\mathcal{T}$ be any well–generated triangulated category. Choose some regular cardinal*

α, for which $\mathcal{T}$ is α–compactly generated. Suppose for that α, the abelian category $\mathcal{E}x\Big(\{\mathcal{T}^\alpha\}^{op}, \mathcal{A}b\Big)$ has enough injectives.

Let H be a covariant functor $H : \mathcal{T} \longrightarrow \mathcal{A}b$. Then H will be representable if and only if it is homological, and takes products in $\mathcal{T}$ to products of abelian groups. In other words, the representable functors $\mathcal{T}(t, -)$ can be characterised as the homological functors respecting products.

Proof: Theorem 8.6.1. □

We can formalise this

DEFINITION 1.20. *A triangulated category $\mathcal{T}$ satisfying [TR5] is said to* satisfy the representability theorem *if the (contravariant) representable functors $\mathcal{T}(-, t)$ are precisely the homological functors $H : \mathcal{T}^{op} \longrightarrow \mathcal{A}b$ taking coproducts to products.*

The content of Theorems 1.17 and 1.19 is that well–generated categories $\mathcal{T}$ satisfy the representability theorem, as do their duals, provided $\mathcal{E}x\Big(\{\mathcal{T}^\alpha\}^{op}, \mathcal{A}b\Big)$ has enough injectives. An easy proposition states

PROPOSITION 1.21. *Let $F : \mathcal{S} \longrightarrow \mathcal{T}$ be a triangulated functor of triangulated categories. Suppose $\mathcal{S}$ satisfies the representability theorem, as in Definition 1.20. The functor F has a right adjoint $G : \mathcal{T} \longrightarrow \mathcal{S}$ if and only if F respects coproducts.*

Proof: Theorem 8.4.4. □

REMARK 1.22. In Remark 1.4 we noted that the deep questions about triangulated categories involve the existence of adjoints. Let $F : \mathcal{S} \longrightarrow \mathcal{T}$ be a triangulated functor. By Corollary 1.3, F has a right adjoint if and only if $A(F) : A(\mathcal{S}) \longrightarrow A(\mathcal{T})$ does. But this amounts to reducing a difficult problem to an impossible one. The categories $A(\mathcal{S})$ are terrible, and the author does not know a single example where one can show directly that $A(F) : A(\mathcal{S}) \longrightarrow A(\mathcal{T})$ has an adjoint.

By contrast, Proposition 1.21 is practical to apply. If $\mathcal{S}$ or $\mathcal{S}^{op}$ is well–generated, F will have a right adjoint if and only if it preserves coproducts.

REMARK 1.23. It should be noted that Franke has independently obtained a representability theorem strongly reminiscent of Theorem 1.17. Franke's theorem also assumes that $\mathcal{T}$ can be written as a union of T^α satisfying suitable hypotheses. But this is where the similarity becomes confusing. It is not clear whether the $\mathcal{T}^\alpha$'s studied here in general satisfy the hypotheses placed on Franke's T^α's. It also is not clear whether there could be some other choice for Franke's T^α's. In his application, to the derived category of a Grothendieck abelian category, Franke's T^α is just the $\mathcal{T}^\alpha$ we have been studying here. See Franke's [**11**].

Franke's method does not generalise to the dual of a well–generated triangulated category.

So far in the Introduction, we have presented the main results of the book. We ordered them in a way that motivated the definitions. We began with Freyd's construction of the universal homological functor $\mathcal{T} \longrightarrow A(\mathcal{T})$. We discussed its properties, and the fact that, in general, $A(\mathcal{T})$ is terrible. Then we spoke about approximations to $A(\mathcal{T})$, in particular approximations of the form $\mathcal{E}x\big(\mathcal{S}^{op}, \mathcal{A}b\big)$, for an α–localising subcategory $\mathcal{S} \subset \mathcal{T}$. We reasoned that, for every regular cardinal α, there is a canonical best choice for $\mathcal{S}$; the largest possible $\mathcal{S}$ is $\mathcal{T}^\alpha$. Thomason's localisation theorem is the statement that $\mathcal{T}^\alpha$ behaves reasonably well with respect to Verdier localisations. Our main theorems give an application of the $\mathcal{T}^\alpha$'s. The first major theorem asserts that the representability theorem holds for $\mathcal{T}$ whenever $\mathcal{T}^\alpha$ is essentially small and generates $\mathcal{T}$. The second asserts that the representability theorem holds for $\mathcal{T}^{op}$ if $\mathcal{T}$ is α–compactly generated, and furthermore $\mathcal{E}x\Big(\{\mathcal{T}^\alpha\}^{op}, \mathcal{A}b\Big)$ has enough injectives.

Now it is time to explain the way the exposition of these facts is organised in the book, and to discuss some of the less major theorems that we prove on the way, or as consequences. The order in which the results are presented in the book is the Bourbaki order. It is the logical order, not the order that would motivate the constructions. Chapters 1 and 2 give the elementary properties of triangulated categories. Chapters 3 and 4 give the definitions of the categories $\mathcal{T}^\alpha$, and their formal properties. This culminates in Thomason's localisation theorem, which asserts $\mathcal{T}^\alpha$ passes to Verdier quotients. This is quite unmotivated. We define the categories $\mathcal{T}^\alpha$, and study their formal properties, before we have any indication that they might be of some use.

Only in Chapter 5 do we treat Freyd's classical theorem, concerning the universal homological functor. In Chapter 6 we finally come around to the categories $\mathcal{E}x\Big(\{\mathcal{T}^\alpha\}^{op}, \mathcal{A}b\Big)$. We develop the elementary properties of the categories $\mathcal{E}x\big(\mathcal{S}^{op}, \mathcal{A}b\big)$, and of the functor $A(\mathcal{T}) \longrightarrow \mathcal{E}x\big(\mathcal{S}^{op}, \mathcal{A}b\big)$. Chapter 6 should help clarify, somewhat belatedly, the point of studying the categories $\mathcal{T}^\alpha$.

In Theorem 1.19, we saw that Brown representability sometimes holds for the dual of $\mathcal{T}$; in particular, it holds if $\mathcal{E}x\big(\mathcal{S}^{op}, \mathcal{A}b\big)$ has enough injectives. It becomes interesting to study whether there are enough injectives.

The general answer is No. Counterexamples may be found in Sections C.4 and D.2. Nevertheless, it is possible that Brown representability for the dual could be proved with less than the existence of injectives. The homological algebra of the categories $\mathcal{E}x\big(\mathcal{S}^{op}, \mathcal{A}b\big)$ is interesting, and its careful study might yield great results. In Chapter 7, I assemble assorted facts I know. These do not really lead anywhere yet, but I thought they might be useful to future researchers. The Chapter may safely be skipped by all but the truly committed.

The category $\mathcal{E}x(\mathcal{S}^{op}, \mathcal{A}b)$ does not satisfy [AB5]. We remind the reader: this means direct limits of exact sequences need not be exact. It follows that $\mathcal{E}x(\mathcal{S}^{op}, \mathcal{A}b)$ is not a Grothendieck abelian category, and hence the classical proofs of the existence of injectives break down. Of course, the proofs must break down, since we know that there are not, in general, enough injectives. But it is interesting to analyse just where the breakdown occurs. We will analyse this for the argument that appears in Grothendieck's Tôhoku paper; see Théorème 1.10.1, on page 135 of [**18**]. I do not know the origin of the argument; Grothendieck said that it had been well–known, and he was merely sketching it. The argument is based on adding cells. We should perhaps remind the reader.

Let $\mathcal{A}$ be an abelian category, x an object of $\mathcal{A}$. We wish to embed x in an injective I. This means that given any extension in $\mathrm{Ext}^1(z, x)$, that is any exact sequence

$$0 \longrightarrow x \longrightarrow y \longrightarrow z \longrightarrow 0,$$

the map $x \longrightarrow I$ should kill it. In other words, the map $x \longrightarrow I$ should factor as

$$x \longrightarrow y \longrightarrow I.$$

This suggests a natural way to try to construct I. If x is not injective, it has an extension

$$0 \longrightarrow x \longrightarrow y \longrightarrow z \longrightarrow 0.$$

If y is not injective, we can repeat the process. We can construct a sequence of monomorphisms

$$x = x_0 \longrightarrow x_1 \longrightarrow x_2 \longrightarrow \cdots$$

and hope that the colimit of the x_i will be injective. This is the process of adding cells, and the proof in [**18**] was based on a slightly refined version of this construction.

The first and most serious problem with this construction is that, for an abelian category not satisfying [AB5], it is not clear that x injects into the $\underrightarrow{\mathrm{colim}}\, x_i$. It might well be that $\underrightarrow{\mathrm{colim}}\, x_i = 0$. We say that an abelian category satisfies [AB4.5] if, for any (transfinite) sequence of monomorphisms as above, the map $x_0 \longrightarrow \underrightarrow{\mathrm{colim}}\, x_i$ is injective.

Actually, for the purpose of the proofs given in the article, it is convenient to give an equivalent statement in terms of the derived functors of the colimit. An abelian category satisfies [AB4.5] if, for any (transfinite) sequence of monomorphisms as above, and for any $n \geq 1$, the nth derived functor of the colimit vanishes. That is,

$$\underrightarrow{\operatorname{colim}}^n \, x_i = 0.$$

We do not prove the equivalence of the two statements; we use the second as a definition, and we use the fact that it implies the first. The converse is true, but of no importance to us. One can easily show (Lemma A.3.15) that if an abelian category satisfies [AB3] (has coproducts) and has enough injectives, then it satisfies [AB4.5].

It is instructive to know that, if $\mathcal{S}$ is sufficiently ridiculous, the category $\mathcal{E}x\big(\mathcal{S}^{op}, \mathcal{A}b\big)$ need not satisfy [AB4.5]. For this reason, in Chapter 6 we begin by defining $\mathcal{E}x\big(\mathcal{S}^{op}, \mathcal{A}b\big)$ for fairly arbitrary $\mathcal{S}$. For the $\mathcal{S}$'s for which we define it, $\mathcal{E}x\big(\mathcal{S}^{op}, \mathcal{A}b\big)$ always satisfies [AB4]; coproducts are exact. See Lemma 6.3.2. However, if $\mathcal{S}$ is the category of normed, non–archimedean, complete topological abelian groups, then $\mathcal{E}x\big(\mathcal{S}^{op}, \mathcal{A}b\big)$ does not satisfy [AB4.5]. See Proposition A.5.12. But in this book, we are mostly interested in the case where the category $\mathcal{S}$ is triangulated. I have no example of a triangulated category $\mathcal{S}$, for which I can show that $\mathcal{E}x\big(\mathcal{S}^{op}, \mathcal{A}b\big)$ does not satisfy [AB4.5]. For a while, I thought I could prove [AB4.5] for such $\mathcal{E}x\big(\mathcal{S}^{op}, \mathcal{A}b\big)$. But there is a gap. Included in Chapter 7, is the part of the argument that is correct.

The study of derived functors of co–Mittag–Leffler sequences in abelian categories is of some independent interest, and the existence of an abelian category $\mathcal{E}x\big(\mathcal{S}^{op}, \mathcal{A}b\big)$, satisfying [AB4] but not [AB4.5], is new and surprising. The reader is referred to Proposition 1 in [**29**], or Lemma 1.15 on page 213 of [**20**], to see just how striking it is. Since the results are about abelian, as opposed to triangulated, categories, they have been put in an appendix; see Appendix A.

The property [AB4.5] is extensively studied, for the abelian categories $\mathcal{E}x\big(\mathcal{S}^{op}, \mathcal{A}b\big)$, in Chapter 7 and Appendix A. The study is inconclusive, but might be helpful to others. This occupies most of Chapter 7. But the final section, Section 7.5, is quite unrelated.

Let $\mathcal{S}$ be a triangulated category closed under coproducts of $< \alpha$ of its objects. The category $\mathcal{E}x\big(\mathcal{S}^{op}, \mathcal{A}b\big)$ is the category of functors $\mathcal{S}^{op} \longrightarrow \mathcal{A}b$, which respect products of fewer than α objects. It is contained in the category $\mathcal{C}at\big(\mathcal{S}^{op}, \mathcal{A}b\big)$, of all additive functors $\mathcal{S}^{op} \longrightarrow \mathcal{A}b$. Let i be the inclusion

$$i : \mathcal{E}x\big(\mathcal{S}^{op}, \mathcal{A}b\big) \longrightarrow \mathcal{C}at\big(\mathcal{S}^{op}, \mathcal{A}b\big).$$

In Section 7.5, we prove that i has a left adjoint j. So far, this is a special case of a theorem of Gabriel and Ulmer [**16**]. But more interestingly, the functor j has left derived functors $L^n j$. And most remarkably, if F is an object of $\mathcal{E}x\big(\mathcal{S}^{op}, \mathcal{A}b\big)$, then iF is an object of $\mathcal{C}at\big(\mathcal{S}^{op}, \mathcal{A}b\big)$, and we prove that, for $n \geq 1$,

$$L^n j\{iF\} \quad = \quad 0.$$

An easy consequence is that, given objects F and G in the abelian category $\mathcal{E}x(\mathcal{S}^{op}, \mathcal{A}b)$, the groups $\mathrm{Ext}^n(F, G)$ agree, whether we compute them in $\mathcal{E}x(\mathcal{S}^{op}, \mathcal{A}b)$ or in the larger $\mathcal{C}at(\mathcal{S}^{op}, \mathcal{A}b)$.

Chapter 8 has the proof of the two Brown representability theorems. It shows how the earlier theory can be used to prove practical theorems. It is now time to give fairly precise statements of what we prove, and explain the consequences.

In Theorems 1.17 and 1.19, we saw that if $\mathcal{T}$ is well–generated, then the representability theorem holds for $\mathcal{T}$, and also for $\mathcal{T}^{op}$, as long as some abelian category has enough injectives. See Definition 1.20 for what it means for a triangulated category $\mathcal{T}$ to satisfy the representability theorem. These are true statements, but we prove more. Now it is time to be precise about what we prove.

To say that $\mathcal{T}$ is well–generated asserts that, for some regular cardinal α, $\mathcal{S} = \mathcal{T}^\alpha$ is essentially small, and the natural homological functor

$$\mathcal{T} \longrightarrow \mathcal{E}x(\mathcal{S}^{op}, \mathcal{A}b)$$

does not annihilate any object. A more precise statement of the theorem we prove would be

THEOREM 1.24. *Suppose $\mathcal{T}$ is a triangulated category satisfying [TR5]. Suppose α is a regular cardinal, and suppose $\mathcal{S} \subset \mathcal{T}$ is an α–localising subcategory. Suppose $\mathcal{S}$ is essentially small, and suppose the natural homological functor*

$$\mathcal{T} \longrightarrow \mathcal{E}x(\mathcal{S}^{op}, \mathcal{A}b)$$

does not annihilate any object, and respects countable *coproducts. Then $\mathcal{T}$ satisfies the representability theorem.*

Note that we do not assume, in the statement of the theorem, that $\mathcal{S} = \mathcal{T}^\alpha$. If $\mathcal{S} = \mathcal{T}^\alpha$, the conditions placed on $\mathcal{S}$ amount to saying that $\mathcal{T}$ is α–compactly generated. For $\mathcal{S} = \mathcal{T}^\alpha$, Theorem 1.24 reduces to Theorem 1.17. The fact that we allow other $\mathcal{S}$'s is a generalisation. Now let us analyse this.

If we assume that the map

$$\mathcal{T} \longrightarrow \mathcal{E}x(\mathcal{S}^{op}, \mathcal{A}b)$$

preserves all coproducts, then the generalisation is in fact very minor. Let us explain why. The point is that if

$$\mathcal{T} \longrightarrow \mathcal{E}x(\mathcal{S}^{op}, \mathcal{A}b)$$

respects all coproducts, then $\mathcal{S} \subset \mathcal{T}^\alpha$. Recall that $\mathcal{T}^\alpha$ is the largest of all the $\mathcal{S}$'s for which coproducts are preserved. It contains all others, in particular it contains $\mathcal{S}$. It turns out that if $\alpha \geq \aleph_1$, $\mathcal{S} \subset \mathcal{T}^\alpha$ is α–localising, and the map

$$\mathcal{T} \longrightarrow \mathcal{E}x\big(\mathcal{S}^{op}, \mathcal{A}b\big)$$

does not annihilate any object, then $\mathcal{S} = \mathcal{T}^\alpha$. We know this because in Theorem 8.3.3 we prove $\mathcal{T} = \langle \mathcal{S} \rangle$, and Thomason's localisation theorem (Theorem 4.4.9) then tells us that $\mathcal{T}^\alpha = \langle \mathcal{S} \rangle^\alpha$. But $\langle \mathcal{S} \rangle^\alpha = \mathcal{S}$, since $\mathcal{S} \subset \mathcal{T}^\alpha$ is α–localising. If $\alpha = \aleph_0$ the statement is more delicate, and what we said is true only up to splitting idempotents. Anyway, the point is that, up to splitting idempotents when $\alpha = \aleph_0$, $\mathcal{S} = \mathcal{T}^\alpha$ is the only choice. The only practical value of the seemingly more general statement about arbitrary $\mathcal{S}$'s, is that it gives us a way to show that some $\mathcal{S}$ is, in fact, equal to $\mathcal{T}^\alpha$.

But the fact that the functor

$$\mathcal{T} \longrightarrow \mathcal{E}x\big(\mathcal{S}^{op}, \mathcal{A}b\big)$$

need only respect countable coproducts seems a genuine relaxation of the hypothesis that $\mathcal{T}$ be well–generated. I use the word "seems" because I know of no example. I know no non–well–generated category to which this applies. Still, we could define a triangulated category $\mathcal{T}$ satisfying [TR5] to be $\aleph_1$*–perfectly generated*, if there exists an essentially small $\aleph_1$–localising subcategory $\mathcal{S} \subset \mathcal{T}$, so that the functor

$$\mathcal{T} \longrightarrow \mathcal{E}x\big(\mathcal{S}^{op}, \mathcal{A}b\big)$$

does not annihilate any object, and preserves countable coproducts. Theorem 1.24 applies, and we would deduce that $\mathcal{T}$ satisfies the representability theorem. For a while I had hopes that maybe the dual of a well–generated triangulated category would be $\aleph_1$–perfectly generated. We will see in Section E.2 that the dual of $D(\mathbb{Q})$ is not $\aleph_1$–perfectly generated. I would like to thank Shelah for pointing out the cardinality argument at the heart of Section E.2.

So we now know the precise statement of Theorem 8.3.3, which asserts that $\aleph_1$–perfectly generated triangulated categories satisfy the representability theorem. The statement for the dual, that is Theorem 8.6.1, is precisely as we quoted it in Theorem 1.19. That is, if $\mathcal{T}$ is α–compactly generated and $\mathcal{E}x\Big(\{\mathcal{T}^\alpha\}^{op}, \mathcal{A}b\Big)$ has enough injectives, then the representability theorem holds for $\mathcal{T}^{op}$. In other words, here we do not get away with countable coproducts. We need to assume the functor

$$\mathcal{T} \longrightarrow \mathcal{E}x\big(\mathcal{S}^{op}, \mathcal{A}b\big)$$

preserves all coproducts.

Now we have made precise the two representability theorems we prove. It is time to briefly review the applications. We have already mentioned the application to adjoints. See Proposition 1.21; if $\mathcal{S}$ is an α–compactly generated triangulated category, and if $\mathcal{E}x\big(\mathcal{S}^{op}, \mathcal{A}b\big)$ has enough injectives, then a triangulated functor $\mathcal{S} \longrightarrow \mathcal{T}$ has a left (respectively right) adjoint if and only if it preserves products (respectively coproducts). Next we want to discuss what follows, still under the hypothesis that the category $\mathcal{E}x(\mathcal{S}^{op}, \mathcal{A}b)$ has enough injectives.

First we should remind the reader. Let $\mathcal{T}$ be a triangulated category satisfying [TR5]. When $\mathcal{S}$ is an α–localising subcategory of $\mathcal{T}$, the homological functor

$$\mathcal{T} \longrightarrow \mathcal{E}x(\mathcal{S}^{op}, \mathcal{A}b)$$

factors uniquely through Freyd's universal homological functor. It factors

$$\mathcal{T} \longrightarrow A(\mathcal{T}) \xrightarrow{\pi} \mathcal{E}x(\mathcal{S}^{op}, \mathcal{A}b)$$

where π is exact. The functor π respects products and has a left adjoint

$$\mathcal{E}x(\mathcal{S}^{op}, \mathcal{A}b) \xrightarrow{F} A(\mathcal{T}),$$

by Proposition 1.5. Also by Proposition 1.5, the unit of adjunction

$$\eta : 1 \longrightarrow \pi F$$

is an isomorphism if π respects all coproducts. The interesting case of this is $\mathcal{S} = \mathcal{T}^\alpha$.

In particular, for $\mathcal{S} = \mathcal{T}^\alpha$, the map π does respect coproducts, the unit of adjunction is an isomorphism, and by Gabriel's theory of localisation (in abelian categories), $\mathcal{E}x(\mathcal{S}^{op}, \mathcal{A}b)$ is a quotient of $A(\mathcal{T})$ by the class of objects annihilated by π. See Section A.2 for a summary of Gabriel's pertinent results.

So much is completely general. But we prove more. If $\mathcal{S} = \mathcal{T}^\alpha$ is essentially small, if $\mathcal{E}x(\mathcal{S}^{op}, \mathcal{A}b)$ has enough injectives, and if $\mathcal{T}$ satisfies the representability theorem, then the functor

$$A(\mathcal{T}) \xrightarrow{\pi} \mathcal{E}x(\mathcal{S}^{op}, \mathcal{A}b)$$

also has a right adjoint

$$\mathcal{E}x(\mathcal{S}^{op}, \mathcal{A}b) \xrightarrow{G} A(\mathcal{T}).$$

This is proved in Corollary 8.5.3. And it means the following. We already knew that $\mathcal{E}x(\mathcal{S}^{op}, \mathcal{A}b)$ is a Gabriel quotient of $A(\mathcal{T})$ by the kernel of π. We knew quite generally that there is a left adjoint to π. But under the hypotheses given above, which hold, for example, if $\alpha = \aleph_0$, the functor π also has a right adjoint G. The quotient is a localizant–colocalizant one.

The existence of enough injectives in $\mathcal{E}x(\mathcal{S}^{op}, \mathcal{A}b)$ implies that $\pi : A(\mathcal{T}) \longrightarrow \mathcal{E}x(\mathcal{S}^{op}, \mathcal{A}b)$ has a right adjoint G. Lemma 8.5.5 establishes that the existence of the right adjoint G implies that the category $\mathcal{E}x(\mathcal{S}^{op}, \mathcal{A}b)$

has a cogenerator. We have implications

$$\left\{\begin{array}{c}\mathcal{E}x\big(\mathcal{S}^{op},\mathcal{A}b\big)\\ \text{has enough}\\ \text{injectives}\end{array}\right\}\Longrightarrow\left\{\begin{array}{c}A(\mathcal{T})\longrightarrow\mathcal{E}x\big(\mathcal{S}^{op},\mathcal{A}b\big)\\ \text{has a right adjoint}\end{array}\right\}\Longrightarrow\left\{\begin{array}{c}\mathcal{E}x\big(\mathcal{S}^{op},\mathcal{A}b\big)\\ \text{has a}\\ \text{cogenerator}\end{array}\right\}$$

And in the counterexamples of Sections C.4 and D.2, we see that in general the category $\mathcal{E}x\big(\mathcal{S}^{op},\mathcal{A}b\big)$ need not have a cogenerator. Thus the right adjoint G need not exist, and $\mathcal{E}x\big(\mathcal{S}^{op},\mathcal{A}b\big)$ may fail to have enough injectives.

Injective objects and right adjoints are abstract and perhaps uninviting. It is therefore illuminating to rephrase everything in terms of phantom maps.

A morphism $f : x \longrightarrow y$ in $\mathcal{T}$ is called α–phantom if its image vanishes in $\mathcal{E}x\big(\mathcal{S}^{op},\mathcal{A}b\big)$. That is,

$$\mathcal{T}(-,f)|_{\mathcal{S}} : \mathcal{T}(-,x)|_{\mathcal{S}} \longrightarrow \mathcal{T}(-,y)|_{\mathcal{S}}$$

is the zero map. In Lemma 8.5.20, we prove that the right adjoint to $A(\mathcal{T}) \longrightarrow \mathcal{E}x\big(\mathcal{S}^{op},\mathcal{A}b\big)$ exists if and only if, for every object $z \in \mathcal{T}$, there is a maximal α–phantom map $y \longrightarrow z$. That is, every α–phantom $x \longrightarrow z$ must factor, non–uniquely, as

$$x \longrightarrow y \longrightarrow z.$$

In Lemma 8.5.17, we see that the category $\mathcal{E}x\big(\mathcal{S}^{op},\mathcal{A}b\big)$ will have enough injectives if and only if, for every object $z \in \mathcal{T}$, the maximal α–phantom map $y \longrightarrow z$ may be so chosen that, in the triangle

$$y \longrightarrow z \longrightarrow t \longrightarrow \Sigma y$$

the object t is orthogonal to the α–phantom maps. Every α–phantom map $x \longrightarrow t$ vanishes.

If the category $\mathcal{E}x\big(\mathcal{S}^{op},\mathcal{A}b\big)$ has enough injectives, there is a right adjoint G to the functor $\pi : A(\mathcal{T}) \longrightarrow \mathcal{E}x\big(\mathcal{S}^{op},\mathcal{A}b\big)$. Let I be an injective cogenerator of $\mathcal{E}x\big(\mathcal{S}^{op},\mathcal{A}b\big)$. We may form the object GI. Since G has an exact left adjoint π, GI must be injective in $A(\mathcal{T})$. That is, GI is really an object in $\mathcal{T} \subset A(\mathcal{T})$; the injective objects are direct summands of objects in $\mathcal{T}$, and as $\mathcal{T}$ satisfies [TR5], idempotents split in $\mathcal{T}$. See Proposition 1.6.8.

We call the object GI a *Brown–Comenetz object*, and denote it $\mathbb{BC}$. The Brown–Comenetz objects are somehow crucial to our proof that the dual of $\mathcal{T}$ satisfies the representability theorem. Since $\mathcal{E}x\big(\mathcal{S}^{op},\mathcal{A}b\big)$ need not have enough injectives, it would be nice to have another proof, which does not so critically hinge on the existence of injectives in $\mathcal{E}x\big(\mathcal{S}^{op},\mathcal{A}b\big)$.

The last chapter before the appendices is Chapter 9. It discusses Bousfield localisation. It is relatively short, and exposes no new results. Let me briefly tell the reader the contents of the Chapter.

Let $\mathcal{T}$ be a triangulated category, $\mathcal{S} \subset \mathcal{T}$ a triangulated subcategory. We say that a Bousfield localisation functor exists for the pair $\mathcal{S} \subset \mathcal{T}$ if the map $\mathcal{T} \longrightarrow \mathcal{T}/\mathcal{S}$ has a right adjoint. Chapter 9 explores in some detail what happens. It turns out that the adjoint

$$\mathcal{T}/\mathcal{S} \xrightarrow{\quad i \quad} \mathcal{T}$$

is always fully faithful, which allows us to think of $\mathcal{T}/\mathcal{S}$ as a subcategory of $\mathcal{T}$. The Verdier quotient

$$\frac{\mathcal{T}}{\mathcal{T}/\mathcal{S}}$$

is naturally isomorphic to $\mathcal{S}$, and the embedding

$$\tfrac{\mathcal{T}}{\mathcal{T}/\mathcal{S}} = \mathcal{S} \longrightarrow \mathcal{T}$$

is left adjoint to the quotient map

$$\mathcal{T} \longrightarrow \tfrac{\mathcal{T}}{\mathcal{T}/\mathcal{S}} = \mathcal{S}.$$

There are some parallels with Gabriel's constructions of quotients of abelian categories. Anyway, we hope the reader finds the exposition of Chapter 9 amusing, even if the results are basically all known. In the second volume, which Voevodsky and the author still promise to write, there will hopefully be more about Bousfield localisation.

There are also some appendices. The appendices contain two types of results. The first is background which the reader will need, and which is assembled here for convenience. There are several facts we want to use about abelian categories, which go beyond the elementary homological algebra that is a prerequisite for the book. These results may be found elsewhere, scattered around the literature. But Appendix A offers the reader a condensed summary. See Section A.1 for locally presentable categories, Section A.2 for Gabriel's treatment of localisation in abelian categories, and Sections A.3 and A.4 for the derived functors of limits.

The remaining material in the appendices is new, often of independent interest. Generally, if a result has no strong, direct bearing on the development of the theory, it is left to an appendix. For example, Appendix A contains more than just a summary of known facts about abelian categories. The treatment of Mittag–Leffler sequences is new, as is [AB4.5]. And the example of the category $\mathcal{S}$ for which $\mathcal{E}x\big(\mathcal{S}^{op}, \mathcal{A}b\big)$ satisfies [AB4] but not [AB4.5] is not only new, it is quite surprising. It goes against the expectations in the literature. Since it is only tangentially related to the main subject of the book, the reader will find it consigned to Section A.5.

Appendix A is about abelian categories, with some of the material being old, some new. The remaining appendices offer results about triangulated categories. These also divide into two types.

In Appendix B, we show that the functor $\mathcal{T} \longrightarrow \mathcal{E}x\Big(\{\mathcal{T}^\alpha\}^{op}, \mathcal{A}b\Big)$ can be characterised by a universal property. An abelian category $\mathcal{A}$ is said to satisfy [AB5$^\alpha$] if α–filtered colimits are exact in $\mathcal{A}$. We prove

THEOREM 1.25. *Let $\mathcal{T}$ be an α–compactly generated triangulated category. The coproduct–preserving homological functors $H : \mathcal{T} \longrightarrow \mathcal{A}$, where $\mathcal{A}$ is an abelian category satisfying [AB5$^\alpha$], factor uniquely, up to canonical equivalence, as*

$$\mathcal{T} \longrightarrow \mathcal{E}x\Big(\{\mathcal{T}^\alpha\}^{op}, \mathcal{A}b\Big) \xrightarrow{\exists !} \mathcal{A},$$

with the functor $\mathcal{E}x\Big(\{\mathcal{T}^\alpha\}^{op}, \mathcal{A}b\Big) \longrightarrow \mathcal{A}$ coproduct–preserving and exact. Natural transformations between coproduct–preserving homological functors $\mathcal{T} \longrightarrow \mathcal{A}$ are in 1–to–1 correspondence with natural transformations of coproduct–preserving exact functors $\mathcal{E}x\Big(\{\mathcal{T}^\alpha\}^{op}, \mathcal{A}b\Big) \longrightarrow \mathcal{A}$.

Proof: Theorem B.2.5. □

This result is the type that perhaps merits inclusion in the body of the book. But we do not really use it elsewhere. It gives more evidence that the construction of the categories $\mathcal{T}^\alpha$ is natural. Beyond this, it does not have a strong bearing on the development of the theory. At least, not yet. For this reason, it was put in an appendix.

I would like to thank Christensen, who kept asking me about such results. Christensen and Strickland, in [**9**], proved the assertion when $\mathcal{T}$ is the category of spectra, and $\alpha = \aleph_0$. Their methods do not seem to generalise, even to other $\mathcal{T}$ but with α still $\aleph_0$. If not for Christensen's prodding, I would probably never have obtained the result.

Now for the remaining three appendices. These are basically examples. In Appendix D, we work out in some detail what the general theory says in the special case, where $\mathcal{T}$ is the homotopy category of spectra. We began the Introduction with this, hence we will not repeat it. There are two remaining Appedices, C and E. These mostly are about pathological behavior. The reader is expected to know a little bit about the derived category to read these examples. The body of the book does not discuss examples, and does not depend on knowing any. But in the appendices, we assume some acquaintance with the derived category.

Appendix C has two results. First it proves that, in general, the objects of Freyd's universal category $A(\mathcal{T})$ have classes, not sets, of subobjects. Very concretely, we show it for the object $\mathbb{Z} \in D(\mathbb{Z}) \subset A\big[D(\mathbb{Z})\big]$. This result is certainly known to the experts, and in fact seems to have been independently rediscovered several times. The earliest reference I could find is Freyd's [**14**].

The second result is that the category $\mathcal{E}x\big(\mathcal{S}^{op}, \mathcal{A}b\big)$ need not have a cogenerator. As we have seen above, this also means that there need not be a right adjoint to $\pi : A(\mathcal{T}) \longrightarrow \mathcal{E}x\big(\mathcal{S}^{op}, \mathcal{A}b\big)$, and that $\mathcal{E}x\big(\mathcal{S}^{op}, \mathcal{A}b\big)$ need not have enough injectives.

Appendix E offers examples of categories which are not well–generated. In Corollary E.1.3, we see that if $\mathcal{T}$ is $\aleph_0$–compactly generated, then $\mathcal{T}^{op}$ cannot be well–generated. For this result, the reader does not need to know any examples. In Section E.3, more specifically Summary E.3.3, we see that if $K(\mathbb{Z})$ is the homotopy category of chain complexes of abelian groups, then neither $K(\mathbb{Z})$ nor $K(\mathbb{Z})^{op}$ is well–generated. Section E.2 treats a condition possibly weaker than well–generation. It shows that the dual of $D(\mathbb{Q})$, the derived category of vector spaces over the field $\mathbb{Q}$, is not even $\aleph_1$–perfectly generated.

One thing should be noted. In Corollary E.1.3, we prove that the dual of an $\aleph_0$–compactly generated $\mathcal{T}$ cannot be well–generated. If $\mathcal{T}$ is the homotopy category of spectra, we deduce that $\mathcal{T}^{op}$ cannot be well–generated. Hence $\mathcal{T}$ and $\mathcal{T}^{op}$ cannot be equivalent. The assertion that the homotopy category is not self–dual is an old theorem of Boardman, [**2**]. What we have here is a generalisation. It is not the best generalisation one can prove, but in the interest of simplicity, it is the one we give.

CHAPTER 1

Definition and elementary properties of triangulated categories

1.1. Pre–triangulated categories

DEFINITION 1.1.1. *Let $\mathcal{C}$ be an additive category and $\Sigma : \mathcal{C} \longrightarrow \mathcal{C}$ be an additive endofunctor of $\mathcal{C}$. Assume throughout that the endofunctor Σ is invertible. A* candidate triangle *in $\mathcal{C}$ (with respect to Σ) is a diagram of the form:*

$$X \xrightarrow{\ u\ } Y \xrightarrow{\ v\ } Z \xrightarrow{\ w\ } \Sigma X$$

such that the composites $v \circ u$, $w \circ v$ and $\Sigma u \circ w$ are the zero morphisms.

A morphism of candidate triangles is a commutative diagram

$$\begin{array}{ccccccc}
X & \xrightarrow{\ u\ } & Y & \xrightarrow{\ v\ } & Z & \xrightarrow{\ w\ } & \Sigma X \\
{\scriptstyle f}\downarrow & & {\scriptstyle g}\downarrow & & {\scriptstyle h}\downarrow & & {\scriptstyle \Sigma f}\downarrow \\
X' & \xrightarrow{\ u'\ } & Y' & \xrightarrow{\ v'\ } & Z' & \xrightarrow{\ w'\ } & \Sigma X'
\end{array}$$

where each row is a candidate triangle.

DEFINITION 1.1.2. A pre–triangulated category $\mathcal{T}$ is an additive category, together with an additive automorphism Σ, and a class of candidate triangles (with respect to Σ) called *distinguished* triangles. The following conditions must hold:

TR0: Any candidate triangle which is isomorphic to a distinguished triangle is a distinguished triangle. The candidate triangle

$$X \xrightarrow{\ 1\ } X \longrightarrow 0 \longrightarrow \Sigma X$$

is distinguished.

TR1: For any morphism $f : X \to Y$ in $\mathcal{T}$ there exists a distinguished triangle of the form

$$X \xrightarrow{\ f\ } Y \longrightarrow Z \longrightarrow \Sigma X$$

TR2: Consider the two candidate triangles

$$X \xrightarrow{\ u\ } Y \xrightarrow{\ v\ } Z \xrightarrow{\ w\ } \Sigma X$$

and

$$Y \xrightarrow{-v} Z \xrightarrow{-w} \Sigma X \xrightarrow{-\Sigma u} \Sigma Y.$$

If one is a distinguished triangle, then so is the other.

TR3: For any commutative diagram of the form

$$\begin{array}{ccccccc} X & \xrightarrow{u} & Y & \xrightarrow{v} & Z & \xrightarrow{w} & \Sigma X \\ {\scriptstyle f}\downarrow & & {\scriptstyle g}\downarrow & & & & \\ X' & \xrightarrow{u'} & Y' & \xrightarrow{v'} & Z' & \xrightarrow{w'} & \Sigma X' \end{array}$$

where the rows are distinguished triangles, there is a morphism $h : Z \to Z'$, not necessarily unique, which makes the diagram

$$\begin{array}{ccccccc} X & \xrightarrow{u} & Y & \xrightarrow{v} & Z & \xrightarrow{w} & \Sigma X \\ {\scriptstyle f}\downarrow & & {\scriptstyle g}\downarrow & & {\scriptstyle h}\downarrow & & {\scriptstyle \Sigma f}\downarrow \\ X' & \xrightarrow{u'} & Y' & \xrightarrow{v'} & Z' & \xrightarrow{w'} & \Sigma X' \end{array}$$

commutative.

REMARK 1.1.3. Parts of Definition 1.1.2 are known to be redundant. For instance, it is not necessary to assume that distinguished triangles are candidate triangles. In other words, we can assume that the distinguished triangles are sequences

$$X \xrightarrow{u} Y \xrightarrow{v} Z \xrightarrow{w} \Sigma X$$

without necessarily postulating that the composites $v \circ u$, $w \circ v$ and $\Sigma u \circ w$ vanish. It follows from the other axioms that the composites $v \circ u$, $w \circ v$ and $\Sigma u \circ w$ must be zero. Just consider the diagram

$$\begin{array}{ccccccc} X & \xrightarrow{1} & X & \longrightarrow & 0 & \longrightarrow & \Sigma X \\ {\scriptstyle 1}\downarrow & & {\scriptstyle u}\downarrow & & & & \\ X & \xrightarrow{u} & Y & \xrightarrow{v} & Z & \xrightarrow{w} & \Sigma X \end{array}$$

The bottom row is a distinguished triangle by hypothesis, the top by [TR0]. But by [TR3] the diagram may be completed to a commutative

$$\begin{array}{ccccccc} X & \xrightarrow{1} & X & \longrightarrow & 0 & \longrightarrow & \Sigma X \\ {\scriptstyle 1}\downarrow & & {\scriptstyle u}\downarrow & & \downarrow & & {\scriptstyle 1}\downarrow \\ X & \xrightarrow{u} & Y & \xrightarrow{v} & Z & \xrightarrow{w} & \Sigma X \end{array}$$

and we deduce that $v \circ u = 0$. The vanishing of $w \circ v$ and $\Sigma u \circ w$ follows from the above and axiom [TR2].

Similarly, it is not necessary to assume that the category $\mathcal{T}$ is additive; something slightly less suffices. It suffices to assume that the category $\mathcal{T}$ is pointed (there is a zero object), and that the *Hom* sets are abelian groups. The fact that finite coproducts and products exist and agree follows from the other axioms.

NOTATION 1.1.4. Let $\mathcal{T}$ be a pre–triangulated category. If we speak of "triangles" in $\mathcal{T}$, we mean distinguished triangles. When we mean candidate triangles, the adjective will always be explicitly used.

REMARK 1.1.5. If $\mathcal{T}$ is a pre–triangulated category, then clearly so is its dual $\mathcal{T}^{op}$. For $\mathcal{T}^{op}$, the functor Σ gets replaced by Σ^{-1}.

PROPOSITION 1.1.6. *Let $\mathcal{T}$ be a pre–triangulated category. Then the functor Σ preserves products and coproducts. Let us state this precisely. Suppose $\{X_\lambda, \lambda \in \Lambda\}$ is a set of objects of $\mathcal{T}$, and suppose the categorical coproduct $\coprod_{\lambda\in\Lambda} X_\lambda$ exists in $\mathcal{T}$. Then the natural map*

$$\coprod_{\lambda\in\Lambda} \{\Sigma X_\lambda\} \longrightarrow \Sigma\left\{\coprod_{\lambda\in\Lambda} X_\lambda\right\}$$

is an isomorphism. In other words, the natural maps

$$\Sigma X_\lambda \longrightarrow \Sigma\left\{\coprod_{\lambda\in\Lambda} X_\lambda\right\}$$

give $\Sigma\left\{\coprod_{\lambda\in\Lambda} X_\lambda\right\}$ the structure of a coproduct in the category $\mathcal{T}$. Similarly, if $\prod_{\lambda\in\Lambda} X_\lambda$ exists in $\mathcal{T}$, then the natural maps

$$\Sigma\left\{\prod_{\lambda\in\Lambda} X_\lambda\right\} \longrightarrow \Sigma X_\lambda$$

give $\Sigma\left\{\prod_{\lambda\in\Lambda} X_\lambda\right\}$ the structure of a product in the category $\mathcal{T}$.

Proof: The point is that Σ, being invertible, has both a right and a left adjoint, namely Σ^{-1}. There are natural isomorphisms

$$Hom\,(\Sigma X, Y) \simeq Hom\,\left(X, \Sigma^{-1}Y\right)$$

and

$$Hom\,(X, \Sigma Y) \simeq Hom\,\left(\Sigma^{-1}X, Y\right),$$

induced by Σ^{-1}. A functor possessing a left adjoint respects products, a functor possessing a right adjoint respects coproducts. Thus Σ respects both. □

Because $\mathcal{T}^{op}$ is a pre–triangulated category with Σ^{-1} playing the role of Σ, it follows that Σ^{-1} also respects products and coproducts.

DEFINITION 1.1.7. *Let $\mathcal{T}$ be a pre–triangulated category. Let H be a functor from $\mathcal{T}$ to some abelian category $\mathcal{A}$. The functor H is called* homological *if, for every (distinguished) triangle*

$$X \xrightarrow{\quad u \quad} Y \xrightarrow{\quad v \quad} Z \xrightarrow{\quad w \quad} \Sigma X$$

the sequence

$$H(X) \xrightarrow{H(u)} H(Y) \xrightarrow{H(v)} H(Z)$$

is exact in the abelian category $\mathcal{A}$.

REMARK 1.1.8. Because of axiom [TR2], it follows that the sequence above can be continued indefinitely in both directions. In other words, the infinite sequence

$$H(\Sigma^{-1}Z) \xrightarrow{H(\Sigma^{-1}w)} H(X) \xrightarrow{H(u)} H(Y) \xrightarrow{H(v)} H(Z) \xrightarrow{H(w)} H(\Sigma X)$$

is exact everywhere.

REMARK 1.1.9. A homological functor on the pre–triangulated category $\mathcal{T}^{op}$ is called a *cohomological* functor on $\mathcal{T}$. Thus, a cohomological functor is a contravariant functor $H : \mathcal{T} \to \mathcal{A}$ such that, for any triangle

$$X \xrightarrow{\quad u \quad} Y \xrightarrow{\quad v \quad} Z \xrightarrow{\quad w \quad} \Sigma X$$

the sequence

$$H(Z) \xrightarrow{H(v)} H(Y) \xrightarrow{H(u)} H(X)$$

is exact in the abelian category $\mathcal{A}$.

LEMMA 1.1.10. *Let $\mathcal{T}$ be a pre–triangulated category, U be an object of $\mathcal{T}$. Then the representable functor $Hom(U, -)$ is homological.*

Proof: Suppose we are given a triangle

$$X \xrightarrow{\quad u \quad} Y \xrightarrow{\quad v \quad} Z \xrightarrow{\quad w \quad} \Sigma X.$$

We need to show the exactness of the sequence

$$Hom(U, X) \longrightarrow Hom(U, Y) \longrightarrow Hom(U, Z)$$

We know in any case that the composite is zero. Let $f \in Hom(U, Y)$ map to zero in $Hom(U, Z)$. That is, let $f : U \to Y$ be such that the composite

$$U \xrightarrow{\quad f \quad} Y \xrightarrow{\quad v \quad} Z$$

is zero. Then we have a commutative diagram

$$\begin{array}{ccccccc}
U & \longrightarrow & 0 & \longrightarrow & \Sigma U & \xrightarrow{-1} & \Sigma U \\
{\scriptstyle f}\downarrow & & \downarrow & & & & \downarrow{\scriptstyle \Sigma f} \\
Y & \xrightarrow{-v} & Z & \xrightarrow{-w} & \Sigma X & \xrightarrow{-\Sigma u} & \Sigma Y.
\end{array}$$

The bottom row is a triangle by [TR2], the top row by [TR0] and [TR2]. There is therefore, by [TR3], a map $h : U \to X$ such that $\Sigma h : \Sigma U \to \Sigma X$ makes the diagram above commute. But this means in particular that the square

$$\begin{array}{ccc} \Sigma U & \xrightarrow{-1} & \Sigma U \\ {\scriptstyle \Sigma h}\downarrow & & {\scriptstyle \Sigma f}\downarrow \\ \Sigma X & \xrightarrow{-\Sigma u} & \Sigma Y \end{array}$$

commutes, and hence $f = u \circ h$. Thus, we have produced an $h \in Hom(U, X)$ mapping to $f \in Hom(U, Y)$.

□

REMARK 1.1.11. Recall that the dual of a pre–triangulated category is pre–triangulated. It follows from Lemma 1.1.10, applied to the dual of $\mathcal{T}$, that the functor $Hom(-, U)$ is cohomological.

DEFINITION 1.1.12. *Let $H : \mathcal{T} \to \mathcal{A}$ be a homological functor. The functor H is called* decent *if*

1.1.12.1. *The abelian category $\mathcal{A}$ satisfies $AB4^*$; that is products exist, and the product of exact sequences is exact.*

1.1.12.2. *The functor H respects products. For any collection $\{X_\lambda, \lambda \in \Lambda\}$ of objects $X_\lambda \in \mathcal{T}$ whose product exists in $\mathcal{T}$, the natural map*

$$H\left(\prod_{\lambda \in \Lambda} X_\lambda\right) \longrightarrow \prod_{\lambda \in \Lambda} H(X_\lambda)$$

is an isomorphism.

EXAMPLE 1.1.13. The functor $Hom(U, -) : \mathcal{T} \to \mathcal{A}b$ is a decent homological functor. It is homological by Lemma 1.1.10, the abelian category $\mathcal{A}b$ of all abelian groups satisfies $AB4^*$, and $Hom(U, -)$ preserves products.

DEFINITION 1.1.14. *Let $\mathcal{T}$ be a pre–triangulated category. A candidate triangle*

$$X \xrightarrow{\ u\ } Y \xrightarrow{\ v\ } Z \xrightarrow{\ w\ } \Sigma X$$

is called a pre–triangle *if, for every decent homological functor $H : \mathcal{T} \to \mathcal{A}$, the long sequence*

$$H(\Sigma^{-1}Z) \xrightarrow{H(\Sigma^{-1}w)} H(X) \xrightarrow{H(u)} H(Y) \xrightarrow{H(v)} H(Z) \xrightarrow{H(w)} H(\Sigma X)$$

is exact.

EXAMPLE 1.1.15. Every (distinguished) triangle is a pre–triangle. A direct summand of a pre–triangle is a pre–triangle. The next little lemma will show that an arbitrary product of pre–triangles is a pre–triangle.

CAUTION 1.1.16. There are pre–triangles which are not distinguished. See for example the discussion of Case 2, pages 232-234 of [**22**]. An example of a pre–triangle which is not a triangle is the mapping cone on the map of triangles in the middle of page 234, loc. cit.

LEMMA 1.1.17. *Let Λ be an index set, and suppose that for every $\lambda \in \Lambda$ we are given a pre–triangle*

$$X_\lambda \longrightarrow Y_\lambda \longrightarrow Z_\lambda \longrightarrow \Sigma X_\lambda.$$

Suppose further that the three products

$$\prod_{\lambda\in\Lambda} X_\lambda, \qquad \prod_{\lambda\in\Lambda} Y_\lambda, \qquad \prod_{\lambda\in\Lambda} Z_\lambda$$

exist in $\mathcal{T}$. The sequence

$$\prod_{\lambda\in\Lambda} X_\lambda \longrightarrow \prod_{\lambda\in\Lambda} Y_\lambda \longrightarrow \prod_{\lambda\in\Lambda} Z_\lambda \longrightarrow \prod_{\lambda\in\Lambda} \{\Sigma X_\lambda\}$$

is identified, using Proposition 1.1.6, with

$$\prod_{\lambda\in\Lambda} X_\lambda \longrightarrow \prod_{\lambda\in\Lambda} Y_\lambda \longrightarrow \prod_{\lambda\in\Lambda} Z_\lambda \longrightarrow \Sigma\left\{\prod_{\lambda\in\Lambda} X_\lambda\right\}.$$

We assert that this candidate triangle is a pre–triangle. Thus, the product of pre–triangles is a pre–triangle.

Proof: Let $H : \mathcal{T} \to \mathcal{A}$ be a decent homological functor. Because for each $\lambda \in \Lambda$ the sequence

$$X_\lambda \longrightarrow Y_\lambda \longrightarrow Z_\lambda \longrightarrow \Sigma X_\lambda$$

is a pre–triangle, applying H we get a long exact sequence in $\mathcal{A}$

$$H\left(\Sigma^{-1} Z_\lambda\right) \longrightarrow H(X_\lambda) \longrightarrow H(Y_\lambda) \longrightarrow H(Z_\lambda) \longrightarrow H(\Sigma X_\lambda)$$

and because $\mathcal{A}$ is assumed to satisfy $AB4^*$, the product of these sequences is exact. But we are assuming H decent, and in particular by 1.1.12.2, the maps

$$H\left(\prod_{\lambda\in\Lambda} X_\lambda\right) \longrightarrow \prod_{\lambda\in\Lambda} H(X_\lambda)$$

$$H\left(\prod_{\lambda\in\Lambda} Y_\lambda\right) \longrightarrow \prod_{\lambda\in\Lambda} H(Y_\lambda)$$

$$H\left(\prod_{\lambda\in\Lambda} Z_\lambda\right) \longrightarrow \prod_{\lambda\in\Lambda} H(Z_\lambda)$$

are all isomorphisms. This means that the functor H, applied to the sequence

$$\prod_{\lambda\in\Lambda} X_\lambda \longrightarrow \prod_{\lambda\in\Lambda} Y_\lambda \longrightarrow \prod_{\lambda\in\Lambda} Z_\lambda \longrightarrow \Sigma\left\{\prod_{\lambda\in\Lambda} X_\lambda\right\}$$

gives a long exact sequence. This being true for all decent H, we deduce that the sequence is a pre–triangle. □

LEMMA 1.1.18. *Let H be a decent homological functor $\mathcal{T} \longrightarrow \mathcal{A}$. Let the diagram*

$$\begin{array}{ccccccc} X & \xrightarrow{u} & Y & \xrightarrow{v} & Z & \xrightarrow{w} & \Sigma X \\ {\scriptstyle f}\downarrow & & {\scriptstyle g}\downarrow & & {\scriptstyle h}\downarrow & & {\scriptstyle \Sigma f}\downarrow \\ X' & \xrightarrow{u'} & Y' & \xrightarrow{v'} & Z' & \xrightarrow{w'} & \Sigma X' \end{array}$$

be a morphism of pre–triangles. Suppose that for every $n \in \mathbb{Z}$, $H(\Sigma^n f)$ and $H(\Sigma^n g)$ are isomorphisms. Then $H(\Sigma^n h)$ are all isomorphisms.

Proof: Without loss, we are reduced to proving $H(h)$ an isomorphism. But then the diagram

$$\begin{array}{ccccccccc} H(X) & \xrightarrow{H(u)} & H(Y) & \xrightarrow{H(v)} & H(Z) & \xrightarrow{H(w)} & H(\Sigma X) & \xrightarrow{H(\Sigma u)} & H(\Sigma Y) \\ {\scriptstyle H(f)}\downarrow & & {\scriptstyle H(g)}\downarrow & & {\scriptstyle H(h)}\downarrow & & {\scriptstyle H(\Sigma f)}\downarrow & & {\scriptstyle H(\Sigma g)}\downarrow \\ H(X') & \xrightarrow{H(u')} & H(Y') & \xrightarrow{H(v')} & H(Z') & \xrightarrow{H(w')} & H(\Sigma X') & \xrightarrow{H(\Sigma u')} & H(\Sigma Y') \end{array}$$

is a commutative diagram in the abelian category $\mathcal{A}$ with exact rows. By the 5–lemma, we deduce that $H(h)$ is an isomorphism. □

LEMMA 1.1.19. *In the morphism of pre–triangles*

$$\begin{array}{ccccccc} X & \xrightarrow{u} & Y & \xrightarrow{v} & Z & \xrightarrow{w} & \Sigma X \\ {\scriptstyle f}\downarrow & & {\scriptstyle g}\downarrow & & {\scriptstyle h}\downarrow & & {\scriptstyle \Sigma f}\downarrow \\ X' & \xrightarrow{u'} & Y' & \xrightarrow{v'} & Z' & \xrightarrow{w'} & \Sigma X' \end{array}$$

if f and g are isomorphisms, then for any decent homological functor H, $H(h)$ is an isomorphism.

Proof: If f and g are isomorphisms, so are $\Sigma^n f$ and $\Sigma^n g$ for any n. Hence Lemma 1.1.18 allows us to deduce that $H(h)$ is an isomorphism. □

PROPOSITION 1.1.20. *If in the morphism of pre–triangles*

$$\begin{array}{ccccccc} X & \xrightarrow{u} & Y & \xrightarrow{v} & Z & \xrightarrow{w} & \Sigma X \\ {\scriptstyle f}\downarrow & & {\scriptstyle g}\downarrow & & {\scriptstyle h}\downarrow & & {\scriptstyle \Sigma f}\downarrow \\ X' & \xrightarrow{u'} & Y' & \xrightarrow{v'} & Z' & \xrightarrow{w'} & \Sigma X' \end{array}$$

both f and g are isomorphisms, then so is h.

Proof: By Lemma 1.1.19, we already know that for any decent homological functor $H : \mathcal{T} \longrightarrow \mathcal{A}$, $H(h)$ is an isomorphism. By Example 1.1.13, all representable functors $Hom(U, -)$ are decent, for $U \in \mathcal{T}$. We know therefore that the natural map

$$Hom(U, h) : Hom(U, Z) \longrightarrow Hom(U, Z')$$

is an isomorphism for every U. But then the map

$$Hom(-, h) : Hom(-, Z) \longrightarrow Hom(-, Z')$$

is an isomorphism. It follows from Yoneda's Lemma that h is an isomorphism. □

REMARK 1.1.21. Let $u : X \longrightarrow Y$ be given. By [TR1] it may be completed to a triangle. Let

$$X \xrightarrow{u} Y \xrightarrow{v} Z \xrightarrow{w} \Sigma X$$

and

$$X \xrightarrow{u} Y \xrightarrow{v'} Z' \xrightarrow{w'} \Sigma X$$

be two distinguished triangles "completing" u. We have a diagram

$$\begin{array}{ccccccc} X & \xrightarrow{u} & Y & \xrightarrow{v} & Z & \xrightarrow{w} & \Sigma X \\ {\scriptstyle 1}\downarrow & & {\scriptstyle 1}\downarrow & & & & {\scriptstyle 1}\downarrow \\ X & \xrightarrow{u} & Y & \xrightarrow{v'} & Z' & \xrightarrow{w'} & \Sigma X \end{array}$$

which by [TR3] may be completed to a morphism of triangles

$$\begin{array}{ccccccc} X & \xrightarrow{u} & Y & \xrightarrow{v} & Z & \xrightarrow{w} & \Sigma X \\ {\scriptstyle 1}\downarrow & & {\scriptstyle 1}\downarrow & & {\scriptstyle h}\downarrow & & {\scriptstyle 1}\downarrow \\ X & \xrightarrow{u} & Y & \xrightarrow{v'} & Z' & \xrightarrow{w'} & \Sigma X \end{array}$$

and since $1 : X \longrightarrow X$ and $1 : Y \longrightarrow Y$ are clearly isomorphisms, Proposition 1.1.20 says that h is an isomorphism. It follows that Z is well defined up to isomorphism. In fact, the entire triangle is well defined up to isomorphism. But this isomorphism is not in general canonical.

1.2. Corollaries of Proposition 1.1.20

In this section, we will group together some corollaries of Proposition 1.1.20, which have in common that they concern products and coproducts.

PROPOSITION 1.2.1. *Let $\mathcal{T}$ be a pre–triangulated category and Λ any index set. Suppose for every $\lambda \in \Lambda$ we are given a (distinguished) triangle*

$$X_\lambda \longrightarrow Y_\lambda \longrightarrow Z_\lambda \longrightarrow \Sigma X_\lambda$$

in $\mathcal{T}$. Suppose the three products

$$\prod_{\lambda\in\Lambda} X_\lambda, \qquad \prod_{\lambda\in\Lambda} Y_\lambda, \qquad \prod_{\lambda\in\Lambda} Z_\lambda$$

exist in $\mathcal{T}$. We know by Lemma 1.1.17 that the product is a pre–triangle

$$\prod_{\lambda\in\Lambda} X_\lambda \longrightarrow \prod_{\lambda\in\Lambda} Y_\lambda \longrightarrow \prod_{\lambda\in\Lambda} Z_\lambda \longrightarrow \Sigma\left\{\prod_{\lambda\in\Lambda} X_\lambda\right\}$$

We assert that it is a distinguished triangle.

Proof: By [TR1], the map

$$\prod_{\lambda\in\Lambda} X_\lambda \longrightarrow \prod_{\lambda\in\Lambda} Y_\lambda$$

can be completed to a triangle

$$\prod_{\lambda\in\Lambda} X_\lambda \longrightarrow \prod_{\lambda\in\Lambda} Y_\lambda \longrightarrow Q \longrightarrow \Sigma\left\{\prod_{\lambda\in\Lambda} X_\lambda\right\}.$$

For each $\lambda \in \Lambda$, we get a diagram where the rows are triangles

$$\begin{array}{ccccccc}
\prod_{\lambda\in\Lambda} X_\lambda & \longrightarrow & \prod_{\lambda\in\Lambda} Y_\lambda & \longrightarrow & Q & \longrightarrow & \Sigma\left\{\prod_{\lambda\in\Lambda} X_\lambda\right\} \\
\downarrow & & \downarrow & & & & \downarrow \\
X_\lambda & \longrightarrow & Y_\lambda & \longrightarrow & Z_\lambda & \longrightarrow & \Sigma X_\lambda
\end{array}$$

By [TR3] we may complete this to a morphism of triangles

$$\begin{array}{ccccccc}
\prod_{\lambda\in\Lambda} X_\lambda & \longrightarrow & \prod_{\lambda\in\Lambda} Y_\lambda & \longrightarrow & Q & \longrightarrow & \Sigma\left\{\prod_{\lambda\in\Lambda} X_\lambda\right\} \\
\downarrow & & \downarrow & & \downarrow & & \downarrow \\
X_\lambda & \longrightarrow & Y_\lambda & \longrightarrow & Z_\lambda & \longrightarrow & \Sigma X_\lambda
\end{array}$$

Taking the product of all these maps, we get a morphism

$$\begin{array}{ccccccc}
\prod_{\lambda\in\Lambda} X_\lambda & \longrightarrow & \prod_{\lambda\in\Lambda} Y_\lambda & \longrightarrow & Q & \longrightarrow & \Sigma\left\{\prod_{\lambda\in\Lambda} X_\lambda\right\} \\
{\scriptstyle 1}\downarrow & & {\scriptstyle 1}\downarrow & & \downarrow & & {\scriptstyle 1}\downarrow \\
\prod_{\lambda\in\Lambda} X_\lambda & \longrightarrow & \prod_{\lambda\in\Lambda} Y_\lambda & \longrightarrow & \prod_{\lambda\in\Lambda} Z_\lambda & \longrightarrow & \Sigma\left\{\prod_{\lambda\in\Lambda} X_\lambda\right\}
\end{array}$$

Both rows are pre–triangles. The top row because it is a triangle, the bottom row by Lemma 1.1.17. It follows from Proposition 1.1.20 that this map is an isomorphism of the top row (a distinguished triangle) with the bottom, which is therefore a triangle. □

REMARK 1.2.2. Dually, the coproduct of distinguished triangles is distinguished.

PROPOSITION 1.2.3. *Let $\mathcal{T}$ be a pre–triangulated category. Let*

$$X \longrightarrow Y \longrightarrow Z \longrightarrow \Sigma X$$

$$X' \longrightarrow Y' \longrightarrow Z' \longrightarrow \Sigma X'$$

be candidate triangles. Suppose the direct sum

$$X \oplus X' \longrightarrow Y \oplus Y' \longrightarrow Z \oplus Z' \longrightarrow \Sigma X \oplus \Sigma X'$$

is a distinguished triangle. Then so are the summands.

Proof: The situation being symmetric, it suffices to prove that

$$X \longrightarrow Y \longrightarrow Z \longrightarrow \Sigma X$$

is a triangle. Since it is the direct summand of a pre–triangle, it is in any case a pre–triangle, by Example 1.1.15. Let

$$X \longrightarrow Y \longrightarrow Q \longrightarrow \Sigma X$$

be a distinguished triangle. The diagram

$$\begin{array}{ccccccc}
X & \longrightarrow & Y & \longrightarrow & Q & \longrightarrow & \Sigma X \\
\begin{pmatrix}1\\0\end{pmatrix}\downarrow & & \begin{pmatrix}1\\0\end{pmatrix}\downarrow & & & & \begin{pmatrix}1\\0\end{pmatrix}\downarrow \\
X \oplus X' & \longrightarrow & Y \oplus Y' & \longrightarrow & Z \oplus Z' & \longrightarrow & \Sigma X \oplus \Sigma X'
\end{array}$$

may be completed to a morphism of triangles

$$\begin{array}{ccccccc}
X & \longrightarrow & Y & \longrightarrow & Q & \longrightarrow & \Sigma X \\
\begin{pmatrix}1\\0\end{pmatrix}\downarrow & & \begin{pmatrix}1\\0\end{pmatrix}\downarrow & & \downarrow & & \begin{pmatrix}1\\0\end{pmatrix}\downarrow \\
X \oplus X' & \longrightarrow & Y \oplus Y' & \longrightarrow & Z \oplus Z' & \longrightarrow & \Sigma X \oplus \Sigma X'
\end{array}$$

If we compose this with the projection

$$\begin{array}{ccccccc}
X \oplus X' & \longrightarrow & Y \oplus Y' & \longrightarrow & Z \oplus Z' & \longrightarrow & \Sigma X \oplus \Sigma X' \\
{\scriptstyle (\,1\ \ 0\,)}\big\downarrow & & {\scriptstyle (\,1\ \ 0\,)}\big\downarrow & & {\scriptstyle (\,1\ \ 0\,)}\big\downarrow & & {\scriptstyle (\,1\ \ 0\,)}\big\downarrow \\
X & \longrightarrow & Y & \longrightarrow & Z & \longrightarrow & \Sigma X
\end{array}$$

we get a morphism of pre–triangles

$$\begin{array}{ccccccc}
X & \longrightarrow & Y & \longrightarrow & Q & \longrightarrow & \Sigma X \\
{\scriptstyle 1}\big\downarrow & & {\scriptstyle 1}\big\downarrow & & {\scriptstyle h}\big\downarrow & & {\scriptstyle 1}\big\downarrow \\
X & \longrightarrow & Y & \longrightarrow & Z & \longrightarrow & \Sigma X
\end{array}$$

where the bottom is a pre–triangle because it is the direct summand of a triangle, and the top is a triangle. Once again, Proposition 1.1.20 implies that h is an isomorphism. Thus the bottom row is isomorphic to the top row, which is a distinguished triangle. By [TR0], the bottom row is also a triangle. □

There is one special case of the above which we will have occasion to use later, especially in Section 1.4. Observe first

LEMMA 1.2.4. *Suppose we are given a candidate triangle*

$$X \longrightarrow A \oplus Y \xrightarrow{\begin{pmatrix} 1 & \alpha \\ \beta & \gamma \end{pmatrix}} A \oplus Z \longrightarrow \Sigma X$$

Then this candidate triangle is isomorphic to a direct sum of candidate triangles

$$\begin{array}{ccccccc}
0 & \longrightarrow & A & \longrightarrow & A & \longrightarrow & 0 \\
X & \longrightarrow & Y & \longrightarrow & Z & \longrightarrow & \Sigma X
\end{array}$$

Proof: Consider the diagram

$$\begin{array}{ccccccc}
 & & A & & & & \\
 & & {\scriptstyle \begin{pmatrix} 1 \\ 0 \end{pmatrix}}\Big\downarrow & & & & \\
X & \longrightarrow & A \oplus Y & \xrightarrow{\begin{pmatrix} 1 & \alpha \\ \beta & \gamma \end{pmatrix}} & A \oplus Z & \longrightarrow & \Sigma X \\
 & & & & \big\downarrow{\scriptstyle (\,1\ \ 0\,)} & & \\
 & & & & A & &
\end{array}$$

Since the composite

$$A \xrightarrow{\begin{pmatrix} 1 \\ 0 \end{pmatrix}} A \oplus Y \xrightarrow{\begin{pmatrix} 1 & \alpha \\ \beta & \gamma \end{pmatrix}} A \oplus Z \xrightarrow{\begin{pmatrix} 1 & 0 \end{pmatrix}} A$$

is the map $1 : A \longrightarrow A$, we deduce that the composite of the maps of candidate triangles

$$\begin{array}{ccccccc}
0 & \longrightarrow & A & \xrightarrow{1} & A & \longrightarrow & 0 \\
\Big\downarrow & & \Big\downarrow \begin{pmatrix} 1 \\ 0 \end{pmatrix} & & \Big\downarrow & & \Big\downarrow \\
X & \longrightarrow & A \oplus Y & \xrightarrow{\begin{pmatrix} 1 & \alpha \\ \beta & \gamma \end{pmatrix}} & A \oplus Z & \longrightarrow & \Sigma X \\
\Big\downarrow & & \Big\downarrow & & \Big\downarrow \begin{pmatrix} 1 & 0 \end{pmatrix} & & \Big\downarrow \\
0 & \longrightarrow & A & \xrightarrow{1} & A & \longrightarrow & 0
\end{array}$$

is the identity. Thus the triangle

$$0 \longrightarrow A \xrightarrow{1} A \longrightarrow 0$$

is a direct summand of

$$X \longrightarrow A \oplus Y \xrightarrow{\begin{pmatrix} 1 & \alpha \\ \beta & \gamma \end{pmatrix}} A \oplus Z \longrightarrow \Sigma X.$$

The other direct summand may be computed, for example, as the kernel of the map

$$\begin{array}{ccccccc}
X & \longrightarrow & A \oplus Y & \xrightarrow{\begin{pmatrix} 1 & \alpha \\ \beta & \gamma \end{pmatrix}} & A \oplus Z & \longrightarrow & \Sigma X \\
\Big\downarrow & & \Big\downarrow & & \Big\downarrow \begin{pmatrix} 1 & 0 \end{pmatrix} & & \Big\downarrow \\
0 & \longrightarrow & A & \xrightarrow{1} & A & \longrightarrow & 0
\end{array}$$

It is a candidate triangle

$$X \longrightarrow Y \longrightarrow Z \longrightarrow \Sigma X$$

and the maps are all very explicitly computable. □

REMARK 1.2.5. If we start with a distinguished triangle

$$X \longrightarrow A \oplus Y \xrightarrow{\begin{pmatrix} 1 & \alpha \\ \beta & \gamma \end{pmatrix}} A \oplus Z \longrightarrow \Sigma X$$

then, by Proposition 1.2.3, the direct summand

$$X \longrightarrow Y \longrightarrow Z \longrightarrow \Sigma X$$

is also distinguished.

Next we give two rather trivial corollaries, which nevertheless are useful.

COROLLARY 1.2.6. *The map $f : X \to Y$ is an isomorphism if and only if, for some Z (necessarily isomorphic to zero), the candidate triangle*

$$X \xrightarrow{f} Y \xrightarrow{0} Z \xrightarrow{0} \Sigma X$$

is distinguished.

Proof: If f is an isomorphism, then the diagram below

$$\begin{array}{ccccccc} X & \xrightarrow{1} & X & \longrightarrow & 0 & \longrightarrow & \Sigma X \\ {\scriptstyle 1}\downarrow & & {\scriptstyle f}\downarrow & & \downarrow & & {\scriptstyle 1}\downarrow \\ X & \xrightarrow{f} & Y & \longrightarrow & 0 & \longrightarrow & \Sigma X \end{array}$$

defines an isomorphism of candidate triangles. The top is distinguished, hence so is the bottom. Thus, we can take $Z = 0$.

Conversely, assume

$$X \xrightarrow{f} Y \xrightarrow{0} Z \xrightarrow{0} \Sigma X$$

is a distinguished triangle. It is the sum of the two candidate triangles

$$X \xrightarrow{f} Y \longrightarrow 0 \longrightarrow \Sigma X$$

and

$$0 \longrightarrow 0 \longrightarrow Z \longrightarrow 0,$$

which by Proposition 1.2.3 must both be triangles. But then the diagram

$$\begin{array}{ccccccc} X & \xrightarrow{1} & X & \longrightarrow & 0 & \longrightarrow & \Sigma X \\ {\scriptstyle 1}\downarrow & & {\scriptstyle f}\downarrow & & \downarrow & & {\scriptstyle 1}\downarrow \\ X & \xrightarrow{f} & Y & \longrightarrow & 0 & \longrightarrow & \Sigma X \end{array}$$

gives a morphism of triangles. We know that $1 : X \to X$ and $1 : 0 \to 0$ are isomorphisms. Proposition 1.1.20 implies that the morphism $f : X \to Y$ also is. Now in the morphism of triangles

$$\begin{array}{ccccccc} X & \xrightarrow{1} & X & \longrightarrow & 0 & \longrightarrow & \Sigma X \\ {\scriptstyle 1}\downarrow & & {\scriptstyle f}\downarrow & & \downarrow & & {\scriptstyle 1}\downarrow \\ X & \xrightarrow{f} & Y & \xrightarrow{0} & Z & \xrightarrow{0} & \Sigma X \end{array}$$

we know that $1 : X \to X$ and $f : X \to Y$ are isomorphisms, hence so is $0 : 0 \to Z$. Thus, Z is isomorphic to zero. □

Corollary 1.2.7. *Any triangle of the form*

$$X \longrightarrow Y \longrightarrow Z \xrightarrow{0} \Sigma X$$

is isomorphic to

$$X \longrightarrow X \oplus Z \longrightarrow Z \xrightarrow{0} \Sigma X;$$

that is, if the map $Z \to \Sigma X$ vanishes, then the triangle splits.

Proof: By Corollary 1.2.6, for any isomorphism $f : Z \longrightarrow Z$, there is a triangle

$$Z \xrightarrow{f} Z \longrightarrow 0 \longrightarrow \Sigma Z.$$

In particular, we may take $f = -1$. By [TR2] it then follows that

$$0 \longrightarrow Z \xrightarrow{1} Z \longrightarrow 0$$

is a distinguished triangle. By [TR0], so is

$$X \xrightarrow{1} X \longrightarrow 0 \longrightarrow \Sigma X.$$

From Proposition 1.2.1, we learn that so is the direct sum

$$X \longrightarrow X \oplus Z \longrightarrow Z \xrightarrow{0} \Sigma X.$$

But now the diagram

$$\begin{array}{ccccccc} X & \longrightarrow & X \oplus Z & \longrightarrow & Z & \xrightarrow{0} & \Sigma X \\ {\scriptstyle 1}\downarrow & & & & {\scriptstyle 1}\downarrow & & {\scriptstyle 1}\downarrow \\ X & \longrightarrow & Y & \longrightarrow & Z & \xrightarrow{0} & \Sigma X \end{array}$$

can be completed to a morphism of triangles, which must be an isomorphism by Proposition 1.1.20. □

It is often necessary to know not only that Y is isomorphic to $X \oplus Z$, but also to give an explicit isomorphism. We observe

Lemma 1.2.8. *Let us be given a triangle*

$$X \xrightarrow{u} Y \xrightarrow{v} Z \xrightarrow{w} \Sigma X.$$

If $v' : Z \longrightarrow Y$ is a map such that

$$Z \xrightarrow{v'} Y \xrightarrow{v} Z$$

composes to the identity on Z, then the map of triangles

$$\begin{array}{ccccccc} X & \longrightarrow & X\oplus Z & \longrightarrow & Z & \longrightarrow & \Sigma X \\ {\scriptstyle 1}\downarrow & & \downarrow{\scriptstyle (u\ v')} & & {\scriptstyle 1}\downarrow & & {\scriptstyle 1}\downarrow \\ X & \xrightarrow{u} & Y & \xrightarrow{v} & Z & \xrightarrow{w} & \Sigma X \end{array}$$

is an isomorphism.

Proof: First let us establish that there is a map of triangles

$$\begin{array}{ccccccc} X & \longrightarrow & X\oplus Z & \longrightarrow & Z & \xrightarrow{0} & \Sigma X \\ {\scriptstyle 1}\downarrow & & \downarrow{\scriptstyle (u\ v')} & & {\scriptstyle 1}\downarrow & & {\scriptstyle 1}\downarrow \\ X & \xrightarrow{u} & Y & \xrightarrow{v} & Z & \xrightarrow{w} & \Sigma X. \end{array}$$

We need to show the squares commutative, and perhaps the square

$$\begin{array}{ccc} Z & \xrightarrow{0} & \Sigma X \\ {\scriptstyle 1}\downarrow & & {\scriptstyle 1}\downarrow \\ Z & \xrightarrow{w} & \Sigma X \end{array}$$

needs a little reflection; we must prove that the composite

$$\begin{array}{ccc} Z & & \\ {\scriptstyle 1}\downarrow & & \\ Z & \xrightarrow{w} & \Sigma X \end{array}$$

vanishes. We are given that the identity on Z may be factored as

$$Z \xrightarrow{v'} Y \xrightarrow{v} Z,$$

from which we conclude that the composite

$$\begin{array}{ccc} Z & & \\ {\scriptstyle 1}\downarrow & & \\ Z & \xrightarrow{w} & \Sigma X \end{array}$$

may be rewritten

$$\begin{array}{ccccc} Z & & & & \\ {\scriptstyle v'}\downarrow & & & & \\ Y & \xrightarrow{v} & Z & \xrightarrow{w} & \Sigma X. \end{array}$$

Since $wv = 0$, we do have a map of triangles

$$\begin{array}{ccccccc} X & \longrightarrow & X \oplus Z & \longrightarrow & Z & \stackrel{0}{\longrightarrow} & \Sigma X \\ {\scriptstyle 1}\downarrow & & \downarrow{\scriptstyle (u \;\; v')} & & {\scriptstyle 1}\downarrow & & {\scriptstyle 1}\downarrow \\ X & \stackrel{u}{\longrightarrow} & Y & \stackrel{v}{\longrightarrow} & Z & \stackrel{w}{\longrightarrow} & \Sigma X. \end{array}$$

It is a map of triangles, where two of the vertical maps are isomorphisms; by Proposition 1.1.20, so is the third. □

REMARK 1.2.9. Dually, given a triangle

$$X \stackrel{u}{\longrightarrow} Y \stackrel{v}{\longrightarrow} Z \stackrel{w}{\longrightarrow} \Sigma X$$

and a map $u' : Y \longrightarrow X$ so that

$$X \stackrel{u}{\longrightarrow} Y \stackrel{u'}{\longrightarrow} X$$

is the identity, then the map of triangles

$$\begin{array}{ccccccc} X & \stackrel{u}{\longrightarrow} & Y & \stackrel{v}{\longrightarrow} & Z & \stackrel{w}{\longrightarrow} & \Sigma X \\ {\scriptstyle 1}\downarrow & & \downarrow{\scriptstyle \begin{pmatrix} u' \\ v \end{pmatrix}} & & {\scriptstyle 1}\downarrow & & {\scriptstyle 1}\downarrow \\ X & \longrightarrow & X \oplus Z & \longrightarrow & Z & \longrightarrow & \Sigma X \end{array}$$

is an isomorphism. In other words, given a triangle

$$X \stackrel{u}{\longrightarrow} Y \stackrel{v}{\longrightarrow} Z \stackrel{w}{\longrightarrow} \Sigma X,$$

we get a canonical isomorphism

$$Y \simeq X \oplus Z$$

whenever we give either a factoring of the identity on Z as

$$Z \stackrel{v'}{\longrightarrow} Y \stackrel{v}{\longrightarrow} Z$$

or a factoring of the identity on X as

$$X \stackrel{u}{\longrightarrow} Y \stackrel{u'}{\longrightarrow} X.$$

LEMMA 1.2.10. *Suppose we have a triangle*

$$X \stackrel{u}{\longrightarrow} Y \stackrel{v}{\longrightarrow} Z \stackrel{w}{\longrightarrow} \Sigma X,$$

a factoring of the identity on Z as

$$Z \stackrel{v'}{\longrightarrow} Y \stackrel{v}{\longrightarrow} Z,$$

and a factoring of the identity on X as

$$X \stackrel{u}{\longrightarrow} Y \stackrel{u'}{\longrightarrow} X.$$

Then we have two isomorphisms

$$Y \simeq X \oplus Z,$$

one from each factoring. These two isomorphisms will agree if and only if the composite

$$Z \xrightarrow{v'} Y \xrightarrow{u'} X$$

vanishes.

Proof: By Lemma 1.2.8, the map $v' : Z \longrightarrow Y$ gives an isomorphism

$$X \oplus Z \xrightarrow{\begin{pmatrix} u & v' \end{pmatrix}} Y,$$

and by the dual of Lemma 1.2.8, the map $u' : Y \longrightarrow X$ gives an isomorphism

$$Y \xrightarrow{\begin{pmatrix} u' \\ v \end{pmatrix}} X \oplus Z.$$

We need to decide whether the isomorphisms agree, meaning whether they are inverse to each other. To do this, it suffices to check whether

$$X \oplus Z \xrightarrow{\begin{pmatrix} u & v' \end{pmatrix}} Y \xrightarrow{\begin{pmatrix} u' \\ v \end{pmatrix}} X \oplus Z$$

composes to the identity on $X \oplus Z$. But the composite is clearly

$$X \oplus Z \xrightarrow{\begin{pmatrix} u'u & u'v' \\ vu & vv' \end{pmatrix}} X \oplus Z.$$

On the other hand, we know that $u'u = 1$ and $vv' = 1$, and $vu = 0$ since it is the composite of two maps in a triangle. This makes the matrix

$$X \oplus Z \xrightarrow{\begin{pmatrix} 1 & u'v' \\ 0 & 1 \end{pmatrix}} X \oplus Z,$$

and it will be the identity precisely if $u'v'$ vanishes. □

1.3. Mapping cones, and the definition of triangulated categories

DEFINITION 1.3.1. Let $\mathfrak{T}$ be a pre–triangulated category. Suppose that we are given a morphism of candidate triangles

$$\begin{array}{ccccccc}
X & \xrightarrow{u} & Y & \xrightarrow{v} & Z & \xrightarrow{w} & \Sigma X \\
{\scriptstyle f}\downarrow & & {\scriptstyle g}\downarrow & & {\scriptstyle h}\downarrow & & {\scriptstyle \Sigma f}\downarrow \\
X' & \xrightarrow{u'} & Y' & \xrightarrow{v'} & Z' & \xrightarrow{w'} & \Sigma X'
\end{array}$$

There is a way to form a new candidate triangle out of this data. It is the diagram

$$Y \oplus X' \xrightarrow{\begin{pmatrix} -v & 0 \\ g & u' \end{pmatrix}} Z \oplus Y' \xrightarrow{\begin{pmatrix} -w & 0 \\ h & v' \end{pmatrix}} \Sigma X \oplus Z' \xrightarrow{\begin{pmatrix} -\Sigma u & 0 \\ \Sigma f & w' \end{pmatrix}} \Sigma Y \oplus \Sigma X'.$$

This new candidate triangle is called the *mapping cone* on a map of candidate triangles.

DEFINITION 1.3.2. Two maps of candidate triangles

$$\begin{array}{ccccccc} X & \xrightarrow{u} & Y & \xrightarrow{v} & Z & \xrightarrow{w} & \Sigma X \\ {\scriptstyle f}\downarrow & & {\scriptstyle g}\downarrow & & {\scriptstyle h}\downarrow & & {\scriptstyle \Sigma f}\downarrow \\ X' & \xrightarrow{u'} & Y' & \xrightarrow{v'} & Z' & \xrightarrow{w'} & \Sigma X' \end{array}$$

and

$$\begin{array}{ccccccc} X & \xrightarrow{u} & Y & \xrightarrow{v} & Z & \xrightarrow{w} & \Sigma X \\ {\scriptstyle f'}\downarrow & & {\scriptstyle g'}\downarrow & & {\scriptstyle h'}\downarrow & & {\scriptstyle \Sigma f'}\downarrow \\ X' & \xrightarrow{u'} & Y' & \xrightarrow{v'} & Z' & \xrightarrow{w'} & \Sigma X' \end{array}$$

are called *homotopic* if they differ by a homotopy; that is, if there exist Θ, Φ and Ψ below

$$\begin{array}{ccccccc} X & \xrightarrow{u} & Y & \xrightarrow{v} & Z & \xrightarrow{w} & \Sigma X \\ & {\scriptstyle \Theta}\swarrow & & {\scriptstyle \Phi}\swarrow & & {\scriptstyle \Psi}\swarrow & \\ X' & \xrightarrow{u'} & Y' & \xrightarrow{v'} & Z' & \xrightarrow{w'} & \Sigma X' \end{array},$$

with

$$f - f' = \Theta u + \Sigma^{-1}\{w'\Psi\} \qquad g - g' = \Phi v + u'\Theta \qquad h - h' = \Psi w + v'\Phi.$$

LEMMA 1.3.3. *Up to isomorphism, the mapping cone depends not on the morphism of candidate triangles*

$$\begin{array}{ccccccc} X & \xrightarrow{u} & Y & \xrightarrow{v} & Z & \xrightarrow{w} & \Sigma X \\ {\scriptstyle f}\downarrow & & {\scriptstyle g}\downarrow & & {\scriptstyle h}\downarrow & & {\scriptstyle \Sigma f}\downarrow \\ X' & \xrightarrow{u'} & Y' & \xrightarrow{v'} & Z' & \xrightarrow{w'} & \Sigma X' \end{array}$$

but only on the homotopy equivalence class of this morphism. If the map above is homotopic *to the map*

$$\begin{array}{ccccccc}
X & \xrightarrow{u} & Y & \xrightarrow{v} & Z & \xrightarrow{w} & \Sigma X \\
{\scriptstyle f'}\downarrow & & {\scriptstyle g'}\downarrow & & {\scriptstyle h'}\downarrow & & {\scriptstyle \Sigma f'}\downarrow \\
X' & \xrightarrow{u'} & Y' & \xrightarrow{v'} & Z' & \xrightarrow{w'} & \Sigma X'
\end{array}$$

as in Definition 1.3.2, then the mapping cones are isomorphic candidate triangles.

Proof: The diagram below is commutative

$$\begin{array}{ccccccc}
Y \oplus X' & \xrightarrow{\begin{pmatrix} -v & 0 \\ g & u' \end{pmatrix}} & Z \oplus Y' & \xrightarrow{\begin{pmatrix} -w & 0 \\ h & v' \end{pmatrix}} & \Sigma X \oplus Z' & \xrightarrow{\begin{pmatrix} -\Sigma u & 0 \\ \Sigma f & w' \end{pmatrix}} & \Sigma Y \oplus \Sigma X' \\
\left(\begin{array}{c|c} 1 & 0 \\ \Theta & 1 \end{array}\right)\Big\downarrow & & \left(\begin{array}{c|c} 1 & 0 \\ \Phi & 1 \end{array}\right)\Big\downarrow & & \left(\begin{array}{c|c} 1 & 0 \\ \Psi & 1 \end{array}\right)\Big\downarrow & & \left(\begin{array}{c|c} 1 & 0 \\ \Sigma\Theta & 1 \end{array}\right)\Big\downarrow \\
Y \oplus X' & \xrightarrow[\begin{pmatrix} -v & 0 \\ g' & u' \end{pmatrix}]{} & Z \oplus Y' & \xrightarrow[\begin{pmatrix} -w & 0 \\ h' & v' \end{pmatrix}]{} & \Sigma X \oplus Z' & \xrightarrow[\begin{pmatrix} -\Sigma u & 0 \\ \Sigma f' & w' \end{pmatrix}]{} & \Sigma Y \oplus \Sigma X'
\end{array}$$

and the vertical maps are isomorphisms; hence we have an isomorphism of the top row with the bottom row. □

The following is an elementary fact of homological algebra, whose proof we leave to the reader.

LEMMA 1.3.4. *Suppose $F : C \to D$ and $F' : C \to D$ are two morphisms of candidate triangles. Suppose F and F' are homotopic. Then for any map $G : C' \to C$ and any map $H : D \to D'$, the composites $H \circ F \circ G$ and $H \circ F' \circ G$ are homotopic.* □

DEFINITION 1.3.5. *A candidate triangle C is called* contractible *if the identity map $1 : C \to C$ is homotopic to the zero map $0 : C \to C$.*

LEMMA 1.3.6. *If C is a contractible candidate triangle, then any map from $F : C \to D$ or $F' : D \to C$ of candidate triangles is homotopic to the zero map.*

Proof: The two cases being dual, we can restrict attention to F. But $F = F \circ 1_C$, and since C is contractible, 1_C is homotopic to 0. But then F is homotopic to $F \circ 0 = 0$. □

LEMMA 1.3.7. *If C is a contractible candidate triangle, then C is a pre–triangle.*

Proof: We need to show that if H is any decent homological functor, and C is the contractible candidate triangle

$$X \longrightarrow Y \longrightarrow Z \longrightarrow \Sigma X,$$

then the long sequence

$$H(\Sigma^{-1}Z) \longrightarrow H(X) \longrightarrow H(Y) \longrightarrow H(Z) \longrightarrow H(\Sigma X)$$

is exact. In fact, this is true not only for decent homological functors H, but for any additive functor. The point is that the identity on C is homotopic to the zero map. There are Θ, Φ and Ψ as below

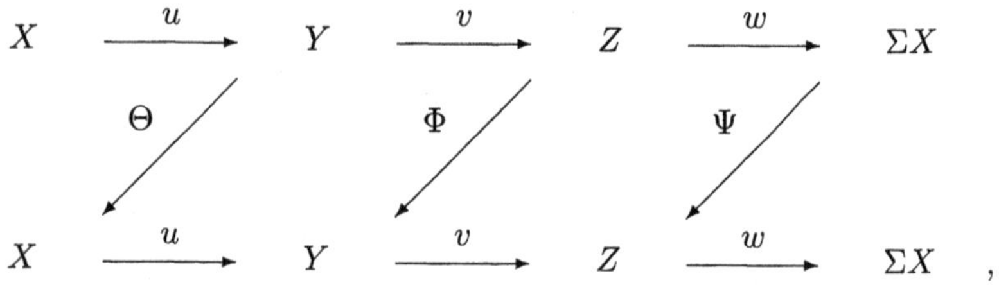

with

$$1_X = \Theta u + \Sigma^{-1}\{w\Psi\} \qquad 1_Y = \Phi v + u\Theta \qquad 1_Z = \Psi w + v\Phi.$$

Applying any additive functor H to this, we deduce that the identity on the sequence

$$H(\Sigma^{-1}Z) \longrightarrow H(X) \longrightarrow H(Y) \longrightarrow H(Z) \longrightarrow H(\Sigma X)$$

is chain homotopic to the zero map; hence the sequence is exact. □

PROPOSITION 1.3.8. *Let C be a contractible candidate triangle. Then C is a distinguished triangle.*

Proof: If C is the sequence

$$X \xrightarrow{\ u\ } Y \xrightarrow{\ v\ } Z \xrightarrow{\ w\ } \Sigma X$$

then we can complete $u : X \to Y$ to a triangle

$$X \xrightarrow{\ u\ } Y \xrightarrow{\ v'\ } Q \xrightarrow{\ w'\ } \Sigma X$$

Because C is contractible, there are the three maps

$$\begin{array}{ccccccc} X & \xrightarrow{u} & Y & \xrightarrow{v} & Z & \xrightarrow{w} & \Sigma X \\ & \swarrow{\scriptstyle\Theta} & & \swarrow{\scriptstyle\Phi} & & \swarrow{\scriptstyle\Psi} & \\ X & \xrightarrow{u} & Y & \xrightarrow{v} & Z & \xrightarrow{w} & \Sigma X \end{array},$$

giving the homotopy of 1_C to the zero map. Consider the map $w\Psi$. Clearly, $\Sigma u \circ \{w\Psi\} = 0$, since $\Sigma u \circ w = 0$. But because $Hom(\Sigma X, -)$ is a homological functor, when we apply it to the triangle

$$X \xrightarrow{u} Y \xrightarrow{v'} Q \xrightarrow{w'} \Sigma X$$

we get an exact sequence. Since Σu kills $w\Psi$, there must be a map $\Psi' : \Sigma X \to Q$ with $w'\Psi' = w\Psi$.

Now form the map of pre–triangles

$$\begin{array}{ccccccc} X & \xrightarrow{u} & Y & \xrightarrow{v} & Z & \xrightarrow{w} & \Sigma X \\ {\scriptstyle 1}\downarrow & & {\scriptstyle 1}\downarrow & & {\scriptstyle \Psi'w+v'\Phi}\downarrow & & {\scriptstyle 1}\downarrow \\ X & \xrightarrow{u} & Y & \xrightarrow{v'} & Q & \xrightarrow{w'} & \Sigma X \end{array}$$

with Ψ' as above. The reader will easily show that it is a map of pre–triangles; the diagram commutes. But as $1 : X \to X$ and $1 : Y \to Y$ are isomorphisms, so is $\Psi'w + v'\Phi$ by Proposition 1.1.20. We deduce that the top candidate triangle is isomorphic to the bottom, and the bottom is a distinguished triangle. □

From now on, when we speak of contractible candidate triangles, we will call them *contractible triangles.* Since we know by Proposition 1.3.8 that they are all distinguished, there is no risk of confusion. Remember that, in this book, the word "triangle", with no adjective preceding it, means distinguished triangle.

LEMMA 1.3.9. *Let the diagram*

$$\begin{array}{ccccccc} X & \xrightarrow{u} & Y & \xrightarrow{v} & Z & \xrightarrow{w} & \Sigma X \\ {\scriptstyle f}\downarrow & & {\scriptstyle g}\downarrow & & {\scriptstyle h}\downarrow & & {\scriptstyle \Sigma f}\downarrow \\ X' & \xrightarrow{u'} & Y' & \xrightarrow{v'} & Z' & \xrightarrow{w'} & \Sigma X' \end{array}$$

be a map of pre–triangles in the pre–triangulated category $\mathcal{T}$. Then the mapping cone is a pre–triangle.

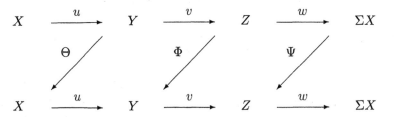

Proof: Let H be a decent homological functor. We need to show that H takes the mapping cone to an exact sequence. Because each row is a pre–triangle, we have two exact sequences

$$H(\Sigma^{-1}Z) \longrightarrow H(X) \longrightarrow H(Y) \longrightarrow H(Z) \longrightarrow H(\Sigma X)$$

and

$$H(\Sigma^{-1}Z') \longrightarrow H(X') \longrightarrow H(Y') \longrightarrow H(Z') \longrightarrow H(\Sigma X')$$

and a map between them. The mapping cone on this map of exact sequences is exact. But it agrees with what we get if we apply H to the candidate triangle

$$Y \oplus X' \xrightarrow{\begin{pmatrix} -v & 0 \\ g & u' \end{pmatrix}} Z \oplus Y' \xrightarrow{\begin{pmatrix} -w & 0 \\ h & v' \end{pmatrix}} \Sigma X \oplus Z' \xrightarrow{\begin{pmatrix} -\Sigma u & 0 \\ \Sigma f & w' \end{pmatrix}} \Sigma Y \oplus \Sigma X'.$$

Hence the candidate triangle is a pre–triangle. □

Now suppose we are given a morphism of triangles

$$\begin{array}{ccccccc} X & \xrightarrow{u} & Y & \xrightarrow{v} & Z & \xrightarrow{w} & \Sigma X \\ {\scriptstyle f}\downarrow & & {\scriptstyle g}\downarrow & & {\scriptstyle h}\downarrow & & {\scriptstyle \Sigma f}\downarrow \\ X' & \xrightarrow{u'} & Y' & \xrightarrow{v'} & Z' & \xrightarrow{w'} & \Sigma X' \end{array}$$

Then the mapping cone is a pre–triangle by Lemma 1.3.9. It turns out that it need not always be a triangle. Let us first analyse the trivial cases.

LEMMA 1.3.10. *The mapping cone on the zero map between triangles is a triangle.*

Proof: Consider the zero map

$$\begin{array}{ccccccc} X & \xrightarrow{u} & Y & \xrightarrow{v} & Z & \xrightarrow{w} & \Sigma X \\ {\scriptstyle 0}\downarrow & & {\scriptstyle 0}\downarrow & & {\scriptstyle 0}\downarrow & & {\scriptstyle 0}\downarrow \\ X' & \xrightarrow{u'} & Y' & \xrightarrow{v'} & Z' & \xrightarrow{w'} & \Sigma X' \end{array}$$

The mapping cone is the sequence

$$Y \oplus X' \xrightarrow{\begin{pmatrix} -v & 0 \\ 0 & u' \end{pmatrix}} Z \oplus Y' \xrightarrow{\begin{pmatrix} -w & 0 \\ 0 & v' \end{pmatrix}} \Sigma X \oplus Z' \xrightarrow{\begin{pmatrix} -\Sigma u & 0 \\ 0 & w' \end{pmatrix}} \Sigma Y \oplus \Sigma X'.$$

This is nothing other than the direct sum of the sequences

$$X' \xrightarrow{u'} Y' \xrightarrow{v'} Z' \xrightarrow{w'} \Sigma X'$$

and

$$Y \xrightarrow{-v} Z \xrightarrow{-w} \Sigma X \xrightarrow{-\Sigma u} \Sigma Y.$$

The first row is a triangle by hypothesis. The second row is a triangle by [TR2], and because

$$X \xrightarrow{u} Y \xrightarrow{v} Z \xrightarrow{w} \Sigma X$$

is. The direct sum is therefore a triangle, by Proposition 1.2.1. □

Because the mapping cone does not change, up to isomorphism, if we replace a map by a homotopic one, we immediately deduce

COROLLARY 1.3.11. *Given a map of triangles*

$$\begin{array}{ccccccc} X & \xrightarrow{u} & Y & \xrightarrow{v} & Z & \xrightarrow{w} & \Sigma X \\ {\scriptstyle f}\downarrow & & {\scriptstyle g}\downarrow & & {\scriptstyle h}\downarrow & & {\scriptstyle \Sigma f}\downarrow \\ X' & \xrightarrow{u'} & Y' & \xrightarrow{v'} & Z' & \xrightarrow{w'} & \Sigma X' \end{array}$$

if the map is homotopic to zero, then the mapping cone is a triangle. □

COROLLARY 1.3.12. *If either*

$$X \xrightarrow{u} Y \xrightarrow{v} Z \xrightarrow{w} \Sigma X$$

or

$$X' \xrightarrow{u'} Y' \xrightarrow{v'} Z' \xrightarrow{w'} \Sigma X'$$

is a contractible triangle, and the other is a triangle, then the mapping cone on any map

$$\begin{array}{ccccccc} X & \xrightarrow{u} & Y & \xrightarrow{v} & Z & \xrightarrow{w} & \Sigma X \\ {\scriptstyle f}\downarrow & & {\scriptstyle g}\downarrow & & {\scriptstyle h}\downarrow & & {\scriptstyle \Sigma f}\downarrow \\ X' & \xrightarrow{u'} & Y' & \xrightarrow{v'} & Z' & \xrightarrow{w'} & \Sigma X' \end{array}$$

is a triangle.

Proof: By Lemma 1.3.6, the map is homotopic to the zero map. Then by Corollary 1.3.11, the mapping cone is a triangle. □

This is as far as one gets without further assumptions. Now we come to the main definition of this section:

DEFINITION 1.3.13. *Let $\mathcal{T}$ be a pre–triangulated category. Then $\mathcal{T}$ is* triangulated *if it satisfies the further hypothesis*

TR4': *Given any diagram*

$$\begin{array}{ccccccc} X & \xrightarrow{u} & Y & \xrightarrow{v} & Z & \xrightarrow{w} & \Sigma X \\ {\scriptstyle f}\downarrow & & {\scriptstyle g}\downarrow & & & & {\scriptstyle \Sigma f}\downarrow \\ X' & \xrightarrow{u'} & Y' & \xrightarrow{v'} & Z' & \xrightarrow{w'} & \Sigma X' \end{array}$$

where the rows are triangles, there is, by [TR3], a way to choose an $h : Z \to Z'$ to make the diagram commutative. This h may be chosen so that the mapping cone

$$Y \oplus X' \xrightarrow{\begin{pmatrix} -v & 0 \\ g & u' \end{pmatrix}} Z \oplus Y' \xrightarrow{\begin{pmatrix} -w & 0 \\ h & v' \end{pmatrix}} \Sigma X \oplus Z' \xrightarrow{\begin{pmatrix} -\Sigma u & 0 \\ \Sigma f & w' \end{pmatrix}} \Sigma Y \oplus \Sigma X'$$

is a triangle.

DEFINITION 1.3.14. *A morphism of triangles will be called* good *if its mapping cone is a triangle.*

REMARK 1.3.15. [TR4'] can be restated as saying that any diagram

$$\begin{array}{ccccccc} X & \xrightarrow{u} & Y & \xrightarrow{v} & Z & \xrightarrow{w} & \Sigma X \\ {\scriptstyle f}\downarrow & & {\scriptstyle g}\downarrow & & & & \downarrow{\scriptstyle \Sigma f} \\ X' & \xrightarrow{u'} & Y' & \xrightarrow{v'} & Z' & \xrightarrow{w'} & \Sigma X' \end{array}$$

where the rows are distinguished triangles, may be completed to a good morphism of triangles.

The author does not know an example of a pre–triangulated category which is not triangulated.

1.4. Elementary properties of triangulated categories

We begin with the definition of homotopy cartesian squares.

DEFINITION 1.4.1. *Let $\mathcal{T}$ be a triangulated category. Then a commutative square*

$$\begin{array}{ccc} Y & \xrightarrow{f} & Z \\ {\scriptstyle g}\downarrow & & \downarrow{\scriptstyle g'} \\ Y' & \xrightarrow[f']{} & Z' \end{array}$$

is called homotopy cartesian *if there is a distinguished triangle*

$$Y \xrightarrow{\begin{pmatrix} g \\ -f \end{pmatrix}} Y' \oplus Z \xrightarrow{\begin{pmatrix} f' & g' \end{pmatrix}} Z' \xrightarrow{\partial} \Sigma Y$$

for some $\partial : Z' \to \Sigma Y$.

NOTATION 1.4.2. If

$$\begin{array}{ccc} Y & \xrightarrow{f} & Z \\ {\scriptstyle g}\downarrow & & \downarrow{\scriptstyle g'} \\ Y' & \xrightarrow[f']{} & Z' \end{array}$$

is a homotopy cartesian square, we call Y the *homotopy pullback* of

$$\begin{array}{ccc} & & Z \\ & & \downarrow{\scriptstyle g'} \\ Y' & \xrightarrow[f']{} & Z' \end{array}$$

and Z' the *homotopy pushout* of

$$\begin{array}{ccc} Y & \xrightarrow{f} & Z \\ {\scriptstyle g}\downarrow & & \\ Y' & & \end{array}$$

It follows from [TR1] that any diagram

$$\begin{array}{ccc} Y & \xrightarrow{f} & Z \\ {\scriptstyle g}\downarrow & & \\ Y' & & \end{array}$$

has a homotopy pushout; the morphism

$$Y \longrightarrow Y' \oplus Z$$

can be completed to a triangle

$$Y \xrightarrow{\begin{pmatrix} g \\ -f \end{pmatrix}} Y' \oplus Z \longrightarrow Z' \longrightarrow \Sigma Y$$

and this triangle defines a homotopy cartesian square

$$\begin{array}{ccc} Y & \xrightarrow{f} & Z \\ {\scriptstyle g}\downarrow & & \downarrow{\scriptstyle g'} \\ Y' & \xrightarrow[f']{} & Z'. \end{array}$$

By Remark 1.1.21, the homotopy pushout is unique up to non–canonical isomorphism. Also, any commutative square

$$\begin{array}{ccc} Y & \xrightarrow{f} & Z \\ {\scriptstyle g}\downarrow & & \downarrow \\ Y' & \longrightarrow & P \end{array}$$

corresponds to a map $Y' \oplus Z \to P$ so that the composite

$$Y \xrightarrow{\begin{pmatrix} g \\ -f \end{pmatrix}} Y' \oplus Z \longrightarrow P$$

vanishes. But $Hom(-, P)$ is a cohomological functor, and in particular takes the triangle

$$Y \xrightarrow{\begin{pmatrix} g \\ -f \end{pmatrix}} Y' \oplus Z \longrightarrow Z' \longrightarrow \Sigma Y$$

to a long exact sequence. We are given a map in $Hom(Y' \oplus Z, P)$ whose image in $Hom(Y, P)$ vanishes. It must therefore come from $Hom(Z', P)$. There is a map $Z' \to P$ (non–unique) which maps the square

$$\begin{array}{ccc} Y & \xrightarrow{f} & Z \\ {\scriptstyle g}\downarrow & & \downarrow{\scriptstyle g'} \\ Y' & \xrightarrow[f']{} & Z' \end{array}$$

to the square

$$\begin{array}{ccc} Y & \xrightarrow{f} & Z \\ {\scriptstyle g}\downarrow & & \downarrow \\ Y' & \longrightarrow & P. \end{array}$$

Dually, homotopy pullbacks always exist and are unique up to non–canonical isomorphism. And given a commutative square

$$\begin{array}{ccc} P & \longrightarrow & Z \\ \downarrow & & \downarrow{\scriptstyle g'} \\ Y' & \xrightarrow[f']{} & Z' \end{array}$$

there is always a map $P \to Y$ mapping this square to the homotopy pullback square

$$\begin{array}{ccc} Y & \xrightarrow{f} & Z \\ {\scriptstyle g}\downarrow & & \downarrow{\scriptstyle g'} \\ Y' & \xrightarrow[f']{} & Z'. \end{array}$$

LEMMA 1.4.3. *Let the following be a commutative diagram with triangles for rows*

$$\begin{array}{ccccccc} X & \xrightarrow{f} & Y & \longrightarrow & Z & \longrightarrow & \Sigma X \\ {\scriptstyle 1}\downarrow & & {\scriptstyle g}\downarrow & & & & {\scriptstyle 1}\downarrow \\ X & \xrightarrow{gf} & Y' & \longrightarrow & Z' & \longrightarrow & \Sigma X. \end{array}$$

It may be completed to a morphism of triangles

$$\begin{array}{ccccccc} X & \xrightarrow{f} & Y & \longrightarrow & Z & \longrightarrow & \Sigma X \\ {\scriptstyle 1}\downarrow & & {\scriptstyle g}\downarrow & & \downarrow & & {\scriptstyle 1}\downarrow \\ X & \xrightarrow{gf} & Y' & \longrightarrow & Z' & \xrightarrow{w'} & \Sigma X, \end{array}$$

so that

$$\begin{array}{ccc} Y & \longrightarrow & Z \\ \downarrow & & \downarrow \\ Y' & \longrightarrow & Z' \end{array}$$

is homotopy cartesian. In fact, the differential $\partial : Z' \to \Sigma Y$ may be chosen to be the composite

$$Z' \xrightarrow{w'} \Sigma X \xrightarrow{\Sigma f} \Sigma Y.$$

Proof: By [TR4'] we may complete

$$\begin{array}{ccccccc} X & \xrightarrow{f} & Y & \longrightarrow & Z & \longrightarrow & \Sigma X \\ {\scriptstyle 1}\downarrow & & {\scriptstyle g}\downarrow & & & & {\scriptstyle 1}\downarrow \\ X & \xrightarrow{gf} & Y' & \longrightarrow & Z' & \longrightarrow & \Sigma X. \end{array}$$

to a good morphism of triangles

$$\begin{array}{ccccccc} X & \xrightarrow{f} & Y & \longrightarrow & Z & \longrightarrow & \Sigma X \\ {\scriptstyle 1}\downarrow & & {\scriptstyle g}\downarrow & & \downarrow & & {\scriptstyle 1}\downarrow \\ X & \xrightarrow{gf} & Y' & \longrightarrow & Z' & \xrightarrow{w'} & \Sigma X. \end{array}$$

Then the mapping cone is a triangle

$$X \oplus Y \longrightarrow Y' \oplus Z \longrightarrow \Sigma X \oplus Z' \longrightarrow \Sigma X \oplus \Sigma Y.$$

An elementary computation [see Lemma 1.2.4] allows us to show that this triangle is isomorphic to the direct sum of the two candidate triangles

$$X \longrightarrow 0 \longrightarrow \Sigma X \longrightarrow \Sigma X$$

and

$$Y \longrightarrow Y' \oplus Z \longrightarrow Z' \longrightarrow \Sigma Y.$$

By Proposition 1.2.3, each direct summand of a triangle is a triangle. In particular,

$$Y \longrightarrow Y' \oplus Z \longrightarrow Z' \longrightarrow \Sigma Y$$

must be a distinguished triangle. It is also easy to compute that the differential is as in the statement of the Lemma. □

LEMMA 1.4.4. *Let*

$$\begin{array}{ccc} Y & \longrightarrow & Z \\ {\scriptstyle g}\downarrow & & \downarrow{\scriptstyle h} \\ Y' & \longrightarrow & Z' \end{array}$$

be a homotopy cartesian square. If

$$Y \xrightarrow{\ g\ } Y' \longrightarrow Y'' \longrightarrow \Sigma Y$$

is a triangle, then there is a triangle

$$Z \xrightarrow{\ h\ } Z' \longrightarrow Y'' \longrightarrow \Sigma Z$$

which completes the homotopy cartesian square to a map of triangles

$$\begin{array}{ccccccc} Y & \xrightarrow{\ g\ } & Y' & \longrightarrow & Y'' & \longrightarrow & \Sigma Y \\ \downarrow & & \downarrow & & {\scriptstyle 1}\downarrow & & \downarrow \\ Z & \xrightarrow{\ h\ } & Z' & \longrightarrow & Y'' & \longrightarrow & \Sigma Z. \end{array}$$

That is, the differential $Z' \to \Sigma Y$ is the composite

$$Z' \longrightarrow Y'' \longrightarrow \Sigma Y.$$

Proof: We know that the square

$$\begin{array}{ccc} Y & \longrightarrow & Z \\ {\scriptstyle g}\downarrow & & \downarrow{\scriptstyle h} \\ Y' & \longrightarrow & Z' \end{array}$$

is homotopy cartesian, in other words we have a triangle

$$Y \longrightarrow Y' \oplus Z \longrightarrow Z' \longrightarrow \Sigma Y.$$

But then the diagram

$$\begin{CD} Y @>>> Y'\oplus Z @>>> Z' @>>> \Sigma Y \\ @V{1}VV @VVV @. @V{1}VV \\ Y @>>> Y' @>>> Y'' @>>> \Sigma Y \end{CD}$$

may be completed to a good morphism of triangles

$$\begin{CD} Y @>>> Y'\oplus Z @>>> Z' @>>> \Sigma Y \\ @V{1}VV @VVV @VVV @V{1}VV \\ Y @>>> Y' @>>> Y'' @>>> \Sigma Y \end{CD}$$

and in particular the mapping cone is a triangle. The mapping cone can easily be written as a direct sum of the three candidate triangles

$$\begin{CD} Y' @>{1}>> Y' @>>> 0 @>>> \Sigma Y' \end{CD}$$

$$\begin{CD} Y @>>> 0 @>>> \Sigma Y @>{1}>> \Sigma Y \end{CD}$$

$$\begin{CD} Z @>>> Z' @>>> Y'' @>>> \Sigma Z \end{CD}$$

which must therefore all be triangles; in particular,

$$\begin{CD} Z @>>> Z' @>>> Y'' @>>> \Sigma Z \end{CD}$$

is a distinguished triangle. Computing the various maps, we deduce the map of triangles

$$\begin{CD} Y @>{g}>> Y' @>>> Y'' @>>> \Sigma Y \\ @VVV @VVV @V{1}VV @VVV \\ Z @>{h}>> Z' @>>> Y'' @>>> \Sigma Z. \end{CD}$$

of the Lemma. □

REMARK 1.4.5. Combining Lemmas 1.4.3 and 1.4.4, we have that good maps of triangles

$$\begin{CD} X @>{f}>> Y @>>> Z @>>> \Sigma X \\ @V{1}VV @V{g}VV @V{h}VV @V{1}VV \\ X @>{gf}>> Y' @>>> Z' @>>> \Sigma X \end{CD}$$

are closely related to homotopy cartesian squares

$$\begin{CD} Y @>>> Z \\ @V{g}VV @V{h}VV \\ Y' @>>> Z'. \end{CD}$$

One can pass back and forth from one to the other, of course not uniquely. By Lemma 1.4.3 a good map gives a homotopy cartesian square, and by Lemma 1.4.4, a homotopy cartesian square

$$\begin{array}{ccc} Y & \longrightarrow & Z \\ {\scriptstyle g}\downarrow & & {\scriptstyle h}\downarrow \\ Y' & \longrightarrow & Z' \end{array}$$

and a triangle

$$X \xrightarrow{f} Y \longrightarrow Z \longrightarrow \Sigma X$$

give a good morphism of triangles

$$\begin{array}{ccccccc} X & \xrightarrow{f} & Y & \longrightarrow & Z & \longrightarrow & \Sigma X \\ {\scriptstyle 1}\downarrow & & {\scriptstyle g}\downarrow & & \downarrow & & {\scriptstyle 1}\downarrow \\ X & \xrightarrow{gf} & Y' & \longrightarrow & Z' & \longrightarrow & \Sigma X. \end{array}$$

Putting together Lemmas 1.4.3 and 1.4.4 slightly differently, we deduce

PROPOSITION 1.4.6. *Let $\mathfrak{T}$ be a triangulated category. Let $f : X \to Y$ and $g : Y \to Y'$ be two composable morphisms. Let us be given triangles*

$$\begin{array}{ccccccc} X & \xrightarrow{f} & Y & \longrightarrow & Z & \longrightarrow & \Sigma X \\ X & \xrightarrow{gf} & Y' & \longrightarrow & Z' & \longrightarrow & \Sigma X \\ Y & \xrightarrow{g} & Y' & \longrightarrow & Y'' & \longrightarrow & \Sigma Y. \end{array}$$

Then we can complete this to a commutative diagram

$$\begin{array}{ccccccc} X & \xrightarrow{f} & Y & \longrightarrow & Z & \longrightarrow & \Sigma X \\ {\scriptstyle 1}\downarrow & & {\scriptstyle g}\downarrow & & \downarrow & & {\scriptstyle 1}\downarrow \\ X & \xrightarrow{gf} & Y' & \longrightarrow & Z' & \longrightarrow & \Sigma X \\ \downarrow & & \downarrow & & \downarrow & & \downarrow \\ 0 & \longrightarrow & Y'' & \xrightarrow{1} & Y'' & \longrightarrow & 0 \\ \downarrow & & \downarrow & & \downarrow & & \downarrow \\ \Sigma X & \xrightarrow{\Sigma f} & \Sigma Y & \longrightarrow & \Sigma Z & \longrightarrow & \Sigma^2 X \end{array}$$

where the first and second row and second column are our given three triangles, and every row and column in the diagram is a distinguished triangle.

Furthermore, the square

$$\begin{array}{ccc} Y & \longrightarrow & Z \\ \downarrow & & \downarrow \\ Y' & \longrightarrow & Z' \end{array}$$

is homotopy cartesian, with differential being given by the equal composites

$$Z' \longrightarrow \Sigma X \longrightarrow \Sigma Y,$$

$$Z' \longrightarrow Y'' \longrightarrow \Sigma Y.$$

Proof: By Lemma 1.4.3, the diagram

$$\begin{array}{ccccccc} X & \xrightarrow{f} & Y & \longrightarrow & Z & \longrightarrow & \Sigma X \\ {\scriptstyle 1}\downarrow & & {\scriptstyle g}\downarrow & & & & {\scriptstyle 1}\downarrow \\ X & \xrightarrow{gf} & Y' & \longrightarrow & Z' & \longrightarrow & \Sigma X \end{array}$$

may be completed to a morphism of triangles

$$\begin{array}{ccccccc} X & \xrightarrow{f} & Y & \longrightarrow & Z & \longrightarrow & \Sigma X \\ {\scriptstyle 1}\downarrow & & {\scriptstyle g}\downarrow & & \downarrow & & {\scriptstyle 1}\downarrow \\ X & \xrightarrow{gf} & Y' & \longrightarrow & Z' & \longrightarrow & \Sigma X, \end{array}$$

so that

$$\begin{array}{ccc} Y & \longrightarrow & Z \\ \downarrow & & \downarrow \\ Y' & \longrightarrow & Z' \end{array}$$

is homotopy cartesian. By Lemma 1.4.4, the homotopy cartesian square

$$\begin{array}{ccc} Y & \longrightarrow & Z \\ \downarrow & & \downarrow \\ Y' & \longrightarrow & Z' \end{array}$$

and the triangle

$$Y \longrightarrow Y' \longrightarrow Y'' \longrightarrow \Sigma Y$$

may be completed to a map of triangles

$$\begin{array}{ccccccc} Y & \xrightarrow{g} & Y' & \longrightarrow & Y'' & \longrightarrow & \Sigma Y \\ \downarrow & & \downarrow & & {\scriptstyle 1}\downarrow & & \downarrow \\ Z & \xrightarrow{h} & Z' & \longrightarrow & Y'' & \longrightarrow & \Sigma Z, \end{array}$$

and the diagram

$$\begin{array}{ccccccc}
X & \xrightarrow{f} & Y & \longrightarrow & Z & \longrightarrow & \Sigma X \\
{\scriptstyle 1}\downarrow & & {\scriptstyle g}\downarrow & & \downarrow & & {\scriptstyle 1}\downarrow \\
X & \xrightarrow{gf} & Y' & \longrightarrow & Z' & \longrightarrow & \Sigma X \\
\downarrow & & \downarrow & & \downarrow & & \downarrow \\
0 & \longrightarrow & Y'' & \xrightarrow{1} & Y'' & \longrightarrow & 0 \\
\downarrow & & \downarrow & & \downarrow & & \downarrow \\
\Sigma X & \xrightarrow{\Sigma f} & \Sigma Y & \longrightarrow & \Sigma Z & \longrightarrow & \Sigma^2 X
\end{array}$$

just assembles all this information together. □

REMARK 1.4.7. Proposition 1.4.6 is generally known as [TR4], or the Octahedral Axiom. The diagram whose existence the Proposition asserts is known as an octahedron. The reason for this name is that we do in fact have 8 triangles which can be assembled to an octahedron. The rows and the columns of the diagram give 4 distinguished triangles. But there are also 4 commutative triangles

$$\begin{array}{ccc}
X & \xrightarrow{f} & Y \\
{\scriptstyle 1}\downarrow & & {\scriptstyle g}\downarrow \\
X & \xrightarrow{gf} & Y'
\end{array}
\qquad\text{and}\qquad
\begin{array}{ccc}
Z & \longrightarrow & \Sigma X \\
\downarrow & & {\scriptstyle 1}\downarrow \\
Z' & \longrightarrow & \Sigma X
\end{array}$$

$$\begin{array}{ccc}
Y' & \longrightarrow & Y'' \\
\downarrow & & {\scriptstyle 1}\downarrow \\
Z' & \xrightarrow{h} & Y''
\end{array}
\qquad\text{and}\qquad
\begin{array}{ccc}
Y'' & \longrightarrow & \Sigma Y \\
{\scriptstyle 1}\downarrow & & \downarrow \\
Y'' & \longrightarrow & \Sigma Z.
\end{array}$$

These 8 triangles make an octahedron, and the reader is referred elsewhere for the study of its symmetries.

It may be shown that a pre–triangulated category satisfies [TR4] if and only if it satisfies [TR4']. We have shown the "if". The proof that any pre–triangulated category satisfying [TR4] also satisfies [TR4'] may be found in [**22**], Theorem 1.8.

1.5. Triangulated subcategories

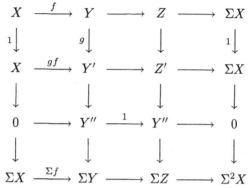

DEFINITION 1.5.1. *Let $\mathcal{T}$ be a triangulated category. A full additive subcategory $\mathcal{S}$ in $\mathcal{T}$ is called a* triangulated subcategory *if every object isomorphic to an object of $\mathcal{S}$ is in $\mathcal{S}$, if $\Sigma\mathcal{S} = \mathcal{S}$, and if for any distinguished*

triangle

$$X \longrightarrow Y \longrightarrow Z \longrightarrow \Sigma X$$

such that the objects X and Y are in $\mathcal{S}$, the object Z is also in $\mathcal{S}$.

REMARK 1.5.2. From [TR2] we easily deduce that if $\mathcal{S}$ is a triangulated subcategory of $\mathcal{T}$ and

$$X \longrightarrow Y \longrightarrow Z \longrightarrow \Sigma X$$

is a triangle in $\mathcal{T}$, then if any two of the objects X, Y or Z are in $\mathcal{S}$, so is the third.

DEFINITION 1.5.3. *Let $\mathcal{T}$ be a triangulated category, $\mathcal{S}$ a triangulated subcategory. We define a collection of morphisms $Mor_{\mathcal{S}} \subset \mathcal{T}$ by the following rule. A morphism $f : X \to Y$ belongs to $Mor_{\mathcal{S}}$ if and only if, in some triangle*

$$X \xrightarrow{\ f\ } Y \longrightarrow Z \longrightarrow \Sigma X,$$

the object Z lies in $\mathcal{S}$.

REMARK 1.5.4. Note that it is irrelevant which particular triangle

$$X \xrightarrow{\ f\ } Y \longrightarrow Z \longrightarrow \Sigma X$$

we take. By Remark 1.1.21 the object Z is unique up to isomorphism, and by Definition 1.5.1, $\mathcal{S}$ contains all objects in $\mathcal{T}$ isomorphic to its objects.

LEMMA 1.5.5. *Every isomorphism $f : X \to Y$ is in $Mor_{\mathcal{S}}$.*

Proof: Let $f : X \to Y$ be an isomorphism. By Corollary 1.2.6, the diagram

$$X \xrightarrow{\ f\ } Y \longrightarrow 0 \longrightarrow \Sigma X$$

is a triangle in $\mathcal{T}$. But since $\mathcal{S}$ is, among other things, an additive subcategory of $\mathcal{T}$, 0 must be in $\mathcal{S}$. Thus f is in $Mor_{\mathcal{S}}$. □

LEMMA 1.5.6. *Let $f : X \to Y$ and $g : Y \to Y'$ be two morphisms in $\mathcal{T}$. If any two of $f : X \to Y$, $g : Y \to Y'$ and $gf : X \to Y'$ lie in $Mor_{\mathcal{S}}$, then so does the third.*

Proof: By Proposition 1.4.6, there is a diagram of triangles

$$\begin{array}{ccccccc}
X & \xrightarrow{f} & Y & \longrightarrow & Z & \longrightarrow & \Sigma X \\
{\scriptstyle 1}\downarrow & & {\scriptstyle g}\downarrow & & \downarrow & & {\scriptstyle 1}\downarrow \\
X & \xrightarrow{gf} & Y' & \longrightarrow & Z' & \longrightarrow & \Sigma X \\
\downarrow & & \downarrow & & \downarrow & & \downarrow \\
0 & \longrightarrow & Y'' & \xrightarrow{1} & Y'' & \longrightarrow & 0 \\
\downarrow & & \downarrow & & \downarrow & & \downarrow \\
\Sigma X & \xrightarrow{\Sigma f} & \Sigma Y & \longrightarrow & \Sigma Z & \longrightarrow & \Sigma^2 X
\end{array}$$

Now f lies in $Mor_{\mathcal{S}}$ if and only if Z lies in $\mathcal{S}$, gf lies in $Mor_{\mathcal{S}}$ if and only if Z' lies in $\mathcal{S}$, and g lies in $Mor_{\mathcal{S}}$ if and only if Y'' lies in $\mathcal{S}$. From the triangle

$$Z \longrightarrow Z' \longrightarrow Y'' \longrightarrow \Sigma Z$$

we learn that if any two of Z, Z' and Y'' lie in $\mathcal{S}$, then so does the third.□

LEMMA 1.5.7. *If $\mathcal{S}$ is a triangulated subcategory of $\mathcal{T}$, then there is a subcategory of $\mathcal{T}$ whose objects are all the objects of $\mathcal{T}$, and whose morphisms are the ones in $Mor_{\mathcal{S}}$.*

Proof: Let X be an object of $\mathcal{T}$. From Lemma 1.5.5 we learn that the identity morphisms $1 : X \to X$, being isomorphisms, lie in $Mor_{\mathcal{S}}$.

But by Lemma 1.5.6 we know that the composite of two morphisms in $Mor_{\mathcal{S}}$ is again in $Mor_{\mathcal{S}}$. Thus $Mor_{\mathcal{S}}$ is a subcategory of $\mathcal{T}$. □

LEMMA 1.5.8. *Let the square*

$$\begin{array}{ccc}
Y & \xrightarrow{f} & Z \\
{\scriptstyle g}\downarrow & & \downarrow{\scriptstyle g'} \\
Y' & \xrightarrow[f']{} & Z'
\end{array}$$

be a homotopy cartesian square. Then f is in $Mor_{\mathcal{S}}$ if and only if f' is, and g is in $Mor_{\mathcal{S}}$ if and only if g' is. Another way to phrase this is that homotopy pushout and homotopy pullback of morphisms in $Mor_{\mathcal{S}}$ give morphisms in $Mor_{\mathcal{S}}$.

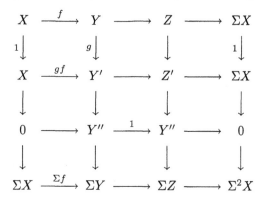

Proof: The two statement being transposes of each other, we will prove only that g is in $Mor_{\mathcal{S}}$ if and only if g' is. By Lemma 1.4.4 the homotopy cartesian square above may be completed to a morphism of triangles

$$\begin{array}{ccccccc} Y & \xrightarrow{g} & Y' & \longrightarrow & Y'' & \longrightarrow & \Sigma Y \\ {\scriptstyle f}\downarrow & & \downarrow{\scriptstyle f'} & & {\scriptstyle 1}\downarrow & & {\scriptstyle \Sigma f}\downarrow \\ Z & \xrightarrow[g']{} & Z' & \longrightarrow & Y'' & \longrightarrow & \Sigma Z. \end{array}$$

Now Y'' will lie in $\mathcal{S}$ precisely if both maps g and g' are in $Mor_{\mathcal{S}}$. This proves that g is in $Mor_{\mathcal{S}}$ if and only if g' is. □

1.6. Direct sums and products, and homotopy limits and colimits

DEFINITION 1.6.1. Let α be an infinite cardinal. A triangulated category $\mathcal{T}$ is said to satisfy [TR5(α)] if, in addition to the other axioms, the following holds.

TR5(α): For any set Λ of cardinality $< \alpha$, and any collection $\{X_\lambda, \lambda \in \Lambda\}$ of objects of $\mathcal{T}$, the coproduct $\coprod_{\lambda \in \Lambda} X_\lambda$ exists in $\mathcal{T}$. If $\mathcal{T}$ satisfies [TR5(α)] for all infinite cardinals α, we say $\mathcal{T}$ satisfies [TR5].

DEFINITION 1.6.2. *If the dual triangulated category $\mathcal{T}^{op}$ satisfies the condition [TR5(α)], we say $\mathcal{T}$ satisfies [TR5*(α)]. If the dual satisfies [TR5], we say that $\mathcal{T}$ satisfies [TR5*].*

REMARK 1.6.3. It follows from Proposition 1.2.1 and Remark 1.2.2 that coproducts and products of triangles are triangles. More precisely, if $\mathcal{T}$ satisfies [TR5*], then given any set of triangles, the product exists, and is a triangle by Proposition 1.2.1. Dually, if $\mathcal{T}$ satisfies [TR5], the coproduct of any collection of triangles exists and is a triangle.

DEFINITION 1.6.4. *Suppose $\mathcal{T}$ is a triangulated category, and assume it satisfies [TR5($\aleph_1$)]. That is, countable coproducts exist in $\mathcal{T}$. Let*

$$X_0 \xrightarrow{j_1} X_1 \xrightarrow{j_2} X_2 \xrightarrow{j_3} \cdots$$

be a sequence of objects and morphisms in $\mathcal{T}$. The homotopy colimit *of the sequence, denoted $\underrightarrow{\mathrm{Hocolim}}\, X_i$, is by definition given, up to non–canonical isomorphism, by the triangle*

$$\coprod_{i=0}^{\infty} X_i \xrightarrow{1-shift} \coprod_{i=0}^{\infty} X_i \longrightarrow \underrightarrow{\mathrm{Hocolim}}\, X_i \longrightarrow \Sigma\left\{\coprod_{i=0}^{\infty} X_i\right\}.$$

Here, the shift map $\coprod_{i=0}^{\infty} X_i \xrightarrow{shift} \coprod_{i=0}^{\infty} X_i$ is understood to be the direct sum of $j_{i+1} : X_i \to X_{i+1}$. In other words, the map $\{1 - shift\}$ is the infinite *matrix*

$$\begin{pmatrix} 1_{X_0} & 0 & 0 & 0 & \cdots \\ -j_1 & 1_{X_1} & 0 & 0 & \cdots \\ 0 & -j_2 & 1_{X_2} & 0 & \cdots \\ 0 & 0 & -j_3 & 1_{X_3} & \cdots \\ \vdots & \vdots & \vdots & \vdots & \end{pmatrix}$$

LEMMA 1.6.5. *If we have two sequences*

$$X_0 \longrightarrow X_1 \longrightarrow X_2 \longrightarrow \cdots$$

and

$$Y_0 \longrightarrow Y_1 \longrightarrow Y_2 \longrightarrow \cdots$$

then, non–canonically,

$$\underrightarrow{\text{Hocolim}}\ \{X_i \oplus Y_i\} = \left\{\underrightarrow{\text{Hocolim}}\ X_i\right\} \oplus \left\{\underrightarrow{\text{Hocolim}}\ Y_i\right\}.$$

Proof: Because the direct sum of two triangles is a triangle by Proposition 1.1.20, there is a triangle

$$\begin{array}{ccc} \left\{\coprod_{i=0}^{\infty} X_i\right\} \oplus \left\{\coprod_{i=0}^{\infty} Y_i\right\} & \xrightarrow{1-shift} & \left\{\coprod_{i=0}^{\infty} X_i\right\} \oplus \left\{\coprod_{i=0}^{\infty} Y_i\right\} \\ & & \downarrow \\ & & \left\{\underrightarrow{\text{Hocolim}}\ X_i\right\} \oplus \left\{\underrightarrow{\text{Hocolim}}\ Y_i\right\} \end{array}$$

and this triangle identifies

$$\left\{\underrightarrow{\text{Hocolim}}\ X_i\right\} \oplus \left\{\underrightarrow{\text{Hocolim}}\ Y_i\right\} \quad = \quad \underrightarrow{\text{Hocolim}}\ \{X_i \oplus Y_i\}.$$

□

LEMMA 1.6.6. *Let X be an object of $\mathcal{T}$, and let*

$$X \xrightarrow{\ 1\ } X \xrightarrow{\ 1\ } X \xrightarrow{\ 1\ } \cdots$$

be the sequence where all the maps are identities on X. Then

$$\underrightarrow{\text{Hocolim}}\, X = X,$$

even canonically.

Proof: The point is that the map

$$\coprod_{i=0}^{\infty} X \xrightarrow{1-shift} \coprod_{i=0}^{\infty} X$$

is split. Perhaps a simpler way to say this is that the map

$$X \oplus \left\{ \coprod_{i=0}^{\infty} X \right\} \xrightarrow{\left(i_0 \quad \{1-shift\} \right)} \coprod_{i=0}^{\infty} X$$

is an isomorphism, where $i_0 : X \longrightarrow \coprod_{i=0}^{\infty} X$ is the inclusion into the zeroth summand. In other words, the candidate triangle

$$\coprod_{i=0}^{\infty} X \xrightarrow{1-shift} \coprod_{i=0}^{\infty} X \xrightarrow{pr} X \xrightarrow{0} \Sigma \left\{ \coprod_{i=0}^{\infty} X \right\}$$

where $pr : \coprod_{i=0}^{\infty} X \longrightarrow X$ is the map which is 1 on every summand, is isomorphic to the sum of the two triangles

$$\coprod_{i=0}^{\infty} X \xrightarrow{1} \coprod_{i=0}^{\infty} X \longrightarrow 0 \longrightarrow \Sigma \left\{ \coprod_{i=0}^{\infty} X \right\}$$

and

$$0 \longrightarrow X \xrightarrow{1} X \longrightarrow 0.$$

Hence X is identified as $\underrightarrow{\text{Hocolim}}\, X$. □

LEMMA 1.6.7. *If in the sequence*

$$X_0 \xrightarrow{0} X_1 \xrightarrow{0} X_2 \xrightarrow{0} \cdots$$

all the maps are zero, then $\underrightarrow{\text{Hocolim}}\, X_i = 0$.

Proof: The point is that then the shift map in

$$\coprod_{i=0}^{\infty} X_i \xrightarrow{1-shift} \coprod_{i=0}^{\infty} X_i$$

vanishes. But by [TR0] there is a triangle

$$\coprod_{i=0}^{\infty} X_i \xrightarrow{1} \coprod_{i=0}^{\infty} X_i \longrightarrow 0 \longrightarrow \Sigma \left\{ \coprod_{i=0}^{\infty} X_i \right\}$$

and this identifies 0 as $\underrightarrow{\text{Hocolim}}\, X_i$. □

PROPOSITION 1.6.8. *Suppose $\mathcal{T}$ is a triangulated category satisfying [TR5($\aleph_1$)]. Let X be an object of $\mathcal{T}$, and suppose $e : X \to X$ is idempotent; that is, $e^2 = e$. Then e splits in $\mathcal{T}$. There are morphisms f and g below*

$$X \xrightarrow{f} Y \xrightarrow{g} X$$

with $gf = e$ and $fg = 1_Y$.

Proof: Cosider the two sequences

$$X \xrightarrow{\ e\ } X \xrightarrow{\ e\ } X \xrightarrow{\ e\ } \cdots$$

and

$$X \xrightarrow{1-e} X \xrightarrow{1-e} X \xrightarrow{1-e} \cdots$$

Let Y be the homotopy colimit of the first, and Z the homotopy colimit of the second. We will denote this by writing $Y = \underrightarrow{\mathrm{Hocolim}}\,(e)$ and $Z = \underrightarrow{\mathrm{Hocolim}}\,(1-e)$.

By Lemma 1.6.5, $Y \oplus Z$ is the homotopy colimit of the direct sum of the two sequences, that is of

$$X\oplus X \xrightarrow{\begin{pmatrix} e & 0 \\ 0 & 1-e \end{pmatrix}} X\oplus X \xrightarrow{\begin{pmatrix} e & 0 \\ 0 & 1-e \end{pmatrix}} X\oplus X \xrightarrow{\begin{pmatrix} e & 0 \\ 0 & 1-e \end{pmatrix}} \cdots$$

But the following is a map of sequences

$$\begin{array}{ccccccc}
X\oplus X & \xrightarrow{\begin{pmatrix} e & 0 \\ 0 & 1-e \end{pmatrix}} & X\oplus X & \xrightarrow{\begin{pmatrix} e & 0 \\ 0 & 1-e \end{pmatrix}} & X\oplus X & \xrightarrow{\begin{pmatrix} e & 0 \\ 0 & 1-e \end{pmatrix}} & \\
\begin{pmatrix} e & 1-e \\ 1-e & e \end{pmatrix}\Big\downarrow & & \begin{pmatrix} e & 1-e \\ 1-e & e \end{pmatrix}\Big\downarrow & & \begin{pmatrix} e & 1-e \\ 1-e & e \end{pmatrix}\Big\downarrow & & \\
X\oplus X & \xrightarrow[\begin{pmatrix} 1 & 0 \\ 0 & 0 \end{pmatrix}]{} & X\oplus X & \xrightarrow[\begin{pmatrix} 1 & 0 \\ 0 & 0 \end{pmatrix}]{} & X\oplus X & \xrightarrow[\begin{pmatrix} 1 & 0 \\ 0 & 0 \end{pmatrix}]{} &
\end{array}$$

and in fact, the vertical maps are isomorphisms. The map

$$\begin{pmatrix} e & 1-e \\ 1-e & e \end{pmatrix} : X\oplus X \longrightarrow X \oplus X$$

is its own inverse; its square is easily computed to be the identity.

It follows that the homotopy limits of the two sequences are the same. Thus $Y \oplus Z$ is the homotopy limit of the bottom row, and the bottom row decomposes as the direct sum of the two sequences

$$X \xrightarrow{\ 1\ } X \xrightarrow{\ 1\ } X \xrightarrow{\ 1\ } \cdots$$

and

$$X \xrightarrow{\ 0\ } X \xrightarrow{\ 0\ } X \xrightarrow{\ 0\ } \cdots$$

By Lemma 1.6.6, the homotopy colimit of the first sequence is X, while by Lemma 1.6.7, the homotopy colimit of the second sequence is 0. The homotopy colimit of the sum, which is $Y \oplus Z$, is therefore isomorphic to $X \oplus 0 = X$.

More concretely, consider the maps of sequences

$$\begin{array}{ccccccc} X & \xrightarrow{e} & X & \xrightarrow{e} & X & \xrightarrow{e} & \cdots \\ {\scriptstyle e}\downarrow & & {\scriptstyle e}\downarrow & & {\scriptstyle e}\downarrow & & \\ X & \xrightarrow{1} & X & \xrightarrow{1} & X & \xrightarrow{1} & \cdots \end{array}$$

and

$$\begin{array}{ccccccc} X & \xrightarrow{1-e} & X & \xrightarrow{1-e} & X & \xrightarrow{1-e} & \cdots \\ {\scriptstyle 1-e}\downarrow & & {\scriptstyle 1-e}\downarrow & & {\scriptstyle 1-e}\downarrow & & \\ X & \xrightarrow{1} & X & \xrightarrow{1} & X & \xrightarrow{1} & \cdots \end{array}$$

What we have shown is that the induced maps on homotopy colimits, that is $g : Y \to X$ and $g' : Z \to X$ can be chosen so that the sum $Y \oplus Z \to X$ is an isomorphism.

In the sequence

$$X \xrightarrow{e} X \xrightarrow{e} X \xrightarrow{e} \cdots$$

defining Y as the homotopy colimit, we get a map $f : X \to Y$, just the map from a finite term to the colimit. In the sequence

$$X \xrightarrow{1} X \xrightarrow{1} X \xrightarrow{1} \cdots$$

the map from the finite terms to the homotopy colimit is the identity. We deduce a commutative square

$$\begin{array}{ccc} X & \xrightarrow{f} & Y \\ {\scriptstyle e}\downarrow & & \downarrow{\scriptstyle g} \\ X & \xrightarrow{1} & X. \end{array}$$

Similarly, from the other sequence we deduce a commutative square

$$\begin{array}{ccc} X & \xrightarrow{f'} & Z \\ {\scriptstyle 1-e}\downarrow & & \downarrow{\scriptstyle g'} \\ X & \xrightarrow{1} & X. \end{array}$$

In other words, we conclude in total that $e = gf$ and $1 - e = g'f'$. The composite

$$X \xrightarrow{\begin{pmatrix} f \\ f' \end{pmatrix}} Y \oplus Z \xrightarrow{\begin{pmatrix} g & g' \end{pmatrix}} X$$

is $e+(1-e) = 1$. Since we know that the map $Y \oplus Z \to X$ is an isomorphism, it follows that the map $X \to Y \oplus Z$ above is its (two-sided) inverse. The composite in the other order is also the identity. In particular, $fg = 1_Y$ and $f'g' = 1_Z$. □

REMARK 1.6.9. Dually, if $\mathcal{T}$ satisfies [TR5*($\aleph_1$)], then idempotents also split.

REMARK 1.6.10. In this entire section, we have used nothing more than countable coproducts and products.

1.7. Some weak "functoriality" for homotopy limits and colimits

In the last section we saw the definition and elementary properties of the homotopy colimit of a sequence. We also saw how homotopy colimits can be useful; for example, they prove that idempotents split. See Proposition 1.6.8. Since homotopy colimits are defined by a triangle, they are not in any reasonable sense functorial. But they do have some good properties. Let us mention one here.

LEMMA 1.7.1. *Let $\mathcal{T}$ be a triangulated category satisfying [TR5($\aleph_1$)]. Suppose we are given a sequence of objects and morphisms in $\mathcal{T}$*

$$X_0 \longrightarrow X_1 \longrightarrow X_2 \longrightarrow \cdots$$

Suppose we take any increasing sequence of integers

$$0 \le i_0 < i_1 < i_2 < i_3 < \cdots$$

Then we can form the subsequence

$$X_{i_0} \longrightarrow X_{i_1} \longrightarrow X_{i_2} \longrightarrow \cdots$$

The two sequences have isomorphic homotopy colimits.

Proof: For each integer n, define j_n to be the smallest i_m such that $n \le i_m$. Then there is a map of sequences

$$\begin{array}{ccccccc} X_0 & \longrightarrow & X_1 & \longrightarrow & X_2 & \longrightarrow & \cdots \\ \downarrow & & \downarrow & & \downarrow & & \\ X_{j_0} & \longrightarrow & X_{j_1} & \longrightarrow & X_{j_2} & \longrightarrow & \cdots \end{array}$$

This seems much more complicated than it actually is. If we start with the sequence

$$0 < 2 < 5 < \cdots$$

our map of sequences is simply

$$\begin{array}{ccccccccccccc} X_0 & \longrightarrow & X_1 & \longrightarrow & X_2 & \longrightarrow & X_3 & \longrightarrow & X_4 & \longrightarrow & X_5 & \longrightarrow \\ \downarrow & & \downarrow & & \downarrow & & \downarrow & & \downarrow & & \downarrow & \\ X_0 & \longrightarrow & X_2 & \longrightarrow & X_2 & \longrightarrow & X_5 & \longrightarrow & X_5 & \longrightarrow & X_5 & \longrightarrow \end{array}$$

The map of sequences yields a commutative square

$$\begin{array}{ccc}
\displaystyle\coprod_{n=0}^{\infty} X_n & \xrightarrow{1-shift} & \displaystyle\coprod_{n=0}^{\infty} X_n \\
\downarrow & & \downarrow \\
\displaystyle\coprod_{n=0}^{\infty} X_{j_n} & \xrightarrow{1-shift} & \displaystyle\coprod_{n=0}^{\infty} X_{j_n}
\end{array}$$

and the point is that this commutative square is homotopy cartesian (see Definition 1.4.1). In fact, the zero map is the differential. In the sequence

$$\coprod_{n=0}^{\infty} X_n \longrightarrow \left\{\coprod_{n=0}^{\infty} X_n\right\} \oplus \left\{\coprod_{n=0}^{\infty} X_{j_n}\right\} \longrightarrow \coprod_{n=0}^{\infty} X_{j_n}$$

both maps are split, expressing the middle as the direct sum of the two outside terms. We leave it to the reader to check this fact.

By Lemma 1.4.4, the homotopy commutative square may be completed to a morphism of triangles

$$\begin{array}{ccccccc}
\displaystyle\coprod_{n=0}^{\infty} X_n & \xrightarrow{1-shift} & \displaystyle\coprod_{n=0}^{\infty} X_n & \longrightarrow & Y & \longrightarrow & \Sigma\left\{\displaystyle\coprod_{n=0}^{\infty} X_n\right\} \\
\downarrow & & \downarrow & & \downarrow & & \downarrow \\
\displaystyle\coprod_{n=0}^{\infty} X_{j_n} & \xrightarrow{1-shift} & \displaystyle\coprod_{n=0}^{\infty} X_{j_n} & \longrightarrow & Y & \longrightarrow & \Sigma\left\{\displaystyle\coprod_{n=0}^{\infty} X_{j_n}\right\}
\end{array}$$

and Y is identified as both $\underrightarrow{\mathrm{Hocolim}}\, X_n$ and $\underrightarrow{\mathrm{Hocolim}}\, X_{j_n}$.

Now we need to identify $\underrightarrow{\mathrm{Hocolim}}\, X_{j_n}$ with $\underrightarrow{\mathrm{Hocolim}}\, X_{i_n}$; recall that the sequence

$$X_{j_0} \longrightarrow X_{j_1} \longrightarrow X_{j_2} \longrightarrow \cdots$$

is obtained from the sequence

$$X_{i_0} \longrightarrow X_{i_1} \longrightarrow X_{i_2} \longrightarrow \cdots$$

by repeating many of the terms. The reader can check that the triangle

$$\coprod_{n=0}^{\infty} X_{j_n} \xrightarrow{1-shift} \coprod_{n=0}^{\infty} X_{j_n} \longrightarrow Y \longrightarrow \Sigma\left\{\coprod_{n=0}^{\infty} X_{j_n}\right\}$$

is isomorphic to the direct sum of two candidate triangles

$$\begin{array}{ccccccc} \displaystyle\coprod_{n=0}^{\infty} X_{i_n} & \xrightarrow{1-shift} & \displaystyle\coprod_{n=0}^{\infty} X_{i_n} & \longrightarrow & Y & \longrightarrow & \Sigma\left\{\displaystyle\coprod_{n=0}^{\infty} X_{i_n}\right\} \\ \displaystyle\coprod_{n=0}^{\infty} X_{i_n}^{\oplus a_n} & \longrightarrow & \displaystyle\coprod_{n=0}^{\infty} X_{i_n}^{\oplus a_n} & \longrightarrow & 0 & \longrightarrow & \Sigma\left\{\displaystyle\coprod_{n=0}^{\infty} X_{i_n}^{\oplus a_n}\right\} \end{array}$$

where $a_n = i_n - i_{n-1} - 1$, with i_{-1} being defined to be -1. The second distinguished triangle above corresponds to the "repeated terms". By Proposition 1.2.3 both candidate triangles are triangles, and in particular the triangle

$$\coprod_{n=0}^{\infty} X_{i_n} \xrightarrow{1-shift} \coprod_{n=0}^{\infty} X_{i_n} \longrightarrow Y \longrightarrow \Sigma\left\{\coprod_{n=0}^{\infty} X_{i_n}\right\}$$

allows us to identify Y as $\underrightarrow{\mathrm{Hocolim}}\, X_{i_n}$. □

REMARK 1.7.2. Lemma 1.7.1 is in fact very useful. Given a sequence

$$X_0 \longrightarrow X_1 \longrightarrow X_2 \longrightarrow \cdots$$

the lemma tells us that any two subsequences have isomorphic homotopy colimits.

1.8. History of the results in Chapter 1

Triangulated categories were defined independently and around the same time by Puppe and Verdier. Puppe works with the homotopy category as the main example. His axiomatic description is that of a pre–triangulated category. [TR4] is missing. Verdier's main example was the derived category, and Verdier discovered [TR4], which in the homotopy theory literature is sometimes referred to as the Verdier axiom.

Puppe's work may be found in [**27**]. Verdier's original work was in his PhD thesis, which was only published very recently in [**36**]. There is an account of some of the work in Chapter 1 of Hartshorne's [**19**], and a later and more general treatment in Verdier's [**35**].

Because Verdier's thesis remained unpublished for so long, some of his results were independently rediscovered later, and are generally attributed to the rediscoverers.

The idea that one should systematically study coproducts in triangulated categories originated later, and from the homotopy theorists. For example, the proof of Theorem 1.2.1, which asserts that the product of triangles is a triangle, may be found in Margolis' book [**21**]. The notion of the homotopy colimit of a sequence (Definition 1.6.4) is basically Milnor's mapping telescope. The use of mapping telescopes to split idempotents, as in Proposition 1.6.8, is a fairly standard application, although I do not know a classical reference.

The only part of the Chapter which is not completely classical is the presentation of [TR4]. Verdier's version of the axiom is the existence of octahedra, and it is given in Proposition 1.5.6. In Proposition 1.5.6, with complete disregard to the history of the subject, we show that a pre–triangulated category satisfying [TR4'] must also satisfy [TR4]. But [TR4'] came long after [TR4], and we do not give here the proof that any pre–triangulated category satisfying [TR4] also satisfies [TR4']. This is a true fact, for which the reader is referred to the author's [**22**]. Classically, triangulated categories were defined to be pre–triangulated categories satisfying [TR4].

The discussion of mapping cones on maps of triangles, their basic properties, and whether they are distinguished triangles, is all a departure from the classical literature; the classical literature says nothing about any of this. The reader will find a much amplified version of this discussion in [**22**].

CHAPTER 2

Triangulated functors and localizations of triangulated categories

2.1. Verdier localization and thick subcategories

DEFINITION 2.1.1. *Let $\mathcal{D}_1$, $\mathcal{D}_2$ be triangulated categories. A triangulated functor $F : \mathcal{D}_1 \to \mathcal{D}_2$ is an additive functor $F : \mathcal{D}_1 \to \mathcal{D}_2$ together with natural isomorphisms*

$$\phi_X : F(\Sigma(X)) \longrightarrow \Sigma(F(X))$$

such that for any distinguished triangle

$$X \xrightarrow{\quad u \quad} Y \xrightarrow{\quad v \quad} Z \xrightarrow{\quad w \quad} \Sigma X$$

in $\mathcal{D}_1$ the candidate triangle

$$F(X) \xrightarrow{F(u)} F(Y) \xrightarrow{F(v)} F(Z) \xrightarrow{\phi_X \circ F(w)} \Sigma(F(X))$$

is a distinguished triangle in $\mathcal{D}_2$.

We remind the reader of the definition of triangulated subcategories (see Section 1.5)

DEFINITION 1.5.1 *Let $\mathcal{D}$ be a triangulated category. A full additive subcategory $\mathcal{C}$ in $\mathcal{D}$ is called a* triangulated subcategory *if every object isomorphic to an object of $\mathcal{C}$ is in $\mathcal{C}$, and the inclusion $\mathcal{C} \hookrightarrow \mathcal{D}$ is a triangulated functor, as in Definition 2.1.1. We assume further that*

$$\phi_X : 1(\Sigma(X)) \longrightarrow \Sigma(1(X))$$

is the identity on ΣX.

REMARK 2.1.2. To say that the inclusion functor is triangulated comes down concretely to saying that triangles in $\mathcal{C}$ are also triangles in $\mathcal{D}$. Because every morphism can be completed to a triangle and any two triangles on the same morphism are isomorphic (see Remark 1.1.21), the triangles in $\mathcal{C}$ must just be the triangles in $\mathcal{D}$ whose objects happen to lie in $\mathcal{C}$. And if a morphism lies in $\mathcal{C}$, so must the triangle on it.

DEFINITION 2.1.3. *Let $F : \mathcal{D} \to \mathcal{T}$ be a triangulated functor. The* kernel *of F is defined to be the full subcategory $\mathcal{C}$ of $\mathcal{D}$ whose objects map to objects of $\mathcal{T}$ isomorphic to 0. That is,*

$$\mathcal{C} = \{x \in Ob(\mathcal{D}) | F(x) \text{ is isomorphic to } 0\}.$$

LEMMA 2.1.4. *Let $F : \mathcal{D} \to \mathcal{T}$ be a triangulated functor. Then the kernel $\mathcal{C}$ of F is a triangulated subcategory of $\mathcal{D}$.*

Proof: An object $x \in \mathcal{D}$ is in the kernel if and only if $F(x)$ is isomorphic to 0. But $F(x)$ is isomorphic to 0 if and only if so is $\Sigma F(x) = F(\Sigma x)$. Thus Σx is in the kernel if and only if x is. Also, if

$$x \longrightarrow y \longrightarrow z \longrightarrow \Sigma x$$

is a triangle in $\mathcal{D}$, then

$$F(x) \longrightarrow F(y) \longrightarrow F(z) \longrightarrow F(\Sigma x)$$

is a triangle in $\mathcal{T}$. If $F(x)$ and $F(y)$ are isomorphic to 0, then by Remark 1.1.21 the above triangle must be isomorphic in $\mathcal{T}$ to

$$0 \longrightarrow 0 \longrightarrow 0 \longrightarrow 0$$

and in particular, $F(z)$ is isomorphic to 0. Thus, if x and y are in $\mathcal{C}$, so is z. □

More is in fact true. One can prove

LEMMA 2.1.5. *Let $F : \mathcal{D} \to \mathcal{T}$ be a triangulated functor. Let $\mathcal{C} \subset \mathcal{D}$ be the kernel of F. If $x \oplus y$ is an object of $\mathcal{C}$, then so are the direct summands x and y.*

Proof: Since F is an additive functor, $F(x \oplus y) = F(x) \oplus F(y)$. But then if $F(x \oplus y)$ is isomorphic to zero, so are $F(x)$ and $F(y)$, since they are direct summands of 0. □

DEFINITION 2.1.6. *A subcategory $\mathcal{C}$ of a triangulated category $\mathcal{D}$ is called* thick *if it is triangulated, and it contains all direct summands of its objects.*

REMARK 2.1.7. Rephrased in terms of Definition 2.1.6, Lemma 2.1.5 says that the kernel of a triangulated functor is thick.

The main theorem of this section, due essentially to Verdier, is

THEOREM 2.1.8. *Let $\mathcal{D}$ be a triangulated category, $\mathcal{C} \subset \mathcal{D}$ a triangulated subcategory (not necessarily thick). Then there is a universal functor $F : \mathcal{D} \to \mathcal{T}$ with $\mathcal{C} \subset ker(F)$. In other words, there exists a triangulated category $\mathcal{D}/\mathcal{C}$, and a triangulated functor $F_{univ} : \mathcal{D} \longrightarrow \mathcal{D}/\mathcal{C}$ so that $\mathcal{C}$ is in the kernel of F_{univ}, and F_{univ} is universal with this property. If $F : \mathcal{D} \to \mathcal{T}$*

is a triangulated functor whose kernel contains $\mathcal{C}$, then it factors uniquely as

$$\mathcal{D} \xrightarrow{F_{univ}} \mathcal{D}/\mathcal{C} \longrightarrow \mathcal{T}.$$

REMARK 2.1.9. The quotient category $\mathcal{D}/\mathcal{C}$ is called the Verdier quotient of $\mathcal{D}$ by $\mathcal{C}$; the natural map $F_{univ} : \mathcal{D} \longrightarrow \mathcal{D}/\mathcal{C}$ is called the Verdier localisation map.

REMARK 2.1.10. Note that we are not assuming the subcategory $\mathcal{C} \subset \mathcal{D}$ to be thick, and we do not conclude that $\mathcal{C}$ is the kernel of $F_{univ} : \mathcal{D} \longrightarrow \mathcal{D}/\mathcal{C}$. We only claim that $\mathcal{C}$ is *contained* in the kernel. By Remark 2.1.7 the kernel of F_{univ} is thick, and it contains $\mathcal{C}$. It turns out to be the smallest thick subcategory containing $\mathcal{C}$. We will learn a very precise description of the kernel in the course of the proof. The kernel is the full subcategory of all objects which are direct summands in $\mathcal{D}$ of the objects of $\mathcal{C}$. See Lemma 2.1.33. In other words, up to splitting idempotents in $\mathcal{C}$, the kernel is just $\mathcal{C}$.

The proof of Theorem 2.1.8 will proceed by a series of easy lemmas, in which we will define the category $\mathcal{D}/\mathcal{C}$ and prove its universality. The objects of $\mathcal{D}/\mathcal{C}$ are simple; they are just the objects of $\mathcal{D}$, and the functor F_{univ} is the identity on objects.

Now we turn to defining the morphisms in $\mathcal{D}/\mathcal{C}$. Recall Definition 1.5.3. For the triangulated subcategory $\mathcal{C} \subset \mathcal{D}$, we defined a category $Mor_{\mathcal{C}} \subset \mathcal{D}$. A morphism $f : X \to Y$ lies in $Mor_{\mathcal{C}}$ if and only if in the triangle

$$X \xrightarrow{f} Y \longrightarrow Z \longrightarrow \Sigma X$$

the object Z lies in $\mathcal{C}$. We refer the reader to Section 1.5 for the elementary properties of $Mor_{\mathcal{C}}$.

Note that a triangulated functor F takes an object Z to zero if and only if it takes the distinguished triangle

$$X \xrightarrow{f} Y \longrightarrow Z \longrightarrow \Sigma X$$

to a triangle isomorphic to the image of

$$X \xrightarrow{1} X \longrightarrow 0 \longrightarrow \Sigma X;$$

in other words $F(f) : F(X) \to F(Y)$ must be an isomorphism. Therefore, in the category $\mathcal{D}/\mathcal{C}$ all the morphisms in $Mor_{\mathcal{C}}$ will become invertible. It is natural therefore to define

DEFINITION 2.1.11. *For any two objects X, Y in $\mathcal{D}$ let $\alpha(X, Y)$ be the class of diagrams of the form*

$$\begin{array}{ccc} & Z & \\ {}^{f}\swarrow & & \searrow{}^{g} \\ X & & Y \end{array}$$

such that f belongs to $Mor_{\mathcal{C}}$. Consider a relation $R(X,Y)$ on $\alpha(X,Y)$ such that $[(Z,f,g),(Z',f',g')]$ belongs to $R(X,Y)$ if and only if there is an element (Z'',f'',g'') in $\alpha(X,Y)$ and morphisms

$$u : Z'' \to Z$$

$$v : Z'' \to Z'$$

which make the diagram

$$\begin{array}{ccccc} & & Z' & & \\ & \overset{f'}{\swarrow} & v\uparrow & \overset{g'}{\searrow} & \\ X & \overset{f''}{\longleftarrow} & Z'' & \overset{g''}{\longrightarrow} & Y \\ & \overset{f}{\nwarrow} & u\downarrow & \overset{g}{\nearrow} & \\ & & Z & & \end{array}$$

commute.

LEMMA 2.1.12. *With the notation as in Definition 2.1.11, u and v must be in $Mor_{\mathcal{C}}$.*

Proof: The two cases being symmetric, it suffices to consider u. But $fu = f''$, and f and f'' are in $Mor_{\mathcal{C}}$. By Lemma 1.5.6, it follows that u is in $Mor_{\mathcal{C}}$. □

REMARK 2.1.13. Elements of $\alpha(X,Y)$ should be thought of as maps gf^{-1} in $\mathcal{D}/\mathcal{C}$, which would be well defined because in $\mathcal{D}/\mathcal{C}$ any $f \in Mor_{\mathcal{C}}$ is invertible. The existence of a diagram as in $R(X,Y)$, that is

$$\begin{array}{ccccc} & & Z' & & \\ & \overset{f'}{\swarrow} & v\uparrow & \overset{g'}{\searrow} & \\ X & \overset{f''}{\longleftarrow} & Z'' & \overset{g''}{\longrightarrow} & Y \\ & \overset{f}{\nwarrow} & u\downarrow & \overset{g}{\nearrow} & \\ & & Z & & \end{array}$$

says in particular

$$\begin{aligned} gf^{-1} &= guu^{-1}f^{-1} \\ &= g''\{f''\}^{-1} \\ &= g'vv^{-1}\{f'\}^{-1} \\ &= g'\{f'\}^{-1} \end{aligned}$$

and it is therefore natural to identify (Z,f,g) with (Z',f',g').

LEMMA 2.1.14. *The relation $R(X,Y)$ is an equivalence relation.*

Proof: The only thing we need to check is that R is transitive. Let (Z_i, f_i, g_i), $i = 1, 2, 3$ be three elements in $\alpha(X,Y)$ such that

$$\big[(Z_1, f_1, g_1), (Z_2, f_2, g_2)\big] \qquad \text{and} \qquad \big[(Z_2, f_2, g_2), (Z_3, f_3, g_3)\big]$$

both belong to $R(X,Y)$. Then there are elements (Z,p,q), (Z',p',q') in $\alpha(X,Y)$ and morphisms

$$u : Z \to Z_1, \qquad v : Z \to Z_2$$

$$u' : Z' \to Z_2 \qquad v' : Z' \to Z_3$$

which make the corresponding diagrams

$$\begin{array}{ccccc} & & Z_1 & & \\ & \overset{f_1}{\swarrow} & u \uparrow & \overset{g_1}{\searrow} & \\ X & \longleftarrow & Z & \longrightarrow & Y \\ & \overset{f_2}{\nwarrow} & v \downarrow & \overset{g_2}{\nearrow} & \\ & & Z_2 & & \end{array}$$

and

$$\begin{array}{ccccc} & & Z_2 & & \\ & \overset{f_2}{\swarrow} & u' \uparrow & \overset{g_2}{\searrow} & \\ X & \longleftarrow & Z' & \longrightarrow & Y \\ & \overset{f_3}{\nwarrow} & v' \downarrow & \overset{g_3}{\nearrow} & \\ & & Z_3 & & \end{array}$$

commutative. Consider a homotopy pullback diagram as in Notation 1.4.2

$$\begin{array}{ccc} Z'' & \xrightarrow{\ w\ } & Z \\ {\scriptstyle w'}\big\downarrow & & \big\downarrow{\scriptstyle v} \\ Z' & \xrightarrow{\ u'\ } & Z_2 \end{array}$$

Lemma 2.1.12 says that v and u' are in $Mor_{\mathcal{C}}$, and from Lemma 1.5.8 we deduce that so are w and w'. One can easily see now that $(Z'', f_2 \circ v \circ w, g_2 \circ v \circ w)$ is an element of $\alpha(X,Y)$, and that the morphisms morphisms

$$Z'' \to Z \to Z_1$$

$$Z'' \to Z' \to Z_3$$

make the diagram

$$\begin{array}{ccccc} & & Z_1 & & \\ & \overset{f_1}{\swarrow} & \uparrow & \overset{g_1}{\searrow} & \\ X & \longleftarrow & Z'' & \longrightarrow & Y \\ & \overset{f_3}{\nwarrow} & \downarrow & \overset{g_3}{\nearrow} & \\ & & Z_3 & & \end{array}$$

commutative. Therefore the pair $[(Z_1, f_1, g_1), (Z_3, f_3, g_3)]$ must belong to $R(X,Y)$. □

DEFINITION 2.1.15. *We denote by* $''Hom''_{\mathcal{D}/\mathcal{C}}(X,Y)$ *the class of equivalence classes in* $\alpha(X,Y)$ *with respect to* $R(X,Y)$.

For an element (W_1, f_1, g_1) in $\alpha(X,Y)$ and an element (W_2, f_2, g_2) in $\alpha(Y,Z)$, consider the diagram

$$\begin{array}{ccccc} W_3 & \xrightarrow{u} & W_2 & \xrightarrow{g_2} & Z \\ {\scriptstyle v}\downarrow & & {\scriptstyle f_2}\downarrow & & \\ W_1 & \xrightarrow{g_1} & Y & & \\ {\scriptstyle f_1}\downarrow & & & & \\ X & & & & \end{array}$$

where the square

$$\begin{array}{ccc} W_3 & \xrightarrow{u} & W_2 \\ {\scriptstyle v}\downarrow & & {\scriptstyle f_2}\downarrow \\ W_1 & \xrightarrow{g_1} & Y \end{array}$$

is obtained by homotopy pullback. Since f_2 is in $Mor_{\mathcal{C}}$, by Lemma 1.5.8 so is its homotopy pullback v. Since f_1 is also in $Mor_{\mathcal{C}}$, so is the composite $f_1 v$. We deduce therefore

LEMMA 2.1.16. *There is a map*

$$\alpha(X,Y) \times \alpha(Y,Z) \to {''Hom''_{\mathcal{D}/\mathcal{C}}}(X,Z),$$

sending an element (W_1, f_1, g_1) *in* $\alpha(X,Y)$ *and an element* (W_2, f_2, g_2) *in* $\alpha(Y,Z)$ *to* $(W_3, f_1 \circ v, g_2 \circ u)$ *in* $''Hom''_{\mathcal{D}/\mathcal{C}}(X,Z)$.

Proof: Since $f_1 \circ v$ is in $Mor_{\mathcal{C}}$, the triple $(W_3, f_1 \circ v, g_2 \circ u)$ is indeed in $''Hom''_{\mathcal{D}/\mathcal{C}}(X, Z)$. And since the homotopy pullback square

$$\begin{array}{ccc} W_3 & \xrightarrow{u} & W_2 \\ {\scriptstyle v}\downarrow & & \downarrow{\scriptstyle f_2} \\ W_1 & \xrightarrow{g_1} & Y \end{array}$$

is unique up to isomorphism in $\mathcal{D}$, and all isomorphisms are in $Mor_{\mathcal{C}}$, the image in $''Hom''_{\mathcal{D}/\mathcal{C}}(X, Z)$ is well–defined. □

Next comes a little lemma which is useful in identifying these products.

LEMMA 2.1.17. *Given an element (W_1, f_1, g_1) in $\alpha(X, Y)$, an element (W_2, f_2, g_2) in $\alpha(Y, Z)$, and a commutative diagram*

$$\begin{array}{ccccc} P & \xrightarrow{u'} & W_2 & \xrightarrow{g_2} & Z \\ {\scriptstyle v'}\downarrow & & \downarrow{\scriptstyle f_2} & & \\ W_1 & \xrightarrow{g_1} & Y & & \\ {\scriptstyle f_1}\downarrow & & & & \\ X & & & & \end{array}$$

with v' in $mor_{\mathcal{C}}$, the elements $(W_3, f_1 v, g_2 u)$ and $(P, f_1 v', g_2 u')$ agree in $''Hom''_{\mathcal{D}/\mathcal{C}}(X, Z)$, where $(W_3, f_1 v, g_2 u)$ is obtained from a diagram

$$\begin{array}{ccccc} W_3 & \xrightarrow{u} & W_2 & \xrightarrow{g_2} & Z \\ {\scriptstyle v}\downarrow & & \downarrow{\scriptstyle f_2} & & \\ W_1 & \xrightarrow{g_1} & Y & & \\ {\scriptstyle f_1}\downarrow & & & & \\ X & & & & \end{array}$$

where the square

$$\begin{array}{ccc} W_3 & \xrightarrow{u} & W_2 \\ {\scriptstyle v}\downarrow & & \downarrow{\scriptstyle f_2} \\ W_1 & \xrightarrow{g_1} & Y \end{array}$$

is a homotopy pullback.

Proof: The point is that, as in Notation 1.4.2, there is a map from the commutative square to the homotopy pullback square; there is a morphism

$P \to W_3$ in $\mathcal{D}$ which maps the square

$$\begin{array}{ccc} P & \xrightarrow{u'} & W_2 \\ {\scriptstyle v'}\big\downarrow & & \big\downarrow{\scriptstyle f_2} \\ W_1 & \xrightarrow{g_1} & Y \end{array}$$

to the square

$$\begin{array}{ccc} W_3 & \xrightarrow{u} & W_2 \\ {\scriptstyle v}\big\downarrow & & \big\downarrow{\scriptstyle f_2} \\ W_1 & \xrightarrow{g_1} & Y. \end{array}$$

This map gives the desired diagram, which establishes the equivalence of (W_3, f_1v, g_2u) and (P, f_1v', g_2u') modulo the relation $R(X, Z)$. □

We leave it to the reader to check the next Lemma

LEMMA 2.1.18. *The map*

$$\alpha(X,Y) \times \alpha(Y,Z) \to {''Hom''_{\mathcal{D}/\mathcal{C}}}(X,Z)$$

is consistent with equivalence relations $R(X,Y)$ and $R(Y,Z)$, and thus gives a pairing

$${''Hom''_{\mathcal{D}/\mathcal{C}}}(X,Y) \times {''Hom''_{\mathcal{D}/\mathcal{C}}}(Y,Z) \to {''Hom''_{\mathcal{D}/\mathcal{C}}}(X,Z).$$

It is easy to check that the element $(X, 1, 1)$ in ${''Hom''_{\mathcal{D}/\mathcal{C}}}(X,X)$ is a two–sided identity for this composition. It is also easy to check that the composition satisfies the associative law. Let us actually do this check, by way of illusatration.

LEMMA 2.1.19. *The composition map*

$${''Hom''_{\mathcal{D}/\mathcal{C}}}(X,Y) \times {''Hom''_{\mathcal{D}/\mathcal{C}}}(Y,Z) \to {''Hom''_{\mathcal{D}/\mathcal{C}}}(X,Z)$$

satisfies the associative law.

Proof: Let us be given (W, f, g) in $\alpha(X, X')$, (W', f', g') in $\alpha(X', X'')$ and (W'', f'', g'') in $\alpha(X'', X''')$. Let

$$\begin{array}{ccccccc}
U & \longrightarrow & V' & \longrightarrow & W'' & \longrightarrow & X''' \\
{\scriptstyle \overline{f}''}\big\downarrow & & {\scriptstyle \hat{f}''}\big\downarrow & & {\scriptstyle f''}\big\downarrow & & \\
V & \longrightarrow & W' & \longrightarrow & X'' & & \\
{\scriptstyle \hat{f}'}\big\downarrow & & {\scriptstyle f'}\big\downarrow & & & & \\
W & \longrightarrow & X' & & & & \\
{\scriptstyle f}\big\downarrow & & & & & & \\
X & & & & & &
\end{array}$$

be a diagram in which all the small squares are homotopy pullback squares. That is, the three squares

$$\begin{array}{ccc}
U & \longrightarrow & V' \\
{\scriptstyle \overline{f}''}\big\downarrow & & {\scriptstyle \hat{f}''}\big\downarrow \\
V & \longrightarrow & W'
\end{array}
\qquad
\begin{array}{ccc}
V' & \longrightarrow & W'' \\
{\scriptstyle \hat{f}''}\big\downarrow & & {\scriptstyle f''}\big\downarrow \\
W' & \longrightarrow & X''
\end{array}
\qquad
\begin{array}{ccc}
V & \longrightarrow & W' \\
{\scriptstyle \hat{f}'}\big\downarrow & & {\scriptstyle f'}\big\downarrow \\
W & \longrightarrow & X'
\end{array}$$

are homotopy cartesian. Then since f' is in $Mor_{\mathcal{C}}$, so is its homotopy pullback $\hat{f}'$. Since f'' is in $Mor_{\mathcal{C}}$, so is also its homotopy pullback $\hat{f}''$, and so is $\overline{f}''$, being the homotopy pullback of $\hat{f}''$. We deduce that all the vertical maps in the diagram are in $Mor_{\mathcal{C}}$.

But now just from the commutativity of the diagram, the fact that all the vertical maps are in $Mor_{\mathcal{C}}$, and from Lemma 2.1.17, we deduce that by reading parts of the diagram we get the composites in the two orders of (W, f, g) in $\alpha(X, X')$, (W', f', g') in $\alpha(X', X'')$ and (W'', f'', g'') in $\alpha(X'', X''')$. Since both come out to be

$$\begin{array}{ccc}
U & \longrightarrow & X''' \\
\big\downarrow & & \\
X & &
\end{array}$$

the associative law follows. $\square$

DEFINITION 2.1.20. *With* $"Hom"_{\mathcal{D}/\mathcal{C}}(X, Y)$ *for the morphisms from* X *to* Y, $\mathcal{D}/\mathcal{C}$ *is therefore a category. From now on, we write* $Hom_{\mathcal{D}/\mathcal{C}}(X, Y)$ *instead of* $"Hom"_{\mathcal{D}/\mathcal{C}}(X, Y)$ *for the morphisms in this category. Define a functor* $F_{univ} : \mathcal{D} \to \mathcal{D}/\mathcal{C}$ *to be the identity on objects, and to take*

$f : X \to Y$ to

$$\begin{array}{ccc} X & \xrightarrow{f} & Y \\ {\scriptstyle 1}\downarrow & & \\ X & & \end{array}$$

LEMMA 2.1.21. *Let $f : X \to Y$ be a morphism in $Mor_{\mathcal{C}}$. In the category $\mathcal{D}/\mathcal{C}$, the morphisms*

$$\begin{array}{ccc} X & \xrightarrow{f} & Y \\ {\scriptstyle 1}\downarrow & & \\ X & & \end{array} \qquad\qquad \begin{array}{ccc} X & \xrightarrow{1} & X \\ {\scriptstyle f}\downarrow & & \\ Y & & \end{array}$$

are inverse to each other.

Proof: Consider the diagram

$$\begin{array}{ccccc} X & \xrightarrow{1} & X & \xrightarrow{1} & X \\ {\scriptstyle 1}\downarrow & & {\scriptstyle f}\downarrow & & \\ X & \xrightarrow{f} & Y & & \\ {\scriptstyle 1}\downarrow & & & & \\ X & & & & \end{array}$$

It shows that the composite $X \to Y \to X$, in the category $\mathcal{D}/\mathcal{C}$, is the identity. But the diagram

$$\begin{array}{ccccc} X & \xrightarrow{1} & X & \xrightarrow{f} & Y \\ {\scriptstyle 1}\downarrow & & {\scriptstyle 1}\downarrow & & \\ X & \xrightarrow{1} & X & & \\ {\scriptstyle f}\downarrow & & & & \\ Y & & & & \end{array}$$

computes the composite $Y \to X \to Y$ in $\mathcal{D}/\mathcal{C}$ to be

$$\begin{array}{ccc} X & \xrightarrow{f} & Y \\ {\scriptstyle f}\downarrow & & \\ Y & & \end{array}$$

and the map $f : X \to Y$ shows that this is equivalent to the identity

$$\begin{array}{ccc} Y & \xrightarrow{1} & Y \\ {\scriptstyle 1}\big\downarrow & & \\ Y & & \end{array}$$

□

LEMMA 2.1.22. *For any map in $\mathcal{D}/\mathcal{C}$ of the form*

$$\begin{array}{ccc} W & \xrightarrow{g} & Y \\ {\scriptstyle f}\big\downarrow & & \\ X & & \end{array}$$

one can write it as the composite of the two maps

$$\begin{array}{ccc} W & \xrightarrow{1} & W \\ {\scriptstyle f}\big\downarrow & & \\ X & & \end{array} \qquad\qquad \begin{array}{ccc} W & \xrightarrow{g} & Y \\ {\scriptstyle 1}\big\downarrow & & \\ W & & \end{array}$$

Proof: The diagram

$$\begin{array}{ccccc} W & \xrightarrow{1} & W & \xrightarrow{g} & Y \\ {\scriptstyle 1}\big\downarrow & & {\scriptstyle 1}\big\downarrow & & \\ W & \xrightarrow{1} & W & & \\ {\scriptstyle f}\big\downarrow & & & & \\ X & & & & \end{array}$$

amounts to the proof; it computes the composite. □

REMARK 2.1.23. Lemma 2.1.21 asserts that if $f : X \to Y$ is in $Mor_{\mathcal{C}}$, then $F_{univ}(f)$ is invertible in $\mathcal{D}/\mathcal{C}$. Lemma 2.1.22 states that every morphism in $\mathcal{D}/\mathcal{C}$ is of the form $F_{univ}(g)F_{univ}(f)^{-1}$, with $f \in Mor_{\mathcal{C}}$ and $g \in \mathcal{D}$. Finally, Lemma 2.1.17 asserts that to compose two morphisms $F_{univ}(g_1)F_{univ}(f_1)^{-1} : X \longrightarrow Y$ and $F_{univ}(g_2)F_{univ}(f_2)^{-1} : Y \longrightarrow Z$, we find a commutative diagram in $\mathcal{D}$ where all the vertical maps are in $Mor_{\mathcal{C}}$

$$\begin{array}{ccccc} P & \xrightarrow{u'} & W_2 & \xrightarrow{g_2} & Z \\ {\scriptstyle v'}\big\downarrow & & {\scriptstyle f_2}\big\downarrow & & \\ W_1 & \xrightarrow{g_1} & Y & & \\ {\scriptstyle f_1}\big\downarrow & & & & \\ X & & & & \end{array}$$

and then

$$F_{univ}(g_2)F_{univ}(f_2)^{-1}F_{univ}(g_1)F_{univ}(f_1)^{-1} = \\ F_{univ}(g_1)F_{univ}(u')F_{univ}(v')^{-1}F_{univ}(f_1)^{-1}.$$

In other words, to give two composable morphisms $X \longrightarrow Y$ and $Y \longrightarrow Z$ in $\mathcal{D}/\mathcal{C}$ and their composite, is nothing other than to give $X' \longrightarrow Y'$ and $Y' \longrightarrow Z'$ in $\mathcal{D}$ and maps $X' \to X$, $Y' \to Y$ and $Z' \to Z$ in $Mor_{\mathcal{C}}$ expressing the isomorphism of the composable pair in $\mathcal{D}/\mathcal{C}$ with a composable pair in $\mathcal{D}$. We may choose $Z' \to Z$ to be the identity.

PROPOSITION 2.1.24. *The functor $F_{univ} : \mathcal{D} \longrightarrow \mathcal{D}/\mathcal{C}$ is universal for all functors $F : \mathcal{D} \longrightarrow \mathcal{T}$ which take all morphisms in $Mor_{\mathcal{C}}$ to invertible morphisms.*

Proof: This is really immediate from the construction. If $\mathcal{T}$ is any category, and $F : \mathcal{D} \longrightarrow \mathcal{T}$ is a functor sending $Mor_{\mathcal{C}}$ to invertible maps, then F extends to any diagram in $\alpha(X, Y)$, and obviously sends two diagrams equivalent modulo $R(X, Y)$ to the same. □

REMARK 2.1.25. The universal property of F_{univ} is self–dual. Thus the same construction on the dual category $\mathcal{D}^{op}$ will lead to $\{\mathcal{D}/\mathcal{C}\}^{op}$. We deduce that morphisms in $Hom_{\mathcal{D}/\mathcal{C}}(X, Y)$ can also be described as diagrams

$$\begin{array}{ccc} & & Y \\ & & {\scriptstyle f'}\big\downarrow \\ X & \xrightarrow{g'} & W' \end{array}$$

where f' is in $Mor_{\mathcal{C}}$. The equivalence relation on such diagrams is the dual of $R(X, Y)$. Such a diagram is to be thought of as $F_{univ}(f')^{-1}F_{univ}(g')$.

LEMMA 2.1.26. *Let f and g be two morphisms $X \to Y$ in $\mathcal{D}$. Then the following are equivalent*

2.1.26.1. $F_{univ}(f) = F_{univ}(g)$.

2.1.26.2. *There exists a map* $\alpha : W \to X$ *in* $Mor_{\mathcal{C}}$ *with* $f\alpha = g\alpha$.

2.1.26.3. *The map* $f - g : X \to Y$ *factors as*

$$X \to C \to Y$$

with $C \in \mathcal{C}$.

Proof: Let us first prove the equivalence of 2.1.26.1 and 2.1.26.2. The morphisms $F_{univ}(f)$ and $F_{univ}(g)$ will agree in $\mathcal{D}/\mathcal{C}$ if and only if the diagrams

$$\begin{array}{ccc} X & \xrightarrow{f} & Y \\ {\scriptstyle 1}\downarrow & & \\ X & & \end{array} \qquad\qquad \begin{array}{ccc} X & \xrightarrow{g} & Y \\ {\scriptstyle 1}\downarrow & & \\ X & & \end{array}$$

are equivalent. This will happen if and only if there is an object $W \in \mathcal{D}$, and maps $\alpha_1 : W \to X$ and $\alpha_2 : W \to X$ in $Mor_{\mathcal{C}}$, rendering commutative the squares

$$\begin{array}{ccc} W & \xrightarrow{\alpha_1} & X \\ {\scriptstyle \alpha_2}\downarrow & & \downarrow{\scriptstyle 1} \\ X & \xrightarrow{1} & X \end{array} \qquad\qquad \begin{array}{ccc} W & \xrightarrow{\alpha_1} & X \\ {\scriptstyle \alpha_2}\downarrow & & \downarrow{\scriptstyle f} \\ X & \xrightarrow{g} & Y. \end{array}$$

But the commutativity of the first square implies $\alpha_1 = \alpha_2 = \alpha$, and form the second square we learn that $f\alpha = g\alpha$.

Now let us prove the equivalence of 2.1.26.2 and 2.1.26.3. 2.1.26.2 will hold if and only if for some $\alpha : W \to X$ in $Mor_{\mathcal{C}}$, $(f - g)\alpha = 0$. Consider now the triangle

$$W \xrightarrow{\alpha} X \longrightarrow C \longrightarrow \Sigma W.$$

Because $Hom(-, Y)$ is a cohomological functor, $(f - g)\alpha = 0$ if and only if $f - g$ factors through C. But $\alpha \in Mor_{\mathcal{C}}$ if and only if $C \in \mathcal{C}$. Consequently there exists an $\alpha \in Mor_{\mathcal{C}}$ with $(f - g)\alpha = 0$ if and only if $f - g$ factors through $C \in \mathcal{C}$. □

LEMMA 2.1.27. *Any commutative square in* $\mathcal{D}/\mathcal{C}$ *isomorphic to the image of a commutative square in* $\mathcal{D}$. *More precisely, if*

$$\begin{array}{ccc} W & \longrightarrow & X \\ \downarrow & & \downarrow \\ Y & \longrightarrow & Z \end{array}$$

is a commutative square in $\mathcal{D}/\mathcal{C}$, there is a commutative square in $\mathcal{D}$

$$\begin{array}{ccc} W' & \longrightarrow & X' \\ \downarrow & & \downarrow \\ Y' & \longrightarrow & Z' \end{array}$$

and maps in $Mor_{\mathcal{C}}$ $W' \to W$, $X' \to X$, $Y' \to Y$ and $Z' \to Z$ which, being isomorphisms in $\mathcal{D}/\mathcal{C}$, express the isomorphism of the two diagrams.

Proof: From Remark 2.1.23 we know that we can lift the composites $W \to X \to Z$ and $W \to Y \to Z$ to $W_1 \to X' \to Z$ and $W_2 \to Y' \to Z$. Replacing W_1 and W_2 by the homotopy pullback

$$\begin{array}{ccc} W_3 & \longrightarrow & W_1 \\ \downarrow & & \downarrow \\ W_2 & \longrightarrow & W \end{array}$$

we may assume $W_1 = W_2$. But now the commutativity in $\mathcal{D}/\mathcal{C}$ of the diagram

$$\begin{array}{ccc} W_3 & \longrightarrow & X' \\ \downarrow & & \downarrow \\ Y' & \longrightarrow & Z \end{array}$$

means, by Lemma 2.1.26, that there is a map $W' \to W_3$ in $Mor_{\mathcal{C}}$ which equalises the two composites in $\mathcal{D}$. Hence a commutative diagram in $\mathcal{D}$

$$\begin{array}{ccc} W' & \longrightarrow & X' \\ \downarrow & & \downarrow \\ Y' & \longrightarrow & Z. \end{array}$$

Thus we may even take $Z' \to Z$ to be the identity. □

Next we want to prove that $\mathcal{D}/\mathcal{C}$ is an additive category.

LEMMA 2.1.28. *The object $0 \in \mathcal{D}$ is a terminal and initial object in $\mathcal{D}/\mathcal{C}$.*

Proof: The two statements being dual, we may restrict ourselves to proving 0 terminal. Let X be an object of $\mathcal{D}/\mathcal{C}$, that is an object of $\mathcal{D}$. The diagram

$$\begin{array}{ccc} X & \longrightarrow & 0 \\ {\scriptstyle 1}\downarrow & & \\ X & & \end{array}$$

exhibits a morphism $X \to 0$ in $\mathcal{D}/\mathcal{C}$. Given any other,

$$\begin{array}{ccc} P & \longrightarrow & 0 \\ {\scriptstyle f}\big\downarrow & & \\ X & & \end{array}$$

then the map $f : P \to X$ shows that

$$\begin{array}{ccc} P & \longrightarrow & 0 \\ {\scriptstyle f}\big\downarrow & & \\ X & & \end{array} \qquad \begin{array}{ccc} X & \longrightarrow & 0 \\ {\scriptstyle 1}\big\downarrow & & \\ X & & \end{array}$$

are equivalent. Hence there is only one map $X \to 0$ in $\mathcal{D}/\mathcal{C}$. □

LEMMA 2.1.29. *Let X and Y be two objects of $\mathcal{D}/\mathcal{C}$, that is objects of $\mathcal{D}$. Then the direct sum $X \oplus Y$ in $\mathcal{D}$ is a biproduct in $\mathcal{D}/\mathcal{C}$. It satisfies the universal properties both of product and coproduct.*

Proof: There are maps in $\mathcal{D}$

$$\begin{array}{ccccc} & & X \oplus Y & & \\ & {\scriptstyle p_1}\swarrow & & \searrow{\scriptstyle p_2} & \\ X & & & & Y \end{array} \qquad \begin{array}{ccccc} X & & & & Y \\ & {\scriptstyle i_1}\searrow & & \swarrow{\scriptstyle i_2} & \\ & & X \oplus Y & & \end{array}$$

which give $X \oplus Y$ the structure of a product (respectively coproduct) in $\mathcal{D}$. We will show that these maps work also in $\mathcal{D}/\mathcal{C}$.

The two statements being dual, we may restrict ourselves to showing the statement about coproducts. We need to show that $X \oplus Y$ is the coproduct of X and Y in $\mathcal{D}/\mathcal{C}$. Giving two morphisms $X \to Q$ and $Y \to Q$ in $\mathcal{D}/\mathcal{C}$ is the same as giving equivalence classes of diagrams

$$\begin{array}{ccc} P & \xrightarrow{\ f\ } & Q \\ {\scriptstyle \alpha}\big\downarrow & & \\ X & & \end{array} \qquad \begin{array}{ccc} P' & \xrightarrow{\ g\ } & Q \\ {\scriptstyle \alpha'}\big\downarrow & & \\ Y & & \end{array}$$

Since α and α' lie in $Mor_{\mathcal{C}}$, they fit into triangles

$$P \xrightarrow{\ \alpha\ } X \longrightarrow Z \longrightarrow \Sigma P$$

$$P' \xrightarrow{\ \alpha'\ } Y \longrightarrow Z' \longrightarrow \Sigma P'$$

where Z and Z' are in $\mathcal{C}$. The direct sum of these triangles is a triangle by Proposition 1.2.1. That is, we have a triangle

$$P \oplus P' \xrightarrow{\ \alpha\oplus\alpha'\ } X \oplus Y \longrightarrow Z \oplus Z' \longrightarrow \Sigma P \oplus \Sigma P'.$$

But $Z \oplus Z'$ is in $\mathcal{C}$, and hence $\alpha \oplus \alpha'$ is in $Mor_{\mathcal{C}}$. This makes

$$\begin{array}{ccc} P \oplus P' & \xrightarrow{\begin{pmatrix} f & g \end{pmatrix}} & Q \\ \big\downarrow & & \\ X \oplus Y & & \end{array}$$

a well–defined representative for a morphism in $\mathcal{D}/\mathcal{C}$. The composite of it with

$$X \xrightarrow{i_1} X \oplus Y$$

is computed by the commutative diagram

$$\begin{array}{ccccc} P & \longrightarrow & P \oplus P' & \xrightarrow{\begin{pmatrix} f & g \end{pmatrix}} & Q \\ {\scriptstyle\alpha}\big\downarrow & & \big\downarrow & & \\ X & \xrightarrow{i_1} & X \oplus Y & & \end{array}$$

to be just

$$\begin{array}{ccc} P & \xrightarrow{f} & Q \\ {\scriptstyle\alpha}\big\downarrow & & \\ X & & \end{array}$$

and similarly for the composite with $i_2 : Y \longrightarrow X \oplus Y$. Hence a pair of maps $X \to Q$ and $Y \to Q$ in $\mathcal{D}/\mathcal{C}$ does factor through the object $X \oplus Y$. Now we need to show the uniqueness of the factorisation.

For the uniqueness, it is handier to work with the dual description of morphisms in $\mathcal{D}/\mathcal{C}$, as in Remark 2.1.25. Suppose therefore that we are given two morphisms in $\mathcal{D}/\mathcal{C}$

$$\begin{array}{ccc} & & Q \\ & & \big\downarrow \\ X \oplus Y & \xrightarrow{f} & P \end{array} \qquad\qquad \begin{array}{ccc} & & Q \\ & & \big\downarrow \\ X \oplus Y & \xrightarrow{g} & P' \end{array}$$

so that the composites with i_1 and i_2 agree. First, we may assume that $P = P'$ and the vertical maps $Q \to P$ agree, by replacing P and P' by the homotopy pushout

$$\begin{array}{ccc} Q & \longrightarrow & P \\ \big\downarrow & & \big\downarrow \\ P' & \longrightarrow & N \end{array}$$

Thus, we may assume we have two diagrams

$$\begin{array}{ccc} & & Q \\ & & \big\downarrow{\scriptstyle \alpha} \\ X \oplus Y & \xrightarrow{f} & P \end{array} \qquad\qquad \begin{array}{ccc} & & Q \\ & & \big\downarrow{\scriptstyle \alpha} \\ X \oplus Y & \xrightarrow{g} & P \end{array}$$

so that the composites with i_1 and i_2 induce the same morphism in $\mathcal{D}/\mathcal{C}$; that is

$$F_{univ}(\alpha)^{-1}F_{univ}(fi_1) = F_{univ}(\alpha)^{-1}F_{univ}(gi_1),$$

and

$$F_{univ}(\alpha)^{-1}F_{univ}(fi_2) = F_{univ}(\alpha)^{-1}F_{univ}(gi_2).$$

Multiplying by $F_{univ}(\alpha)$, this means

$$F_{univ}(fi_1) = F_{univ}(gi_1) \qquad \text{and} \qquad F_{univ}(fi_2) = F_{univ}(gi_2).$$

By Lemma 2.1.26, this means that $(f-g)i_1$ factors through $C \in \mathcal{C}$ and $(f-g)i_2$ factors through $C' \in \mathcal{C}$. This means that $f-g$ factors through $C \oplus C'$, and again by Lemma 2.1.26 we deduce that $F_{univ}(f) = F_{univ}(g)$. But then

$$F_{univ}(\alpha)^{-1}F_{univ}(f) = F_{univ}(\alpha)^{-1}F_{univ}(g),$$

giving the uniqueness. □

LEMMA 2.1.30. *The category $\mathcal{D}/\mathcal{C}$ is an additive category, and the functor $F_{univ} : \mathcal{D} \longrightarrow \mathcal{D}/\mathcal{C}$ is an additive functor.*

Proof: The category $\mathcal{D}/\mathcal{C}$ is a pointed category by Lemma 2.1.28 (there is an object which is simultaneously initial and terminal). Also, $\mathcal{D}/\mathcal{C}$ has finite biproducts, by Lemma 2.1.29. Furthermore, the functor $F_{univ} : \mathcal{D} \longrightarrow \mathcal{D}/\mathcal{C}$ respects biproducts and the 0 object. There is therefore a well–defined addition on $\mathcal{D}/\mathcal{C}$, and the functor $F_{univ} : \mathcal{D} \longrightarrow \mathcal{D}/\mathcal{C}$ respects the addition. Given $f : X \longrightarrow Y$ and $g : X \longrightarrow Y$, then $f + g : X \longrightarrow Y$ is given by either of the composites

$$X \xrightarrow{\Delta} X \oplus X \xrightarrow{\begin{pmatrix} f & g \end{pmatrix}} Y \qquad\qquad X \xrightarrow{\begin{pmatrix} f \\ g \end{pmatrix}} Y \oplus Y \xrightarrow{\Delta} Y$$

which agree. It remains only to show that this addition gives a group law, rather than just a commutative monoid; we need to show that a map $X \to Y$ in $\mathcal{D}/\mathcal{C}$ has an additive inverse.

Let us be given a map in $\mathcal{D}/\mathcal{C}$, it can be expressed as

$$F_{univ}(\alpha)^{-1}F_{univ}(f).$$

But

$$\begin{aligned} F_{univ}(\alpha)^{-1}F_{univ}(f) + F_{univ}(\alpha)^{-1}F_{univ}(-f) &= \\ &= F_{univ}(\alpha)^{-1}F_{univ}\Big(f+(-f)\Big) \\ &= F_{univ}(\alpha)^{-1}F_{univ}(0) \\ &= 0 \end{aligned}$$

□

Next we need to discuss the triangulated structure on $\mathcal{D}/\mathcal{C}$. There is a slight subtlety introduced by [TR0]. Recall that [TR0] asserts, among other things, that any object isomorphic to a distinguished triangle is a distinguished triangle. Before we continue the proof of Theorem 2.1.8, it will be helpful to study isomorphisms in $\mathcal{D}/\mathcal{C}$.

LEMMA 2.1.31. *If a morphism in $\mathcal{D}/\mathcal{C}$ of the form*

$$\begin{array}{ccc} P & \xrightarrow{f} & X \\ {\scriptstyle\alpha}\downarrow & & \\ X & & \end{array}$$

is in the equivalence class of the identity $1 : X \to X$, then $f \in Mor_{\mathcal{C}}$.

Proof: An equivalence of the above morphism with

$$\begin{array}{ccc} X & \xrightarrow{1} & X \\ {\scriptstyle 1}\downarrow & & \\ X & & \end{array}$$

will consist of two commutative squares

$$\begin{array}{ccc} W & \xrightarrow{u} & P \\ {\scriptstyle u'}\downarrow & & \downarrow{\scriptstyle\alpha} \\ X & \xrightarrow{1} & X \end{array} \qquad \begin{array}{ccc} W & \xrightarrow{u} & P \\ {\scriptstyle u'}\downarrow & & \downarrow{\scriptstyle f} \\ X & \xrightarrow{1} & X \end{array}$$

where in the first square, all the morphisms are in $Mor_{\mathcal{C}}$; that is, u and u' are morphisms in $Mor_{\mathcal{C}}$. But the second square tells us that $u' = fu$. From Lemma 1.5.6 we deduce that f must also be in $Mor_{\mathcal{C}}$. □

LEMMA 2.1.32. *A morphism in $\mathcal{D}/\mathcal{C}$ of the form*

$$\begin{array}{ccc} P & \xrightarrow{g} & Y \\ {\scriptstyle\alpha}\downarrow & & \\ X & & \end{array}$$

will be invertible if and only if there exist morphisms f and h in $\mathcal{D}$ so that gf and hg are both in $Mor_{\mathcal{C}}$.

Proof: The sufficiency is obvious. If there exist f and h so that gf and hg are both in $Mor_{\mathcal{C}}$, then $F_{univ}(hg)$ and $F_{univ}(gf)$ are both invertible. This forces $F_{univ}(g)$ to have both a right and a left inverse in $\mathcal{D}/\mathcal{C}$, hence be invertible. But then the diagram above, which stands for the morphism $F_{univ}(g)F_{univ}(\alpha)^{-1}$, is also invertible.

Next we wish to prove the necessity. Let us suppose, therefore, that $F_{univ}(g)F_{univ}(\alpha)^{-1}$ is invertible in $\mathcal{D}/\mathcal{C}$. Then $F_{univ}(g)$ must also be invertible. We wish to show that there exist f and h with hg and gf in $Mor_{\mathcal{C}}$. The statements being dual, it suffices to produce f.

Let the diagram

$$\begin{array}{ccc} Q & \xrightarrow{f} & P \\ {\scriptstyle\beta}\downarrow & & \\ Y & & \end{array}$$

be a right inverse in $\mathcal{D}/\mathcal{C}$ to $F_{univ}(g) : F_{univ}(P) \longrightarrow F_{univ}(Y)$. Then the composite

$$\begin{array}{ccccc} Q & \xrightarrow{f} & P & \xrightarrow{g} & Y \\ {\scriptstyle\beta}\downarrow & & & & \\ Y & & & & \end{array}$$

is in the equivalence class of the identity. From Lemma 2.1.31 we learn that gf must therefore be in $Mor_{\mathcal{C}}$. □

LEMMA 2.1.33. *The morphism $X \to 0$ in $\mathcal{D}$ becomes an isomorphism in $\mathcal{D}/\mathcal{C}$ if and only if there exists a $Y \in \mathcal{D}$ with $X \oplus Y \in \mathcal{C}$.*

Proof: Suppose the zero map $g : X \to 0$ is a morphism in $\mathcal{D}$ such that F_{univ} takes it to an invertible map. By Lemma 2.1.32, there exists $h : 0 \to \Sigma Y$ so that the composite

$$X \xrightarrow{g} 0 \xrightarrow{h} \Sigma Y$$

is in $Mor_{\mathcal{C}}$. But then we have a triangle

$$X \xrightarrow{0} \Sigma Y \longrightarrow \Sigma(X \oplus Y) \longrightarrow \Sigma X,$$

and since $0 : X \to \Sigma Y$ is in $Mor_{\mathcal{C}}$, $\Sigma(X \oplus Y)$ must be in $\mathcal{C}$. Since $\mathcal{C}$ is triangulated, $X \oplus Y$ is also in $\mathcal{C}$.

Conversely, suppose there exists a $Y \in \mathcal{D}$ with $X \oplus Y$ in $\mathcal{C}$. Define $h : 0 \to \Sigma Y$ and $f : 0 \to X$ to be the zero maps (the only possibilities).

Then $gf : 0 \to 0$ is an isomorphism, while $hg : X \to \Sigma Y$ is the zero map, and fits in a triangle

$$X \xrightarrow{0} \Sigma Y \longrightarrow \Sigma(X \oplus Y) \longrightarrow \Sigma X,$$

with $\Sigma(X \oplus Y) \in \mathcal{C}$. Hence both hg and gf are in $Mor_{\mathcal{C}}$, forcing g to be invertible in $\mathcal{D}/\mathcal{C}$. □

REMARK 2.1.34. In Corollary 4.5.12 we will see that not only does there exist some Y with $X \oplus Y \in \mathcal{C}$, but that Y may always be chosen to be ΣX. It is always true that $X \oplus \Sigma X$ lies in $\mathcal{C}$, if X is isomorphic to 0 in $\mathcal{D}/\mathcal{C}$.

PROPOSITION 2.1.35. *Let $g : Y \to Y'$ be a morphism in $\mathcal{D}$. Then $F_{univ}(g)$ is an isomorphism if and only if in any triangle*

$$Y \xrightarrow{g} Y' \longrightarrow Z \longrightarrow \Sigma Y$$

the object $Z \in \mathcal{D}$ is a direct summand of an object of $\mathcal{C}$; there is an object $Z' \in \mathcal{D}$ so that $Z \oplus Z' \in \mathcal{C}$.

Proof: Suppose there is an object Z' with $Z \oplus Z' \in \mathcal{C}$. To prove one implication, we want to show that $F_{univ}(g)$ is an isomorphism. By Lemma 2.1.32, this is equivalent to producing h and f with gf and hg in $Mor_{\mathcal{C}}$.

Starting with the two triangles

$$\begin{array}{ccccccc} Y & \xrightarrow{g} & Y' & \longrightarrow & Z & \longrightarrow & \Sigma Y \\ 0 & \longrightarrow & Z' & \longrightarrow & Z' & \longrightarrow & 0 \end{array}$$

we can form the direct sum, which is a triangle by Proposition 1.2.1,

$$Y \xrightarrow{\begin{pmatrix} g \\ 0 \end{pmatrix}} Y' \oplus Z' \longrightarrow Z \oplus Z' \longrightarrow \Sigma Y.$$

Since $Z \oplus Z'$ is in $\mathcal{C}$, the map $Y \longrightarrow Y' \oplus Z'$ is in $Mor_{\mathcal{C}}$. But it factors as

$$Y \xrightarrow{g} Y' \xrightarrow{h} Y' \oplus Z'.$$

This exhibits h with $hg \in Mor_{\mathcal{C}}$. But by the dual argument, there is also an f with gf in $Mor_{\mathcal{C}}$. Thus $F_{univ}(g)$ is invertible.

Now we need the converse. Suppose therefore that $F_{univ}(g)$ is invertible. Then there exists an $h : Y' \to Y''$ with $hg \in Mor_{\mathcal{C}}$. Now consider the morphism of triangles

$$\begin{array}{ccccccc} Y & \xrightarrow{g} & Y' & \xrightarrow{\alpha} & Z & \longrightarrow & \Sigma Y \\ {\scriptstyle hg}\downarrow & & \begin{pmatrix} h \\ \alpha \end{pmatrix}\downarrow & & {\scriptstyle 1}\downarrow & & {\scriptstyle \Sigma hg}\downarrow \\ Y'' & \longrightarrow & Y'' \oplus Z & \longrightarrow & Z & \xrightarrow{0} & \Sigma Y'' \end{array}$$

The bottom triangle is contractible, and by Corollary 1.3.12 all maps into contractible triangles are good. Hence the above map is a good map of triangles, and from the proof of Lemma 1.4.3 we learn that the square

$$\begin{array}{ccc} Y & \xrightarrow{\ g\ } & Y' \\ {\scriptstyle hg}\downarrow & & \downarrow{\scriptstyle \begin{pmatrix} h \\ \alpha \end{pmatrix}} \\ Y'' & \longrightarrow & Y'' \oplus Z \end{array}$$

is homotopy cartesian. But hg lies in $Mor_{\mathcal{C}}$, and Lemma 1.5.8 allows us to conclude that the map

$$Y' \xrightarrow{\begin{pmatrix} h \\ \alpha \end{pmatrix}} Y'' \oplus Z$$

lies in $Mor_{\mathcal{C}}$. Among other things, it follows that the map is an isomorphism in $\mathcal{D}/\mathcal{C}$. But since $g : Y \to Y'$ also is, so is the composite

$$Y \xrightarrow{\ g\ } Y' \xrightarrow{\begin{pmatrix} h \\ \alpha \end{pmatrix}} Y'' \oplus Z,$$

and since $\alpha g = 0$ (the composite of two maps in a triangle), we have that

$$Y \xrightarrow{\begin{pmatrix} hg \\ 0 \end{pmatrix}} Y'' \oplus Z$$

induces an isomorphism in the additive category $\mathcal{D}/\mathcal{C}$. Then the composite

$$Y \xrightarrow{\ hg\ } Y'' \xrightarrow{\begin{pmatrix} 1 \\ 0 \end{pmatrix}} Y'' \oplus Z$$

is an isomorphism in $\mathcal{D}/\mathcal{C}$, as is the first map hg. But then the second map

$$Y'' \xrightarrow{\begin{pmatrix} 1 \\ 0 \end{pmatrix}} Y'' \oplus Z$$

is also an isomorphism in $\mathcal{D}/\mathcal{C}$. The composite

$$Y'' \xrightarrow{\begin{pmatrix} 1 \\ 0 \end{pmatrix}} Y'' \oplus Z \xrightarrow{\begin{pmatrix} 1 & 0 \end{pmatrix}} Y''$$

is the identity on Y'', and since the first map is an isomorphism, the second must be its (two–sided) inverse. The composite

$$Y'' \oplus Z \xrightarrow{\begin{pmatrix} 1 & 0 \end{pmatrix}} Y'' \xrightarrow{\begin{pmatrix} 1 \\ 0 \end{pmatrix}} Y'' \oplus Z$$

must therefore be the identity on $Y'' \oplus Z$. In other words, the two maps

$$Y'' \oplus Z \xrightarrow{\begin{pmatrix} 1 & 0 \\ 0 & 1 \end{pmatrix}} Y'' \oplus Z \qquad \text{and} \qquad Y'' \oplus Z \xrightarrow{\begin{pmatrix} 1 & 0 \\ 0 & 0 \end{pmatrix}} Y'' \oplus Z$$

must agree; that is the two maps

$$Z \xrightarrow{\ 1\ } Z \qquad \text{and} \qquad Z \xrightarrow{\ 0\ } Z$$

agree in $\mathcal{D}/\mathcal{C}$. This means that the maps

$$Z \longrightarrow 0 \qquad \text{and} \qquad 0 \longrightarrow Z$$

are inverse to each other. In particular, the map $Z \to 0$ must be an isomorphism in $\mathcal{D}/\mathcal{C}$. By Lemma 2.1.33 we conclude that there is a $Z' \in \mathcal{D}$ with $Z \oplus Z' \in \mathcal{C}$. □

LEMMA 2.1.36. *Let us be given a commutative diagram in $\mathcal{D}$ whose rows are triangles*

$$\begin{array}{ccccccc} X & \xrightarrow{f} & Y & \longrightarrow & Z & \longrightarrow & \Sigma X \\ {\scriptstyle 1}\downarrow & & {\scriptstyle g}\downarrow & & & & {\scriptstyle 1}\downarrow \\ X & \xrightarrow{gf} & Y' & \longrightarrow & Z' & \longrightarrow & \Sigma X. \end{array}$$

Suppose $F_{univ}(g)$ is invertible. Then it is possible to extend the diagram to

$$\begin{array}{ccccccc} X & \xrightarrow{f} & Y & \longrightarrow & Z & \longrightarrow & \Sigma X \\ {\scriptstyle 1}\downarrow & & {\scriptstyle g}\downarrow & & {\scriptstyle h}\downarrow & & {\scriptstyle 1}\downarrow \\ X & \xrightarrow{gf} & Y' & \longrightarrow & Z' & \longrightarrow & \Sigma X \end{array}$$

where $F_{univ}(h)$ is also invertible.

Proof: By Proposition 1.4.6, it is possible to extend to a diagram

$$\begin{array}{ccccccc} X & \xrightarrow{f} & Y & \longrightarrow & Z & \longrightarrow & \Sigma X \\ {\scriptstyle 1}\downarrow & & {\scriptstyle g}\downarrow & & {\scriptstyle h}\downarrow & & {\scriptstyle 1}\downarrow \\ X & \xrightarrow{gf} & Y' & \longrightarrow & Z' & \longrightarrow & \Sigma X \\ \downarrow & & \downarrow & & \downarrow & & \downarrow \\ 0 & \longrightarrow & Y'' & \xrightarrow{1} & Y'' & \longrightarrow & 0 \\ \downarrow & & \downarrow & & \downarrow & & \downarrow \\ \Sigma X & \xrightarrow{\Sigma f} & \Sigma Y & \longrightarrow & \Sigma Z & \longrightarrow & \Sigma^2 X \end{array}$$

and since $F_{univ}(g)$ is invertible, Proposition 2.1.35 for the triangle

$$Y \xrightarrow{g} Y' \longrightarrow Y'' \longrightarrow \Sigma Y$$

implies that Y'' is a direct summand of an object in $\mathfrak{C}$, which by Proposition 2.1.35 for the triangle

$$Z \xrightarrow{h} Z' \longrightarrow Y'' \longrightarrow \Sigma Z$$

implies that $F_{univ}(h)$ is also an isomorphism. □

REMARK 2.1.37. The dual statement is that given a diagram with g deleted, that is

$$\begin{array}{ccccccc} X & \xrightarrow{f} & Y & \longrightarrow & Z & \longrightarrow & \Sigma X \\ {\scriptstyle 1}\downarrow & & & & {\scriptstyle h}\downarrow & & {\scriptstyle 1}\downarrow \\ X & \xrightarrow{gf} & Y' & \longrightarrow & Z' & \longrightarrow & \Sigma X, \end{array}$$

and if h is an isomorphism in $\mathfrak{D}/\mathfrak{C}$, then $g : Y \to Y'$ can be chosen to also be an isomorphism in $\mathfrak{D}/\mathfrak{C}$.

LEMMA 2.1.38. *Given two triangles in* $\mathfrak{D}$

$$X \longrightarrow Y \longrightarrow Z \longrightarrow \Sigma X$$

$$X' \longrightarrow Y' \longrightarrow Z' \longrightarrow \Sigma X'$$

and a commutative square in $\mathfrak{D}/\mathfrak{C}$ *with vertical isomorphisms*

$$\begin{array}{ccc} X & \longrightarrow & Y \\ \downarrow & & \downarrow \\ X' & \longrightarrow & Y' \end{array}$$

it may be extended to an isomorphism in $\mathfrak{D}/\mathfrak{C}$ *of the triangles.*

Proof: The isomorphism $Y \to Y'$ is a map in $\mathfrak{D}/\mathfrak{C}$, that is it may be represented as a diagram

$$\begin{array}{ccc} Y'' & \xrightarrow{f} & Y \\ {\scriptstyle \alpha}\downarrow & & \\ Y' & & \end{array}$$

where $\alpha \in Mor_{\mathfrak{C}}$ and $F_{univ}(f)$ is invertible. In other words, both α and f are invertible in $\mathfrak{D}/\mathfrak{C}$. But then the diagrams in $\mathfrak{D}$ whose rows are triangles

$$\begin{array}{ccccccc} X'' & \longrightarrow & Y'' & \longrightarrow & Z' & \longrightarrow & \Sigma X'' \\ & & {\scriptstyle \alpha}\downarrow & & {\scriptstyle 1}\downarrow & & \\ X' & \longrightarrow & Y' & \longrightarrow & Z' & \longrightarrow & \Sigma X' \end{array}$$

and

$$\begin{array}{ccccccc}
\overline{X}'' & \longrightarrow & Y'' & \longrightarrow & Z & \longrightarrow & \Sigma\overline{X}'' \\
 & & {\scriptstyle f}\big\downarrow & & {\scriptstyle 1}\big\downarrow & & \\
X & \longrightarrow & Y & \longrightarrow & Z & \longrightarrow & \Sigma X
\end{array}$$

can be completed to isomorphisms in $\mathcal{D}/\mathcal{C}$ by the dual of Lemma 2.1.36 (see Remark 2.1.37). We may therefore assume that $Y = Y'$ and that in the commutative square

$$\begin{array}{ccc}
X & \longrightarrow & Y \\
\big\downarrow & & {\scriptstyle 1}\big\downarrow \\
X' & \longrightarrow & Y
\end{array}$$

the vertical map $Y \to Y$ is the identity. But now the isomorphism $X \to X'$ in $\mathcal{D}/\mathcal{C}$ may be represented by

$$\begin{array}{ccc}
X'' & \xrightarrow{\ f\ } & X \\
{\scriptstyle \alpha}\big\downarrow & & \\
X' & &
\end{array}$$

and we deduce a square in $\mathcal{D}$, which commutes in $\mathcal{D}/\mathcal{C}$,

$$\begin{array}{ccc}
X'' & \xrightarrow{\ f\ } & X \\
{\scriptstyle \alpha}\big\downarrow & & \big\downarrow \\
X' & \longrightarrow & Y.
\end{array}$$

By Lemma 2.1.26, replacing X'' by W with a map $W \to X''$ in $Mor_{\mathcal{C}}$, we may assume the square commutes in $\mathcal{D}$.

But now the diagrams

$$\begin{array}{ccccccc}
X'' & \longrightarrow & Y & \longrightarrow & Z'' & \longrightarrow & \Sigma X'' \\
{\scriptstyle f}\big\downarrow & & {\scriptstyle 1}\big\downarrow & & & & {\scriptstyle \Sigma f}\big\downarrow \\
X & \longrightarrow & Y & \longrightarrow & Z & \longrightarrow & \Sigma X
\end{array}$$

and

$$\begin{array}{ccccccc}
X'' & \longrightarrow & Y & \longrightarrow & Z'' & \longrightarrow & \Sigma X'' \\
{\scriptstyle \alpha}\big\downarrow & & {\scriptstyle 1}\big\downarrow & & & & {\scriptstyle \Sigma\alpha}\big\downarrow \\
X' & \longrightarrow & Y & \longrightarrow & Z' & \longrightarrow & \Sigma X'
\end{array}$$

can be extended to isomorphisms in $\mathcal{D}/\mathcal{C}$, completing the proof. □

Proof of Theorem 2.1.8 Now it only remains to show that $\mathcal{D}/\mathcal{C}$ is a triangulated category, that $F_{univ} : \mathcal{D} \longrightarrow \mathcal{D}/\mathcal{C}$ is a triangulated functor, and that F_{univ} is universal.

Define the suspension functor on $\mathcal{D}/\mathcal{C}$ to be just the suspension functor of $\mathcal{D}$ on objects, and the suspension functor on diagrams defining morphisms for morphisms. Let $\Phi_{univ} : \Sigma F_{univ} \longrightarrow F_{univ}\Sigma$ be the identity.

Define the distinguished triangles in $\mathcal{D}/\mathcal{C}$ to be all candidate triangles isomorphic to

$$F_{univ}(X) \longrightarrow F_{univ}(Y) \longrightarrow F_{univ}(Z) \longrightarrow \Sigma F_{univ}(X)$$

where

$$X \longrightarrow Y \longrightarrow Z \longrightarrow \Sigma X$$

is a distinguished triangle in $\mathcal{D}$. Then [TR0] and [TR2] are obvious for $\mathcal{D}/\mathcal{C}$. To prove [TR1], note first that any morphism in $\mathcal{D}/\mathcal{C}$ may be written in the form $F_{univ}(u)F_{univ}(f)^{-1}$, where $f : P \to X$ is in $Mor_{\mathcal{C}}$ and $u : P \to Y$ is any morphism in $\mathcal{D}$. By [TR1] for $\mathcal{D}$, complete u to a triangle in $\mathcal{D}$

$$P \xrightarrow{\ u\ } Y \xrightarrow{\ v\ } Z \xrightarrow{\ w\ } \Sigma P.$$

Then

$$\begin{array}{ccccc} F_{univ}(X) & \xrightarrow{F_{univ}(u)F_{univ}(f)^{-1}} & F_{univ}(Y) & \xrightarrow{F_{univ}(v)} & F_{univ}(Z) \\ & & & & \Big\downarrow {\scriptstyle F_{univ}(\Sigma f)F_{univ}(w)} \\ & & & & \Sigma F_{univ}(X) \end{array}$$

is isomorphic in $\mathcal{D}/\mathcal{C}$ to

$$F_{univ}(P) \xrightarrow{F_{univ}(u)} F_{univ}(Y) \xrightarrow{F_{univ}(v)} F_{univ}(Z) \xrightarrow{F_{univ}(w)} \Sigma F_{univ}(X)$$

and hence is a distinguished triangle. It remains only to prove [TR3] and [TR4']. Since [TR4'] is stronger, we will prove only it.

We need to show that, given a diagram in $\mathcal{D}/\mathcal{C}$ where the rows are triangles

$$\begin{array}{ccccccc} X & \longrightarrow & Y & \longrightarrow & Z & \longrightarrow & \Sigma X \\ \downarrow & & \downarrow & & & & \downarrow \\ X' & \longrightarrow & Y' & \longrightarrow & Z' & \longrightarrow & \Sigma X' \end{array}$$

then there is a way to choose $Z \to Z'$ making the above a good map of triangles; that is, the mapping cone is a triangle. We remind the reader that [TR3] is weaker, asserting only that the diagram can be made commutative.

Observe first that by Lemma 2.1.27, the commutative square

$$\begin{array}{ccc} X & \longrightarrow & Y \\ \downarrow & & \downarrow \\ X' & \longrightarrow & Y' \end{array}$$

can be lifted from $\mathcal{D}/\mathcal{C}$ to $\mathcal{D}$. Choose a square in $\mathcal{D}$ isomorphic to it as in Lemma 2.1.27. Then in that square, which we will denote

$$\begin{array}{ccc} \overline{X} & \longrightarrow & \overline{Y} \\ \downarrow & & \downarrow \\ \overline{X}' & \longrightarrow & \overline{Y}' \end{array}$$

the rows can be extended to triangles in $\mathcal{D}$, and the diagram

$$\begin{array}{ccccccc} \overline{X} & \longrightarrow & \overline{Y} & \longrightarrow & \overline{Z} & \longrightarrow & \Sigma\overline{X} \\ \downarrow & & \downarrow & & & & \downarrow \\ \overline{X}' & \longrightarrow & \overline{Y}' & \longrightarrow & \overline{Z}' & \longrightarrow & \Sigma\overline{X}' \end{array}$$

can be extended to a good morphism of triangles in $\mathcal{D}$, hence also in $\mathcal{D}/\mathcal{C}$. But by Lemma 2.1.38 the commutative square with vertical isomorphisms

$$\begin{array}{ccc} \overline{X} & \longrightarrow & \overline{Y} \\ \downarrow & & \downarrow \\ X & \longrightarrow & Y \end{array}$$

extends to an isomorphism of candidate triangles in $\mathcal{D}/\mathcal{C}$

$$\begin{array}{ccccccc} \overline{X} & \longrightarrow & \overline{Y} & \longrightarrow & \overline{Z} & \longrightarrow & \Sigma\overline{X} \\ \downarrow & & \downarrow & & \downarrow & & \downarrow \\ X & \longrightarrow & Y & \longrightarrow & Z & \longrightarrow & X \end{array}$$

and similarly, the commutative square

$$\begin{array}{ccc} \overline{X}' & \longrightarrow & \overline{Y}' \\ \downarrow & & \downarrow \\ X' & \longrightarrow & Y' \end{array}$$

extends to an isomorphism of candidate triangles in $\mathcal{D}/\mathcal{C}$

$$\begin{array}{ccccccc} \overline{X}' & \longrightarrow & \overline{Y}' & \longrightarrow & \overline{Z}' & \longrightarrow & \Sigma\overline{X}' \\ \downarrow & & \downarrow & & \downarrow & & \downarrow \\ X' & \longrightarrow & Y' & \longrightarrow & Z' & \longrightarrow & X' \end{array}$$

and hence we have defined a good morphism, as required,

$$\begin{array}{ccccccc} X & \longrightarrow & Y & \longrightarrow & Z & \longrightarrow & X \\ \downarrow & & \downarrow & & \downarrow & & \downarrow \\ X' & \longrightarrow & Y' & \longrightarrow & Z' & \longrightarrow & X'. \end{array}$$

Thus $\mathcal{D}/\mathcal{C}$ is triangulated.

The fact that F_{univ} is a triangulated functor is obvious; by the definition of triangles in $\mathcal{D}/\mathcal{C}$, F_{univ} takes triangles to triangles. The fact that $\mathcal{C}$ is contained in the kernel of F_{univ} follows, for instance, from Lemma 2.1.33. The universality is also obvious, since a triangulated functor taking $\mathcal{C}$ to zero must take all maps in $Mor_{\mathcal{C}}$ to isomorphisms, and by Proposition 2.1.24 F_{univ} is the universal functor with this property. □

REMARK 2.1.39. Let $\mathcal{D}$ be a triangulated category, $\mathcal{C}$ a triangulated subcategory. By Remark 2.1.7, the kernel of $F_{univ} : \mathcal{D} \to \mathcal{D}/\mathcal{C}$ is a thick subcategory of $\mathcal{D}$, that is a triangulated category containing all direct summands of its objects. From Lemma 2.1.33 we learn that the kernel contains $\mathcal{C}$, and can be described as the full subcategory whose objects are the direct summands of objects in $\mathcal{C}$. We will call this category the *thick closure* of $\mathcal{C}$, and denote it $\widehat{\mathcal{C}}$.

Because the kernel of a triangulated functor is always triangulated, we deduce that for any triangulated subcategory $\mathcal{C} \subset \mathcal{D}$, the subcategory $\widehat{\mathcal{C}}$ is triangulated. The triangulated subcategory $\mathcal{C} \subset \mathcal{D}$ is thick if and only if $\mathcal{C} = \widehat{\mathcal{C}}$.

2.2. Sets and classes

In Section 2.1, we gave a proof of Verdier's localisation theorem. Given a triangulated category $\mathcal{D}$ and a triangulated subcategory $\mathcal{C}$, one can form a triangulated quotient $\mathcal{D}/\mathcal{C}$, with a universal property. But there is here a technical point which deserves mention.

Suppose $\mathcal{D}$ is a category with small *Hom*–sets. That is, for any objects x and y in $\mathcal{D}$, we require that $Hom_{\mathcal{D}}(x, y)$ should be a set. Then of course $\mathcal{C}$, being a subcategory, also has small *Hom*–sets. However, the quotient category $\mathcal{D}/\mathcal{C}$ in general *does not!*

Recall that the class $Hom_{\mathcal{D}/\mathcal{C}}(x, y)$ is defined by taking the class of all diagrams

$$\begin{array}{ccc} p & \xrightarrow{\alpha} & y \\ {\scriptstyle f}\downarrow & & \\ x & & \end{array}$$

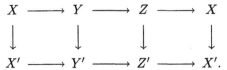

with $f \in Mor_{\mathcal{C}}$, modulo a suitable equivalence. There is no reason in general to expect there to only be a small set of diagrams as above; usually the collection of such diagrams forms a class. It may happen that even after identifying equivalent diagrams, we end up with a class of equivalence classes. See, for example, Exercise 1, Chapter 6, pp 131-132 in [**12**].

Note that since $f : p \longrightarrow x$ is assumed in $Mor_{\mathcal{C}}$, there is a triangle

$$p \xrightarrow{\ f\ } x \xrightarrow{\ g\ } z \longrightarrow \Sigma p$$

with $z \in \mathcal{C}$. If it so happened that $\mathcal{C}$ is a small category (that is, a small set of objects), then we would deduce that there is only a small set of choices for z, and for each z only a small set of choices for $g : x \longrightarrow z$. This gives a small set of choices (up to isomorphism) of $f : p \longrightarrow x$, and for each of these a small set of choices for $\alpha : p \longrightarrow y$. Thus we end up with a small set of equivalence classes. We conclude

PROPOSITION 2.2.1. *Let $\mathcal{D}$ be a triangulated category with small Hom–sets. Let $\mathcal{C}$ be a small triangulated subcategory of $\mathcal{D}$. Then $\mathcal{D}/\mathcal{C}$, the Verdier quotient of Section 2.1, is a category with small Hom–sets.* □

This is true, but not very useful. In fact, one mostly cares about subcategories $\mathcal{C} \subset \mathcal{D}$ which are not small, and frequently one needs criteria which guarantee that $\mathcal{D}/\mathcal{C}$ has small *Hom*–sets. In the coming Chapters, we will, among other things, develop criteria on $\mathcal{D}$ and $\mathcal{C}$ which guarantee the smallness of the *Hom*–sets in $\mathcal{D}/\mathcal{C}$.

2.3. History of the results in Chapter 2

The main result of the Chapter, Theorem 2.1.8, is due to Verdier. Verdier, in [**35**], constructed the quotient $\mathcal{T}/\mathcal{S}$ if $\mathcal{S}$ is a thick subcategory of $\mathcal{T}$. We allow $\mathcal{S}$ to be an arbitrary triangulated subcategory. The quotient of $\mathcal{T}$ by $\mathcal{S}$ then agrees with the quotient of $\mathcal{T}$ by the thick closure of $\mathcal{S}$. The advantage of treating the more general case is that it allows us to analyse the thick closure, and prove that it consists precisely of direct summands of objects of $\mathcal{S}$. In Parshall–Scott's preprint for [**26**], the authors also constructed the quotient by an arbitrary triangulated $\mathcal{S}$. In the published version this was left out.

The proof here is somewhat different from Verdier's. This is due to an observation of Rickard's that Verdier's original characterisation of thick subcategories was unnecessarily opaque. Rickard characterises thick subcategories as subcategories containing direct summands of their objects (Definition 2.1.6). We do not give here Verdier's definition, nor the fact that the two are equivalent. The interested reader is referred to Rickard's [**28**].

In the literature, the treatment of set theoretic problems that arise in constructing $\mathcal{T}/\mathcal{S}$ is patchy at best. In Section 2.2, we remind the reader that even if $\mathcal{S}$ and $\mathcal{T}$ are categories with small Hom–sets, $\mathcal{T}/\mathcal{S}$ need not be. This has led to many mistakes in the literature. Subtle points of this nature tend to arise more in the homotopy theoretic literature. The literature on derived categories traditionally ignores this difficulty completely, and usually does so quite safely.

CHAPTER 3

Perfection of classes

3.1. Cardinals

Let us briefly review some standard definitions for large cardinals. A cardinal α is called *singular* if α can be written as a sum of fewer than α cardinals, all smaller than α.

Let $\aleph_n$ be the nth infinite cardinal. Thus, $\aleph_0$ is the zeroth, that is the countable cardinal, $\aleph_1$ the next, and so on. The cardinal $\aleph_\omega$ is defined to be the smallest cardinal bigger than $\aleph_n$ for all n. Clearly,

$$\aleph_\omega = \sum_{n=1}^{\infty} \aleph_n$$

is a countable union of cardinals, each strictly smaller than $\aleph_\omega$. Hence $\aleph_\omega$ is an example of a singular cardinal.

A cardinal which is not singular is called *regular.* If α is a cardinal and β is the successor of α, then β is regular. The successor of a cardinal α is the smallest cardinal exceeding it. If β is the successor of α, a sum of fewer than β cardinals, each less than β, is a sum of less than or equal to α cardinals, each less than or equal to α. This is bounded by $\alpha \times \alpha = \alpha$.

3.2. Generated subcategories

DEFINITION 3.2.1. *Let $\mathcal{T}$ be a triangulated category satisfying [TR5]. Let β be an infinite cardinal. Let S be a class of objects of $\mathcal{T}$. Then $\langle S\rangle^\beta$ will stand for the smallest $\mathcal{S}$, $\mathcal{S}$ a triangulated subcategory of $\mathcal{T}$ satisfying*

3.2.1.1. *The objects of S lie in $\mathcal{S}$.*

3.2.1.2. *Any coproduct of fewer than β objects of $\mathcal{S}$ lies in $\mathcal{S}$.*

3.2.1.3. *The subcategory $\mathcal{S} \subset \mathcal{T}$ is thick.*

REMARK 3.2.2. The subcategory $\langle S\rangle^\beta$ is well–defined. It is the intersection of all the subcategories $\mathcal{S}$ of $\mathcal{T}$ satisfying 3.2.1.1, 3.2.1.2 and 3.2.1.3.

The key fact we want to prove in this section is that, as long as S is a set, for all infinite β the subcategory $\langle S\rangle^\beta$ is essentially small. That is, there is only a set of isomorphism classes of objects of $\langle S\rangle^\beta$. The first observation is

LEMMA 3.2.3. *If $\beta \le \gamma$ and $S \subset \mathcal{T}$ is a set of objects, then $\langle S\rangle^{\beta} \subset \langle S\rangle^{\gamma}$.*

Proof: $\langle S\rangle^{\gamma}$ satisfies 3.2.1.1 and 3.2.1.3, which do not involve γ, as well as 3.2.1.2 for γ; that is the coproduct of fewer than γ objects in $\langle S\rangle^{\gamma}$ is in $\langle S\rangle^{\gamma}$. But then the coproduct of fewer than β objects of $\langle S\rangle^{\gamma}$ certainly must be in $\langle S\rangle^{\gamma}$. But $\langle S\rangle^{\beta}$ is minimal with this property; hence $\langle S\rangle^{\beta} \subset \langle S\rangle^{\gamma}$. □

LEMMA 3.2.4. *Let S be a set of objects in a triangulated category $\mathcal{T}$. Suppose $\mathcal{T}$ has small Hom–sets. Then the smallest triangulated subcategory of $\mathcal{T}$ containing S, denoted $\overline{T}(S)$, is essentially small. Furthermore, in the course of the proof we will construct a small category $T(S)$ equivalent to $\overline{T}(S)$, containing S.*

Proof: Define $T_1(S)$ to be the full subcategory whose objects are $S \cup \{0\}$. Because S is a set the category $T_1(S)$ has only a set of objects, and since $\mathcal{T}$ has small *Hom*–sets, $T_1(S)$ is small. Now inductively define $T_n(S)$. Suppose $T_1(S), T_2(S), \ldots, T_n(S)$ have already been defined, and are small. To define $T_{n+1}(S)$ choose, for every morphism $f : x \to y$ in $T_n(S)$, one object of $\mathcal{T}$ in the isomorphism class of z in the triangle

$$x \xrightarrow{\ f\ } y \longrightarrow z \longrightarrow \Sigma x$$

Call this object C_f. Let $T_{n+1}(S)$ be the smallest full subcategory containing $T_n(S)$ and all the C_f's.

Now put $T(S) = \bigcup_{n=1}^{\infty} T_n(S)$. Given any morphism $f : x \to y$ in $T(S)$, it lies in some $T_n(S)$. But then $T_{n+1}(S)$ contains an object C_f which fits in a triangle

$$x \xrightarrow{\ f\ } y \longrightarrow C_f \longrightarrow \Sigma x$$

Thus the full subcategory of $\mathcal{T}$ whose objects are all the isomorphs of the objects of $T(S)$ is triangulated. Call it $\overline{T}(S)$. Since $\overline{T}(S)$ is equivalent to $T(S)$, $\overline{T}(S)$ is essentially small. It is easy to see that $\overline{T}(S)$ is in fact the smallest triangulated subcategory containing S. □

PROPOSITION 3.2.5. *Let $\mathcal{T}$ be triangulated category with small Hom–sets, satisfying [TR5]. Suppose S is some set of objects in $\mathcal{T}$. Let β be an infinite cardinal. Then the category $\langle S\rangle^{\beta}$ is essentially small.*

Proof: By Lemma 3.2.3, if $\beta \le \gamma$, then $\langle S\rangle^{\beta} \subset \langle S\rangle^{\gamma}$. It therefore clearly suffices to show that if β is an infinite cardinal and γ is the successor of β, then $\langle S\rangle^{\gamma}$ is essentially small. But γ, being a successor cardinal, is regular (see Section 3.1). Furthermore, $\gamma > \beta \ge \aleph_0$ ($\aleph_0$ is the first infinite cardinal). Replacing β by its successor γ, we may therefore assume β regular, and $\beta > \aleph_0$.

Now observe that any subcategory $\mathcal{S}$ satisfying 3.2.1.2 automatically satisfies 3.2.1.3; after all $\beta > \aleph_0$ together with 3.2.1.2 implies that the category $\mathcal{S}$ is closed with respect to countable coproducts. But then by Proposition 1.6.8 all idempotents in $\mathcal{S}$ split, and the subcategory $\mathcal{S}$ must be thick.

In other words, for $\beta > \aleph_0$ 3.2.1.3 is redundant; $\langle S \rangle^\beta$ is the smallest subcategory satisfying 3.2.1.1 and 3.2.1.2. Now we will show $\langle S \rangle^\beta$ essentially small. But first a little notation.

Let S be a set of objects in $\mathcal{T}$, containing the zero object. Let M be a set of cardinality β. Consider the set

$$\prod_{\mu \in M} S,$$

of all sequences of objects in S of length β. Consider the subset

$$G \subset \prod_{\mu \in M} S,$$

consisting of all sequences such that the cardinality of the set of non–zero terms is $< \beta$. For each sequence in G, choose a representative in the isomorphism class of the coproduct of the sequence. Let $CP(S)$ be the full subcategory of $\mathcal{T}$, whose objects are the union of S and our choices of isomorphism representatives of coproducts of length $< \beta$. Then $CP(S)$ is small, contains S, and contains an object isomorphic to the coproduct of any $< \beta$ objects of S.

Now we proceed by transfinite induction. Inductively define S_i, for all ordinals i. Set $S_0 = T(S)$ (a small category equivalent to the smallest triangulated subcategory containing S, as in Lemma 3.2.4). For a successor ordinal $i+1$, define S_{i+1} to be $T(CP(S_i))$; that is, first choose something isomorphic to any coproduct of fewer than β objects in S_i and throw it in, then close with respect to triangles. For i a limit ordinal, define S_i to be the union

$$S_i = \bigcup_{j<i} S_j.$$

Clearly, each S_i is small. I assert that the full subcategory $\overline{S}_\beta$ of $\mathcal{T}$, whose objects are all the isomorphs of objects of S_β, is triangulated and satisfies 3.2.1.1 and 3.2.1.2.

First, 3.2.1.1 is obvious; by construction $S \subset S_0 \subset S_\beta$.

The fact that $\overline{S}_\beta$ is triangulated is proved by induction. $S_0 = T(S)$, and clearly $\overline{S}_0 = \overline{T}(S)$ is triangulated. For any successor ordinal, $S_{i+1} = T(CP(S_i))$, in other words, S_{i+1} is T of something, and hence $\overline{S}_{i+1}$ must be triangulated, by the T construction. But if i is a limit ordinal, and for all $j < i$ $\overline{S}_j$ is triangulated, then from the fact that

$$\overline{S}_i = \bigcup_{j<i} \overline{S}_j$$

is an increasing union of triangulated subcategories, we deduce that $\overline{S}_i$ is triangulated. By transfinite induction, $\overline{S}_i$ are triangulated for all i. In particular, $\overline{S}_\beta$ is triangulated.

Now suppose $\{X_\lambda; \lambda \in \Lambda\}$ is a collection of fewer than β objects in S_β. Recall that β, being a cardinal, is certainly a limit ordinal. Therefore

$$S_\beta = \bigcup_{j<\beta} S_j$$

and $X_\lambda \in S_\beta$ means that for some $j_\lambda < \beta$, $X_\lambda \in S_{j_\lambda}$. For each λ, choose a j_λ.

Let γ be the smallest ordinal larger than all the j_λ's, $\lambda \in \Lambda$. Now each $j_\lambda < \beta$, and there are fewer than β of them (the set Λ has cardinality less than β). Therefore the sum of the j_λ's is a sum of fewer than β cardinals, each smaller than β. But β is regular. It follows that the cardinality of γ must be strictly less than β. In other words, $\gamma < \beta$.

We deduce that all the X_λ's lie in S_γ. But then $S_{\gamma+1} \subset S_\beta$ contains an object isomorphic to the coproduct. Hence $\overline{S}_\beta$ contains all coproducts of fewer than β of its objects. In other words, $\overline{S}_\beta$ satisfies 3.2.1.1 and 3.2.1.2.

Since $\overline{S}_\beta$ satisfies 3.2.1.1 and 3.2.1.2, it contains $\langle S\rangle^\beta$. But since $\overline{S}_\beta$ is equivalent to the small category S_β, it is essentially small. It follows that the smaller subcategory $\langle S\rangle^\beta$ is essentially small. □

DEFINITION 3.2.6. *Let β be an infinite cardinal. Let $\mathcal{T}$ be a triangulated category satisfying [TR5]. A subcategory $\mathcal{S} \subset \mathcal{T}$ is called* β–localising *if it is thick and closed with respect to the formation of coproducts of fewer than β of its objects. That means that the coproduct of fewer than β objects of $\mathcal{S}$ exists, as a coproduct in $\mathcal{T}$, and is an object of $\mathcal{S}$. A triangulated subcategory $\mathcal{S} \subset \mathcal{T}$ is called* localising *if it is β–localising for all β. Equivalently, $\mathcal{S}$ is localising if it is closed under the formation of all small $\mathcal{T}$–coproducts of its objects. That is, if $\{X_\lambda, \lambda \in \Lambda\}$ is a family of objects in $\mathcal{S}$, then the $\mathcal{T}$–coproduct*

$$\coprod_{\lambda\in\Lambda} X_\lambda$$

is an object of $\mathcal{S}$.

REMARK 3.2.7. If $\beta > \aleph_0$, then any *triangulated* subcategory $\mathcal{S} \subset \mathcal{T}$ closed under the formation of coproducts of fewer than β of its objects is automatically thick, hence localising. It is redundant to assume $\mathcal{S}$ thick. The point is that if $\beta > \aleph_0$, then $\mathcal{S}$ contains all countable coproducts of its objects. By Proposition 1.6.8 every idempotent in $\mathcal{S}$ splits, and hence $\mathcal{S}$ contains all direct summands of its objects.

EXAMPLE 3.2.8. Let $\mathcal{T}$ be a triangulated category satisfying [TR5]. Let S be a class of objects of $\mathcal{T}$. Then $\langle S\rangle^{\beta}$ is β–localising, whereas

$$\langle S\rangle = \bigcup_{\beta} \langle S\rangle^{\beta}$$

is localising, as in Definition 3.2.6. The statement for $\langle S\rangle^{\beta}$ is really part of its definition, as the smallest β–localising subcategory with certain properties. The statement for $\langle S\rangle$ requires a little proof. Let us give the proof.

Let $\{X_\lambda, \lambda \in \Lambda\}$ be any collection of objects in $\langle S\rangle$. Let the cardinality of Λ be β, and suppose that for each $\lambda \in \Lambda$, the object X_λ lies in $\langle S\rangle^{\beta_\lambda}$, for some cardinal β_λ. Let γ be a cardinal greater than the sum of β and the β_λ's. The X_λ's are all in $\langle S\rangle^{\gamma}$, and since $\gamma > \beta$, the coproduct of the X_λ's, that is of β objects of $\langle S\rangle^{\gamma}$, must lie in $\langle S\rangle^{\gamma}$. Hence in $\langle S\rangle$, which is therefore localising.

In fact, $\langle S\rangle$ is the smallest localising subcategory containing S. Any localising subcategory containing S will satisfy 3.2.1.1, 3.2.1.2 and 3.2.1.3 for all infinite β. Therefore it must contain $\langle S\rangle^{\beta}$ for all β, and hence $\langle S\rangle$, the union.

DEFINITION 3.2.9. *Let $\mathcal{T}$ be a triangulated category satisfying [TR5]. Let S be a class of objects of $\mathcal{T}$. Then the union of all the $\langle S\rangle^{\beta}$'s, that is*

$$\bigcup_{\beta} \langle S\rangle^{\beta}$$

will be denoted $\langle S\rangle$. Note that even when S is a small set and $\mathcal{T}$ has small Hom–sets, in which case Proposition 3.2.5 tells us that the categories $\langle S\rangle^{\beta}$ are all essentially small, the category $\langle S\rangle$ will usually be gigantic. It is called the localising subcategory generated by S.

LEMMA 3.2.10. *Let β be an infinite cardinal. Let $\mathcal{T}$ be a triangulated category closed under the formation of coproducts of fewer than β of its objects. Let $\mathcal{S}$ be a β–localising subcategory of $\mathcal{T}$. Then $\mathcal{T}/\mathcal{S}$ is closed with respect to the formation of coproducts of fewer than β of its objects, and the universal functor $F : \mathcal{T} \longrightarrow \mathcal{T}/\mathcal{S}$ preserves coproducts.*

Proof: (cf. proof of Lemma 2.1.29). Since the objects of $\mathcal{T}$ and $\mathcal{T}/\mathcal{S}$ are the same, it suffices to show that the coproduct in $\mathcal{T}$ of fewer than β objects is also the coproduct in $\mathcal{T}/\mathcal{S}$. Let Λ be a set of cardinality less than β, $\{X_\lambda, \lambda \in \Lambda\}$ a collection of objects of $\mathcal{T}$. Form their coproduct in $\mathcal{T}$, which we know exists,

$$\coprod_{\lambda\in\Lambda} X_\lambda.$$

We need to show this is also a coproduct in $\mathcal{T}/\mathcal{S}$.

Let Y be an arbitrary object of $\mathcal{T}$. We need to show that any collection of morphisms in $\mathcal{T}/\mathcal{S}$

$$\{X_\lambda \longrightarrow Y, \lambda \in \Lambda\}$$

factors uniquely through $\coprod_{\lambda\in\Lambda} X_\lambda$. That is, we need to show the existence and uniqueness of a factorisation.

The morphisms $X_\lambda \longrightarrow Y$ in $\mathcal{T}/\mathcal{S}$ can be represented by diagrams

$$\begin{array}{ccc} P_\lambda & \longrightarrow & Y \\ {\scriptstyle f_\lambda}\big\downarrow & & \\ X_\lambda & & \end{array}$$

where $f_\lambda : P_\lambda \longrightarrow X_\lambda$ are in $Mor_{\mathcal{S}}$. That is, in the triangle

$$P_\lambda \xrightarrow{f_\lambda} X_\lambda \longrightarrow Z_\lambda \longrightarrow \Sigma P_\lambda$$

the object Z_λ must be in $\mathcal{S}$. But now the coproduct is a triangle

$$\coprod_{\lambda\in\Lambda} P_\lambda \xrightarrow{\coprod_{\lambda\in\Lambda} f_\lambda} \coprod_{\lambda\in\Lambda} X_\lambda \longrightarrow \coprod_{\lambda\in\Lambda} Z_\lambda \longrightarrow \coprod_{\lambda\in\Lambda} \Sigma P_\lambda$$

and $\coprod_{\lambda\in\Lambda} Z_\lambda$, being the coproduct of fewer than β objects in $\mathcal{S}$, must be in $\mathcal{S}$. The map

$$\coprod_{\lambda\in\Lambda} P_\lambda \xrightarrow{\coprod_{\lambda\in\Lambda} f_\lambda} \coprod_{\lambda\in\Lambda} X_\lambda$$

therefore lies in $Mor_{\mathcal{C}}$. The diagram

$$\begin{array}{ccc} \coprod_{\lambda\in\Lambda} P_\lambda & \longrightarrow & Y \\ {\scriptstyle \coprod_{\lambda\in\Lambda} f_\lambda}\big\downarrow & & \\ \coprod_{\lambda\in\Lambda} X_\lambda & & \end{array}$$

represents a morphism in $\mathcal{T}/\mathcal{S}$ of the form $\coprod_{\lambda\in\Lambda} X_\lambda \longrightarrow Y$. Its composites with the inclusions

$$i_\lambda : X_\lambda \longrightarrow \coprod_{\lambda\in\Lambda} X_\lambda$$

are computed by the commutative diagrams

$$\begin{array}{ccccc} P_\lambda & \longrightarrow & \coprod_{\lambda\in\Lambda} P_\lambda & \longrightarrow & Y \\ {\scriptstyle f_\lambda}\downarrow & & \coprod_{\lambda\in\Lambda} f_\lambda \downarrow & & \\ X_\lambda & \longrightarrow & \coprod_{\lambda\in\Lambda} X_\lambda & & \end{array}$$

to be the given maps

$$\begin{array}{ccc} P_\lambda & \longrightarrow & Y \\ {\scriptstyle f_\lambda}\downarrow & & \\ X_\lambda & & \end{array}$$

We have therefore factored $X_\lambda \longrightarrow Y$ through $\coprod_{\lambda\in\Lambda} X_\lambda$.

Now for the uniqueness of the factorisation. Suppose we are given a map in $\mathcal{T}/\mathcal{S}$ of the form

$$\phi : \coprod_{\lambda\in\Lambda} X_\lambda \longrightarrow Y$$

so that the composites $\phi \circ i_\lambda$ with every

$$i_\lambda : X_\lambda \longrightarrow \coprod_{\lambda\in\Lambda} X_\lambda$$

vanish. We need to show the map ϕ vanishes. Represent ϕ as a diagram

$$\begin{array}{ccc} & & Y \\ & & \downarrow{\scriptstyle g} \\ \coprod_{\lambda\in\Lambda} X_\lambda & \longrightarrow & Q \end{array}$$

with $g : Y \longrightarrow Q$ in $Mor_{\mathcal{S}}$. Then for each λ, the composite with i_λ is represented by

$$\begin{array}{ccc} & & Y \\ & & \downarrow{\scriptstyle g} \\ X_\lambda & \longrightarrow & Q \end{array}$$

and must vanish in $\mathcal{T}/\mathcal{S}$. By Lemma 2.1.26, there must then exist $Z_\lambda \in \mathcal{S}$ so that $X_\lambda \longrightarrow Q$ factors as

$$X_\lambda \longrightarrow Z_\lambda \longrightarrow Q.$$

But then

$$\coprod_{\lambda\in\Lambda} X_\lambda \longrightarrow Q$$

factors as

$$\coprod_{\lambda\in\Lambda} X_\lambda \longrightarrow \coprod_{\lambda\in\Lambda} Z_\lambda \longrightarrow Q$$

and $\coprod_{\lambda\in\Lambda} Z_\lambda$, being a coproduct of fewer than β objects of $\mathcal{S}$, must lie in $\mathcal{S}$. Lemma 2.1.26 tells us that the map

$$\begin{array}{ccc} & & Y \\ & & \big\downarrow{\scriptstyle g} \\ \coprod_{\lambda\in\Lambda} X_\lambda & \longrightarrow & Q \end{array}$$

must then vanish in $\mathcal{T}/\mathcal{S}$. □

The following is an immediate corollary.

COROLLARY 3.2.11. *Let $\mathcal{T}$ be a triangulated category satisfying [TR5]. Let $\mathcal{S}$ be a localising subcategory. Then $\mathcal{T}/\mathcal{S}$ satisfies [TR5], and the universal functor $\mathcal{T} \longrightarrow \mathcal{T}/\mathcal{S}$ preserves coproducts.*

3.3. Perfect classes

DEFINITION 3.3.1. *Let $\mathcal{T}$ be a triangulated category satisfying [TR5]. Let β be an infinite cardinal. A class of objects $S \subset \mathcal{T}$ is called β–*perfect *if the following hold*

3.3.1.1. *S contains* 0.

3.3.1.2. *Suppose $\{X_\lambda, \lambda \in \Lambda\}$ is a collection of objects in $\mathcal{T}$. Suppose the cardinality of Λ is less than β. Let k be an object of S. Then any map*

$$k \longrightarrow \coprod_{\lambda\in\Lambda} X_\lambda$$

factors as

$$k \longrightarrow \coprod_{\lambda\in\Lambda} k_\lambda \longrightarrow \coprod_{\lambda\in\Lambda} X_\lambda$$

with $k_\lambda \in S$. More precisely there is, for each $\lambda \in \Lambda$, an object $k_\lambda \in S$ and a map $f_\lambda : k_\lambda \longrightarrow X_\lambda$, so that the map factors as

$$k \longrightarrow \coprod_{\lambda\in\Lambda} k_\lambda \xrightarrow{\coprod_{\lambda\in\Lambda} f_\lambda} \coprod_{\lambda\in\Lambda} X_\lambda.$$

3.3.1.3. *Suppose again that Λ is a set of cardinality $< \beta$. Suppose k and the k_λ's, $\lambda \in \Lambda$, are objects of S and the X_λ's are any objects of $\mathcal{T}$, and the composite*

$$k \longrightarrow \coprod_{\lambda\in\Lambda} k_\lambda \xrightarrow{\coprod_{\lambda\in\Lambda} f_\lambda} \coprod_{\lambda\in\Lambda} X_\lambda$$

vanishes. Then it is possible to factor each $f_\lambda : k_\lambda \longrightarrow X_\lambda$ as

$$k_\lambda \xrightarrow{g_\lambda} l_\lambda \xrightarrow{h_\lambda} X_\lambda$$

so that $l_\lambda \in S$, and the composite

$$k \longrightarrow \coprod_{\lambda\in\Lambda} k_\lambda \xrightarrow{\coprod_{\lambda\in\Lambda} g_\lambda} \coprod_{\lambda\in\Lambda} l_\lambda$$

already vanishes.

We begin with a trivial lemma.

LEMMA 3.3.2. *Let $\mathcal{T}$ be a triangulated category satisfying [TR5]. Suppose β is an infinite cardinal, and T is a β-perfect class in $\mathcal{T}$. Suppose that $S \subset T$ is an equivalent class; that is, any object of T is isomorphic to some object of S. Then S is also β-perfect.*

Proof: Trivial. □

LEMMA 3.3.3. *Let $\mathcal{T}$ be a triangulated category satisfying [TR5]. Let β be an infinite cardinal. Let S be a β-perfect class of $\mathcal{T}$. Let T be the class of all objects in $\mathcal{T}$ which are direct summands of objects of S. Then T is also β-perfect.*

Proof: Suppose we are given an object $k \in T$, a set Λ of cardinality $< \beta$, a family of X_λ's in $\mathcal{T}$ and a map

$$k \longrightarrow \coprod_{\lambda\in\Lambda} X_\lambda.$$

Because $k \in T$, there must exist some $k' \in \mathcal{T}$ so that $k \oplus k' \in S$; T is defined to be the class of all direct summands of objects of S. Consider the composite

$$k \oplus k' \xrightarrow{\begin{pmatrix} 1 & 0 \end{pmatrix}} k \longrightarrow \coprod_{\lambda\in\Lambda} X_\lambda.$$

Since $k \oplus k' \in S$ and S is β–perfect, this factors

$$k \oplus k' \longrightarrow \coprod_{\lambda\in\Lambda} k_\lambda \xrightarrow{\coprod_{\lambda\in\Lambda} f_\lambda} \coprod_{\lambda\in\Lambda} X_\lambda$$

with $k_\lambda \in S \subset T$. But then

$$k \xrightarrow{\begin{pmatrix} 1 \\ 0 \end{pmatrix}} k \oplus k' \longrightarrow \coprod_{\lambda\in\Lambda} k_\lambda \xrightarrow{\coprod_{\lambda\in\Lambda} f_\lambda} \coprod_{\lambda\in\Lambda} X_\lambda$$

is a factorisation of the original map

$$k \longrightarrow \coprod_{\lambda\in\Lambda} X_\lambda.$$

Suppose now that we are given a vanishing composite

$$k \longrightarrow \coprod_{\lambda\in\Lambda} k_\lambda \xrightarrow{\coprod_{\lambda\in\Lambda} f_\lambda} \coprod_{\lambda\in\Lambda} X_\lambda$$

with $k, k_\lambda \in T$. Choose k' so that $k \oplus k' \in S$, and for each λ choose k'_λ with $k_\lambda \oplus k'_\lambda \in S$. Then we have a vanishing composite

$$\begin{array}{c} k \oplus k' \xrightarrow{\begin{pmatrix} 1 & 0 \end{pmatrix}} k \longrightarrow \coprod_{\lambda\in\Lambda} k_\lambda \xrightarrow{\coprod_{\lambda\in\Lambda} \begin{pmatrix} 1 \\ 0 \end{pmatrix}} \coprod_{\lambda\in\Lambda} k_\lambda \oplus k'_\lambda \\ \Big\downarrow \coprod_{\lambda\in\Lambda} \begin{pmatrix} f_\lambda & 0 \end{pmatrix} \\ \coprod_{\lambda\in\Lambda} X_\lambda \end{array}$$

Because S is β–perfect, for each λ the map

$$\begin{pmatrix} f_\lambda & 0 \end{pmatrix} : k_\lambda \oplus k'_\lambda \longrightarrow X_\lambda$$

must factor as

$$k_\lambda \oplus k'_\lambda \longrightarrow l_\lambda \longrightarrow X_\lambda$$

with $l_\lambda \in S$, where the composite

$$k \oplus k' \xrightarrow{\begin{pmatrix} 1 & 0 \end{pmatrix}} k \longrightarrow \coprod_{\lambda\in\Lambda} k_\lambda \xrightarrow{\coprod_{\lambda\in\Lambda} \begin{pmatrix} 1 \\ 0 \end{pmatrix}} \coprod_{\lambda\in\Lambda} k_\lambda \oplus k'_\lambda \longrightarrow \coprod_{\lambda\in\Lambda} l_\lambda$$

already vanishes. But then

$$\begin{array}{ccccccccc} k & \xrightarrow{\begin{pmatrix} 1 \\ 0 \end{pmatrix}} & k \oplus k' & \xrightarrow{\begin{pmatrix} 1 & 0 \end{pmatrix}} & k & \longrightarrow & \coprod_{\lambda \in \Lambda} k_\lambda & \xrightarrow{\coprod_{\lambda \in \Lambda} \begin{pmatrix} 1 \\ 0 \end{pmatrix}} & \coprod_{\lambda \in \Lambda} k_\lambda \oplus k'_\lambda \\ &&&&&&&& \downarrow \\ &&&&&&&& \coprod_{\lambda \in \Lambda} l_\lambda \end{array}$$

also vanishes, completing the proof of the β–perfection of T. □

DEFINITION 3.3.4. *Let $\mathcal{T}$ be a triangulated category satisfying [TR5]. Let $\mathcal{S} \subset \mathcal{T}$ be a triangulated subcategory. Let β be an infinite cardinal. An object $k \in \mathcal{T}$ is called β–*good *if the following holds.*

> 3.3.4.1. *Let $\{X_\lambda, \lambda \in \Lambda\}$ be a family of objects of $\mathcal{T}$. Suppose the cardinality of Λ is less than β. Then any map*
>
> $$k \longrightarrow \coprod_{\lambda \in \Lambda} X_\lambda$$
>
> *has a factorisation*
>
> $$k \longrightarrow \coprod_{\lambda \in \Lambda} k_\lambda \xrightarrow{\coprod_{\lambda \in \Lambda} f_\lambda} \coprod_{\lambda \in \Lambda} X_\lambda$$
>
> *with $k_\lambda \in \mathcal{S}$.*

More intuitively, 3.3.1.2 sort of holds for the single object k; it holds where the k_λ may be taken in $\mathcal{S}$.

LEMMA 3.3.5. *Let $\mathcal{T}$ be a triangulated category satisfying [TR5], and let $\mathcal{S}$ be a triangulated subcategory. If k is a β–good object of $\mathcal{T}$, then the following automatically holds.*

> 3.3.5.1. *Suppose Λ is an index set of cardinality $< \beta$. Given a vanishing composite*
>
> $$k \longrightarrow \coprod_{\lambda \in \Lambda} k_\lambda \xrightarrow{\coprod_{\lambda \in \Lambda} f_\lambda} \coprod_{\lambda \in \Lambda} X_\lambda$$
>
> *with $k_\lambda \in \mathcal{S}$, then it is possible to factor each $f_\lambda : k_\lambda \longrightarrow X_\lambda$ as*
>
> $$k_\lambda \xrightarrow{g_\lambda} l_\lambda \xrightarrow{h_\lambda} X_\lambda$$

so that $l_\lambda \in \mathcal{S}$, and the composite

$$k \longrightarrow \coprod_{\lambda\in\Lambda} k_\lambda \xrightarrow{\coprod_{\lambda\in\Lambda} g_\lambda} \coprod_{\lambda\in\Lambda} l_\lambda$$

also vanishes.

Proof: Assume that Λ is a set whose cardinality is less than β. Suppose we are given a vanishing composite

$$k \longrightarrow \coprod_{\lambda\in\Lambda} k_\lambda \xrightarrow{\coprod_{\lambda\in\Lambda} f_\lambda} \coprod_{\lambda\in\Lambda} X_\lambda$$

with $k_\lambda \in \mathcal{S}$.

For each λ, consider the triangle

$$Y_\lambda \longrightarrow k_\lambda \xrightarrow{f_\lambda} X_\lambda \longrightarrow \Sigma Y_\lambda.$$

Summing these triangles, we obtain a triangle

$$\coprod_{\lambda\in\Lambda} Y_\lambda \longrightarrow \coprod_{\lambda\in\Lambda} k_\lambda \xrightarrow{\coprod_{\lambda\in\Lambda} f_\lambda} \coprod_{\lambda\in\Lambda} X_\lambda \longrightarrow \coprod_{\lambda\in\Lambda} \Sigma Y_\lambda.$$

Since the composite

$$k \longrightarrow \coprod_{\lambda\in\Lambda} k_\lambda \xrightarrow{\coprod_{\lambda\in\Lambda} f_\lambda} \coprod_{\lambda\in\Lambda} X_\lambda$$

vanishes, the map

$$k \longrightarrow \coprod_{\lambda\in\Lambda} k_\lambda$$

must factor as

$$k \longrightarrow \coprod_{\lambda\in\Lambda} Y_\lambda \longrightarrow \coprod_{\lambda\in\Lambda} k_\lambda.$$

But then the hypothesis that k is good guarantees that the map

$$k \longrightarrow \coprod_{\lambda\in\Lambda} Y_\lambda$$

factors as

$$k \longrightarrow \coprod_{\lambda\in\Lambda} j_\lambda \longrightarrow \coprod_{\lambda\in\Lambda} Y_\lambda$$

with $j_\lambda \in \mathcal{S}$. The composite

$$j_\lambda \longrightarrow Y_\lambda \longrightarrow k_\lambda \xrightarrow{f_\lambda} X_\lambda$$

clearly vanishes, and if we define l_λ by forming the triangle

$$j_\lambda \longrightarrow k_\lambda \xrightarrow{g_\lambda} l_\lambda \longrightarrow \Sigma j_\lambda,$$

then clearly $k_\lambda \longrightarrow X_\lambda$ factors as

$$k_\lambda \xrightarrow{g_\lambda} l_\lambda \xrightarrow{h_\lambda} X_\lambda.$$

Since j_λ and k_λ are in $\mathcal{S}$ and $\mathcal{S}$ is triangulated, l_λ must also be in $\mathcal{S}$. And the composite

$$k \longrightarrow \coprod_{\lambda\in\Lambda} k_\lambda \longrightarrow \coprod_{\lambda\in\Lambda} l_\lambda$$

must vanish, since it is

$$k \longrightarrow \coprod_{\lambda\in\Lambda} j_\lambda \longrightarrow \coprod_{\lambda\in\Lambda} k_\lambda \longrightarrow \coprod_{\lambda\in\Lambda} l_\lambda$$

and

$$j_\lambda \longrightarrow k_\lambda \xrightarrow{g_\lambda} l_\lambda$$

are two morphisms in a triangle. □

REMARK 3.3.6. From Lemma 3.3.5 we learn that, if $\mathcal{S}$ is a triangulated subcategory of $\mathcal{T}$ all of whose objects are β–good, then $\mathcal{S}$ is a β–perfect class; more precisely, the collection of all objects of $\mathcal{S}$ is a β–perfect class. For any $k \in \mathcal{S}$, if it satisfies 3.2.1.2 then 3.2.1.3 is automatic. The next couple of Lemmas allow us to cut the work even shorter. One need not check that every object of $\mathcal{S}$ is β–good. It is enough to check all the objects of a large enough generating class.

Before the next Lemmas, we should perhaps remind the reader of the notation of Section 3.2. Let S be a class of objects of $\mathcal{T}$. Then $\overline{T}(S)$ is the smallest triangulated subcategory containing S. For an infinite cardinal α, $\langle S\rangle^\alpha$ is the smallest α–localising subcategory containing S. More explicitly, $\langle S\rangle^\alpha$ is the minimal thick subcategory $\mathcal{S} \subset \mathcal{T}$ such that

3.2.1.1: $\mathcal{S}$ contains S.

3.2.1.2: $\mathcal{S}$ is closed under the formation of coproducts of fewer than α of its objects.

Recall that by Remark 3.2.7, if $\alpha > \aleph_0$ it is redundant to assume $\mathcal{S}$ is thick; 3.2.1.2 already tells us that idempotents split in $\mathcal{S}$.

LEMMA 3.3.7. *Let α and β be infinite cardinals. Let $\mathcal{T}$ be a triangulated category satisfying [TR5]. Let S be a class of objects of $\mathcal{T}$. Suppose that every object $k \in S$ is β–good, as an object of the triangulated subcategory $\langle S \rangle^{\alpha} \subset \mathcal{T}$. Then the objects of $\langle S \rangle^{\alpha}$ form a β–perfect class.*

A similar Lemma, whose proof is nearly identical, is

LEMMA 3.3.8. *Let β be an infinite cardinal. Let $\mathcal{T}$ be a triangulated category satisfying [TR5]. Let S be a class of objects of $\mathcal{T}$. Suppose that every object $k \in S$ is β–good as an object of the triangulated subcategory $\overline{T}(S) \subset \mathcal{T}$. Then the objects of $\overline{T}(S)$ form a β–perfect class.*

Proofs of Lemmas 3.3.7 and 3.3.8. Before we start the proof, let us make one observation. Assume Lemma 3.3.8; that is assume $\overline{T}(S)$ is β–perfect. By Lemma 3.3.3, the collection of all direct summands of objects of $\overline{T}(S)$ is also β–perfect; but this is precisely $\langle S \rangle^{\aleph_0}$, the thick closure of $\overline{T}(S)$. Without loss, we may therefore assume $\alpha > \aleph_0$ in Lemma 3.3.7; the case $\alpha = \aleph_0$ is an immediate consequence of Lemma 3.3.8.

The proofs of the two Lemmas are so nearly the same, that we will give them together. Consider the following two full subcategories $\mathcal{R} \subset \mathcal{T}$ and $\mathcal{S} \subset \mathcal{T}$, given by

$$\mathcal{S} = \{k \in \langle S \rangle^{\alpha} | k \text{ is } \beta\text{–good}\},$$

$$\mathcal{R} = \{k \in \overline{T}(S) | k \text{ is } \beta\text{–good}\}.$$

It suffices to show that $\mathcal{R}$ contains $\overline{T}(S)$ and $\mathcal{S}$ contains $\langle S \rangle^{\alpha}$. To show that $\mathcal{R}$ contains $\overline{T}(S)$, it suffices to prove that $\mathcal{R}$ is triangulated and contains S. To show that $\mathcal{S}$ contains $\langle S \rangle^{\alpha}$, it suffices to prove that $\mathcal{S}$ is a thick subcategory of $\mathcal{T}$ satisfying 3.2.1.1 and 3.2.1.2. Since the argument for $\mathcal{R}$ is similar and slightly simpler, we leave it to the reader. We assume therefore that $\alpha > \aleph_0$, and we will prove that $\mathcal{S}$ is triangulated, and satisfies 3.2.1.1 and 3.2.1.2. Since $\alpha > \aleph_0$, $\mathcal{S}$ would then be closed under the formation of countable coproducts, hence idempotents would split in $\mathcal{S}$. It is redundant to prove the thickness.

We need to prove three things, of which 3.2.1.1 is the hypothesis of the Lemma. By hypothesis we know that all the objects in S are β–good as objects of $\langle S \rangle^{\alpha}$. Hence $S \subset \mathcal{S}$, $\mathcal{S}$ being the subcategory of all β–good objects.

Proof that $\mathcal{S}$ satisfies 3.2.1.2. We need to show that, if $\{k_\mu, \mu \in M\}$ is a collection of fewer than α objects of $\mathcal{S}$, then the coproduct is in $\mathcal{S}$. Let $\{X_\lambda, \lambda \in \Lambda\}$ be a collection of objects of $\mathcal{T}$, with Λ of cardinality $< \beta$. Suppose that we are given a map

$$\coprod_{\mu \in M} k_\mu \longrightarrow \coprod_{\lambda \in \Lambda} X_\lambda.$$

That is, for each $\mu \in M$, we have a map

$$k_\mu \longrightarrow \coprod_{\lambda \in \Lambda} X_\lambda.$$

Because $k_\mu \in \mathcal{S}$, this map factors as

$$k_\mu \longrightarrow \coprod_{\lambda \in \Lambda} k_{\{\lambda,\mu\}} \xrightarrow{\coprod_{\lambda \in \Lambda} f_{\{\lambda,\mu\}}} \coprod_{\lambda \in \Lambda} X_\lambda$$

with $k_{\{\lambda,\mu\}} \in \langle S \rangle^\alpha$. It follows that our map

$$\coprod_{\mu \in M} k_\mu \longrightarrow \coprod_{\lambda \in \Lambda} X_\lambda$$

factors as

$$\coprod_{\mu \in M} k_\mu \longrightarrow \coprod_{\lambda \in \Lambda} \coprod_{\mu \in M} k_{\{\lambda,\mu\}} \xrightarrow{\coprod_{\lambda \in \Lambda, \mu \in M} f_{\{\lambda,\mu\}}} \coprod_{\lambda \in \Lambda} X_\lambda$$

and as $\langle S \rangle^\alpha$ is closed under the formation of coproducts of fewer than α of its objects,

$$\coprod_{\mu \in M} k_{\{\lambda,\mu\}} \in \langle S \rangle^\alpha.$$

□

Proof that $\mathcal{S}$ is triangulated. Note that the category $\langle S \rangle^\alpha$ is triangulated, in particular closed under suspension. It follows that factoring

$$k \longrightarrow \coprod_{\lambda \in \Lambda} X_\lambda$$

as a composite

$$k \longrightarrow \coprod_{\lambda \in \Lambda} k_\lambda \xrightarrow{\coprod_{\lambda \in \Lambda} f_\lambda} \coprod_{\lambda \in \Lambda} X_\lambda$$

with $k_\lambda \in \langle S \rangle^\alpha$ is the same as factoring

$$\Sigma k \longrightarrow \coprod_{\lambda \in \Lambda} \Sigma X_\lambda$$

as a composite

$$\Sigma k \longrightarrow \coprod_{\lambda \in \Lambda} \Sigma k_\lambda \xrightarrow{\coprod_{\lambda \in \Lambda} \Sigma f_\lambda} \coprod_{\lambda \in \Lambda} \Sigma X_\lambda$$

with $\Sigma k_\lambda \in \langle S \rangle^\alpha$. Thus $\mathcal{S}$ satisfies $\Sigma\mathcal{S} = \mathcal{S}$.

Now suppose $f : k \to l$ is a morphism in $\mathcal{S}$. Complete it to a triangle

$$k \xrightarrow{\ f\ } l \longrightarrow m \longrightarrow \Sigma k.$$

We know that k and l are in $\mathcal{S}$. We must prove that so is m.

Let $\{X_\lambda; \lambda \in \Lambda\}$ be a set of objects of $\mathcal{T}$, with Λ of cardinality $< \beta$. We deduce an exact sequence

$$Hom\left(m, \coprod_{\lambda\in\Lambda} X_\lambda\right) \longrightarrow Hom\left(l, \coprod_{\lambda\in\Lambda} X_\lambda\right) \longrightarrow Hom\left(k, \coprod_{\lambda\in\Lambda} X_\lambda\right)$$

Suppose now that we are given a map

$$m \xrightarrow{\ h\ } \coprod_{\lambda\in\Lambda} X_\lambda$$

By the exact sequence, this gives a map

$$l \longrightarrow \coprod_{\lambda\in\Lambda} X_\lambda$$

so that the composite

$$k \longrightarrow l \longrightarrow \coprod_{\lambda\in\Lambda} X_\lambda$$

vanishes. But l is in $\mathcal{S}$, meaning it is β–good; hence there exists a factorisation

$$l \longrightarrow \coprod_{\lambda\in\Lambda} l_\lambda \xrightarrow{\ \coprod_{\lambda\in\Lambda} f_\lambda\ } \coprod_{\lambda\in\Lambda} X_\lambda$$

with $l_\lambda \in \langle S \rangle^\alpha$. Furthermore, the composite

$$k \longrightarrow l \longrightarrow \coprod_{\lambda\in\Lambda} l_\lambda \xrightarrow{\ \coprod_{\lambda\in\Lambda} f_\lambda\ } \coprod_{\lambda\in\Lambda} X_\lambda$$

vanishes, and hence since $k \in \mathcal{S}$, we deduce by Lemma 3.3.5 that

$$l_\lambda \xrightarrow{\ f_\lambda\ } X_\lambda$$

factors as

$$l_\lambda \xrightarrow{\ g_\lambda\ } m_\lambda \xrightarrow{\ h_\lambda\ } X_\lambda$$

so that the composite

$$k \longrightarrow l \longrightarrow \coprod_{\lambda\in\Lambda} l_\lambda \xrightarrow{\ \coprod_{\lambda\in\Lambda} g_\lambda\ } \coprod_{\lambda\in\Lambda} m_\lambda$$

vanishes. But then the map

$$l \longrightarrow \coprod_{\lambda \in \Lambda} m_\lambda$$

factors through m; there is a map

$$m \xrightarrow{\ g\ } \coprod_{\lambda \in \Lambda} m_\lambda$$

giving it. Now the composite

$$m \xrightarrow{\ g\ } \coprod_{\lambda \in \Lambda} m_\lambda \xrightarrow{\coprod_{\lambda \in \Lambda} h_\lambda} \coprod_{\lambda \in \Lambda} X_\lambda$$

need not agree with the given map

$$m \xrightarrow{\ h\ } \coprod_{\lambda \in \Lambda} X_\lambda.$$

However, by construction, the composites with $l \to m$ agree. The difference therefore factors as a map

$$m \longrightarrow \Sigma k \longrightarrow \coprod_{\lambda \in \Lambda} X_\lambda.$$

But $\Sigma k \in \mathcal{S}$, and hence the map

$$\Sigma k \longrightarrow \coprod_{\lambda \in \Lambda} X_\lambda$$

factors as

$$\Sigma k \longrightarrow \coprod_{\lambda \in \Lambda} m'_\lambda \xrightarrow{\coprod_{\lambda \in \Lambda} h'_\lambda} \coprod_{\lambda \in \Lambda} X_\lambda$$

with $m'_\lambda \in \langle S \rangle^\alpha$, and the map h factors as

$$m \longrightarrow \coprod_{\lambda \in \Lambda} \{m_\lambda \oplus m'_\lambda\} \xrightarrow{\coprod_{\lambda \in \Lambda} (\ h_\lambda \quad h'_\lambda\ )} \coprod_{\lambda \in \Lambda} X_\lambda$$

with $\{m_\lambda \oplus m'_\lambda\} \in \langle S \rangle^\alpha$. □

THEOREM 3.3.9. *Let α and β be infinite cardinals. Let $\mathcal{T}$ be a triangulated category satisfying [TR5]. Let $\{S_i, i \in I\}$ be a family of β–perfect classes of $\mathcal{T}$. Recall that if $\cup S_i$ is the union*

$$\bigcup_{i \in I} S_i,$$

then $\overline{T}(\cup S_i)$ *is the smallest triangulated subcategory of* $\mathcal{T}$ *containing* $\cup S_i$, *and* $\langle \cup S_i \rangle^\alpha$ *is the smallest thick,* α*–localising subcategory* $\mathcal{S} \subset \mathcal{T}$ *containing* $\cup S_i$.

Our theorem asserts that for any β*–perfect* S_i*'s as above, the collection of objects of* $\overline{T}(\cup S_i)$ *is a* β*–perfect class. Furthermore, for any infinite* α, *the objects of* $\langle \cup S_i \rangle^\alpha$ *also form a* β*–perfect class.*

Proof: Since the proofs are virtually identical, we will prove the statement about $\langle \cup S_i \rangle^\alpha$. By Lemma 3.3.7, it suffices to show that any $k \in \cup_{i\in I} S_i$ is β–good. More explicitly, let $\{X_\lambda, \lambda \in \Lambda\}$ be a collection of objects of $\mathcal{T}$, where Λ has cardinality $< \beta$. Given any map

$$k \longrightarrow \coprod_{\lambda\in\Lambda} X_\lambda$$

we must show there is a factorisation

$$k \longrightarrow \coprod_{\lambda\in\Lambda} k_\lambda \xrightarrow{\coprod_{\lambda\in\Lambda} f_\lambda} \coprod_{\lambda\in\Lambda} X_\lambda$$

with $k_\lambda \in \langle \cup S_i \rangle^\alpha$.

So take $k \in \cup_{i\in I} S_i$. For some $i \in I$, k must lie in S_i. But then a map

$$k \longrightarrow \coprod_{\lambda\in\Lambda} X_\lambda$$

as above must factor as

$$k \longrightarrow \coprod_{\lambda\in\Lambda} k_\lambda \xrightarrow{\coprod_{\lambda\in\Lambda} f_\lambda} \coprod_{\lambda\in\Lambda} X_\lambda$$

with $k_\lambda \in S_i$, hence $k_\lambda \in \langle \cup S_i \rangle^\alpha$. □

COROLLARY 3.3.10. *Let* β *be an infinite cardinal. Let* $\mathcal{T}$ *be a triangulated category satisfying [TR5]. Let* $\mathcal{S}$ *be a triangulated subcategory. The collection of* β*–perfect classes* $S_i \subset \mathcal{S}$ *of* $\mathcal{T}$ *has a unique maximal member* S. *Furthermore, any* β*–perfect class* R *of* $\mathcal{T}$, *whose objects lie in* $\mathcal{S}$, *is contained in the maximal class* S.

Proof: Take the collection of $\{S_i, i \in I\}$ to be the class of all β–perfect classes in $\mathcal{T}$, all of whose objects lie in $\mathcal{S}$. Then by Theorem 3.3.9, the objects of the category $\overline{T}(\cup S_i)$ form an β–perfect class. On the other hand, $\mathcal{S}$ is triangulated and contains $\cup S_i$, hence also $\overline{T}(\cup S_i)$. Thus $\overline{T}(\cup S_i)$ is contained in $\mathcal{S}$, is β–perfect, and contains all the β–perfect S_i's. Putting $S = \overline{T}(\cup S_i)$, we clearly have that S is maximal. □

DEFINITION 3.3.11. *Let β be an infinite cardinal. Let $\mathcal{T}$ be a triangulated category satisfying [TR5]. Let $\mathcal{S}$ be a triangulated subcategory. Then the full subcategory whose object class is the maximal β–perfect class S of Corollary 3.3.10 will be called $\mathcal{S}_\beta$.*

COROLLARY 3.3.12. *Let $\mathcal{S}$ be a triangulated subcategory of a triangulated category $\mathcal{T}$. Suppose $\mathcal{T}$ satisfies [TR5]. Let β be an infinite cardinal. Then $\mathcal{S}_\beta$ is triangulated.*

Proof: $\mathcal{S}$ is a triangulated subcategory containing $\mathcal{S}_\beta$, and hence $\mathcal{S}$ contains the minimal triangulated subcategory containing $\mathcal{S}_\beta$. That is,

$$\mathcal{S}_\beta \subset \overline{T}(\mathcal{S}_\beta) \subset \mathcal{S}.$$

But $\mathcal{S}_\beta$ is a β–perfect class. By Theorem 3.3.9, so is $\overline{T}(\mathcal{S}_\beta)$. Because $\mathcal{S}_\beta$ is the maximal β–perfect class, it must contain $\overline{T}(\mathcal{S}_\beta)$, hence is equal to it, hence is triangulated. □

COROLLARY 3.3.13. *Let $\mathcal{S}$ be a thick subcategory of a triangulated category $\mathcal{T}$ satisfying [TR5]. Let β be an infinite cardinal. Then the category $\mathcal{S}_\beta$ is thick.*

Proof: Let T be the class of all objects isomorphic to direct summands of objects of $\mathcal{S}_\beta$. Since $\mathcal{S}$ contains $\mathcal{S}_\beta$ and is thick, it must contain T. We have

$$\mathcal{S}_\beta \subset T \subset \mathcal{S}.$$

But T is β–perfect by Lemma 3.3.3, and by the maximality of $\mathcal{S}_\beta$ we must have $T \subset \mathcal{S}_\beta$. Hence the two are equal, and $\mathcal{S}_\beta$ is thick. □

COROLLARY 3.3.14. *Let α and β be infinite cardinals. Suppose $\mathcal{S}$ is an α–localising subcategory of a triangulated category $\mathcal{T}$ satisfying [TR5]. That is, $\mathcal{S}$ is a thick subcategory closed under the formation of coproducts of fewer than α of its objects. Then the subcategory $\mathcal{S}_\beta$ is also α–localising.*

Proof: $\mathcal{S}$ is α–localising and contains $\mathcal{S}_\beta$, and hence contains $\langle \mathcal{S}_\beta \rangle^\alpha$, the smallest α–localising subcategory containing $\mathcal{S}_\beta$. We get an inclusion

$$\mathcal{S}_\beta \subset \langle \mathcal{S}_\beta \rangle^\alpha \subset \mathcal{S}.$$

On the other hand, $\mathcal{S}_\beta$ is a β–perfect class. By Theorem 3.3.9, so is $\langle \mathcal{S}_\beta \rangle^\alpha$. By the maximality of $\mathcal{S}_\beta$ we must have

$$\langle \mathcal{S}_\beta \rangle^\alpha \subset \mathcal{S}_\beta.$$

Hence the two are equal and $\mathcal{S}_\beta$ is α–localising. □

REMARK 3.3.15. If S is γ–perfect, and $\gamma > \beta$, then S is also β–perfect. We deduce that for any $\mathcal{S} \subset \mathcal{T}$, $\mathcal{S}_\gamma$, being β–perfect, must be contained in the maximal β–perfect class $\mathcal{S}_\beta$. If $\mathcal{R} \subset \mathcal{S} \subset \mathcal{T}$, then $\mathcal{R}_\beta$ is a β–perfect class in $\mathcal{S}$, hence contained in the maximal $\mathcal{S}_\beta$.

EXAMPLE 3.3.16. Let $\mathcal{T}$ be a triangulated category. Let k be any object of $\mathcal{T}$. Then the class $S = \{0, k\}$ of only two objects is $\aleph_0$ perfect. Given any set Λ of cardinality $< \aleph_0$ (that is, a finite set), and a map

$$k \longrightarrow \coprod_{\lambda\in\Lambda} X_\lambda$$

we can factor it as

$$k \xrightarrow{\ \Delta\ } \coprod_{\lambda\in\Lambda} k \xrightarrow{\coprod_{\lambda\in\Lambda} f_\lambda} \coprod_{\lambda\in\Lambda} X_\lambda$$

where Δ is the diagonal map from k to k^n. If the composite vanishes, then in fact each f_λ must vanish, and the map factors as

$$k \xrightarrow{\ \Delta\ } \coprod_{\lambda\in\Lambda} k \longrightarrow \coprod_{\lambda\in\Lambda} 0 \longrightarrow \coprod_{\lambda\in\Lambda} X_\lambda.$$

It follows that if $\mathcal{S}$ is any triangulated subcategory of $\mathcal{T}$, then $\mathcal{S}_{\aleph_0} = \mathcal{S}$. The case $\beta = \aleph_0$ is the trivial case for perfection.

3.4. History of the results in Chapter 3

The results of Section 3.2 are very standard. The definition of localising subcategories (Definition 3.2.6) is probably due to Bousfield, [**4**], [**6**] and [**5**]. Let $\mathcal{T}$ be a triangulated category satisfying [TR5]. As in Definition 3.2.1, let $\langle S\rangle^\beta$ be the smallest β–localising subcategory of $\mathcal{T}$ containing the set S of objects. The fact that $\langle S\rangle^\beta$ is essentially small (Proposition 3.2.5) is obvious. The fact that quotients by localising subcategories respect coproducts (Corollary 3.2.11) may be found in Bökstedt–Neeman [**3**].

Section 3.3 is completely new. This book introduces the notion of perfect classes to imitate some standard constructions involving transfinite induction on the number of cells of a complex. Whatever the motivation, the definition is new, and in the rest of the book, we will attempt to explain what one can do with it.

CHAPTER 4

Small objects, and Thomason's localisation theorem

4.1. Small objects

DEFINITION 4.1.1. *Let $\mathcal{T}$ be a triangulated category satisfying [TR5] (that is, coproducts exist). Let α be an infinite cardinal. An object $k \in \mathcal{T}$ is called α–*small *if, for any collection $\{X_\lambda; \lambda \in \Lambda\}$ of objects of $\mathcal{T}$, any map*

$$k \longrightarrow \coprod_{\lambda \in \Lambda} X_\lambda$$

factors through some coproduct of cardinality strictly less than α. In other words, there exists a subset $\Lambda' \subset \Lambda$, where the cardinality of Λ' is strictly less than α, and the map above factors as

$$k \longrightarrow \coprod_{\lambda \in \Lambda'} X_\lambda \longrightarrow \coprod_{\lambda \in \Lambda} X_\lambda.$$

EXAMPLE 4.1.2. The special case where $\alpha = \aleph_0$ is of great interest. An object $k \in \mathcal{T}$ is called $\aleph_0$*–small* if for any infinite coproduct in $\mathcal{T}$, say the coproduct of a family $\{X_\lambda; \lambda \in \Lambda\}$ of objects of $\mathcal{T}$, any map

$$k \longrightarrow \coprod_{\lambda \in \Lambda} X_\lambda$$

factors through a finite coproduct. That is, there is a finite subset

$$\{X_1, X_2, \ldots, X_n\} \subset \{X_\lambda; \lambda \in \Lambda\}$$

and a factorisation

$$k \longrightarrow \coprod_{i=1}^{n} X_i \longrightarrow \coprod_{\lambda \in \Lambda} X_\lambda.$$

Expressing this still another way, the natural map

$$\coprod_{\lambda \in \Lambda} Hom(k, X_\lambda) \longrightarrow Hom\left(k, \coprod_{\lambda \in \Lambda} X_\lambda\right)$$

is an isomorphism.

DEFINITION 4.1.3. *Let α be an infinite cardinal. Let $\mathcal{T}$ be a triangulated category satisfying [TR5]. The full subcategory whose objects are all the α–small objects of $\mathcal{T}$ will be denoted $\mathcal{T}^{(\alpha)}$.*

LEMMA 4.1.4. *Let α be an infinite cardinal. Let $\mathcal{T}$ be a triangulated category satisfying [TR5]. Then the subcategory $\mathcal{T}^{(\alpha)} \subset \mathcal{T}$ is triangulated.*

Proof: To begin with, observe that $k \in \mathcal{T}^{(\alpha)}$ if and only if $\Sigma k \in \mathcal{T}^{(\alpha)}$. This comes about from the identity

$$\begin{aligned} Hom\left(\Sigma k, \coprod_{\lambda\in\Lambda} X_\lambda\right) &= Hom\left(k, \Sigma^{-1}\left\{\coprod_{\lambda\in\Lambda} X_\lambda\right\}\right) \\ &= Hom\left(k, \coprod_{\lambda\in\Lambda} \Sigma^{-1} X_\lambda\right) \end{aligned}$$

where the second equality is the fact that the suspension functor respects coproducts, i.e. Proposition 1.1.6.

Let $k \to l$ be a morphism in $\mathcal{T}^{(\alpha)}$. It may be completed to a triangle in $\mathcal{T}$. There is a triangle in $\mathcal{T}$

$$k \to l \to m \to \Sigma k.$$

We know that k and l are α–small. To show that $\mathcal{T}^{(\alpha)}$ is triangulated, we need to show that so is m.

Let $\{X_\lambda; \lambda \in \Lambda\}$ be a set of objects of $\mathcal{T}$. Because

$$Hom\left(-, \coprod_{\lambda\in\Lambda} X_\lambda\right)$$

is a cohomological functor on $\mathcal{T}$, we deduce an exact sequence

$$Hom\left(m, \coprod_{\lambda\in\Lambda} X_\lambda\right) \longrightarrow Hom\left(l, \coprod_{\lambda\in\Lambda} X_\lambda\right) \longrightarrow Hom\left(k, \coprod_{\lambda\in\Lambda} X_\lambda\right)$$

Suppose now that we are given a map

$$m \xrightarrow{\ h\ } \coprod_{\lambda\in\Lambda} X_\lambda$$

By the exact sequence, this gives a map

$$l \longrightarrow \coprod_{\lambda\in\Lambda} X_\lambda$$

so that the composite

$$k \longrightarrow l \longrightarrow \coprod_{\lambda\in\Lambda} X_\lambda$$

vanishes. But l is α–small; hence there exists a subset $\Lambda' \subset \Lambda$ of cardinality $< \alpha$, so that the map

$$l \longrightarrow \coprod_{\lambda\in\Lambda} X_\lambda$$

factors as

$$l \longrightarrow \coprod_{\lambda\in\Lambda'} X_\lambda \longrightarrow \coprod_{\lambda\in\Lambda} X_\lambda.$$

Of course, we know that the composite

$$k \longrightarrow l \longrightarrow \coprod_{\lambda\in\Lambda'} X_\lambda \longrightarrow \coprod_{\lambda\in\Lambda} X_\lambda$$

vanishes. On the other hand, the map

$$\coprod_{\lambda\in\Lambda'} X_\lambda \longrightarrow \coprod_{\lambda\in\Lambda} X_\lambda$$

is the inclusion of a direct summand, hence a monomorphism in $\mathcal{T}$. We deduce that the composite

$$k \longrightarrow l \longrightarrow \coprod_{\lambda\in\Lambda'} X_\lambda$$

is already zero, and therefore that

$$l \longrightarrow \coprod_{\lambda\in\Lambda'} X_\lambda$$

factors through m. We therefore have produced a map

$$m \xrightarrow{\ g\ } \coprod_{\lambda\in\Lambda'} X_\lambda$$

so that the composite

$$m \xrightarrow{\ g\ } \coprod_{\lambda\in\Lambda'} X_\lambda \longrightarrow \coprod_{\lambda\in\Lambda} X_\lambda$$

differs from the map

$$m \xrightarrow{\ h\ } \coprod_{\lambda\in\Lambda} X_\lambda$$

by a map vanishing on l. On the other hand, the exact sequence

$$Hom\left(\Sigma k, \coprod_{\lambda\in\Lambda} X_\lambda\right) \longrightarrow Hom\left(m, \coprod_{\lambda\in\Lambda} X_\lambda\right) \longrightarrow Hom\left(l, \coprod_{\lambda\in\Lambda} X_\lambda\right)$$

tells us that the difference factors as

$$m \longrightarrow \Sigma k \longrightarrow \coprod_{\lambda\in\Lambda} X_\lambda.$$

But $k \in \mathcal{T}^{(\alpha)}$ implies $\Sigma k \in \mathcal{T}^{(\alpha)}$, and we deduce that there is a subset $\Lambda'' \subset \Lambda$, of cardinality $< \alpha$, so that the map

$$\Sigma k \longrightarrow \coprod_{\lambda \in \Lambda} X_\lambda$$

factors as

$$\Sigma k \longrightarrow \coprod_{\lambda \in \Lambda''} X_\lambda \longrightarrow \coprod_{\lambda \in \Lambda} X_\lambda.$$

Putting this all together, one easily deduces that the given map

$$m \xrightarrow{h} \coprod_{\lambda \in \Lambda} X_\lambda$$

factors as

$$m \longrightarrow \coprod_{\lambda \in \Lambda' \cup \Lambda''} X_\lambda \longrightarrow \coprod_{\lambda \in \Lambda} X_\lambda$$

and because α is infinite and Λ' and Λ'' are of cardinality $< \alpha$, the cardinality of $\Lambda' \cup \Lambda''$ is also $< \alpha$. Therefore $\mathcal{T}^{(\alpha)}$ is triangulated. □

LEMMA 4.1.5. *Suppose α is a regular cardinal. That is, α is not the sum of fewer than α cardinals, each of which is less than α. Then $\mathcal{T}^{(\alpha)}$ is α–localising. That is, the coproduct of fewer than α objects of $\mathcal{T}^{(\alpha)}$ is an object of $\mathcal{T}^{(\alpha)}$.*

Proof: Let $\{k_\mu, \mu \in M\}$ be a collection of objects in $\mathcal{T}^{(\alpha)}$, where the cardinality of M is less than α. Let $\{X_\lambda, \lambda \in \Lambda\}$ be an arbitrary collection of objects of $\mathcal{T}$. Suppose we are given a map

$$\coprod_{\mu \in M} k_\mu \longrightarrow \coprod_{\lambda \in \Lambda} X_\lambda.$$

This means that, for every $\mu \in M$, we are given a map

$$k_\mu \longrightarrow \coprod_{\lambda \in \Lambda} X_\lambda.$$

Because k_μ is α–small, for each μ there exists an $\Lambda_\mu \subset \Lambda$, with cardinality $< \alpha$, so that the map

$$k_\mu \longrightarrow \coprod_{\lambda \in \Lambda} X_\lambda$$

factors as

$$k_\mu \longrightarrow \coprod_{\lambda \in \Lambda_\mu} X_\lambda \longrightarrow \coprod_{\lambda \in \Lambda} X_\lambda.$$

Thus the map

$$\coprod_{\mu \in M} k_\mu \longrightarrow \coprod_{\lambda \in \Lambda} X_\lambda$$

factors as

$$\coprod_{\mu \in M} k_\mu \longrightarrow \coprod_{\lambda \in \cup_{\mu \in M} \Lambda_\mu} X_\lambda \longrightarrow \coprod_{\lambda \in \Lambda} X_\lambda,$$

and the cardinality of $\cup_{\mu \in M} \Lambda_\mu$ is bounded by the sum of the cardinalities of Λ_μ over all $\mu \in M$, which is a sum of fewer than α cardinals, each less than α. Because α is regular, this sum is less than α. □

LEMMA 4.1.6. *Let α be an infinite cardinal. The category $\mathcal{T}^{(\alpha)}$ is thick.*

Proof: By Lemma 4.1.4 we know that $\mathcal{T}^{(\alpha)}$ is a triangulated subcategory. To prove it thick, we need to show that any direct summand of an object in $\mathcal{T}^{(\alpha)}$ is in $\mathcal{T}^{(\alpha)}$.

Let k, l be objects of $\mathcal{T}$ and assume that the direct sum $k \oplus l$ is α–small. We wish to show that k is α–small. Take any map

$$k \xrightarrow{h} \coprod_{\lambda \in \Lambda} X_\lambda$$

and consider the map

$$k \oplus m \xrightarrow{\begin{pmatrix} h & 0 \end{pmatrix}} \coprod_{\lambda \in \Lambda} X_\lambda$$

Because $k \oplus m$ is small, there is a subset $\Lambda' \subset \Lambda$ of cardinality $< \alpha$, so that the above factors as

$$k \oplus m \longrightarrow \coprod_{\lambda \in \Lambda'} X_\lambda \longrightarrow \coprod_{\lambda \in \Lambda} X_\lambda$$

and hence the given map

$$k \xrightarrow{h} \coprod_{\lambda \in \Lambda} X_\lambda$$

factors as

$$k \xrightarrow{\begin{pmatrix} 1 \\ 0 \end{pmatrix}} k \oplus m \longrightarrow \coprod_{\lambda \in \Lambda'} X_\lambda \longrightarrow \coprod_{\lambda \in \Lambda} X_\lambda$$

and in particular, it factors through a coproduct of fewer than α terms. □

REMARK 4.1.7. If α is a regular cardinal greater than $\aleph_0$, Lemma 4.1.6 is redundant. By Lemma 4.1.5, $\mathcal{T}^{(\alpha)}$ is α–localising. But since $\alpha > \aleph_0$, Remark 3.2.7 tells us that idempotents split in $\mathcal{T}^{(\alpha)}$, and $\mathcal{T}^{(\alpha)}$ must be thick.

4.2. Compact objects

Let $\mathcal{T}$ be a triangulated category satisfying [TR5]. In Section 4.1 we learned how to construct, for each infinite cardinal α, a triangulated subcategory $\mathcal{T}^{(\alpha)}$ of α–small objects in $\mathcal{T}$. In Section 3.3, we learned that given any triangulated subcategory $\mathcal{S} \subset \mathcal{T}$ and an infinite cardinal β, there is a way to construct a triangulated subcategory $\mathcal{S}_\beta \subset \mathcal{S}$. In this section, the idea will be to combine the constructions and study $\{\mathcal{T}^{(\alpha)}\}_\beta$.

LEMMA 4.2.1. *Let $\mathcal{T}$ be a triangulated category satisfying [TR5]. Let α be an infinite cardinal. Let S be an α–perfect class of α–small objects. Then S is also β–perfect for all infinite β.*

Proof: Suppose k is an object in S, and $\{X_\lambda, \lambda \in \Lambda\}$ a family of fewer than β objects of $\mathcal{T}$. Because k is α–small, any map

$$k \longrightarrow \coprod_{\lambda \in \Lambda} X_\lambda$$

factors as

$$k \longrightarrow \coprod_{\lambda \in \Lambda'} X_\lambda \longrightarrow \coprod_{\lambda \in \Lambda} X_\lambda$$

with the cardinality of Λ' being $\leq \alpha$. Since S is α–perfect, the map

$$k \longrightarrow \coprod_{\lambda \in \Lambda'} X_\lambda$$

factors as

$$k \longrightarrow \coprod_{\lambda \in \Lambda'} k_\lambda \xrightarrow{\coprod_{\lambda \in \Lambda'} f_\lambda} \coprod_{\lambda \in \Lambda'} X_\lambda$$

with $k_\lambda \in S$. For $\lambda \notin \Lambda'$, define $k_\lambda = 0$. Then we deduce a factorisation

$$k \longrightarrow \coprod_{\lambda \in \Lambda} k_\lambda \xrightarrow{\coprod_{\lambda \in \Lambda} f_\lambda} \coprod_{\lambda \in \Lambda} X_\lambda,$$

where of course most of the k_λ's vanish, but in any case they are all in S.

Suppose now that we have a vanishing composite

$$k \longrightarrow \coprod_{\lambda\in\Lambda} k_\lambda \xrightarrow{\coprod_{\lambda\in\Lambda} f_\lambda} \coprod_{\lambda\in\Lambda} X_\lambda$$

with k and k_λ all in S and Λ of cardinality $< \beta$. Because k is α–small, the map

$$k \longrightarrow \coprod_{\lambda\in\Lambda} k_\lambda$$

must factor as

$$k \longrightarrow \coprod_{\lambda\in\Lambda'} k_\lambda \longrightarrow \coprod_{\lambda\in\Lambda} k_\lambda$$

where the cardinality of Λ' is $< \alpha$. The composite

$$k \longrightarrow \coprod_{\lambda\in\Lambda'} k_\lambda \xrightarrow{\coprod_{\lambda\in\Lambda'} f_\lambda} \coprod_{\lambda\in\Lambda'} X_\lambda$$

vanishes, and since S is α–perfect, we deduce that for each $\lambda \in \Lambda'$, the map $f_\lambda : k_\lambda \longrightarrow X_\lambda$ factors as

$$k_\lambda \xrightarrow{g_\lambda} l_\lambda \xrightarrow{h_\lambda} X_\lambda$$

so that $l_\lambda \in S$ and the composite

$$k \longrightarrow \coprod_{\lambda\in\Lambda'} k_\lambda \xrightarrow{\coprod_{\lambda\in\Lambda'} g_\lambda} \coprod_{\lambda\in\Lambda'} l_\lambda$$

already vanishes. For $\lambda \notin \Lambda'$, define $g_\lambda : k_\lambda \longrightarrow l_\lambda$ to be the identity. Then we still have the vanishing of

$$k \longrightarrow \coprod_{\lambda\in\Lambda} k_\lambda \xrightarrow{\coprod_{\lambda\in\Lambda} g_\lambda} \coprod_{\lambda\in\Lambda} l_\lambda.$$

□

DEFINITION 4.2.2. *Let $\mathcal{T}$ be a triangulated category satisfying [TR5]. Let α be an infinite cardinal. Define a triangulated subcategory $\mathcal{T}^\alpha$ by*

$$\mathcal{T}^\alpha = \left\{\mathcal{T}^{(\alpha)}\right\}_\alpha.$$

LEMMA 4.2.3. *If $\alpha < \beta$ are infinite cardinals, then $\mathcal{T}^\alpha \subset \mathcal{T}^\beta$.*

Proof: $\mathcal{T}^\alpha$ is an α–perfect class, whose objects are α–small. We know, from Lemma 4.2.1, that $\mathcal{T}^\alpha$ must be β–perfect. Since its objects are α–small, they are also β–small. Thus $\mathcal{T}^\alpha$ is a β–perfect class in $\mathcal{T}^{(\beta)}$, hence it must be contained in the maximal one, $\left\{\mathcal{T}^{(\beta)}\right\}_\beta$. □

LEMMA 4.2.4. *For every infinite α, $\mathcal{T}^\alpha$ is thick.*

Proof: By Lemma 4.1.6, $\mathcal{T}^{(\alpha)}$ is thick. By Corollary 3.3.13, for every infinite cardinal β, $\left\{\mathcal{T}^{(\alpha)}\right\}_\beta$ is also thick. In particular, letting $\beta = \alpha$, $\mathcal{T}^\alpha$ is thick. □

LEMMA 4.2.5. *Let α be a regular cardinal. Then $\mathcal{T}^\alpha$ is α–localising.*

Proof: α is regular, and hence Lemma 4.1.5 says that $\mathcal{T}^{(\alpha)}$ is α–localising. But then Corollary 3.3.14 asserts that, for any infinite β, $\left\{\mathcal{T}^{(\alpha)}\right\}_\beta$ is also α–localising. Letting $\beta = \alpha$, we deduce that $\mathcal{T}^\alpha$ is α–localising. □

REMARK 4.2.6. In the special case $\alpha = \aleph_0$, all classes are α–perfect; see Example 3.3.16. Thus

$$\left\{\mathcal{T}^{(\aleph_0)}\right\}_{\aleph_0} = \mathcal{T}^{(\aleph_0)}.$$

In other words,

$$\mathcal{T}^{\aleph_0} = \mathcal{T}^{(\aleph_0)}.$$

DEFINITION 4.2.7. *The objects of $\mathcal{T}^\alpha$ will be called the* α–compact *objects of $\mathcal{T}$. In the case $\alpha = \aleph_0$, the objects of $\mathcal{T}^{\aleph_0}$ will be called the* compact *objects. They are β–compact for any infinite β. We will permit ourselves to write $\mathcal{T}^c$ for $\mathcal{T}^{\aleph_0}$; the superscript c stands for compact.*

4.3. Maps factor through $\langle S\rangle^\beta$

REMINDER 4.3.1. Let $\mathcal{T}$ be a triangulated category satisfying [TR5]. Let S be a class of objects of $\mathcal{T}$. We remind the reader of Definition 3.2.9; $\langle S\rangle$ stands for the localising subcategory generated by S, that is

$$\langle S\rangle = \bigcup_\beta \langle S\rangle^\beta.$$

$\langle S\rangle^\beta$ is the smallest thick subcategory containing S and closed with respect to forming the coproducts of fewer than β of its objects; See Definition 3.2.1.

LEMMA 4.3.2. *Let $\mathcal{T}$ be a triangulated category satisfying [TR5]. Let β be a regular cardinal. Suppose S is a class of objects of $\mathcal{T}^\beta$. That is, $S \subset \mathcal{T}^\beta$. Then the subcategory $\langle S\rangle^\beta$ is also contained in $\mathcal{T}^\beta$.*

Proof: If β is regular then, by Lemma 4.2.5, $\mathcal{T}^\beta$ is β–localising. By hypothesis, S is contained in $\mathcal{T}^\beta$. But $\langle S\rangle^\beta$ is the minimal β–localising subcategory containing S, hence $\langle S\rangle^\beta \subset \mathcal{T}^\beta$. □

The main theorem of this section is the following.

THEOREM 4.3.3. *Let $\mathcal{T}$ be a triangulated category satisfying [TR5]. Let β be a regular cardinal. Let S be some class of objects in $\mathcal{T}^\beta$. Let x be a β–compact object of $\mathcal{T}$. Let z be an object of $\langle S\rangle$. Suppose $f : x \longrightarrow z$ is a morphism in $\mathcal{T}$. Then there exists an object $y \in \langle S\rangle^\beta$ so that f factors as*

$$x \longrightarrow y \longrightarrow z.$$

Proof: We define a full subcategory $\mathcal{S}$ of $\mathcal{T}$ as follows. If $Ob(\mathcal{S})$ is the class of objects of $\mathcal{S}$, then

$$Ob(\mathcal{S}) = \left\{ \begin{array}{c} z \in Ob(\mathcal{T}) \,|\, \forall x \in Ob(\mathcal{T}^\beta),\, \forall f : x \longrightarrow z,\, \exists y \in \langle S\rangle^\beta \\ \text{and a factorisation of } f \text{ as } \ x \longrightarrow y \longrightarrow z \end{array} \right\}.$$

It suffices to prove that $\langle S\rangle \subset \mathcal{S}$. To do this, we will show that $\mathcal{S}$ contains S, is triangulated, and contains all coproducts of its objects. Since $\langle S\rangle$ is minimal with these properties (see Example 3.2.8), it will follow that $\langle S\rangle \subset \mathcal{S}$.

The fact that $\mathcal{S}$ contains S is obvious. Take any objects $z \in S$ and $x \in \mathcal{T}^\beta$, and any morphism $x \dashrightarrow z$. Since we know that $z \in S$, clearly $z \in \langle S\rangle^\beta$, the smallest thick category containing S and closed with respect to coproducts of fewer than β of its objects. Put $y = z$, and factor $f : x \longrightarrow z$ as

$$x \xrightarrow{\ \ f\ \ } z \xrightarrow{\ \ 1\ \ } z.$$

Equally clearly, $z \in \mathcal{S}$ if and only if $\Sigma z \in \mathcal{S}$. After all, $x \longrightarrow \Sigma z$ can be factored as

$$x \longrightarrow y \longrightarrow \Sigma z$$

if and only if $\Sigma^{-1}x \longrightarrow z$ can be factored as

$$\Sigma^{-1}x \longrightarrow \Sigma^{-1}y \longrightarrow z$$

and $x \in \mathcal{T}^\beta$ iff $\Sigma^{-1}x \in \mathcal{T}^\beta$, $y \in \langle S\rangle^\beta$ iff $\Sigma^{-1}y \in \langle S\rangle^\beta$.

Suppose now that $\phi : z \longrightarrow z'$ is a morphism in $\mathcal{S}$. Complete it to a triangle

$$z \longrightarrow z' \longrightarrow z'' \longrightarrow \Sigma z.$$

We know that z and z' are in $\mathcal{S}$. To show that $\mathcal{S}$ is triangulated, we must establish that z'' is also in $\mathcal{S}$. Choose any $x \in \mathcal{T}^\beta$, and any map $f : x \longrightarrow z''$. We need to factor it as

$$x \longrightarrow y \longrightarrow z'',$$

with $y \in \langle S \rangle^{\beta}$.

First of all, the composite

$$x \longrightarrow z'' \longrightarrow \Sigma z$$

gives a map from $x \in \mathcal{T}^{\beta}$ to $\Sigma z \in \mathcal{S}$, which must factor as

$$x \longrightarrow y \longrightarrow \Sigma z,$$

with $y \in \langle S \rangle^{\beta}$. Now the composite

$$x \longrightarrow z'' \longrightarrow \Sigma z \longrightarrow \Sigma z'$$

clearly vanishes, and is equal to the composite

$$x \longrightarrow y \longrightarrow \Sigma z \longrightarrow \Sigma z'.$$

Complete $x \longrightarrow y$ to a triangle

$$x \longrightarrow y \longrightarrow C \longrightarrow \Sigma x;$$

then the map $y \longrightarrow \Sigma z'$ must factor through C. We deduce a commutative square

$$\begin{array}{ccc} y & \longrightarrow & C \\ \downarrow & & \downarrow \\ \Sigma z & \longrightarrow & \Sigma z'. \end{array}$$

Now, in the triangle defining C, the other two objects are x and y. By hypothesis, $x \in \mathcal{T}^{\beta}$. By construction, $y \in \langle S \rangle^{\beta}$, and by Lemma 4.3.2, $\langle S \rangle^{\beta} \subset \mathcal{T}^{\beta}$. Because x and y are both in $\mathcal{T}^{\beta}$, so is C. Since $\Sigma z' \in \mathcal{S}$, we deduce that the map $C \longrightarrow \Sigma z'$ factors as

$$C \longrightarrow y' \longrightarrow \Sigma z'$$

with $y' \in \langle S \rangle^{\beta}$. Our commutative square above gets replaced by another,

$$\begin{array}{ccc} y & \longrightarrow & y' \\ \downarrow & & \downarrow \\ \Sigma z & \longrightarrow & \Sigma z', \end{array}$$

where the top row involves only objects in $\langle S \rangle^{\beta}$. Note also that since the composite

$$x \longrightarrow y \longrightarrow C$$

vanishes, so does the longer composite

$$x \longrightarrow y \longrightarrow C \longrightarrow y'.$$

Now complete the commutative square

$$\begin{array}{ccc} y & \longrightarrow & y' \\ \downarrow & & \downarrow \\ \Sigma z & \longrightarrow & \Sigma z' \end{array}$$

to a map of triangles

$$\begin{array}{ccccccc} \Sigma^{-1}y'' & \longrightarrow & y & \longrightarrow & y' & \longrightarrow & y'' \\ \downarrow & & \downarrow & & \downarrow & & \downarrow \\ z'' & \longrightarrow & \Sigma z & \longrightarrow & \Sigma z' & \longrightarrow & \Sigma z'' \end{array}$$

and note that, because y and y' are in $\langle S\rangle^\beta$, so is y'', and because the composite

$$x \longrightarrow y \longrightarrow y'$$

vanishes, the map $x \longrightarrow y$ factors as

$$x \longrightarrow \Sigma^{-1}y'' \longrightarrow y.$$

We deduce a commutative diagram

$$\begin{array}{ccccc} x & \longrightarrow & \Sigma^{-1}y'' & \longrightarrow & y \\ & & \downarrow & & \downarrow \\ & & z'' & \longrightarrow & \Sigma z. \end{array}$$

The composite

$$x \longrightarrow \Sigma^{-1}y'' \longrightarrow z''$$

is not the map $f : x \longrightarrow z''$ we began with. But when we compose

$$x \longrightarrow \Sigma^{-1}y'' \longrightarrow z'' \longrightarrow \Sigma z$$

we do get the given map

$$x \xrightarrow{\ f\ } z'' \longrightarrow \Sigma z.$$

In other words, the difference between the composite

$$x \longrightarrow \Sigma^{-1}y'' \longrightarrow z''$$

and $f : x \longrightarrow z''$ factors as $x \longrightarrow z' \longrightarrow z''$. We know that $z' \in \mathcal{S}$, and hence $x \longrightarrow z'$ must factor as

$$x \longrightarrow \overline{y} \longrightarrow z'$$

with $\overline{y} \in \langle S\rangle^\beta$. But then $f : x \longrightarrow y$ factors as

$$x \longrightarrow \{\overline{y} \oplus \Sigma^{-1}y''\} \longrightarrow z'',$$

and $\overline{y} \oplus \Sigma^{-1}y''$ is in $\langle S\rangle^\beta$. Thus $z'' \in \mathcal{S}$, as required.

It remains to prove that $\mathcal{S}$ is closed with respect to the formation of coproducts of its objects. Let $\{z_\lambda, \lambda \in \Lambda\}$ be a set of objects of $\mathcal{S}$. We wish to show that

$$\coprod_{\lambda \in \Lambda} z_\lambda$$

is an object of $\mathcal{S}$. Rephrasing this again, we wish to show that if $x \in \mathcal{T}^\beta$ and

$$f : x \longrightarrow \coprod_{\lambda \in \Lambda} z_\lambda$$

is any map, then there is a factorisation

$$x \longrightarrow y \longrightarrow \coprod_{\lambda \in \Lambda} z_\lambda$$

with $y \in \langle S \rangle^\beta$.

Take therefore any map

$$f : x \longrightarrow \coprod_{\lambda \in \Lambda} z_\lambda.$$

Now recall that $x \in \mathcal{T}^\beta \subset \mathcal{T}^{(\beta)}$. In particular, x is β–small. Therefore there is a subset $\Lambda' \subset \Lambda$, where the cardinality of Λ' is less than β, so that f factors as

$$x \xrightarrow{\ g\ } \coprod_{\lambda \in \Lambda'} z_\lambda \longrightarrow \coprod_{\lambda \in \Lambda} z_\lambda.$$

Now x is in fact not only β–small, but also β–compact. In other words, x belongs to the β–perfect class $\mathcal{T}^\beta$. Therefore the map g factors as

$$x \longrightarrow \coprod_{\lambda \in \Lambda'} x_\lambda \xrightarrow{\coprod_{\lambda \in \Lambda'} h_\lambda} \coprod_{\lambda \in \Lambda'} z_\lambda$$

for some collection of $h_\lambda : x_\lambda \longrightarrow z_\lambda$, with x_λ in $\mathcal{T}^\beta$.

For each $\lambda \in \Lambda'$, we have $x_\lambda \in \mathcal{T}^\beta$, $z_\lambda \in \mathcal{S}$ and a map $h_\lambda : x_\lambda \longrightarrow z_\lambda$. It follows that, for each λ, we may choose a $y_\lambda \in \langle S \rangle^\beta$ and a factorisation of $h_\lambda : x_\lambda \longrightarrow z_\lambda$ as

$$x_\lambda \longrightarrow y_\lambda \longrightarrow z_\lambda,$$

with $y_\lambda \in \langle S \rangle^\beta$. But then f factorises as

$$x \longrightarrow \coprod_{\lambda \in \Lambda'} y_\lambda \longrightarrow \coprod_{\lambda \in \Lambda} z_\lambda,$$

and $\coprod_{\lambda \in \Lambda'} y_\lambda$, being a coproduct of fewer than β objects of $\langle S \rangle^\beta$, must lie in $\langle S \rangle^\beta$. □

4.4. Maps in the quotient

As we have said in Section 2.2, one of the problems with Verdier's construction of the quotient is that one ends up with a category in which the *Hom*–sets need not be small. Suppose $\mathcal{T}$ is a triangulated category with small *Hom*–sets, and $\mathcal{S}$ is a triangulated subcategory. Then the quotient $\mathcal{T}/\mathcal{S}$ of Theorem 2.1.8 is a category, which in general need not have small *Hom*–sets. Nevertheless, we can already give one criterion that guarantees the smallness of the *Hom*–sets. The criterion is Corollary 4.4.3.

Then we will further explore some of the consequences of the machinery that has been developed so far. We lead up to Thomason's localisation theorem (Theorem 4.4.9), which will give a summary of the results in this Section.

PROPOSITION 4.4.1. *Let $\mathcal{T}$ be a triangulated category. Let β be a regular cardinal. Let S be a subclass of the objects of $\mathcal{T}^\beta$. Let $y \in \mathcal{T}$ be an arbitrary object, $x \in \mathcal{T}$ a β–compact object (i.e. $x \in \mathcal{T}^\beta$). Then*

$$\{\mathcal{T}/\langle S\rangle\}(x,y) = \left\{\mathcal{T}/\langle S\rangle^\beta\right\}(x,y).$$

In other words, the maps $x \longrightarrow y$ are the same in the Verdier quotient categories $\mathcal{T}/\langle S\rangle$ and $\mathcal{T}/\langle S\rangle^\beta$.

Proof: There is a natural map

$$\phi : \left\{\mathcal{T}/\langle S\rangle^\beta\right\}(x,y) \longrightarrow \{\mathcal{T}/\langle S\rangle\}(x,y),$$

and we want to prove it an isomorphism. We need to show it injective and surjective. Let us begin by proving it surjective.

Proof that ϕ is surjective. Let $x \longrightarrow y$ be a morphism in $\mathcal{T}/\langle S\rangle$. That is, an equivalence class of diagrams

$$\begin{array}{ccc} p & \xrightarrow{\ \alpha\ } & y \\ {\scriptstyle f}\big\downarrow & & \\ x & & \end{array}$$

with f in $Mor_{\langle S\rangle}$. In the triangle

$$p \longrightarrow x \longrightarrow z \longrightarrow \Sigma p$$

we must have $z \in \langle S\rangle$. On the other hand, from the hypothesis of the Proposition, $x \in \mathcal{T}^\beta$, $S \subset \mathcal{T}^\beta$ and $z \in \langle S\rangle$. By Theorem 4.3.3, we know that there is a $z' \in \langle S\rangle^\beta$ so that $x \longrightarrow z$ factors as

$$x \longrightarrow z' \longrightarrow z.$$

We deduce a map of triangles

$$\begin{array}{ccccccc} p' & \xrightarrow{fg} & x & \longrightarrow & z' & \longrightarrow & \Sigma p' \\ {\scriptstyle g}\downarrow & & {\scriptstyle 1}\downarrow & & \downarrow & & \downarrow \\ p & \xrightarrow{f} & x & \longrightarrow & z & \longrightarrow & \Sigma p \end{array}$$

The morphism $fg : p' \longrightarrow x$ lies in $Mor_{\langle S \rangle^\beta}$ since $z' \in \langle S \rangle^\beta$. The diagram

$$\begin{array}{ccc} p' & \xrightarrow{\alpha g} & y \\ {\scriptstyle fg}\downarrow & & \\ x & & \end{array}$$

is therefore a morphism in $\mathcal{T}/\langle S \rangle^\beta$, whose image under ϕ is clearly equivalent to the given morphism

$$\begin{array}{ccc} p & \xrightarrow{\alpha} & y \\ {\scriptstyle f}\downarrow & & \\ x & & \end{array}$$

Hence the surjectivity of ϕ.
Proof that ϕ is injective. Let the diagram

$$\begin{array}{ccc} p & \xrightarrow{\alpha} & y \\ {\scriptstyle f}\downarrow & & \\ x & & \end{array}$$

represent a morphism in $\mathcal{T}/\langle S \rangle^\beta$ whose image under ϕ is zero. Because the diagram is a morphism in $\mathcal{T}/\langle S \rangle^\beta$, f must be in $Mor_{\langle S \rangle^\beta}$. In the triangle

$$p \longrightarrow x \longrightarrow z \longrightarrow \Sigma p,$$

we must have $z \in \langle S \rangle^\beta$. On the other hand, β is regular, and Lemma 4.3.2 tells us that $\langle S \rangle^\beta \subset \mathcal{T}^\beta$. We were given that $x \in \mathcal{T}^\beta$, and deduce that p, the third vertex of the triangle, must also be in $\mathcal{T}^\beta$.

We also assume that the image under ϕ of the morphism represented by the diagram

$$\begin{array}{ccc} p & \xrightarrow{\alpha} & y \\ {\scriptstyle f}\downarrow & & \\ x & & \end{array}$$

vanishes. In other words, the diagram is equivalent, in $\mathcal{T}/\langle S\rangle$, to the diagram

$$\begin{array}{ccc} p & \xrightarrow{0} & y \\ {\scriptstyle f}\downarrow & & \\ x & & \end{array}$$

By Lemma 2.1.26, the map $\alpha : p \longrightarrow y$ must factor as

$$p \longrightarrow z \longrightarrow y$$

with $z \in \langle S\rangle$. But we have just shown that $p \in \mathcal{T}^\beta$. As $z \in \langle S\rangle$, we conclude from Theorem 4.3.3 that $p \to z$ factors as

$$p \longrightarrow z' \longrightarrow z$$

with $z' \in \langle S\rangle^\beta$. Since we have factored $p \to y$ through $z' \in \langle S\rangle^\beta$, it follows, again from Lemma 2.1.26, that the class of

$$\begin{array}{ccc} p & \xrightarrow{\alpha} & y \\ {\scriptstyle f}\downarrow & & \\ x & & \end{array}$$

vanishes already in $\mathcal{T}/\langle S\rangle^\beta$. □

An immediate corollary is

COROLLARY 4.4.2. *With the notation as in Proposition 4.4.1, the natural functor*

$$\mathcal{T}^\beta/\langle S\rangle^\beta \longrightarrow \mathcal{T}/\langle S\rangle$$

is fully faithful.

Proof: Proposition 4.4.1 asserts that for all $x \in \mathcal{T}^\beta$, $y \in \mathcal{T}$,

$$\{\mathcal{T}/\langle S\rangle\}(x,y) = \left\{\mathcal{T}/\langle S\rangle^\beta\right\}(x,y).$$

Corollary 4.4.2 is the weaker assertion that this holds if $y \in \mathcal{T}^\beta$ as well. □

All of this becomes useful when we have $\mathcal{T} = \cup_\beta \mathcal{T}^\beta$. We then know

COROLLARY 4.4.3. *Suppose $\mathcal{T}$ is a triangulated category with small Hom–sets, satisfying [TR5]. Suppose $\mathcal{T} = \cup_\beta \mathcal{T}^\beta$; that is, every object of $\mathcal{T}$ is β–compact for some β. Suppose S is a* set *of objects in $\mathcal{T}^\alpha$, for some infinite cardinal α. Then the category $\mathcal{T}/\langle S\rangle$ also has small Hom–sets.*

Proof: Let x and y be objects of $\mathfrak{T}$, we need to show that

$$\{\mathfrak{T}/\langle S\rangle\}(x,y)$$

is a set. But by hypothesis, $\mathfrak{T} = \cup_\beta \mathfrak{T}^\beta$. Therefore x must lie in some $\mathfrak{T}^\beta$. Recall that if $\beta < \gamma$, then by Lemma 4.2.3, $\mathfrak{T}^\beta \subset \mathfrak{T}^\gamma$. We may therefore choose β so that

4.4.3.1. x is β–compact.

4.4.3.2. β is regular.

4.4.3.3. $\beta > \alpha$, where $S \subset \mathfrak{T}^\alpha$, $\alpha \geq \aleph_0$ given.

Then x and S both lie in $\mathfrak{T}^\beta$, and by Proposition 4.4.1,

$$\{\mathfrak{T}/\langle S\rangle\}(x,y) = \left\{\mathfrak{T}/\langle S\rangle^\beta\right\}(x,y).$$

On the other hand, S was a set, and Proposition 3.2.5 gives us that the category $\langle S\rangle^\beta$ is essentially small. But by Proposition 2.2.1, the Verdier quotient of $\mathfrak{T}$ by a small (or essentially small) category has small *Hom*–sets. Therefore $\left\{\mathfrak{T}/\langle S\rangle^\beta\right\}(x,y)$ is a small set; $\{\mathfrak{T}/\langle S\rangle\}(x,y)$, being equal to it, is also a set. □

The situation of Corollary 4.4.3 is worth analysing more closely.

LEMMA 4.4.4. *Suppose $\mathfrak{T}$ is a triangulated category satisfying [TR5]. Suppose $\mathfrak{T} = \cup_\beta \mathfrak{T}^\beta$; that is, every object of $\mathfrak{T}$ is β–compact for some β. Suppose S is a class of objects in $\mathfrak{T}^\alpha$, for some infinite cardinal α. Then for any regular $\beta \geq \alpha$, the image of $\mathfrak{T}^\beta$ in $\mathfrak{T}/\mathfrak{S}$ satisfies*

$$\mathfrak{T}^\beta \subset \{\mathfrak{T}/\langle S\rangle\}^\beta.$$

In other words, every object of $\mathfrak{T}^\beta$ is β-small even in $\mathfrak{T}/\langle S\rangle$, and in fact the class $\mathfrak{T}^\beta$ is a β–perfect class of $\mathfrak{T}/\langle S\rangle$, hence contained in the maximal one inside $\{\mathfrak{T}/\langle S\rangle\}^{(\beta)}$.

Proof: We need to show that $\mathfrak{T}^\beta$ is consists of β–small objects of $\mathfrak{T}/\langle S\rangle$, and that $\mathfrak{T}^\beta$ is β–perfect in $\mathfrak{T}/\langle S\rangle$. Let k be an object of $\mathfrak{T}^\beta$. Let $\{X_\lambda, \lambda \in \Lambda\}$ be a set of objects of $\mathfrak{T}/\langle S\rangle$, which of course are the same as objects of $\mathfrak{T}$. Let us be given a map in $\mathfrak{T}/\langle S\rangle$

$$k \longrightarrow \coprod_{\lambda\in\Lambda} X_\lambda.$$

To show that k is β–small in $\mathfrak{T}/\langle S\rangle$, we need to factor the map as

$$k \longrightarrow \coprod_{\lambda\in\Lambda'} X_\lambda \longrightarrow \coprod_{\lambda\in\Lambda} X_\lambda$$

for some subset $\Lambda' \subset \Lambda$ of cardinality $< \beta$.

To show further that $\mathcal{T}^\beta$ is a β–perfect, note first that the objects of $\mathcal{T}^\beta$ form a subcategory equivalent to a triangulated subcategory. By Corollary 4.4.2, the subcategory

$$\mathcal{T}^\beta/\langle S\rangle^\beta \subset \mathcal{T}/\langle S\rangle$$

is a full subcategory, and since it is closed with respect to the formation of triangles it is equivalent to a triangulated subcategory of $\mathcal{T}/\langle S\rangle$. By Remark 3.3.6, to check that the objects of a triangulated subcategory form a β–perfect class, one needs only show that each object is β–good. In other words, we need to further prove that the map

$$k \longrightarrow \coprod_{\lambda\in\Lambda'} X_\lambda$$

can be factored as

$$k \longrightarrow \coprod_{\lambda\in\Lambda'} k_\lambda \xrightarrow{\coprod_{\lambda\in\Lambda'} f_\lambda} \coprod_{\lambda\in\Lambda'} X_\lambda$$

with $k_\lambda \in \mathcal{T}^\beta$.

Let us begin therefore with a map in $\mathcal{T}/\langle S\rangle$

$$k \longrightarrow \coprod_{\lambda\in\Lambda} X_\lambda.$$

By Corollary 3.2.11, the coproduct $\coprod_{\lambda\in\Lambda} X_\lambda$ exists in $\mathcal{T}/\langle S\rangle$; in fact, it agrees with the coproduct in $\mathcal{T}$. We are given a map in $\mathcal{T}/\langle S\rangle$

$$k \longrightarrow \coprod_{\lambda\in\Lambda} X_\lambda.$$

It is a map in $\mathcal{T}/\langle S\rangle$ from an object in $k \in \mathcal{T}^\beta$ to an object $\coprod_{\lambda\in\Lambda} X_\lambda \in \mathcal{T}$, and by Proposition 4.4.1,

$$\{\mathcal{T}/\langle S\rangle\}\left(k, \coprod_{\lambda\in\Lambda} X_\lambda\right) = \left\{\mathcal{T}/\langle S\rangle^\beta\right\}\left(k, \coprod_{\lambda\in\Lambda} X_\lambda\right).$$

That is, the map

$$k \longrightarrow \coprod_{\lambda\in\Lambda} X_\lambda$$

comes from a morphism in $\mathcal{T}/\langle S\rangle^\beta$. It therefore is represented by a diagram

$$\begin{array}{ccc} p & \longrightarrow & \coprod_{\lambda\in\Lambda} X_\lambda \\ {\scriptstyle f}\big\downarrow & & \\ k & & \end{array}$$

where $f \in Mor_{\langle S \rangle^\beta}$. This means that there is a triangle

$$p \longrightarrow k \longrightarrow z \longrightarrow \Sigma p$$

with $z \in \langle S \rangle^\beta \subset \mathfrak{T}^\beta$. Since k is also assumed in $\mathfrak{T}^\beta$, it follows that $p \in \mathfrak{T}^\beta$. But then the β–smallness of $p \in \mathfrak{T}$ guarantees that the map in $\mathfrak{T}$

$$p \longrightarrow \coprod_{\lambda \in \Lambda} X_\lambda$$

factors as

$$p \longrightarrow \coprod_{\lambda \in \Lambda'} X_\lambda \longrightarrow \coprod_{\lambda \in \Lambda} X_\lambda$$

where $\Lambda' \subset \Lambda$ has cardinality less than β. The β–compactness of p in $\mathfrak{T}$ says that the map in $\mathfrak{T}$

$$p \longrightarrow \coprod_{\lambda \in \Lambda'} X_\lambda$$

factors as

$$p \longrightarrow \coprod_{\lambda \in \Lambda'} k_\lambda \longrightarrow \coprod_{\lambda \in \Lambda'} X_\lambda$$

with $k_\lambda \in \mathfrak{T}^\beta$. In other words, in $\mathfrak{T}/\langle S \rangle$ we factored the map

$$k \longrightarrow \coprod_{\lambda \in \Lambda} X_\lambda$$

as

$$k \longrightarrow \coprod_{\lambda \in \Lambda'} k_\lambda \longrightarrow \coprod_{\lambda \in \Lambda'} X_\lambda \longrightarrow \coprod_{\lambda \in \Lambda} X_\lambda.$$

□

Even the seemingly stupid case, where $\langle S \rangle = \mathfrak{T}$, is worth considering further.

Lemma 4.4.5. *Let $\mathfrak{T}$ be a triangulated category satisfying [TR5]. Let S be a class of objects in $\mathfrak{T}^\alpha$, for some infinite α. Suppose $\langle S \rangle = \mathfrak{T}$. Let β be a regular cardinal $\geq \alpha$. Then the inclusion $\langle S \rangle^\beta \subset \mathfrak{T}^\beta$ is an equality. In other words, every object of $\mathfrak{T}^\beta$ is in $\langle S \rangle^\beta$.*

Proof: Let x be an object of $\mathfrak{T}^\beta$. We need to prove that x is in $\langle S \rangle^\beta$. The identity map $1 : x \longrightarrow x$ is a morphism from $x \in \mathfrak{T}^\beta$ to $x \in \langle S \rangle$ (we are assuming $\langle S \rangle = \mathfrak{T}$). By Theorem 4.3.3, it factors through some object $y \in \langle S \rangle^\beta$. Thus x is a direct summand of $y \in \langle S \rangle^\beta$. But $\langle S \rangle^\beta$ is thick, hence $x \in \langle S \rangle^\beta$. □

PROPOSITION 4.4.6. *Let $\mathcal{T}$ be a triangulated category satisfying [TR5]. Let $T \subset \mathcal{T}^\alpha$ and $S \subset \mathcal{T}^\alpha$ be two classes of α–compact objects, α an infinite cardinal. Suppose that $\langle T \rangle = \mathcal{T}$. Suppose that $\beta \geq \alpha$ is a regular cardinal. Then the inclusion*

$$\mathcal{T}^\beta / \langle S \rangle^\beta \subset \{\mathcal{T}/\langle S \rangle\}^\beta$$

is almost an equivalence; every object of $\{\mathcal{T}/\langle S \rangle\}^\beta$ is isomorphic to a direct summand of something in the image. That is, the β–compact objects of $\mathcal{T}/\langle S \rangle$ are, up to splitting idempotents, the images of β–compact objects of $\mathcal{T}$.

Proof: Note that the map $\mathcal{T} \longrightarrow \mathcal{T}/\langle S \rangle$ takes $\mathcal{T}^\beta$ to β–compact objects of $\mathcal{T}/\langle S \rangle$, by Lemma 4.4.4. Hence there is a well defined map

$$\mathcal{T}^\beta / \langle S \rangle^\beta \longrightarrow \{\mathcal{T}/\langle S \rangle\}^\beta.$$

The fact that this map is fully faithful is a consequence of Corollary 4.4.2. We may therefore view

$$\mathcal{T}^\beta / \langle S \rangle^\beta \subset \{\mathcal{T}/\langle S \rangle\}^\beta$$

as a fully faithful embedding of categories. Let $\overline{\mathcal{T}^\beta / \langle S \rangle^\beta}$ be the thick closure of $\mathcal{T}^\beta / \langle S \rangle^\beta$. Since $\{\mathcal{T}/\langle S \rangle\}^\beta$ is a thick subcategory of $\mathcal{T}/\langle S \rangle$ containing $\mathcal{T}^\beta / \langle S \rangle^\beta$, we deduce

$$\overline{\mathcal{T}^\beta / \langle S \rangle^\beta} \subset \{\mathcal{T}/\langle S \rangle\}^\beta.$$

We need to prove the opposite inclusion,

$$\overline{\mathcal{T}^\beta / \langle S \rangle^\beta} \supset \{\mathcal{T}/\langle S \rangle\}^\beta.$$

Now $\mathcal{T}^\beta$ is a triangulated subcategory of $\mathcal{T}$, and contains all coproducts of fewer that β of its objects. By Lemma 3.2.10, coproducts in $\mathcal{T}$ and $\mathcal{T}/\langle S \rangle$ agree. Therefore $\mathcal{T}^\beta / \langle S \rangle^\beta$, and hence also $\overline{\mathcal{T}^\beta / \langle S \rangle^\beta}$, are closed under the formation of coproducts in $\mathcal{T}/\langle S \rangle$ of fewer than β of their objects. Then $\overline{\mathcal{T}^\beta / \langle S \rangle^\beta}$ is thick, and contains the coproducts of fewer than β of its objects. Since $T \subset \mathcal{T}^\alpha \subset \mathcal{T}^\beta = \mathcal{T}^\beta / \langle S \rangle^\beta$,

$$\overline{\mathcal{T}^\beta / \langle S \rangle^\beta} \supset \langle T \rangle^\beta$$

where $\langle T \rangle^\beta \subset \mathcal{T}/\langle S \rangle$ is the smallest β–localising, thick subcategory containing T.

On the other hand, we assume that $\langle T \rangle = \mathcal{T}$; that is, $\mathcal{T}$ is the smallest localising subcategory of $\mathcal{T}$ containing T. We deduce that $\mathcal{T}/\langle S \rangle$ is the smallest localising subcategory of $\mathcal{T}/\langle S \rangle$ containing T. If we view T as a subclass of $\mathcal{T}/\langle S \rangle$, we still get $\langle T \rangle = \mathcal{T}/\langle S \rangle$. By Lemma 4.4.5, applied to the class T in the category $\mathcal{T}/\langle S \rangle$, we get

$$\langle T \rangle^\beta = \{\mathcal{T}/\langle S \rangle\}^\beta.$$

Hence

$$\overline{\mathcal{T}^\beta/\langle S\rangle^\beta} \supset \{\mathcal{T}/\langle S\rangle\}^\beta.$$

Thus the two subcategories are equal, as stated. □

REMARK 4.4.7. In Proposition 4.4.6, the statement can be improved if β is not only regular but also $\beta > \aleph_0$. In this case, $\mathcal{T}^\beta/\langle S\rangle^\beta$ is closed under the formation of coproducts of countably many objects, and idempotents in $\mathcal{T}^\beta/\langle S\rangle^\beta$ must split. Therefore, if $\beta > \aleph_0$ then every object of $\{\mathcal{T}/\langle S\rangle\}^\beta$ is isomorphic in $\mathcal{T}/\langle S\rangle$ to an object in $\mathcal{T}^\beta/\langle S\rangle^\beta$. There is no need to split idempotents; the categories $\{\mathcal{T}/\langle S\rangle\}^\beta$ and $\mathcal{T}^\beta/\langle S\rangle^\beta$ agree, up to extending $\mathcal{T}^\beta/\langle S\rangle^\beta$ to include every object in $\mathcal{T}/\langle S\rangle$ isomorphic to an object of $\mathcal{T}^\beta/\langle S\rangle^\beta$.

LEMMA 4.4.8. *Let $\mathcal{T}$ be a triangulated category satisfying [TR5]. Let $S \subset \mathcal{T}^\alpha$ be a class of α–compact objects, α an infinite cardinal. Let $\mathcal{S} = \langle S\rangle$ be the localising subcategory generated by S. Suppose that $\beta \geq \alpha$ is a regular cardinal. Then there is an inclusion*

$$\mathcal{S} \cap \mathcal{T}^\beta \subset \mathcal{S}^\beta.$$

Proof: We will show that $\mathcal{S} \cap \mathcal{T}^\beta$ is a β–perfect class of objects in $\mathcal{S}^{(\beta)}$; hence it must be contained in the maximal such, $\mathcal{S}^\beta$.

Let $k \in \mathcal{S} \cap \mathcal{T}^\beta$ be any object. Let $\{X_\lambda, \lambda \in \Lambda\}$ be a family of objects in $\mathcal{S} = \langle S\rangle$. Suppose we are given a map

$$k \longrightarrow \coprod_{\lambda\in\Lambda} X_\lambda.$$

Because $k \in \mathcal{T}^\beta$, it is β–small in $\mathcal{T}$, and hence there must be a subset $\Lambda' \subset \Lambda$ of cardinality $< \beta$, so that the map factors as

$$k \longrightarrow \coprod_{\lambda\in\Lambda'} X_\lambda \longrightarrow \coprod_{\lambda\in\Lambda} X_\lambda.$$

This proves that k is β–small in $\mathcal{S}$. In other words, we have established that $\mathcal{S} \cap \mathcal{T}^\beta \subset \mathcal{S}^{(\beta)}$.

Next we want to show that $\mathcal{S} \cap \mathcal{T}^\beta$ is a perfect class. By Remark 3.3.6, it suffices to show that every object $k \in \mathcal{S} \cap \mathcal{T}^\beta$ is β–good. Assume therefore that Λ is a set of cardinality $< \beta$, and we have a map

$$k \longrightarrow \coprod_{\lambda\in\Lambda} X_\lambda.$$

Because $k \in \mathcal{T}^\beta$, this map must factor as

$$k \longrightarrow \coprod_{\lambda\in\Lambda} k_\lambda \xrightarrow{\coprod_{\lambda\in\Lambda} f_\lambda} \coprod_{\lambda\in\Lambda} X_\lambda$$

with $k_\lambda \in \mathcal{T}^\beta$. On the other hand, $f_\lambda : k_\lambda \longrightarrow X_\lambda$ is a map from an object of $\mathcal{T}^\beta$ to an object of $\langle S\rangle$. We know from Theorem 4.3.3 that it must factor as

$$k_\lambda \longrightarrow k'_\lambda \longrightarrow X_\lambda$$

where $k'_\lambda \in \langle S\rangle^\beta \subset \mathcal{S} \cap \mathcal{T}^\beta$. Thus we have factored

$$k \longrightarrow \coprod_{\lambda\in\Lambda} X_\lambda$$

as

$$k \longrightarrow \coprod_{\lambda\in\Lambda} k'_\lambda \longrightarrow \coprod_{\lambda\in\Lambda} X_\lambda.$$

and k is β–good. □

Summarising the work of this Section, we get

THEOREM 4.4.9. *Let $\mathcal{S}$ be a triangulated category satisfying [TR5], $\mathcal{R} \subset \mathcal{S}$ a localising subcategory. Write $\mathcal{T}$ for the Verdier quotient $\mathcal{S}/\mathcal{R}$.*

Suppose there is an infinite cardinal α, a class of objects $S \subset \mathcal{S}^\alpha$ and another class of objects $R \subset \mathcal{R} \cap \mathcal{S}^\alpha$, so that

$$\mathcal{R} = \langle R\rangle \qquad \textit{and} \qquad \mathcal{S} = \langle S\rangle.$$

Then for any regular $\beta \geq \alpha$,

$$\langle R\rangle^\beta = \mathcal{R}^\beta = \mathcal{R} \cap \mathcal{S}^\beta,$$

$$\langle S\rangle^\beta = \mathcal{S}^\beta.$$

The natural map

$$\mathcal{S}^\beta/\mathcal{R}^\beta \longrightarrow \mathcal{T}$$

factors as

$$\mathcal{S}^\beta/\mathcal{R}^\beta \longrightarrow \mathcal{T}^\beta \subset \mathcal{T},$$

and the functor

$$\mathcal{S}^\beta/\mathcal{R}^\beta \longrightarrow \mathcal{T}^\beta$$

is fully faithful. If $\beta > \aleph_0$, the functor

$$\mathcal{S}^\beta/\mathcal{R}^\beta \longrightarrow \mathcal{T}^\beta$$

is an equivalence of categories. If $\beta = \aleph_0$, then every object of $\mathcal{T}^\beta$ is a direct summand of an object in $\mathcal{S}^\beta/\mathcal{R}^\beta$.

Proof: From Lemma 4.4.8 we deduce an inclusion $\mathcal{R} \cap \mathcal{S}^\beta \subset \mathcal{R}^\beta$. Trivially, we have an inclusion $\langle R \rangle^\beta \subset \mathcal{R} \cap \mathcal{S}^\beta$. Combining, we have inclusions

$$\langle R \rangle^\beta \subset \mathcal{R} \cap \mathcal{S}^\beta \subset \mathcal{R}^\beta.$$

By Lemma 4.4.5, applied to $R \subset \mathcal{R} \cap \mathcal{S}^\beta \subset \mathcal{R}^\beta$, we have

$$\langle R \rangle^\beta = \mathcal{R}^\beta.$$

Hence equality must hold throughout, and we have

$$\langle R \rangle^\beta = \mathcal{R} \cap \mathcal{S}^\beta = \mathcal{R}^\beta.$$

By Lemma 4.4.5, applied to $S \subset \mathcal{S}^\alpha$, we have

$$\langle S \rangle^\beta = \mathcal{S}^\beta.$$

That the natural map

$$\mathcal{S}^\beta/\mathcal{R}^\beta \longrightarrow \mathcal{T}$$

factors as

$$\mathcal{S}^\beta/\mathcal{R}^\beta \longrightarrow \mathcal{T}^\beta \subset \mathcal{T}$$

is the statement that the image of a β–compact object of $\mathcal{S}$ is β–compact in $\mathcal{T}$, that is Lemma 4.4.4. That the functor

$$\mathcal{S}^\beta/\mathcal{R}^\beta \longrightarrow \mathcal{T}^\beta$$

is fully faithful is Corollary 4.4.2. That the functor

$$\mathcal{S}^\beta/\mathcal{R}^\beta \longrightarrow \mathcal{T}^\beta$$

is an equivalence if $\beta > \aleph_0$ is Remark 4.4.7. The statement that, for $\beta = \aleph_0$, every object of $\mathcal{T}^\beta$ is a direct summand of an object of the full subcategory $\mathcal{S}^\beta/\mathcal{R}^\beta$ follows from Proposition 4.4.6. □

4.5. A refinement in the countable case

The classical, most useful case of Thomason's localisation theorem is the case $\beta = \aleph_0$. As stated, the theorem says that $\mathcal{S}^{\aleph_0}/\mathcal{R}^{\aleph_0}$ is embedded fully faithfully in $\mathcal{T}^{\aleph_0}$, and that the embedding is an equivalence up to splitting idempotents. But one gets a refinement, which we will discuss in this section. We begin with some definitions.

DEFINITION 4.5.1. *Let $\mathcal{T}$ be an essentially small category. Define $\mathbb{Z}(\mathcal{T})$ to be the free abelian group on isomorphism classes of the objects of $\mathcal{T}$. The object X of $\mathcal{T}$, viewed as an element in $\mathbb{Z}(\mathcal{T})$, will be denoted $[X]$.*

DEFINITION 4.5.2. *Let $\mathfrak{T}$ be an essentially small additive category. One defines a group $A(\mathfrak{T}) \subset \mathbb{Z}(\mathfrak{T})$ to be the subgroup generated by all*

$$[X \oplus Y] - [X] - [Y]$$

where X, Y are objects of $\mathfrak{T}$.

DEFINITION 4.5.3. *Let $\mathfrak{T}$ be an essentially small additive category. The set of isomorphism classes of objects of $\mathfrak{T}$ forms an abelian semigroup, with addition being given by direct sum. Define $K_A(\mathfrak{T})$ to be the group completion of $\mathfrak{T}$ with respect to this addition. That is, elements of $K_A(\mathfrak{T})$ are equivalence classes of formal differences $[X] - [Y]$, where $[X]$ and $[Y]$ are isomorphism classes of objects of $\mathfrak{T}$. We declare $[X] - [Y]$ equivalent to $[X'] - [Y']$ if, for some object $P \in \mathfrak{T}$, there is an isomorphism*

$$X \oplus Y' \oplus P \simeq X' \oplus Y \oplus P.$$

One proves easily that this is an equivalence relation, and that $K_A(\mathfrak{T})$ is naturally an abelian group.

REMARK 4.5.4. This is usually called the Grothendieck group of the symmetric monoidal category $\mathfrak{T}$. The subscript A here is to distinguish $K_A(\mathfrak{T})$ from $K_0(\mathfrak{T})$, whose definition will come in Definition 4.5.8.

LEMMA 4.5.5. *Let $\mathfrak{T}$ be an essentially small additive category. Then there is a natural isomorphism*

$$\frac{\mathbb{Z}(\mathfrak{T})}{A(\mathfrak{T})} \longrightarrow K_A(\mathfrak{T}).$$

Proof: Define a map

$$\mathbb{Z}(\mathfrak{T}) \longrightarrow K_A(\mathfrak{T})$$

to be the following. It sends an element of $\mathbb{Z}(\mathfrak{T})$, that is a linear combination

$$\sum_{i=1}^{n}[X_i] - \sum_{j=1}^{m}[Y_j],$$

to the element

$$\left[\bigoplus_{i=1}^{n} X_i\right] - \left[\bigoplus_{j=1}^{m} Y_j\right].$$

This is clearly surjective, and the kernel is $A(\mathfrak{T})$; hence the identification

$$\frac{\mathbb{Z}(\mathfrak{T})}{A(\mathfrak{T})} \simeq K_A(\mathfrak{T}).$$

□

DEFINITION 4.5.6. *Let $\mathcal{T}$ be an essentially small triangulated category. One defines a subgroup $T(\mathcal{T}) \subset \mathbb{Z}(\mathcal{T})$ to be the subgroup generated by sums*

$$[Y] - [X] - [Z]$$

where

$$X \longrightarrow Y \longrightarrow Z \longrightarrow \Sigma X$$

is a triangle in $\mathcal{T}$.

LEMMA 4.5.7. *Suppose $\mathcal{T}$ is an essentially small triangulated category. Then $A(\mathcal{T}) \subset T(\mathcal{T})$.*

Proof: Let X and Y be objects of $\mathcal{T}$. Then the following is a triangle

$$X \longrightarrow X \oplus Y \longrightarrow Y \longrightarrow \Sigma X$$

establishing that

$$[X \oplus Y] - [X] - [Y]$$

is in $T(\mathcal{T})$. □

DEFINITION 4.5.8. *Let $\mathcal{T}$ be an essentially small triangulated category. One defines a group $K_0(\mathcal{T})$, the Grothendieck group of $\mathcal{T}$, to be*

$$K_0(\mathcal{T}) = \frac{\mathbb{Z}(\mathcal{T})}{T(\mathcal{T})}.$$

REMARK 4.5.9. In particular, from the triangle

$$X \longrightarrow 0 \longrightarrow \Sigma X \longrightarrow \Sigma X$$

in $\mathcal{T}$, we learn that $[X] + [\Sigma X]$ vanishes in $K_0(\mathcal{T})$.

LEMMA 4.5.10. *Suppose $\mathcal{T}$ is an essentially small triangulated category, $\mathcal{S} \subset \mathcal{T}$ a triangulated subcategory. Suppose $\mathcal{T}$ is the thick closure of $\mathcal{S}$. That is, every object of $\mathcal{T}$ is a direct summand of an object of $\mathcal{S}$. Then in $\mathbb{Z}(\mathcal{T})$, which of course contains $\mathbb{Z}(\mathcal{S})$, one gets*

$$T(\mathcal{T}) = A(\mathcal{T}) + T(\mathcal{S}).$$

Proof: Clearly, $A(\mathcal{T})$ and $T(\mathcal{S})$ are both subgroups of $T(\mathcal{T})$, and hence

$$T(\mathcal{T}) \supset A(\mathcal{T}) + T(\mathcal{S}).$$

The problem is to show the reverse inclusion. Choose any of the generators of $T(\mathcal{T})$, that is

$$[Y] - [X] - [Z]$$

where

$$X \longrightarrow Y \longrightarrow Z \longrightarrow \Sigma X$$

is a triangle in $\mathcal{T}$. By hypothesis, $\mathcal{T}$ is the thick closure of $\mathcal{S}$; there is an object $A \in \mathcal{T}$ so that $X \oplus A \in \mathcal{S}$, and there is an object $B \in \mathcal{T}$ so that $Z \oplus B \in \mathcal{S}$. We deduce a triangle in $\mathcal{T}$

$$X \oplus A \longrightarrow Y \oplus A \oplus B \longrightarrow Z \oplus B \longrightarrow \Sigma\{X \oplus A\},$$

and as $X \oplus A \in \mathcal{S}$ and $Z \oplus B \in \mathcal{S}$ and $\mathcal{S}$ is triangulated, the entire triangle lies in $\mathcal{S}$. Hence

$$[Y \oplus A \oplus B] - [X \oplus A] - [Z \oplus B]$$

is an element of $T(\mathcal{S})$. As

$$[Y \oplus A \oplus B] - [Y] - [A] - [B], \qquad [X \oplus A] - [X] - [A],$$

$$[Z \oplus B] - [Z] - [B]$$

all lie in $A(\mathcal{T})$, the identity

$$\begin{aligned} [Y] - [X] - [Z] &= \{[Y \oplus A \oplus B] - [X \oplus A] - [Z \oplus B]\} \\ &\quad - \{[Y \oplus A \oplus B] - [Y] - [A] - [B]\} \\ &\quad + \{[X \oplus A] - [X] - [A]\} \\ &\quad + \{[Z \oplus B] - [Z] - [B]\} \end{aligned}$$

shows that an arbitrary generator $[Y]-[X]-[Z]$ of $T(\mathcal{T})$ lies in $A(\mathcal{T})+T(\mathcal{S})$. Hence

$$T(\mathcal{T}) \subset A(\mathcal{T}) + T(\mathcal{S}),$$

and we are done. □

PROPOSITION 4.5.11. *Let $\mathcal{T}$ be an essentially small triangulated category. Let $\mathcal{S} \subset \mathcal{T}$ be a triangulated subcategory. Suppose the thick closure of $\mathcal{S}$ is all of $\mathcal{T}$. Then the natural map $K_0(\mathcal{S}) \longrightarrow K_0(\mathcal{T})$ is a monomorphism. Furthermore, if $X \in \mathcal{T}$ is an object so that $[X] \in K_0(\mathcal{T})$ lies in the image of $K_0(\mathcal{S}) \longrightarrow K_0(\mathcal{T})$, then $X \in \mathcal{S}$.*

Proof: We need to show that the map $f_0 : K_0(\mathcal{S}) \longrightarrow K_0(\mathcal{T})$ is injective, and analyse when an element $[X] \in K_0(\mathcal{T})$ lies in the image of f_0. But f_0 is identified as the map

$$\frac{\mathbb{Z}(\mathcal{S})}{T(\mathcal{S})} \quad \longrightarrow \quad \frac{\mathbb{Z}(\mathcal{T})}{T(\mathcal{T})} \quad = \quad \frac{\mathbb{Z}(\mathcal{T})}{A(\mathcal{T}) + T(\mathcal{S})}.$$

We are using the fact that $T(\mathcal{T}) = A(\mathcal{T}) + T(\mathcal{S})$, that is Lemma 4.5.10. It therefore suffices to show that the map

$$f_A : \frac{\mathbb{Z}(\mathcal{S})}{A(\mathcal{S})} \longrightarrow \frac{\mathbb{Z}(\mathcal{T})}{A(\mathcal{T})}$$

is injective. The map f_0 is obtained from f_A by further dividing by $T(\mathcal{S})/A(\mathcal{S})$. In other words, we are reduced to showing the injectivity of

$$f_A : K_A(\mathcal{S}) \longrightarrow K_A(\mathcal{T})$$

and analysing its cokernel, which is isomorphic to the cokernel of $f_0 : K_0(\mathcal{S}) \longrightarrow K_0(\mathcal{T})$.

Let $[X] - [Y]$ be an element of the kernel of f_A; that is, X and Y are objects of $\mathcal{S}$, and in $\mathcal{T}$ there exists an object P and an isomorphism

$$X \oplus P \simeq Y \oplus P.$$

But since $\mathcal{T}$ is the thick closure of $\mathcal{S}$, there must exist an object $P' \in \mathcal{T}$ so that $P \oplus P' \in \mathcal{S}$. Then the fact that there is an isomorphism

$$X \oplus P \oplus P' \simeq Y \oplus P \oplus P'$$

says that $[X] - [Y]$ vanishes already in $K_A(\mathcal{S})$; the kernel of f_A vanishes.

Finally, assume that X is an object of $\mathcal{T}$ so that $[X]$ lies in the image of the map $f_0 : K_0(\mathcal{S}) \longrightarrow K_0(\mathcal{T})$, or equivalently in the image of the map $f_A : K_A(\mathcal{S}) \longrightarrow K_A(\mathcal{T})$. Then there exist B and C in $\mathcal{S}$ and an identity in $K_A(\mathcal{T})$

$$[X] = [B] - [C].$$

This means that there is an object $P \in \mathcal{T}$ and an isomorphism

$$X \oplus C \oplus P \simeq B \oplus P.$$

Find an object $P' \in \mathcal{T}$ so that $P \oplus P'$ lies in $\mathcal{S}$. We have an isomorphism

$$X \oplus C \oplus P \oplus P' \simeq B \oplus P \oplus P'.$$

Replacing $C \in \mathcal{S}$ by $C \oplus P \oplus P' \in \mathcal{S}$ and $B \in \mathcal{S}$ by $B \oplus P \oplus P' \in \mathcal{S}$, we may say that there are objects C and B in $\mathcal{S}$ and an isomorphism

$$X \oplus C \simeq B.$$

Now consider the triangle

$$C \longrightarrow B \longrightarrow X \longrightarrow \Sigma C.$$

It is a triangle in $\mathcal{T}$, but since C and B are in $\mathcal{S}$ and $\mathcal{S}$ is triangulated, the triangle lies in $\mathcal{S}$. Hence $X \in \mathcal{S}$. □

A very useful special case of this is

COROLLARY 4.5.12. *Let $\mathcal{S}$ be a triangulated subcategory of a triangulated category $\mathcal{T}$. Suppose $\mathcal{T}$ is the thick closure of $\mathcal{S}$. Then for any object $X \in \mathcal{T}$, the object $X \oplus \Sigma X$ lies in $\mathcal{S}$.*

Proof: If $\mathcal{T}$ is small, this is an immediate corollary of Lemma 4.5.13, once one observes that $X \oplus \Sigma X$ vanishes in $K_0(\mathcal{T})$. One can reduce the general case to the essentially small case; but since a direct proof is very simple, let us give one.

Since $X \in \mathcal{T}$ and $\mathcal{T}$ is the thick closure of $\mathcal{S}$, there exists Y in $\mathcal{T}$ with $\{X \oplus Y\} \in \mathcal{S}$. The suspension of $\{X \oplus Y\}$ is also in $\mathcal{S}$; that is $\{\Sigma X \oplus \Sigma Y\} \in \mathcal{S}$. There are three distinguished triangles in $\mathcal{T}$

$$\begin{array}{ccccccc} Y & \longrightarrow & 0 & \longrightarrow & \Sigma Y & \longrightarrow & \Sigma Y \\ X & \longrightarrow & X & \longrightarrow & 0 & \longrightarrow & \Sigma X \\ 0 & \longrightarrow & \Sigma X & \longrightarrow & \Sigma X & \longrightarrow & 0 \end{array}$$

The direct sum is, by Proposition 1.2.1, a triangle in $\mathcal{T}$

$$X \oplus Y \longrightarrow X \oplus \Sigma X \longrightarrow \Sigma X \oplus \Sigma Y \longrightarrow \Sigma X \oplus \Sigma Y.$$

But two of the terms, namely $X \oplus Y$ and $\Sigma X \oplus \Sigma Y$, lie in $\mathcal{S}$. Hence so does the third term $X \oplus \Sigma X$. □

One more Lemma before we get to the main point.

LEMMA 4.5.13. *As in Theorem 4.4.9, let $\mathcal{S}$ be a triangulated category satisfying [TR5]. But unlike Theorem 4.4.9, we now insist that $\mathcal{S}$ have small Hom–sets. Let $\mathcal{R} \subset \mathcal{S}$ be a localising subcategory. Write $\mathcal{T}$ for the Verdier quotient $\mathcal{S}/\mathcal{R}$.*

Let β be a regular cardinal. Suppose there is a set *(not just a class, as in Theorem 4.4.9) of objects $S \subset \mathcal{S}^\beta$ and another set of objects $R \subset \mathcal{R} \cap \mathcal{S}^\beta$, so that*

$$\mathcal{R} = \langle R \rangle \qquad \textit{and} \qquad \mathcal{S} = \langle S \rangle.$$

Then the categories $\mathcal{R}^\beta$, $\mathcal{S}^\beta$ and $\mathcal{T}^\beta$ are all essentially small.

Proof: By Theorem 4.4.9, we know that $\mathcal{R}^\beta = \langle R \rangle^\beta$, and $\mathcal{S}^\beta = \langle S \rangle^\beta$. Since we are assuming S and R are sets and $\mathcal{S}$ has small *Hom*–sets, it follows from Proposition 3.2.5 that $\langle R \rangle^\beta$ and $\langle S \rangle^\beta$ are essentially small; in other words, $\mathcal{R}^\beta$ and $\mathcal{S}^\beta$ are essentially small. But then by Proposition 2.2.1, the Verdier quotient $\mathcal{S}^\beta/\mathcal{R}^\beta$ has small *Hom*–sets, and since there is clearly only a set of isomorphism classes of objects, the quotient is essentially small. By Theorem 4.4.9 we know that $\mathcal{S}^\beta/\mathcal{R}^\beta$ is a full subcategory of $\mathcal{T}^\beta$, and up to splitting idempotents, the two categories agree. That means that $\mathcal{T}^\beta$ is obtained from $\mathcal{S}^\beta/\mathcal{R}^\beta$ by splitting some idempotents. Since $\mathcal{S}^\beta/\mathcal{R}^\beta$ is essentially small, it follows that so is $\mathcal{T}^\beta$. □

Now we are ready for the refinement of Theorem 4.4.9 in the countable case.

COROLLARY 4.5.14. *Let $\mathcal{S}$ be a triangulated category with small Hom–sets, satisfying [TR5]. Let $\mathcal{R} \subset \mathcal{S}$ be a localising subcategory. Write $\mathcal{T}$ for the Verdier quotient $\mathcal{S}/\mathcal{R}$.*

Suppose there is a set of objects $S \subset \mathcal{S}^{\aleph_0}$ and another set of objects $R \subset \mathcal{R} \cap \mathcal{S}^{\aleph_0}$, so that

$$\mathcal{R} = \langle R \rangle \qquad \text{and} \qquad \mathcal{S} = \langle S \rangle.$$

Then not only is it true that $\mathcal{T}^{\aleph_0}$ is the thick closure of the full subcategory $\mathcal{S}^{\aleph_0}/\mathcal{R}^{\aleph_0}$, but in fact every object in $X \in \mathcal{T}^{\aleph_0}$ which lies in the image of

$$K_0\left(\mathcal{S}^{\aleph_0}/\mathcal{R}^{\aleph_0}\right) \longrightarrow K_0\left(\mathcal{T}^{\aleph_0}\right)$$

is isomorphic to an object in $\mathcal{S}^{\aleph_0}/\mathcal{R}^{\aleph_0} \subset \mathcal{T}^{\aleph_0}$.

Proof: $\mathcal{T}^{\aleph_0}$ is a triangulated category, the category containing all objects of $\mathcal{T}^{\aleph_0}$ isomorphic to objects of $\mathcal{S}^{\aleph_0}/\mathcal{R}^{\aleph_0}$ is a triangulated subcategory whose thick closure is $\mathcal{T}^{\aleph_0}$. Both categories are essentially small by Lemma 4.5.13. Proposition 4.5.11 applies, and in particular we learn that $X \in \mathcal{T}^{\aleph_0}$ will be isomorphic to an object in $\mathcal{S}^{\aleph_0}/\mathcal{R}^{\aleph_0}$ if and only if $[X]$ lie in the image of the map

$$K_0\left(\mathcal{S}^{\aleph_0}/\mathcal{R}^{\aleph_0}\right) \longrightarrow K_0\left(\mathcal{T}^{\aleph_0}\right).$$

□

REMARK 4.5.15. The most useful case turns out to be the object $X \oplus \Sigma X$. See Corollary 4.5.12 above. Thus for any $X \in \mathcal{T}^{\aleph_0}$, there is $Y \in \mathcal{S}^{\aleph_0}$ so that in $\mathcal{T}$ there is an isomorphism

$$X \oplus \Sigma X \simeq Y.$$

4.6. History of the results in Chapter 4

The short way to describe the history is to say that everything is classical if $\alpha = \aleph_0$. An $\aleph_0$–small object, usually known as a compact object, is such that any map from it into a coproduct factors through a finite coproduct. In Remark 4.2.6 we saw that any $\aleph_0$–small object is $\aleph_0$–compact.

Thomason proved his localisation theorem when $\alpha = \aleph_0$ and $\mathcal{T}$ is the derived category of the category of quasi–coherent sheaves on a semi–separated scheme X. This proof may be found in [**34**]. The theorem is what Thomason calls his "key lemma".

For an arbitrary $\mathcal{T}$, but still with $\alpha = \aleph_0$, there is a proof based on Bousfield localisation in the author's [**23**]. The proof here, not appealing to Bousfield localisation, is new. And, of course, the statement for all α, that is Theorem 4.4.9, is entirely new.

Let α be a regular cardinal. Suppose $\mathcal{T} = \langle \mathcal{T}^{\aleph_0} \rangle$. That is, $\mathcal{T}$ is the smallest localising subcategory containing $\mathcal{T}^{\aleph_0}$. Then by Theorem 4.4.9

$$\mathcal{T}^{\alpha} = \langle \mathcal{T}^{\aleph_0} \rangle^{\alpha}.$$

If $\mathcal{T}$ is the homotopy category of spectra, then $\mathcal{T}^{\aleph_0}$ are the finite CW–complexes. By the above, $\mathcal{T}^{\alpha}$ becomes identified with $\langle \mathcal{T}^{\aleph_0} \rangle^{\alpha}$, that is all the spectra with fewer than α cells. This perhaps explains how the definitions of α–perfection and α–smallness were motivated by the attempt to copy classical arguments, which work by induction over the cardinality of the set of cells in a complex.

CHAPTER 5

The category $A(\mathcal{S})$

5.1. The abelian category $A(\mathcal{S})$

Let $\mathcal{S}$ be an additive category. We do not assume that $\mathcal{S}$ is essentially small. We define

DEFINITION 5.1.1. *The category $\mathcal{C}at\big(\mathcal{S}^{op}, \mathcal{A}b\big)$ has for its objects all the additive functors*

$$F : \mathcal{S}^{op} \longrightarrow \mathcal{A}b.$$

The morphisms in $\mathcal{C}at\big(\mathcal{S}^{op}, \mathcal{A}b\big)$ are the natural transformations.

It is well–known that $\mathcal{C}at\big(\mathcal{S}^{op}, \mathcal{A}b\big)$ is an abelian category. We remind the reader what sequences are exact in $\mathcal{C}at\big(\mathcal{S}^{op}, \mathcal{A}b\big)$. Suppose we are given a sequence

$$0 \longrightarrow F'(-) \longrightarrow F(-) \longrightarrow F''(-) \longrightarrow 0$$

of objects and morphisms in $\mathcal{C}at\big(\mathcal{S}^{op}, \mathcal{A}b\big)$, that is functors and natural transformations $\mathcal{S}^{op} \longrightarrow \mathcal{A}b$. This sequence is exact in $\mathcal{C}at\big(\mathcal{S}^{op}, \mathcal{A}b\big)$ if and only if, for every $s \in \mathcal{S}$, the sequence of abelian groups

$$0 \longrightarrow F'(s) \longrightarrow F(s) \longrightarrow F''(s) \longrightarrow 0$$

is exact.

LEMMA 5.1.2. *Let $\mathcal{S}$ be an additive category. Let $\mathcal{C}at\big(\mathcal{S}^{op}, \mathcal{A}b\big)$ be as in Definition 5.1.1. Then the representable functor $\mathcal{S}(-, s)$ is a projective object in the category $\mathcal{C}at\big(\mathcal{S}^{op}, \mathcal{A}b\big)$.*

Proof: The functor $Y_s(-) = \mathcal{S}(-, s)$ is additive, hence an object of $\mathcal{C}at\big(\mathcal{S}^{op}, \mathcal{A}b\big)$. Let F be any object of $\mathcal{C}at\big(\mathcal{S}^{op}, \mathcal{A}b\big)$. Yoneda's lemma tells us that morphisms in $\mathcal{C}at\big(\mathcal{S}^{op}, \mathcal{A}b\big)$, that is natural transformations

$$Y_s(-) = \mathcal{S}(-, s) \longrightarrow F(-)$$

are in one–to–one correspondence with elements of $F(s)$. Suppose we are given an exact sequence in $\mathcal{C}at\big(\mathcal{S}^{op}, \mathcal{A}b\big)$

$$0 \longrightarrow F'(-) \longrightarrow F(-) \longrightarrow F''(-) \longrightarrow 0.$$

Applying the functor

$$\mathcal{C}at\big(\mathcal{S}^{op}, \mathcal{A}b\big)\Big\{Y_s, -\Big\}$$

to the exact sequence gives

$$0 \longrightarrow F'(s) \longrightarrow F(s) \longrightarrow F''(s) \longrightarrow 0.$$

But this sequence of abelian groups is exact. Hence $Y_s = \mathcal{S}(-, s)$ is a projective object. □

DEFINITION 5.1.3. *Let $\mathcal{S}$ be a triangulated category. Recall that we do not assume $\mathcal{S}$ essentially small. The category $\mathcal{C}at\big(\mathcal{S}^{op}, \mathcal{A}b\big)$, as in Definition 5.1.1, is the category of all additive functors $\mathcal{S}^{op} \longrightarrow \mathcal{A}b$. We define*

$$A(\mathcal{S}) \qquad \subset \qquad \mathcal{C}at\big(\mathcal{S}^{op}, \mathcal{A}b\big)$$

to be the full subcategory of all objects F which admit presentations

$$\mathcal{S}(-, s) \longrightarrow \mathcal{S}(-, t) \longrightarrow F(-) \longrightarrow 0.$$

REMARK 5.1.4. In other words, the objects of $A(\mathcal{S})$ are the objects of $\mathcal{C}at\big(\mathcal{S}^{op}, \mathcal{A}b\big)$ admitting a presentation by nice projective objects, namely the representable ones.

LEMMA 5.1.5. *The functors in $A(\mathcal{S})$ take coproducts of objects in $\mathcal{S}$ to products of abelian groups.*

Proof: Let F be an object of $A(\mathcal{S})$; that is, it admits a presentation

$$\mathcal{S}(-, s) \longrightarrow \mathcal{S}(-, t) \longrightarrow F(-) \longrightarrow 0.$$

It is clear that the representable functors $\mathcal{S}(-, s)$ and $\mathcal{S}(-, t)$ take coproducts to products. Let $\{x_\lambda, \lambda \in \Lambda\}$ be a family of objects of $\mathcal{S}$ whose coproduct exists in $\mathcal{S}$. We have a commutative diagram with exact rows

$$\begin{array}{ccccccc}
\mathcal{S}\left(\coprod_{\lambda\in\Lambda} x_\lambda, s\right) & \longrightarrow & \mathcal{S}\left(\coprod_{\lambda\in\Lambda} x_\lambda, t\right) & \longrightarrow & F\left(\coprod_{\lambda\in\Lambda} x_\lambda\right) & \longrightarrow & 0 \\
\| \downarrow \wr & & \| \downarrow \wr & & \downarrow & & \\
\prod_{\lambda\in\Lambda} \mathcal{S}(x_\lambda, s) & \longrightarrow & \prod_{\lambda\in\Lambda} \mathcal{S}(x_\lambda, t) & \longrightarrow & \prod_{\lambda\in\Lambda} F(x_\lambda) & \longrightarrow & 0
\end{array}$$

and we immediately deduce that F sends coproducts to products. □

LEMMA 5.1.6. *Suppose $F \longrightarrow G$ is a morphism in $A(\mathcal{S})$. Then the cokernel is an object of $A(\mathcal{S})$.*

Proof: We have an exact sequence of functors

$$F(-) \longrightarrow G(-) \longrightarrow H(-) \longrightarrow 0,$$

and we are given that F and G are in $A(\mathcal{S})$; we would like to show that so is H. But F and G admit presentations

$$\mathcal{S}(-,s') \longrightarrow \mathcal{S}(-,t') \longrightarrow F(-) \longrightarrow 0$$

$$\mathcal{S}(-,s) \longrightarrow \mathcal{S}(-,t) \longrightarrow G(-) \longrightarrow 0$$

and we have a map $F \longrightarrow G$, hence a composite

$$\begin{array}{ccc} \mathcal{S}(-,t') & \longrightarrow & F(-) \\ & & \downarrow \\ & & G(-). \end{array}$$

By Lemma 5.1.2, the representable functor $\mathcal{S}(-,t')$ is a projective object, and hence the map $\mathcal{S}(-,t') \longrightarrow G(-)$ factors through the surjection $\mathcal{S}(-,t) \longrightarrow G(-)$. In other words the composite

$$\begin{array}{ccc} \mathcal{S}(-,t') & \longrightarrow & F(-) \\ & & \downarrow \\ & & G(-) \end{array}$$

factors to render commutative the square

$$\begin{array}{ccc} \mathcal{S}(-,t') & \longrightarrow & F(-) \\ \downarrow & & \downarrow \\ \mathcal{S}(-,t) & \longrightarrow & G(-). \end{array}$$

We deduce a commutative diagram with exact rows

$$\begin{array}{ccccccc} & & \mathcal{S}(-,t') & \longrightarrow & F(-) & \longrightarrow & 0 \\ & & \downarrow & & \downarrow & & \\ \mathcal{S}(-,s) & \longrightarrow & \mathcal{S}(-,t) & \longrightarrow & G(-) & \longrightarrow & 0 \end{array}$$

and the cokernel H of the map $F \longrightarrow G$ has a presentation

$$\mathcal{S}(-,s \oplus t') \longrightarrow \mathcal{S}(-,t) \longrightarrow H(-) \longrightarrow 0.$$

□

LEMMA 5.1.7. *Suppose F is an object of $A(\mathcal{S})$, and $\phi : \mathcal{S}(-,x) \longrightarrow F(-)$ is an epimorphism. Then the kernel of ϕ is an object of $A(\mathcal{S})$.*

Proof: Because F is an object of $A(\mathcal{S})$, it has a presentation

$$\mathcal{S}(-,s) \longrightarrow \mathcal{S}(-,t) \longrightarrow F(-) \longrightarrow 0.$$

We are also given a map $\phi : \mathcal{S}(-,x) \longrightarrow F(-)$; that is we have a diagram

$$\begin{array}{ccccccc} & & & & \mathcal{S}(-,x) & & \\ & & & & {\scriptstyle\phi}\downarrow & & \\ \mathcal{S}(-,s) & \longrightarrow & \mathcal{S}(-,t) & \longrightarrow & F(-) & \longrightarrow & 0 \end{array}$$

where the row is exact. The object $\mathcal{S}(-,x)$ is projective in $\mathcal{C}at\big(\mathcal{S}^{op},\mathcal{A}b\big)$ by Lemma 5.1.2, and hence the map $\phi : \mathcal{S}(-,x) \longrightarrow F(-)$ factors as

$$\mathcal{S}(-,x) \longrightarrow \mathcal{S}(-,t) \longrightarrow F(-) \longrightarrow 0.$$

We have a diagram

$$\begin{array}{ccccccc} & & \mathcal{S}(-,x) & & & & \\ & & \downarrow & & & & \\ \mathcal{S}(-,s) & \longrightarrow & \mathcal{S}(-,t) & \longrightarrow & F(-) & \longrightarrow & 0 \end{array}$$

where the bottom row is exact and the composite

$$\begin{array}{ccc} \mathcal{S}(-,x) & & \\ \downarrow & & \\ \mathcal{S}(-,t) & \longrightarrow & F(-) \end{array}$$

is surjective. We deduce that

$$\mathcal{S}(-,x) \oplus \mathcal{S}(-,s) \longrightarrow \mathcal{S}(-,t)$$

is surjective.

But by Yoneda this natural transformation of representable functors is induced by a morphism in $\mathcal{S}$

$$x \oplus s \longrightarrow t.$$

To say that this is surjective implies that

$$\mathcal{S}(t, x \oplus s) \longrightarrow \mathcal{S}(t,t)$$

is epi, and hence $1 : t \longrightarrow t$ lies in the image. The identity on t factors as

$$t \longrightarrow x \oplus s \longrightarrow t.$$

Complete $x \oplus s \longrightarrow t$ to a triangle

$$r \longrightarrow x \oplus s \longrightarrow t \longrightarrow \Sigma r.$$

This triangle is split. We must have an isomorphism $x \oplus s \simeq r \oplus t$. We have a commutative square in $\mathcal{T}$

$$\begin{array}{ccc} r & \longrightarrow & x \\ \downarrow & & \downarrow \\ s & \longrightarrow & t \end{array}$$

which is bicartesian for the split exact structure in $\mathcal{T}$. Anyway, we have a bicartesian diagram of functors

$$\begin{array}{ccc} \mathcal{S}(-,r) & \longrightarrow & \mathcal{S}(-,x) \\ \downarrow & & \downarrow \\ \mathcal{S}(-,s) & \longrightarrow & \mathcal{S}(-,t) \end{array}$$

in the abelian category $\mathcal{C}at\big(\mathcal{S}^{op}, \mathcal{A}b\big)$, and the cokernels of the rows are isomorphic; there is a commutative diagram with exact rows

$$\begin{array}{ccccccc} \mathcal{S}(-,r) & \longrightarrow & \mathcal{S}(-,x) & \longrightarrow & F(-) & \longrightarrow & 0 \\ \downarrow & & \downarrow & & \downarrow \wr & & \\ \mathcal{S}(-,s) & \longrightarrow & \mathcal{S}(-,t) & \longrightarrow & F(-) & \longrightarrow & 0 \end{array}$$

The surjection $\mathcal{S}(-,x) \longrightarrow F(-)$ has been completed to a presentation

$$\mathcal{S}(-,r) \longrightarrow \mathcal{S}(-,x) \longrightarrow F(-) \longrightarrow 0.$$

Now complete $r \longrightarrow x$ to a triangle

$$q \longrightarrow r \longrightarrow x \longrightarrow \Sigma q.$$

We have an exact sequence

$$\mathcal{S}(-,q) \longrightarrow \mathcal{S}(-,r) \xrightarrow{\ \rho\ } \mathcal{S}(-,x) \xrightarrow{\ \theta\ } \mathcal{S}(-,\Sigma q).$$

The functor $F(-)$ is identified as the image of θ, and the kernel $K(-)$ of $\mathcal{S}(-,x) \longrightarrow F(-)$ as the image of ρ. We have an exact sequence

$$\mathcal{S}(-,q) \longrightarrow \mathcal{S}(-,r) \longrightarrow K(-) \longrightarrow 0,$$

which establishes that K is an object of $A(\mathcal{S})$. □

LEMMA 5.1.8. *Suppose*

$$0 \longrightarrow F(-) \longrightarrow G(-) \longrightarrow H(-) \longrightarrow 0$$

is an exact sequence of functors in $\mathcal{C}at(\mathcal{S}^{op}, \mathcal{A}b)$. *Suppose* F *and* H *lie in* $A(\mathcal{S})$. *Then there is a commutative diagram with exact rows*

$$\begin{array}{ccccccccc} 0 & \longrightarrow & \mathcal{S}(-,f) & \longrightarrow & \mathcal{S}(-,g) & \longrightarrow & \mathcal{S}(-,h) & \longrightarrow & 0 \\ & & \downarrow & & \downarrow & & \downarrow & & \\ 0 & \longrightarrow & F(-) & \longrightarrow & G(-) & \longrightarrow & H(-) & \longrightarrow & 0 \end{array}$$

where all the vertical maps are surjective.

Proof: Since F and H lie in $A(\mathcal{S})$, we may certainly choose surjections

$$\begin{array}{ccccccccc} & & \mathcal{S}(-,f) & & & & \mathcal{S}(-,h) & & \\ & & \downarrow & & & & \downarrow & & \\ 0 & \longrightarrow & F(-) & \longrightarrow & G(-) & \longrightarrow & H(-) & \longrightarrow & 0 \end{array}$$

By Lemma 5.1.2, the object $\mathcal{S}(-,h)$ is projective in $\mathcal{C}at(\mathcal{S}^{op}, \mathcal{A}b)$. The map $\mathcal{S}(-,h) \longrightarrow H(-)$ must factor through the surjection $G \longrightarrow H$. Letting $g = f \oplus h$, we have a commutative diagram with exact rows

$$\begin{array}{ccccccccc} 0 & \longrightarrow & \mathcal{S}(-,f) & \longrightarrow & \mathcal{S}(-,g) & \longrightarrow & \mathcal{S}(-,h) & \longrightarrow & 0 \\ & & \downarrow & & \downarrow & & \downarrow & & \\ 0 & \longrightarrow & F(-) & \longrightarrow & G(-) & \longrightarrow & H(-) & \longrightarrow & 0 \end{array}$$

and the fact that the two outside vertical maps are surjective forces the middle to also be. □

LEMMA 5.1.9. *Suppose*

$$0 \longrightarrow F(-) \longrightarrow G(-) \longrightarrow H(-) \longrightarrow 0$$

is an exact sequence of functors in $\mathcal{C}at(\mathcal{S}^{op}, \mathcal{A}b)$. *If any two of* F, G *and* H *lie in* $A(\mathcal{S})$, *then so does the third.*

Proof: If F and G lie in $A(\mathcal{S})$, then so does H by Lemma 5.1.6. It remains to consider the other two cases.

Suppose F and H lie in $A(\mathcal{S})$. By Lemma 5.1.8, we may complete to a diagram with exact rows and surjective vertical maps

$$\begin{array}{ccccccccc} 0 & \longrightarrow & \mathcal{S}(-,f) & \longrightarrow & \mathcal{S}(-,g) & \longrightarrow & \mathcal{S}(-,h) & \longrightarrow & 0 \\ & & \downarrow & & \downarrow & & \downarrow & & \\ 0 & \longrightarrow & F(-) & \longrightarrow & G(-) & \longrightarrow & H(-) & \longrightarrow & 0 \end{array}$$

Taking the kernels in the columns, we have a 3×3 diagram with exact rows and columns

$$\begin{array}{ccccccccc} & & 0 & & 0 & & 0 & & \\ & & \downarrow & & \downarrow & & \downarrow & & \\ 0 & \longrightarrow & F'(-) & \longrightarrow & G'(-) & \longrightarrow & H'(-) & \longrightarrow & 0 \\ & & \downarrow & & \downarrow & & \downarrow & & \\ 0 & \longrightarrow & \mathcal{S}(-,f) & \longrightarrow & \mathcal{S}(-,g) & \longrightarrow & \mathcal{S}(-,h) & \longrightarrow & 0 \\ & & \downarrow & & \downarrow & & \downarrow & & \\ 0 & \longrightarrow & F(-) & \longrightarrow & G(-) & \longrightarrow & H(-) & \longrightarrow & 0 \\ & & \downarrow & & \downarrow & & \downarrow & & \\ & & 0 & & 0 & & 0 & & \end{array}$$

and Lemma 5.1.7, applied to the exact sequences

$$0 \longrightarrow F'(-) \longrightarrow \mathcal{S}(-,f) \longrightarrow F(-) \longrightarrow 0$$

$$0 \longrightarrow H'(-) \longrightarrow \mathcal{S}(-,h) \longrightarrow H(-) \longrightarrow 0,$$

allows us to deduce that F' and H' are in $A(\mathcal{S})$. Applying Lemma 5.1.8 to the exact sequence

$$0 \longrightarrow F'(-) \longrightarrow G'(-) \longrightarrow H'(-) \longrightarrow 0$$

we deduce a commutative diagram with exact rows and surjective columns

$$\begin{array}{ccccccccc} 0 & \longrightarrow & \mathcal{S}(-,f') & \longrightarrow & \mathcal{S}(-,g') & \longrightarrow & \mathcal{S}(-,h') & \longrightarrow & 0 \\ & & \downarrow & & \downarrow & & \downarrow & & \\ 0 & \longrightarrow & F'(-) & \longrightarrow & G'(-) & \longrightarrow & H'(-) & \longrightarrow & 0. \end{array}$$

Putting this all together we have a commutative diagram with exact rows and columns

$$\begin{array}{ccccccccc} 0 & \longrightarrow & \mathcal{S}(-,f') & \longrightarrow & \mathcal{S}(-,g') & \longrightarrow & \mathcal{S}(-,h') & \longrightarrow & 0 \\ & & \downarrow & & \downarrow & & \downarrow & & \\ 0 & \longrightarrow & \mathcal{S}(-,f) & \longrightarrow & \mathcal{S}(-,g) & \longrightarrow & \mathcal{S}(-,h) & \longrightarrow & 0 \\ & & \downarrow & & \downarrow & & \downarrow & & \\ 0 & \longrightarrow & F(-) & \longrightarrow & G(-) & \longrightarrow & H(-) & \longrightarrow & 0 \\ & & \downarrow & & \downarrow & & \downarrow & & \\ & & 0 & & 0 & & 0 & & \end{array}$$

and in particular the middle column is exact

$$\mathcal{S}(-,g') \longrightarrow \mathcal{S}(-,g) \longrightarrow G(-) \longrightarrow 0,$$

meaning that G is in $A(\mathcal{S})$.

It remains to show that if in the exact sequence

$$0 \longrightarrow F(-) \longrightarrow G(-) \longrightarrow H(-) \longrightarrow 0$$

the functors G and H are in $A(\mathcal{S})$, then so is F. Since $G \in A(\mathcal{S})$, we may choose a surjection $\mathcal{S}(-,g) \longrightarrow G(-)$. Consider the diagram with exact rows

$$\begin{array}{ccccccccc}
0 & \longrightarrow & 0 & \longrightarrow & \mathcal{S}(-,g) & \longrightarrow & \mathcal{S}(-,g) & \longrightarrow & 0 \\
 & & \downarrow & & \downarrow & & \downarrow & & \\
0 & \longrightarrow & F(-) & \longrightarrow & G(-) & \longrightarrow & H(-) & \longrightarrow & 0
\end{array}$$

We wish to apply the snake lemma to it. The kernels and cokernels fit in a six–term exact sequence. The kernels and cokernels are computed by the diagram with exact rows and columns

$$\begin{array}{ccccccccc}
 & & & & 0 & & 0 & & \\
 & & & & \downarrow & & \downarrow & & \\
 & & 0 & \longrightarrow & G'(-) & \longrightarrow & H'(-) & & \\
 & & \downarrow & & \downarrow & & \downarrow & & \\
0 & \longrightarrow & 0 & \longrightarrow & \mathcal{S}(-,g) & \longrightarrow & \mathcal{S}(-,g) & \longrightarrow & 0 \\
 & & \downarrow & & \downarrow & & \downarrow & & \\
0 & \longrightarrow & F(-) & \longrightarrow & G(-) & \longrightarrow & H(-) & \longrightarrow & 0 \\
 & & \downarrow & & \downarrow & & \downarrow & & \\
 & & F(-) & \longrightarrow & 0 & \longrightarrow & 0 & & \\
 & & \downarrow & & & & & & \\
 & & 0 & & & & & &
\end{array}$$

Lemma 5.1.7 applied to the exact sequences

$$0 \longrightarrow G'(-) \longrightarrow \mathcal{S}(-,g) \longrightarrow G(-) \longrightarrow 0$$

$$0 \longrightarrow H'(-) \longrightarrow \mathcal{S}(-,g) \longrightarrow H(-) \longrightarrow 0$$

allows us to deduce that G' and H' are in $A(\mathcal{S})$. The snake lemma gives an exact sequence

$$0 \longrightarrow G'(-) \longrightarrow H'(-) \longrightarrow F(-) \longrightarrow 0,$$

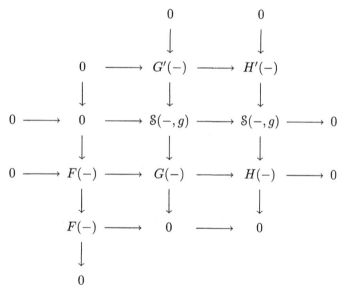

and hence Lemma 5.1.6 establishes that F is also in $A(\mathcal{S})$. □

PROPOSITION 5.1.10. *The full subcategory $A(\mathcal{S}) \subset \mathcal{C}at\big(\mathcal{S}^{op}, \mathcal{A}b\big)$ is an abelian subcategory closed under extensions. That is, if $F \longrightarrow G$ is a morphism of objects in $A(\mathcal{S})$, then the kernel, image and cokernel, computed in the abelian category $\mathcal{C}at\big(\mathcal{S}^{op}, \mathcal{A}b\big)$, lie in $A(\mathcal{S})$. Also, any extension of objects of $A(\mathcal{S})$ is in $A(\mathcal{S})$.*

Proof: Let $f : F \longrightarrow G$ be a morphism in $A(\mathcal{S})$. By Lemma 5.1.6, its cokernel lies in $A(\mathcal{S})$. The image fits in an exact sequence

$$0 \longrightarrow \mathrm{Im}(f) \longrightarrow G \longrightarrow \mathrm{Coker}(f) \longrightarrow 0,$$

and since G and $\mathrm{Coker}(f)$ lie in $A(\mathcal{S})$, Lemma 5.1.9 tells us that so does $\mathrm{Im}(f)$. The kernel fits in the exact sequence

$$0 \longrightarrow \mathrm{Ker}(f) \longrightarrow F \longrightarrow \mathrm{Im}(f) \longrightarrow 0,$$

and since $\mathrm{Im}(f)$ and F lie in $A(\mathcal{S})$, Lemma 5.1.9 tells us that so does $\mathrm{Ker}(f)$.

Finally, if we have an exact sequence

$$0 \longrightarrow F \longrightarrow G \longrightarrow H \longrightarrow 0,$$

with F and H in $A(\mathcal{S})$, Lemma 5.1.9 tells us that G is also in $A(\mathcal{S})$. The subcategory $A(\mathcal{S}) \subset \mathcal{C}at\big(\mathcal{S}^{op}, \mathcal{A}b\big)$ is closed under extensions. □

LEMMA 5.1.11. *The representable functors $\mathcal{S}(-, x)$ lie in $A(\mathcal{S})$, and are projective objects in it. Furthermore, every projective object in $A(\mathcal{S})$ is a direct summand of a representable $\mathcal{S}(-, x)$. If all idempotents in $\mathcal{S}$ split, for example if $\mathcal{S}$ is $\aleph_1$–localising, then the projective objects in $A(\mathcal{S})$ are precisely the representables.*

Proof: The objects $\mathcal{S}(-, s)$ have a presentation

$$\mathcal{S}(-, 0) \longrightarrow \mathcal{S}(-, s) \longrightarrow \mathcal{S}(-, s) \longrightarrow 0,$$

which shows that they lie in $A(\mathcal{S})$. By Lemma 5.1.2, they are projective objects already in the larger category $\mathcal{C}at\big(\mathcal{S}^{op}, \mathcal{A}b\big)$. Hence they must clearly be projective in $A(\mathcal{S})$.

Suppose F is a projective object in $A(\mathcal{S})$. Being an object of $A(\mathcal{S})$, it has a presentation

$$\mathcal{S}(-, s) \longrightarrow \mathcal{S}(-, t) \longrightarrow F(-) \longrightarrow 0.$$

But F is projective, and the identity $1 : F \longrightarrow F$ must factor through the epimorphism $\mathcal{S}(-, t) \longrightarrow F(-)$. That is, F must be a direct summand of $\mathcal{S}(-, t)$. If the category $\mathcal{S}$ is closed under the splitting of idempotents, then clearly F is representable. □

REMARK 5.1.12. We remind the reader: the objects $F \in A(\mathcal{S})$ can be identified with the objects of the larger category $\mathcal{C}at(\mathcal{S}^{op}, \mathcal{A}b)$ which admit projective presentations

$$\mathcal{S}(-,s) \longrightarrow \mathcal{S}(-,t) \longrightarrow F(-) \longrightarrow 0.$$

By Lemma 5.1.5, we know that such functors must take coproducts to products. The reader should be warned that not every functor taking coproducts to products need lie in $A(\mathcal{S})$. Not even for very good $\mathcal{S}$.

Of course in any abelian category, morphisms of objects give rise to morphisms of projective presentations. This leads to another description of the category $A(\mathcal{S})$.

DEFINITION 5.1.13. *Let $B(\mathcal{S})$ be the additive category whose objects are morphisms $\{s \to t\} \in \mathcal{S}$. Morphisms*

$$\{s \to t\} \longrightarrow \{s' \to t'\}$$

in $B(\mathcal{S})$ are equivalence classes of commutative squares

$$\begin{array}{ccc} s & \longrightarrow & t \\ \downarrow & & \downarrow \\ s' & \longrightarrow & t' \end{array}$$

The equivalence relation on morphisms is additive, and a morphism is equivalent to zero if in the diagram

$$\begin{array}{ccc} s & \longrightarrow & t \\ \downarrow & & \downarrow{\scriptstyle \phi} \\ s' & \longrightarrow & t' \end{array}$$

the map $\phi : t \longrightarrow t'$ factors as $t \longrightarrow s' \longrightarrow t'$.

PROPOSITION 5.1.14. *There is a functor $B(\mathcal{S}) \longrightarrow A(\mathcal{S})$ sending the object $\{s \to t\} \in B(\mathcal{S})$ to the cokernel of*

$$\mathcal{S}(-,s) \longrightarrow \mathcal{S}(-,t).$$

This functor is an equivalence of categories.

Proof: Clearly, every object of $A(\mathcal{S})$ is in the image of the functor. And given projective presentations

$$\mathcal{S}(-,s) \longrightarrow \mathcal{S}(-,t) \longrightarrow F(-) \longrightarrow 0$$

$$\mathcal{S}(-,s') \longrightarrow \mathcal{S}(-,t') \longrightarrow F'(-) \longrightarrow 0$$

then maps $F(-) \longrightarrow F'(-)$ are in one–to–one correspondence with homotopy equivalence classes of maps of projective presentations. Precisely, maps $F(-) \longrightarrow F'(-)$ correspond one–to–one to homotopy equivalence classes of chain maps

$$\begin{array}{ccc} \mathcal{S}(-,s) & \longrightarrow & \mathcal{S}(-,t) \\ \downarrow & & \downarrow \\ \mathcal{S}(-,s') & \longrightarrow & \mathcal{S}(-,t') \end{array}$$

that is with chain maps, where we identify two if the difference of the maps $\mathcal{S}(-,t) \longrightarrow \mathcal{S}(-,t')$ factors through $\mathcal{S}(-,s')$. Thus the functor $B(\mathcal{S}) \longrightarrow A(\mathcal{S})$ is fully faithful. □

COROLLARY 5.1.15. *Suppose $\mathcal{S}$ is a triangulated category with small Hom–sets. Then the abelian category $A(\mathcal{S})$ has small Hom–sets.*

Proof: In its description as $B(\mathcal{S})$, this is clear; there is only a set of equivalence classes of diagrams

$$\begin{array}{ccc} s & \longrightarrow & t \\ \downarrow & & \downarrow \\ s' & \longrightarrow & t'. \end{array}$$

□

REMARK 5.1.16. Note that the category $\mathcal{C}at\big(\mathcal{S}^{op}, \mathcal{A}b\big)$ need not have small Hom–sets, even if $\mathcal{S}$ does.

LEMMA 5.1.17. *The Yoneda map $\mathcal{S} \longrightarrow A(\mathcal{S})$, sending s to the representable functor $\mathcal{S}(-,s)$, is a homological functor.*

Proof: $A(\mathcal{S})$ is an exact subcategory of $\mathcal{C}at\big(\mathcal{S}^{op}, \mathcal{A}b\big)$. Exact sequences in the two categories agree. It therefore suffices to show that the functor $\mathcal{S} \longrightarrow \mathcal{C}at\big(\mathcal{S}^{op}, \mathcal{A}b\big)$ is homological. But this is clear: given a triangle in $\mathcal{S}$

$$r \longrightarrow s \longrightarrow t \longrightarrow \Sigma r,$$

the sequence

$$\mathcal{S}(-,r) \longrightarrow \mathcal{S}(-,s) \longrightarrow \mathcal{S}(-,t)$$

is exact by Lemma 1.1.10. □

THEOREM 5.1.18. *Let $\mathcal{S}$ be a triangulated category. The functor $\mathcal{S} \longrightarrow A(\mathcal{S})$ is a universal homological functor. Suppose we are given a homological functor $H : \mathcal{S} \longrightarrow A$, where A is some abelian category. There is, up to canonical isomorphism, a unique exact functor of abelian categories $A(\mathcal{S}) \longrightarrow A$ so that the composite*

$$\mathcal{S} \longrightarrow A(\mathcal{S}) \xrightarrow{\exists !} A$$

is H. Furthermore, any natural transformation of homological functors $\mathcal{S} \longrightarrow A$ factors uniquely through a natural transformation of the asociated exact functors $A(\mathcal{S}) \longrightarrow A$.

Proof: Given an abelian category B with enough projectives, any additive functor on a generating subcategory of projectives extends uniquely, up to canonical isomorphism, to a right exact functor on B. This is standard. In our case, we have $\mathcal{S} \subset A(\mathcal{S})$ is the full subcategory of representable functors. The objects of $\mathcal{S}$ are projective objects when viewed in $A(\mathcal{S})$, and we are given an additive functor $H : \mathcal{S} \longrightarrow A$. Therefore H extends uniquely to a right exact functor

$$A(\mathcal{S}) \xrightarrow{\widetilde{H}} A.$$

Given an object F of $A(\mathcal{S})$ and a projective presentation

$$\mathcal{S}(-,s) \longrightarrow \mathcal{S}(-,t) \longrightarrow F(-) \longrightarrow 0,$$

$\widetilde{H}(F)$ is defined to be the cokernel of

$$H(s) \longrightarrow H(t).$$

If H is assumed not only additive but *homological*, we need to prove that $\widetilde{H}$ is left exact. Note that it is obvious that the composite of $\widetilde{H}$ with $\mathcal{S} \longrightarrow A(\mathcal{S})$ is H.

Suppose first that we are given an exact sequence in $A(\mathcal{S})$ of the special form

$$0 \longrightarrow F'(-) \longrightarrow \mathcal{S}(-,f) \longrightarrow F(-) \longrightarrow 0.$$

We want to begin by showing that $\widetilde{H}$ takes these to exact (as opposed to only right exact) sequences. Since $F' \in A(\mathcal{S})$, we may choose a surjection $\mathcal{S}(-,f') \longrightarrow F'(-)$. In other words, we have a resolution of F

$$\mathcal{S}(-,f') \xrightarrow{\phi} \mathcal{S}(-,f) \longrightarrow F(-) \longrightarrow 0,$$

and F' is identified as the image of ϕ. Complete $f' \longrightarrow f$ to a triangle

$$f'' \longrightarrow f' \longrightarrow f \longrightarrow \Sigma f''.$$

The sequence

$$\mathcal{S}(-,f'') \xrightarrow{\rho} \mathcal{S}(-,f') \xrightarrow{\phi} \mathcal{S}(-,f) \longrightarrow F(-) \longrightarrow 0$$

is exact in $A(\mathcal{S})$, the functor F is the cokernel of ϕ and the functor F' the cokernel of ρ. But the functor H is homological; it takes the triangle

$$f'' \xrightarrow{\rho} f' \xrightarrow{\phi} f \longrightarrow \Sigma f''$$

to an exact sequence

$$H(f'') \xrightarrow{H(\rho)} H(f') \xrightarrow{H(\phi)} H(f)$$

and $\widetilde{H}(F) = \operatorname{coker}(H(\phi))$ while $\widetilde{H}(F') = \operatorname{coker}(H(\rho))$. We immediately deduce an exact sequence

$$0 \longrightarrow \widetilde{H}(F') \longrightarrow H(f) \longrightarrow \widetilde{H}(F) \longrightarrow 0$$

as desired.

Suppose now that we are given a general exact sequence in $A(\mathcal{S})$

$$0 \longrightarrow F(-) \longrightarrow G(-) \longrightarrow K(-) \longrightarrow 0.$$

By Lemma 5.1.8 we can produce in $A(\mathcal{S})$ a 3×3 diagram with exact rows and columns

$$\begin{array}{ccccccccc}
 & & 0 & & 0 & & 0 & & \\
 & & \downarrow & & \downarrow & & \downarrow & & \\
0 & \longrightarrow & F'(-) & \longrightarrow & G'(-) & \longrightarrow & K'(-) & \longrightarrow & 0 \\
 & & \downarrow & & \downarrow & & \downarrow & & \\
0 & \longrightarrow & \mathcal{S}(-,f) & \longrightarrow & \mathcal{S}(-,g) & \longrightarrow & \mathcal{S}(-,k) & \longrightarrow & 0 \\
 & & \downarrow & & \downarrow & & \downarrow & & \\
0 & \longrightarrow & F(-) & \longrightarrow & G(-) & \longrightarrow & K(-) & \longrightarrow & 0 \\
 & & \downarrow & & \downarrow & & \downarrow & & \\
 & & 0 & & 0 & & 0 & &
\end{array}$$

Note that for the sequence of projectives

$$0 \longrightarrow \mathcal{S}(-,f) \longrightarrow \mathcal{S}(-,g) \longrightarrow \mathcal{S}(-,k) \longrightarrow 0$$

to be exact it must actually be split exact. Applying the functor $\widetilde{H}$ to this diagram we get a diagram with exact rows and columns

$$\begin{array}{ccccccccc}
 & & 0 & & 0 & & 0 & & \\
 & & \downarrow & & \downarrow & & \downarrow & & \\
 & & \widetilde{H}(F') & \longrightarrow & \widetilde{H}(G') & \longrightarrow & \widetilde{H}(K') & \longrightarrow & 0 \\
 & & \downarrow & & \downarrow & & \downarrow & & \\
0 & \longrightarrow & H(f) & \longrightarrow & H(g) & \longrightarrow & H(k) & \longrightarrow & 0 \\
 & & \downarrow & & \downarrow & & \downarrow & & \\
 & & \widetilde{H}(F) & \longrightarrow & \widetilde{H}(G) & \longrightarrow & \widetilde{H}(K) & \longrightarrow & 0 \\
 & & \downarrow & & \downarrow & & \downarrow & & \\
 & & 0 & & 0 & & 0 & &
\end{array}$$

The exactness of the columns is the case just discussed. The middle row is exact because it is split exact. The other two rows are *à priori* only right exact.

But the two maps

$$\begin{array}{ccc} \widetilde{H}(F') & & \\ \downarrow & & \\ H(f) & \longrightarrow & H(g) \end{array}$$

are both injective, hence so is their composite. The commutativity of

$$\begin{array}{ccc} \widetilde{H}(F') & \longrightarrow & \widetilde{H}(G') \\ \downarrow & & \downarrow \\ H(f) & \longrightarrow & H(g) \end{array}$$

tells us that the map $\widetilde{H}(F') \longrightarrow \widetilde{H}(G')$ must be injective. Therefore we deduce the exactness of the rows and columns in the commutative diagram

$$\begin{array}{ccccccccc} & & 0 & & 0 & & 0 & & \\ & & \downarrow & & \downarrow & & \downarrow & & \\ 0 & \longrightarrow & \widetilde{H}(F') & \longrightarrow & \widetilde{H}(G') & \longrightarrow & \widetilde{H}(K') & \longrightarrow & 0 \\ & & \downarrow & & \downarrow & & \downarrow & & \\ 0 & \longrightarrow & H(f) & \longrightarrow & H(g) & \longrightarrow & H(k) & \longrightarrow & 0 \\ & & \downarrow & & \downarrow & & \downarrow & & \\ & & \widetilde{H}(F) & \longrightarrow & \widetilde{H}(G) & \longrightarrow & \widetilde{H}(K) & \longrightarrow & 0 \\ & & \downarrow & & \downarrow & & \downarrow & & \\ & & 0 & & 0 & & 0 & & \end{array}$$

and the 3×3 lemma tells us the bottom row must also be exact.

So far, we have proved that a homological functor $H : \mathcal{S} \longrightarrow \mathcal{A}$ extends uniquely, up to canonical isomorphism, to an exact functor $\widetilde{H} : A(\mathcal{S}) \longrightarrow \mathcal{A}$. The statement about the extensions of natural tranformations is easy, and we leave it to the reader. □

REMARK 5.1.19. The universal property of the homological functor $\mathcal{S} \longrightarrow A(\mathcal{S})$ is clearly self–dual. In other words, the dual $\mathcal{S}^{op} \longrightarrow \{A(\mathcal{S})\}^{op}$ must satisfy the same property. Despite appearances, our construction must be self–dual. In the next few lemmas we elaborate on this point.

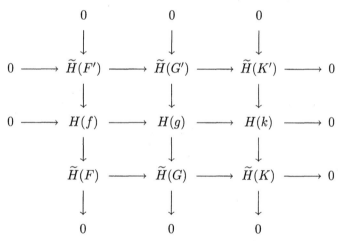

DEFINITION 5.1.20. *Define the category $C(\mathcal{S})$ as follows. The objects are triangles in $\mathcal{S}$*

$$r \longrightarrow s \longrightarrow t \longrightarrow \Sigma r$$

and the morphisms are equivalence classes of morphisms of triangles

$$\begin{array}{ccccccc} r & \longrightarrow & s & \longrightarrow & t & \longrightarrow & \Sigma r \\ \downarrow & & \downarrow & & \downarrow & & \downarrow \\ r' & \longrightarrow & s' & \longrightarrow & t' & \longrightarrow & \Sigma r' \end{array}$$

The equivalence relation is additive. A morphism is equivalent to zero if in the commutative square

$$\begin{array}{ccc} s & \longrightarrow & t \\ \downarrow & & \downarrow \\ s' & \longrightarrow & t' \end{array}$$

the equal composites

$$\begin{array}{ccc} s & & \\ \downarrow & & \\ s' & \longrightarrow & t' \end{array} \qquad\qquad \begin{array}{ccc} s & \longrightarrow & t \\ & & \downarrow \\ & & t' \end{array}$$

both vanish.

LEMMA 5.1.21. *Define a functor $C(\mathcal{S}) \longrightarrow B(\mathcal{S})$ by taking the triangle*

$$r \longrightarrow s \longrightarrow t \longrightarrow \Sigma r$$

to the pair

$$r \longrightarrow s.$$

This functor is an equivalence of categories.

Proof: First let us note that the functor is well–defined. If a morphism in $C(\mathcal{S})$ is equivalent to zero, we must show that its image in $B(\mathcal{S})$ is also equivalent to zero. Take therefore a morphism in $C(\mathcal{S})$ equivalent to zero

$$\begin{array}{ccccccc} r & \longrightarrow & s & \longrightarrow & t & \longrightarrow & \Sigma r \\ \downarrow & & \downarrow & & \downarrow & & \downarrow \\ r' & \longrightarrow & s' & \longrightarrow & t' & \longrightarrow & \Sigma r' \end{array}$$

The fact that it is equivalent to zero means that the composite

$$\begin{array}{ccc} s & & \\ \downarrow & & \\ s' & \longrightarrow & t' \end{array}$$

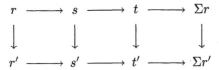

vanishes. But

$$r' \longrightarrow s' \longrightarrow t' \longrightarrow \Sigma r'$$

is a triangle, and hence the map $s \longrightarrow s'$ must factor as $s \longrightarrow r' \longrightarrow s'$. In other words, the square

$$\begin{array}{ccc} r & \longrightarrow & s \\ \downarrow & & \downarrow \\ r' & \longrightarrow & s' \end{array}$$

defines a morphism in $B(\mathcal{S})$ equivalent to zero.

Every object in $B(\mathcal{S})$, that is every $\{r \to s\} \in \mathcal{S}$, can be completed to a triangle

$$r \longrightarrow s \longrightarrow t \longrightarrow \Sigma r$$

and therefore lies in the image of the functor $C(\mathcal{S}) \longrightarrow B(\mathcal{S})$. Any commutative square

$$\begin{array}{ccc} r & \longrightarrow & s \\ \downarrow & & \downarrow \\ r' & \longrightarrow & s' \end{array}$$

can be completed to a morphism of triangles

$$\begin{array}{ccccccc} r & \longrightarrow & s & \longrightarrow & t & \longrightarrow & \Sigma r \\ \downarrow & & \downarrow & & \downarrow & & \downarrow \\ r' & \longrightarrow & s' & \longrightarrow & t' & \longrightarrow & \Sigma r' \end{array}$$

so the functor $C(\mathcal{S}) \longrightarrow B(\mathcal{S})$ is full. But also, if

$$\begin{array}{ccc} r & \longrightarrow & s \\ \downarrow & & \downarrow \\ r' & \longrightarrow & s' \end{array}$$

is equivalent to zero, then the map $s \longrightarrow s'$ factors through $r' \longrightarrow s'$, and in any completion to a morphism of triangles

$$\begin{array}{ccccccc} r & \longrightarrow & s & \longrightarrow & t & \longrightarrow & \Sigma r \\ \downarrow & & \downarrow & & \downarrow & & \downarrow \\ r' & \longrightarrow & s' & \longrightarrow & t' & \longrightarrow & \Sigma r' \end{array}$$

the map

$$\begin{array}{ccc} s & & \\ \downarrow & & \\ s' & \longrightarrow & t' \end{array}$$

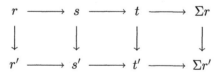

will have to vanish. The functor $C(\mathcal{S}) \longrightarrow B(\mathcal{S})$ is fully faithful, and surjective on objects. Hence it is an equivalence of categories. □

REMARK 5.1.22. The dual of the equivalence $C(\mathcal{S}) \longrightarrow B(\mathcal{S})$ gives a functor $C(\mathcal{S}^{op}) \longrightarrow B(\mathcal{S}^{op})$, which must also be an equivalence. Clearly $C(\mathcal{S}^{op}) = \{C(\mathcal{S})\}^{op}$, the construction being self–dual. We deduce that the functor taking a triangle

$$r \longrightarrow s \longrightarrow t \longrightarrow \Sigma r$$

to the morphism

$$t \longrightarrow \Sigma r$$

is an equivalence of $C(\mathcal{S})$ with $\{B(\mathcal{S}^{op})\}^{op}$. The objects of $C(\mathcal{S})$ of the form

$$0 \longrightarrow s \stackrel{1}{\longrightarrow} s \longrightarrow 0$$

map in $B(\mathcal{S})$ to the projective objects

$$0 \longrightarrow s.$$

Dually, they must map to the projective objects

$$s \longrightarrow 0$$

in $B(\mathcal{S}^{op})$. Since $C(\mathcal{S})$ is naturally the dual of $B(\mathcal{S}^{op})$, the objects

$$0 \longrightarrow s \stackrel{1}{\longrightarrow} s \longrightarrow 0$$

are not only projective objects in $C(\mathcal{S})$, but they are also injective objects(=projective objects in the dual). We conclude

COROLLARY 5.1.23. *The representable functors $\mathcal{S}(-,s)$ are not only projective objects in $A(\mathcal{S})$, but also injective objects. Any projective or injective object is a direct summand of some $\mathcal{S}(-,s)$. If every idempotent in the category $\mathcal{S}$ is split, then the projective(=injective) objects of $A(\mathcal{S})$ are precisely the representable functors $\mathcal{S}(-,s)$. Any object $F \in A(\mathcal{S})$ has both a projective presentation*

$$\mathcal{S}(-,s) \longrightarrow \mathcal{S}(-,t) \longrightarrow F(-) \longrightarrow 0$$

and an injective copresentation

$$\mathcal{S}(-,s') \longleftarrow \mathcal{S}(-,t') \longleftarrow F(-) \longleftarrow 0$$

In other words, the category $A(\mathcal{S}) = B(\mathcal{S}) = C(\mathcal{S})$ is a Frobenius abelian category. It is an abelian category with enough projectives and enough injectives, where it so happens that projectives and injectives coincide.

LEMMA 5.1.24. *Let β be an infinite cardinal. Suppose the category $\mathcal{S}$ satisfies [TR5(β)]. That is, coproducts of fewer than β objects exist in $\mathcal{S}$. Then the category $A(\mathcal{S})$ satisfies [AB3(β)]; that is, it is closed with respect to coproducts of $< \beta$ objects. The universal homological functor*

$\mathcal{S} \longrightarrow A(\mathcal{S})$ respects coproducts of $< \beta$ objects. Furthermore, a homological functor $\mathcal{S} \longrightarrow A$ respects coproducts of of $< \beta$ objects if and only if the induced exact functor $A(\mathcal{S}) \longrightarrow A$ of Theorem 5.1.18 does.

Proof: This is easier to see in the description $B(\mathcal{S})$ or $C(\mathcal{S})$ of the category. Let us do it for $C(\mathcal{S})$.

Suppose that we have a set of $< \beta$ objects in $C(\mathcal{S})$. That is, we have a set Λ, of cardinality $< \beta$, and for every $\lambda \in \Lambda$ an object of $C(\mathcal{S})$, i.e. a triangle in $\mathcal{S}$

$$r_\lambda \longrightarrow s_\lambda \longrightarrow t_\lambda \longrightarrow \Sigma r_\lambda.$$

Since the category $\mathcal{S}$ satisfies [TR5(β)], we can form the coproduct of these triangles

$$\coprod_{\lambda\in\Lambda} r_\lambda \longrightarrow \coprod_{\lambda\in\Lambda} s_\lambda \longrightarrow \coprod_{\lambda\in\Lambda} t_\lambda \longrightarrow \Sigma\left\{\coprod_{\lambda\in\Lambda} r_\lambda\right\}.$$

By Proposition 1.2.1 this is a triangle, that is an object of $C(\mathcal{S})$. The reader will easily see that this object satisfies the universal property of a coproduct in the category $C(\mathcal{S})$.

In the special case of triangles of the form

$$0 \longrightarrow s_\lambda \stackrel{1}{\longrightarrow} s_\lambda \longrightarrow 0,$$

that is objects in the image of the universal homological functor $\mathcal{S} \longrightarrow C(\mathcal{S})$, their coproduct is

$$0 \longrightarrow \coprod_{\lambda\in\Lambda} s_\lambda \stackrel{1}{\longrightarrow} \coprod_{\lambda\in\Lambda} s_\lambda \longrightarrow 0$$

and it immediately follows that the functor $\mathcal{S} \longrightarrow C(\mathcal{S})$ respects coproducts of $< \beta$ objects.

Finally, we need to show that a homological functor $\mathcal{S} \longrightarrow A$, which by Theorem 5.1.18 factorises uniquely as

$$\mathcal{S} \longrightarrow A(\mathcal{S}) \stackrel{\exists!}{\longrightarrow} A,$$

preserves coproducts of $< \beta$ objects if and only if $A(\mathcal{S}) \longrightarrow A$ does. The "if" part is trivial. We know by the above that $\mathcal{S} \longrightarrow A(\mathcal{S})$ preserves coproducts of $< \beta$ objects. If $A(\mathcal{S}) \longrightarrow A$ also does, then so does the composite

$$\mathcal{S} \longrightarrow A(\mathcal{S}) \longrightarrow A.$$

Suppose therefore that $H : \mathcal{S} \longrightarrow A$ is a homological functor preserving coproducts of $< \beta$ objects. We need to show that so does the induced functor $\widetilde{H} : A(\mathcal{S}) \longrightarrow A$. In its realisation $\widetilde{H} : B(\mathcal{S}) \longrightarrow A$, the induced exact functor takes the object $\{r \to s\} \in B(\mathcal{S})$ to the cokernel of the map $H(r) \longrightarrow H(s)$. That is, we have an exact sequence

$$H(r) \longrightarrow H(s) \longrightarrow \widetilde{H}(\{r \longrightarrow s\}) \longrightarrow 0.$$

Suppose Λ is a set of cardinality $< \beta$, and for $\lambda \in \Lambda$, we have objects $\{r_\lambda \to s_\lambda\} \in B(\mathcal{S})$. The coproduct in $B(\mathcal{S})$ is the object

$$\coprod_{\lambda\in\Lambda} r_\lambda \longrightarrow \coprod_{\lambda\in\Lambda} s_\lambda.$$

The functor $B(\mathcal{S}) \longrightarrow A$ takes it to the cokernel of

$$H\left(\coprod_{\lambda\in\Lambda} r_\lambda\right) \longrightarrow H\left(\coprod_{\lambda\in\Lambda} s_\lambda\right);$$

we have an exact sequence

$$H\left(\coprod_{\lambda\in\Lambda} r_\lambda\right) \longrightarrow H\left(\coprod_{\lambda\in\Lambda} s_\lambda\right) \longrightarrow \widetilde{H}\left(\coprod_{\lambda\in\Lambda}\{r_\lambda \to s_\lambda\}\right) \longrightarrow 0.$$

On the other hand, the functor H respects coproducts; in the commutative square below, the vertical maps are isomorphisms

$$\begin{array}{ccc}
\coprod_{\lambda\in\Lambda} H(r_\lambda) & \longrightarrow & \coprod_{\lambda\in\Lambda} H(s_\lambda) \\
\downarrow \wr & & \downarrow \wr \\
H\left(\coprod_{\lambda\in\Lambda} r_\lambda\right) & \longrightarrow & H\left(\coprod_{\lambda\in\Lambda} s_\lambda\right).
\end{array}$$

We deduce a commutative diagram with exact rows

$$\begin{array}{ccccccc}
\coprod_{\lambda\in\Lambda} H(r_\lambda) & \longrightarrow & \coprod_{\lambda\in\Lambda} H(s_\lambda) & \longrightarrow & \coprod_{\lambda\in\Lambda} \widetilde{H}(\{r_\lambda \to s_\lambda\}) & \longrightarrow & 0 \\
\downarrow \wr & & \downarrow \wr & & h\downarrow & & \\
H\left(\coprod_{\lambda\in\Lambda} r_\lambda\right) & \longrightarrow & H\left(\coprod_{\lambda\in\Lambda} s_\lambda\right) & \longrightarrow & \widetilde{H}\left(\coprod_{\lambda\in\Lambda}\{r_\lambda \to s_\lambda\}\right) & \longrightarrow & 0.
\end{array}$$

The exactness of the top row is because coproducts are right exact, in any abelian category $\mathcal{A}$. The 5–lemma now tells us that the map h must be an isomorphism, that is $\widetilde{H}$ preserves coproducts of $< \beta$ objects. □

REMARK 5.1.25. Dually, assume $\mathcal{S}$ is a triangulated category satisfying [TR5*(β)]. Then the category $A(\mathcal{S})$ satisfies [AB3*(β)], and the homological functor $\mathcal{S} \longrightarrow A(\mathcal{S})$ respects products of $< \beta$ objects. A homological functor $\mathcal{S} \longrightarrow A$ preserves products of $< \beta$ objects if and only if the exact functor $A(\mathcal{S}) \longrightarrow A$ does.

5.2. Subobjects and quotient objects in $A(\mathcal{S})$

In any abelian category, it is customary to study the behavior of subobjects and quotient objects. An abelian category is called *well–powered* if any object has a small set of subobjects, or equivalently a small set of quotient objects. Well–powered abelian categories are reasonable. In this sense, the categories $A(\mathcal{S})$ are very unreasonable. In this section, we propose to study the elementary properties of subobjects and quotient objects in the abelian category $A(\mathcal{S})$, where $\mathcal{S}$ is a triangulated category. In Appendix C we will show by example that even the simplest $\mathcal{S}$ may have an $A(\mathcal{S})$ which is not well–powered.

For the purpose of our study, it is convenient to introduce yet another model for the category $A(\mathcal{S})$.

DEFINITION 5.2.1. *Let $\mathcal{S}$ be a triangulated category with small* Hom–*sets. The category $D(\mathcal{S})$ has for its objects the morphisms $\{s \to t\}$ in $\mathcal{S}$. A morphism $\{s \to t\} \longrightarrow \{s' \to t'\}$ in $D(\mathcal{S})$ is an equivalence class of commutative squares in $\mathcal{S}$*

$$\begin{array}{ccc} s & \longrightarrow & t \\ \downarrow & & \downarrow \\ s' & \longrightarrow & t'. \end{array}$$

The equivalence relation on such squares is additive, and a square is defined equivalent to zero if the equal composites

$$\begin{array}{ccc} s & & \\ \downarrow & & \\ s' & \longrightarrow & t' \end{array} \qquad\qquad \begin{array}{ccc} s & \longrightarrow & t \\ & & \downarrow \\ & & t' \end{array}$$

both vanish.

LEMMA 5.2.2. *Let $\mathcal{S}$ be a triangulated category with small* Hom–*sets. There is a natural functor $C(\mathcal{S}) \longrightarrow D(\mathcal{S})$. It takes an object of $C(\mathcal{S})$, that is a triangle in $\mathcal{S}$*

$$r \longrightarrow s \longrightarrow t \longrightarrow \Sigma r$$

to the morphism

$$s \longrightarrow t,$$

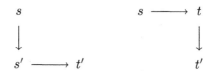

which defines an object in $D(\mathcal{S})$. It takes a morphism in $C(\mathcal{S})$, that is an equivalence class of morphisms of triangles

$$\begin{array}{ccccccc} r & \longrightarrow & s & \longrightarrow & t & \longrightarrow & \Sigma r \\ \downarrow & & \downarrow & & \downarrow & & \downarrow \\ r' & \longrightarrow & s' & \longrightarrow & t' & \longrightarrow & \Sigma r' \end{array}$$

to the square

$$\begin{array}{ccc} s & \longrightarrow & t \\ \downarrow & & \downarrow \\ s' & \longrightarrow & t'. \end{array}$$

This functor is an equivalence of categories.

Proof: The functor is clearly well–defined; equivalent morphisms map to equivalent morphisms. In fact, the definition of the equivalence is the same in both $C(\mathcal{S})$ and $D(\mathcal{S})$. In both cases, a morphism is equivalent to zero if the equal composites

$$\begin{array}{ccc} s & & \\ \downarrow & & \\ s' & \longrightarrow & t' \end{array} \qquad\qquad \begin{array}{ccc} s & \longrightarrow & t \\ & & \downarrow \\ & & t' \end{array}$$

both vanish. Therefore the functor $C(\mathcal{S}) \longrightarrow D(\mathcal{S})$ is clearly faithful.

By [TR1], any morphism $\{s \to t\}$ in $\mathcal{S}$ may be completed to a triangle

$$r \longrightarrow s \longrightarrow t \longrightarrow \Sigma r;$$

hence the functor $C(\mathcal{S}) \longrightarrow D(\mathcal{S})$ is surjective on objects. By [TR3], any commutative diagram with triangles for rows

$$\begin{array}{ccccccc} r & \longrightarrow & s & \longrightarrow & t & \longrightarrow & \Sigma r \\ & & \downarrow & & \downarrow & & \\ r' & \longrightarrow & s' & \longrightarrow & t' & \longrightarrow & \Sigma r' \end{array}$$

may be completed to a morphism of triangles

$$\begin{array}{ccccccc} r & \longrightarrow & s & \longrightarrow & t & \longrightarrow & \Sigma r \\ \downarrow & & \downarrow & & \downarrow & & \downarrow \\ r' & \longrightarrow & s' & \longrightarrow & t' & \longrightarrow & \Sigma r'. \end{array}$$

Thus the functor $C(\mathcal{S}) \longrightarrow D(\mathcal{S})$ is surjective also on morphisms. It is a fully faithful functor, surjective on objects; therefore it gives an equivalence of categories. □

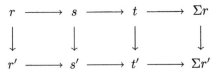

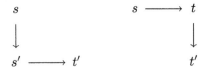

REMARK 5.2.3. By Lemma 5.1.21, there is an equivalence of categories $C(\mathcal{S}) \longrightarrow B(\mathcal{S})$. We remind the reader what this functor does. It takes an object of $C(\mathcal{S})$, that is a triangle in $\mathcal{S}$

$$r \longrightarrow s \longrightarrow t \longrightarrow \Sigma r,$$

to an object of $B(\mathcal{S})$, which was explicitly the morphism in $\mathcal{S}$

$$r \longrightarrow s.$$

By Proposition 5.1.14, there is an equivalence of categories $B(\mathcal{S}) \longrightarrow A(\mathcal{S})$, sending the object

$$r \longrightarrow s$$

to the cokernel of the map

$$\mathcal{S}(-,r) \longrightarrow \mathcal{S}(-,s).$$

Since for the triangle

$$r \longrightarrow s \longrightarrow t \longrightarrow \Sigma r$$

the sequence

$$\mathcal{S}(-,r) \xrightarrow{\alpha} \mathcal{S}(-,s) \xrightarrow{\beta} \mathcal{S}(-,t)$$

is exact, the composite $C(\mathcal{S}) \longrightarrow B(\mathcal{S}) \longrightarrow A(\mathcal{S})$ sends an object of $C(\mathcal{S})$, that is a triangle

$$r \longrightarrow s \longrightarrow t \longrightarrow \Sigma r,$$

to the cokernel of α, which is the same as the image of

$$\mathcal{S}(-,s) \xrightarrow{\beta} \mathcal{S}(-,t).$$

In other words, the equivalence $C(\mathcal{S}) \longrightarrow A(\mathcal{S})$ of Lemma 5.1.21 and Proposition 5.1.14 factors through the equivalence $C(\mathcal{S}) \longrightarrow D(\mathcal{S})$ of Lemma 5.2.2. We deduce an equivalence $D(\mathcal{S}) \longrightarrow A(\mathcal{S})$, taking an object $\{s \to t\}$ of $D(\mathcal{S})$ to the image of the map

$$\mathcal{S}(-,s) \longrightarrow \mathcal{S}(-,t).$$

In Remark 5.1.12 and Proposition 5.1.14, we saw that objects in the the category $B(\mathcal{S})$, that is morphisms $\{r \to s\}$ in $\mathcal{S}$, may be viewed as a projective presentation for the object $F(-)$ of $A(\mathcal{S})$. We have an exact sequence

$$\mathcal{S}(-,r) \longrightarrow \mathcal{S}(-,s) \longrightarrow F(-) \longrightarrow 0.$$

The objects in the category $D(\mathcal{S})$ may be viewed as an object $F(-)$ of $A(\mathcal{S})$, together with an embedding in an injective and a projective mapping onto it. We may think of an object $\{s \to t\}$ in $D(\mathcal{S})$ as

$$\mathcal{S}(-,s) \xrightarrow{\phi} F(-) \xrightarrow{\theta} \mathcal{S}(-,t)$$

where ϕ is surjective and θ injective. The natural functor $D(\mathcal{S}) \longrightarrow A(\mathcal{S})$ sends $\{s \to t\}$ in $D(\mathcal{S})$ to the image of

$$\mathcal{S}(-,s) \longrightarrow \mathcal{S}(-,t).$$

LEMMA 5.2.4. *Let $\mathcal{S}$ be a triangulated category with small* Hom*-sets. Let s be an object of $\mathcal{S}$. Via the universal homological functor $\mathcal{S} \longrightarrow A(\mathcal{S}) = D(\mathcal{S})$, we view s as an object of $D(\mathcal{S})$. Any quotient object of s can be presented as an object $\{s \to t\}$ of $D(\mathcal{S})$, for some t and some morphism $s \longrightarrow t$ in $\mathcal{S}$.*

Proof: Suppose we are given a quotient object of s, that is a surjective map $s \to F$. Choose any embedding of F in an injective object $t \in \mathcal{S}$; then F is the image of $\{s \to t\}$, in $D(\mathcal{S})$. □

LEMMA 5.2.5. *Let $\mathcal{S}$ be a triangulated category with small* Hom*-sets. Let s be an object of $\mathcal{S}$. Let $\{s \to t\}$ and $\{s \to t'\}$ be two quotient objects of s in $D(\mathcal{S})$. These quotients are isomorphic if and only if there exist maps $t \longrightarrow t'$ and $t' \longrightarrow t$ rendering commutative the diagrams*

$$\begin{array}{ccc} s & \longrightarrow & t \\ {\scriptstyle 1}\downarrow & & \downarrow \\ s & \longrightarrow & t' \end{array} \qquad \begin{array}{ccc} s & \longrightarrow & t' \\ {\scriptstyle 1}\downarrow & & \downarrow \\ s & \longrightarrow & t \end{array}$$

Proof: Suppose we are given isomorphic quotients $\{s \to t\}$ and $\{s \to t'\}$ in $D(\mathcal{S})$. That is, we have a quotient $s \longrightarrow F$ in $D(\mathcal{S})$, and two embeddings of it in injectives t and t'. Because t' is injective, the map $F \longrightarrow t'$ factors through the injection $F \longrightarrow t$; we deduce a commutative square

$$\begin{array}{ccc} s & \longrightarrow & t \\ {\scriptstyle 1}\downarrow & & \downarrow \\ s & \longrightarrow & t'. \end{array}$$

By symmetry, we also have a commutative square

$$\begin{array}{ccc} s & \longrightarrow & t' \\ {\scriptstyle 1}\downarrow & & \downarrow \\ s & \longrightarrow & t. \end{array}$$

Now suppose we have two commutative squares

$$\begin{array}{ccc} s & \longrightarrow & t \\ {\scriptstyle 1}\downarrow & & \downarrow \\ s & \longrightarrow & t' \end{array} \qquad \begin{array}{ccc} s & \longrightarrow & t' \\ {\scriptstyle 1}\downarrow & & \downarrow \\ s & \longrightarrow & t \end{array}$$

we need to show that the two quotients of s agree. But these commutative squares define morphisms in $D(\mathcal{S})$ of the quotient objects. The composites of the two morphisms, in both orders, give diagrams

$$\begin{array}{ccc} s & \longrightarrow & t \\ {\scriptstyle 1}\big\downarrow & & \big\downarrow{\scriptstyle 1+\rho} \\ s & \longrightarrow & t \end{array} \qquad \begin{array}{ccc} s & \longrightarrow & t' \\ {\scriptstyle 1}\big\downarrow & & \big\downarrow{\scriptstyle 1+\tau} \\ s & \longrightarrow & t' \end{array}$$

and these differ from the identity morphisms

$$\begin{array}{ccc} s & \longrightarrow & t \\ {\scriptstyle 1}\big\downarrow & & {\scriptstyle 1}\big\downarrow \\ s & \longrightarrow & t \end{array} \qquad \begin{array}{ccc} s & \longrightarrow & t' \\ {\scriptstyle 1}\big\downarrow & & {\scriptstyle 1}\big\downarrow \\ s & \longrightarrow & t' \end{array}$$

by

$$\begin{array}{ccc} s & \longrightarrow & t \\ {\scriptstyle 0}\big\downarrow & & {\scriptstyle \rho}\big\downarrow \\ s & \longrightarrow & t \end{array} \qquad \begin{array}{ccc} s & \longrightarrow & t' \\ {\scriptstyle 0}\big\downarrow & & {\scriptstyle \tau}\big\downarrow \\ s & \longrightarrow & t' \end{array}$$

Since it is clear that

$$\begin{array}{ccc} s & & \\ {\scriptstyle 0}\big\downarrow & & \\ s & \longrightarrow & t \end{array} \qquad \begin{array}{ccc} s & & \\ {\scriptstyle 0}\big\downarrow & & \\ s & \longrightarrow & t' \end{array}$$

vanish, it follows that the composites

$$\begin{array}{ccc} s & \longrightarrow & t \\ {\scriptstyle 1}\big\downarrow & & \big\downarrow{\scriptstyle 1+\rho} \\ s & \longrightarrow & t \end{array} \qquad \begin{array}{ccc} s & \longrightarrow & t' \\ {\scriptstyle 1}\big\downarrow & & \big\downarrow{\scriptstyle 1+\tau} \\ s & \longrightarrow & t' \end{array}$$

are equivalent to the identities. Thus the objects $\{s \to t\}$ and $\{s \to t'\}$ of $D(\mathcal{S})$ are isomorphic to each other in $D(\mathcal{S})$, and the isomorphisms respect the quotient map from s. □

PROPOSITION 5.2.6. *Let $\mathcal{S}$ be a triangulated category with small* Hom–*sets. Let s be an object of $\mathcal{S}$. The quotient objects of s in the category $A(\mathcal{S}) = D(\mathcal{S})$ can be represented as $\{s \to t\} \in D(\mathcal{S})$, and two pairs $\{s \to t\}$ and $\{s \to t'\}$ in $D(\mathcal{S})$ give isomorphic quotients if there are commutative diagrams in $\mathcal{S}$*

$$\begin{array}{ccc} s & \longrightarrow & t \\ {\scriptstyle 1}\big\downarrow & & \big\downarrow \\ s & \longrightarrow & t' \end{array} \qquad \begin{array}{ccc} s & \longrightarrow & t' \\ {\scriptstyle 1}\big\downarrow & & \big\downarrow \\ s & \longrightarrow & t. \end{array}$$

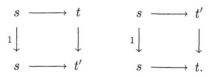

The subobjects of s may be represented as pairs $\{t \to s\} \in D(\mathcal{S})$, and two pairs $\{t \to s\}$ and $\{t' \to s\}$ in $D(\mathcal{S})$ give isomorphic subobjects if there are commutative diagrams in $\mathcal{S}$

$$\begin{array}{ccc} t & \longrightarrow & s \\ \downarrow & & \downarrow{\scriptstyle 1} \\ t' & \longrightarrow & s \end{array} \qquad \begin{array}{ccc} t' & \longrightarrow & s \\ \downarrow & & \downarrow{\scriptstyle 1} \\ t & \longrightarrow & s. \end{array}$$

Proof: For quotients, the existence of a representation $\{s \to t\}$ for a quotient of s is Lemma 5.2.4. The characterisation of isomorphic quotients is Lemma 5.2.5. The statements about subobjects are simply the duals of the statements about quotients. □

REMARK 5.2.7. In Proposition 5.2.6, we described the quotients and subobjects of $s \in \mathcal{S} \subset A(\mathcal{S})$. Since every projective object of $A(\mathcal{S})$ is a direct summand of $s \in \mathcal{S}$, we have described the quotient objects of any projective object of $\mathcal{S}$. Similarly, since every injective object of $A(\mathcal{S})$ is a direct summand of $s \in \mathcal{S}$, we have described the subobjects of injective objects of $A(\mathcal{S})$. Since any object of $A(\mathcal{S})$ may be embedded in an injective and is a quotient of a projective, we have in some sense described all quotients and subobjects of any object of $A(\mathcal{S})$.

CAUTION 5.2.8. An abelian category A is called *well–powered* if, for every object $a \in A$, the class of isomorphism classes of subobjects of a is a small set. Equivalently, the class of isomorphism classes of quotient objects of a is a small set. Subobjects and quotient objects are in 1–to–1 correspondence, with a subobject $b \subset a$ corresponding to the quotient a/b.

If $\mathcal{S}$ is a small category, then so is $A(\mathcal{S})$, and the category is obviously well–powered. But for almost all non–trivial large categories $\mathcal{S}$, the category $A(\mathcal{S})$ is decidedly *not* well–powered. In Appendix C, more precisely in Proposition C.3.2, Corollary C.3.3 and Remark C.3.4, we will show that if $D(\mathbb{Z})$ is the derived category of the category of abelian groups, then $A(D(\mathbb{Z}))$ is not well–powered.

5.3. The functoriality of $A(\mathcal{S})$

Given a triangulated category $\mathcal{T}$, we learned in Section 5.1 how to associate to it an abelian category $A(\mathcal{T})$.

LEMMA 5.3.1. *Let $\mathcal{S}$ and $\mathcal{T}$ be triangulated categories with small* Hom–*sets, $F : \mathcal{S} \longrightarrow \mathcal{T}$ a triangulated functor. Up to canonical isomorphism, there is a unique natural exact functor $A(F) : A(\mathcal{S}) \longrightarrow A(\mathcal{T})$ making* *commutative the diagram*

$$\begin{array}{ccc} \mathcal{S} & \xrightarrow{F} & \mathcal{T} \\ \downarrow & & \downarrow \\ A(\mathcal{S}) & \xrightarrow{A(F)} & A(\mathcal{T}). \end{array}$$

Let β be an infinite cardinal. If $\mathcal{S}$ and $\mathcal{T}$ satisfy [TR5(β)], and the functor $F : \mathcal{S} \longrightarrow \mathcal{T}$ preserves coproducts of $< \beta$ objects, then the induced functor $A(F) : A(\mathcal{S}) \longrightarrow A(\mathcal{T})$ also preserves coproducts of $< \beta$ objects.

Proof: We have a diagram

$$\begin{array}{ccc} \mathcal{S} & \xrightarrow{F} & \mathcal{T} \\ \downarrow & & \downarrow \\ A(\mathcal{S}) & & A(\mathcal{T}) \end{array}$$

and the composite $\mathcal{S} \longrightarrow \mathcal{T} \longrightarrow A(\mathcal{T})$ is a homological functor. By Theorem 5.1.18, any homological functor $\mathcal{S} \longrightarrow A$ factors uniquely through the universal homological functor $\mathcal{S} \longrightarrow A(\mathcal{S})$. There is a unique exact functor $A(F) : A(\mathcal{S}) \longrightarrow A(\mathcal{T})$ making commutative the diagram

$$\begin{array}{ccc} \mathcal{S} & \xrightarrow{F} & \mathcal{T} \\ \downarrow & & \downarrow \\ A(\mathcal{S}) & \xrightarrow{A(F)} & A(\mathcal{T}). \end{array}$$

It remains to prove the statement about coproducts: if the categories $\mathcal{S}$ and $\mathcal{T}$ both satisfy [TR5(β)] and the functor F preserves coproducts of $< \beta$ objects, we need to show that the functor $A(F) : A(\mathcal{S}) \longrightarrow A(\mathcal{T})$ also preserves coproducts of $< \beta$ objects. But in the composite

$$\begin{array}{ccc} \mathcal{S} & \xrightarrow{F} & \mathcal{T} \\ & & \downarrow \\ & & A(\mathcal{T}) \end{array}$$

the functor F preserves coproducts of $< \beta$ objects, by hypothesis. The functor $\mathcal{T} \longrightarrow A(\mathcal{T})$ preserves coproducts because $\mathcal{T}$ satisfies [TR5(β)], and by Lemma 5.1.24. Hence the composite preserves coproducts of $< \beta$ objects. But this composite is equal to

$$\begin{array}{ccc} \mathcal{S} & & \\ \downarrow & & \\ A(\mathcal{S}) & \xrightarrow{A(F)} & A(\mathcal{T}), \end{array}$$

and by Lemma 5.1.24 and the facts that $\mathcal{S}$ satisfies [TR5(β)], the composite preserves coproducts of $< \beta$ objects if and only if $A(F) : A(\mathcal{S}) \longrightarrow A(\mathcal{T})$ does. Hence $A(F) : A(\mathcal{S}) \longrightarrow A(\mathcal{T})$ preserves coproducts of $< \beta$ objects. □

REMARK 5.3.2. Dually, if $\mathcal{S}$ and $\mathcal{T}$ satisfy [TR5*(β)] and F preserves products of $< \beta$ objects, then $A(F) : A(\mathcal{S}) \longrightarrow A(\mathcal{T})$ preserves products of $< \beta$ objects.

REMARK 5.3.3. Let $F : \mathcal{S} \longrightarrow \mathcal{T}$ be a triangulated functor of triangulated categories with small *Hom*–sets. In Lemma 5.3.1 we defined the functor $A(F) : A(\mathcal{S}) \longrightarrow A(\mathcal{T})$. By Proposition 5.1.14 and Lemma 5.1.21 we know that $A(\mathcal{S}) \simeq B(\mathcal{S}) \simeq C(\mathcal{S})$. In terms of B and C, the functors $B(F) : B(\mathcal{S}) \longrightarrow B(\mathcal{T})$ and $C(F) : C(\mathcal{S}) \longrightarrow C(\mathcal{T})$ are very simple to describe explicitly. An object in $B(\mathcal{S})$ (respectively $C(\mathcal{S})$) is a map $\{s \longrightarrow s'\}$ in $\mathcal{S}$ (respectively a triangle $\{s \longrightarrow s' \longrightarrow s'' \longrightarrow \Sigma s\}$ in $\mathcal{S}$). The functor $B(F)$ (respectively $C(F)$) takes this to the morphism $\{Fs \longrightarrow Fs'\}$ in $\mathcal{T}$ (respectively to the triangle $\{Fs \longrightarrow Fs' \longrightarrow Fs'' \longrightarrow \Sigma Fs\}$ in $\mathcal{T}$).

LEMMA 5.3.4. *The assignment $A(-)$ is a lax*[1] *functor from the 2–category of triangulated categories and triangulated functors, to the 2–category of abelian categories and exact functors. It takes a triangulated category $\mathcal{S}$ to the abelian category $A(\mathcal{S})$, takes a triangulated functor F to the exact functor $A(F)$, and takes a natural trasformation ϕ to a natural transformation $A(\phi)$. It respects composition (in the lax sense) and identities.*

Proof: Obvious. □

REMARK 5.3.5. The functors $B(-)$ and $C(-)$ are strict functors of 2–categories, not just lax functors. This follows from the explicit description of $B(F)$ and $C(F)$ in Remark 5.3.3.

Next we wish to say something about the relation between adjoint functors between triangulated categories $\mathcal{S}$ and $\mathcal{T}$, and adjoint functors between $A(\mathcal{S})$ and $A(\mathcal{T})$. But first a lemma.

LEMMA 5.3.6. *Let $\mathcal{S}$ and $\mathcal{T}$ be triangulated categories, $F : \mathcal{S} \longrightarrow \mathcal{T}$ a triangulated functor. Suppose F has an adjoint (left or right) $G : \mathcal{T} \longrightarrow \mathcal{S}$. Then G is also a triangulated functor.*

Proof: We may assume G is a right adjoint, the case of left adjunction being dual. We are given that F commutes with Σ, up to natural isomorphism. That is

$$F\Sigma \quad = \quad \Sigma F.$$

[1]A lax functor between 2–categories only respects composition up to 2–isomorphisms.

Taking right adjoints of this identity and recalling that the right adjoint of Σ is Σ^{-1} (see the proof of Proposition 1.1.6), we deduce

$$\Sigma^{-1}G \quad = \quad G\Sigma^{-1}.$$

Hence G commutes with Σ^{-1}, and therefore also with Σ.

Let $X \longrightarrow Y \longrightarrow Z \longrightarrow \Sigma X$ be a triangle in $\mathcal{T}$; we need to show that $GX \longrightarrow GY \longrightarrow GZ \longrightarrow G\Sigma X$ is a triangle in $\mathcal{S}$. Complete $GX \longrightarrow GY$ to a triangle $GX \longrightarrow GY \longrightarrow C \longrightarrow \Sigma GX$ in $\mathcal{S}$. Because F is triangulated, $FGX \longrightarrow FGY \longrightarrow FC \longrightarrow \Sigma FGX$ is a triangle in $\mathcal{T}$. Let $\varepsilon_X : FGX \longrightarrow X$ and $\varepsilon_Y : FGY \longrightarrow Y$ be the counit of adjunction; it is the map corresponding to the identity $1 : GX \longrightarrow GX$ under the natural isomorphism

$$\mathcal{S}(GX, GX) \quad = \quad \mathcal{T}(FGX, X).$$

By the naturality of ε we have a commutative square in $\mathcal{T}$

$$\begin{array}{ccc} FGX & \longrightarrow & FGY \\ \Big\downarrow{\scriptstyle \varepsilon_X} & & \Big\downarrow{\scriptstyle \varepsilon_Y} \\ X & \longrightarrow & Y \end{array}$$

which we may complete to a morphism of triangles

$$\begin{array}{ccccccc} FGX & \longrightarrow & FGY & \longrightarrow & FC & \longrightarrow & \Sigma FGX \\ {\scriptstyle \varepsilon_X}\Big\downarrow & & {\scriptstyle \varepsilon_Y}\Big\downarrow & & {\scriptstyle \Theta}\Big\downarrow & & \Big\downarrow{\scriptstyle \varepsilon_{\Sigma X}} \\ X & \longrightarrow & Y & \longrightarrow & Z & \longrightarrow & \Sigma X. \end{array}$$

Let R be an object of $\mathcal{S}$. Then Θ defines a map

$$\mathcal{S}(R, C) \xrightarrow{\Theta \circ F(-)} \mathcal{T}(FR, Z),$$

and we deduce a commutative diagram with exact rows:

$$\begin{array}{ccccccccc} \mathcal{S}(R, GX) & \longrightarrow & \mathcal{S}(R, GY) & \longrightarrow & \mathcal{S}(R, C) & \longrightarrow & \mathcal{S}(R, \Sigma GX) & \longrightarrow & \mathcal{S}(R, \Sigma GY) \\ \| \Big\downarrow \wr & & \| \Big\downarrow \wr & & \Big\downarrow & & \| \Big\downarrow \wr & & \| \Big\downarrow \wr \\ \mathcal{T}(FR, X) & \longrightarrow & \mathcal{T}(FR, Y) & \longrightarrow & \mathcal{T}(FR, Z) & \longrightarrow & \mathcal{T}(FR, \Sigma X) & \longrightarrow & \mathcal{T}(FR, \Sigma Y) \end{array}$$

The 5–Lemma implies that $\mathcal{S}(R, C) \longrightarrow \mathcal{T}(FR, Z)$ is an isomorphism, i.e. that $\Theta : FC \longrightarrow Z$ is precisely the counit of adjunction. That is $C = GZ$, and the result follows. □

REMARK 5.3.7. It should be noted that the situation here is better than with abelian categories. If F is an exact functor of abelian categories with a right adjoint G, then G is in general only left exact. If G is the left adjoint of F, then it is in general only right exact. If we think of triangulated functors between triangulated categories as the natural analog of exact functors between abelian categories, then Lemma 5.3.6 is surprising.

LEMMA 5.3.8. *Suppose $F : \mathcal{S} \longrightarrow \mathcal{T}$ is a triangulated functor of triangulated categories. Suppose F has a right adjoint $G : \mathcal{T} \longrightarrow \mathcal{S}$. By Lemma 5.3.6, G is triangulated. By Lemma 5.3.1, both F and G induce exact functors between $A(\mathcal{S})$ and $A(\mathcal{T})$. We assert that these induced functors on abelian categories are also adjoint to each other. We have exact functors*

$$A(\mathcal{S}) \xrightarrow{A(F)} A(\mathcal{T}) \qquad\qquad A(\mathcal{T}) \xrightarrow{A(G)} A(\mathcal{S})$$

and $A(F)$ is left adjoint to $A(G)$.

Proof: Let us use the equivalent categories $B(\mathcal{S}) \simeq A(\mathcal{S})$, $B(\mathcal{T}) \simeq A(\mathcal{T})$. An object in $B(\mathcal{S})$ is a morphism $s \longrightarrow s'$ in $\mathcal{S}$. An object in $B(\mathcal{T})$ is a morphism $t \longrightarrow t'$ in $\mathcal{T}$. The object $B(F)\big[\{s \to s'\}\big]$ is just F applied to $\{s \to s'\}$. A morphism

$$B(F)\big[\{s \to s'\}\big] \longrightarrow \{t \to t'\}$$

is an equivalence class of commutative diagrams in $\mathcal{T}$

$$\begin{array}{ccc} Fs & \longrightarrow & Fs' \\ \downarrow & & \downarrow \\ t & \longrightarrow & t' \end{array}$$

where two diagrams are equivalent if the difference of the maps $Fs' \longrightarrow t'$ factors through $t \longrightarrow t'$. But by adjunction this is the same as an equivalence class of diagrams

$$\begin{array}{ccc} s & \longrightarrow & s' \\ \downarrow & & \downarrow \\ Gt & \longrightarrow & Gt'. \end{array}$$

There is therefore a natural 1–to–1 correspondence of maps

$$\begin{array}{ccc} B(F)\big[\{s \to s'\}\big] & \longrightarrow & \{t \to t'\} \\ \{s \to s'\} & \longrightarrow & B(G)\big[\{t \to t'\}\big]. \end{array}$$

$\square$

PROPOSITION 5.3.9. *Let $\mathcal{S}$ and $\mathcal{T}$ be triangulated categories, closed under splitting idempotents. The triangulated functor $F : \mathcal{S} \longrightarrow \mathcal{T}$ will have a right adjoint $G : \mathcal{T} \longrightarrow \mathcal{S}$ if and only if the exact functor $A(F) : A(\mathcal{S}) \longrightarrow A(\mathcal{T})$ has a right adjoint $A(G) : A(\mathcal{T}) \longrightarrow A(\mathcal{S})$.*

Proof: If $F : \mathcal{S} \longrightarrow \mathcal{T}$ has a right adjoint $G : \mathcal{T} \longrightarrow \mathcal{S}$, we know from Lemma 5.3.8 that $A(F) : A(\mathcal{S}) \longrightarrow A(\mathcal{T})$ has a right adjoint $A(G) : A(\mathcal{T}) \longrightarrow A(\mathcal{S})$. We need the converse. Suppose therefore that $A(F) : A(\mathcal{S}) \longrightarrow A(\mathcal{T})$ has a right adjoint $\widetilde{G} : A(\mathcal{T}) \longrightarrow A(\mathcal{S})$.

The functor $\widetilde{G} : A(\mathcal{T}) \longrightarrow A(\mathcal{S})$ has an exact left adjoint. It therefore takes injectives to injectives. But we are assuming that idempotents split both in $\mathcal{S}$ and in $\mathcal{T}$. By Corollary 5.1.23, the injectives in $A(\mathcal{S})$ (resp. $A(\mathcal{T})$) are $\mathcal{S} \subset A(\mathcal{S})$ (resp. $\mathcal{T} \subset A(\mathcal{T})$). Therefore $\widetilde{G} : A(\mathcal{T}) \longrightarrow A(\mathcal{S})$ restricts to a functor $G : \mathcal{T} \longrightarrow \mathcal{S}$. This functor is clearly right adjoint to F. By Lemma 5.3.6, this adjoint G must be a triangulated functor of triangulated categories. □

REMARK 5.3.10. The dual of Proposition 5.3.9 says that a triangulated functor $G : \mathcal{T} \longrightarrow \mathcal{S}$ will have a left adjoint if and only if the functor $A(G) : A(\mathcal{T}) \longrightarrow A(\mathcal{S})$ does. Thus, looking for adjoints to triangulated functors reduces to looking for adjoints to the associated exact functors.

The problem with this proposition is that it is nearly impossible to apply. Existence theorems for adjoints usually depend on the categories being well–powered. The reader is referred to Caution 5.2.8 and Appendix C, for a discussion of just how far the categories $A(\mathcal{T})$ are from being well–powered.

5.4. History of the results in Chapter 5

The definition of the category $A(\mathcal{S})$, its universal property for homological functors, and the fact that it is a Frobenius abelian category may all be found in Freyd's [**13**], more precisely in Section 3 on pages 127–133. Verdier's thesis, unpublished until very recently, also contains the results. See Sections 3.1 and 3.2, pages 135–144 of [**36**]. I think Freyd had the result first; although one can not be sure. Verdier submitted his thesis in 1967, Freyd's result was already in print by 1966. The treatment we give is more similar to Verdier's. The only result in the Chapter with a claim to originality is Proposition 5.3.9. We remind the reader: Proposition 5.3.9 asserts that a triangulated functor $F : \mathcal{S} \longrightarrow \mathcal{T}$ has an adjoint if and only if the induced functor $A(F) : A(\mathcal{S}) \longrightarrow A(\mathcal{T})$ does.

The analysis of subobjects and quotient objects of an object in $A(\mathcal{T})$, given in Section 5.3, was certainly known to the experts. I have no doubt that either Freyd or Verdier could easily have provided the same treatment. More recently, Strickland was certainly aware of the results, as was Grandis. Part of the purpose of Section 5.3 is to prepare the ground for Appendix C, in which we show that the objects of $A(\mathcal{T})$ may well have classes, not sets, of subobjects.

CHAPTER 6

The category $\mathcal{E}x\big(\mathcal{S}^{op}, \mathcal{A}b\big)$

6.1. $\mathcal{E}x\big(\mathcal{S}^{op}, \mathcal{A}b\big)$ is an abelian category satisfying [AB3] and [AB3*]

Let α be a regular cardinal. Throughout this Chapter, we fix a choice of such a cardinal α. Let $\mathcal{S}$ be a category, satisfying the following hypotheses

HYPOTHESIS 6.1.1. *The category $\mathcal{S}$ is said to satisfy hypothesis 6.1.1 if*

6.1.1.1. *$\mathcal{S}$ is an essentially small additive category.*

6.1.1.2. *The coproduct of fewer than α objects of $\mathcal{S}$ exists in $\mathcal{S}$.*

6.1.1.3. *Homotopy pullback squares exist in $\mathcal{S}$. That is, given a diagram in $\mathcal{S}$*

$$\begin{array}{ccc} & & x \\ & & \downarrow \\ x' & \longrightarrow & y \end{array}$$

it may be completed to a commutative square

$$\begin{array}{ccc} p & \longrightarrow & x \\ \downarrow & & \downarrow \\ x' & \longrightarrow & y \end{array}$$

so that any commutative square

$$\begin{array}{ccc} s & \longrightarrow & x \\ \downarrow & & \downarrow \\ x' & \longrightarrow & y \end{array}$$

is induced by a (non–unique) map $s \longrightarrow p$. The object p is called the homotopy pullback of the diagram

$$\begin{array}{ccc} & & x \\ & & \downarrow \\ x' & \longrightarrow & y \end{array}$$

and the homotopy pullback square

$$\begin{array}{ccc} p & \longrightarrow & x \\ \downarrow & & \downarrow \\ x' & \longrightarrow & y \end{array}$$

is unique up to (non–canonical) isomorphism.

6.1.1.4. *Coproducts of fewer than α homotopy pullback squares are homotopy pullback squares. In other words, let Λ be a set of cardinality $< \alpha$. If for every $\lambda \in \Lambda$ we are given a homotopy pullback square*

$$\begin{array}{ccc} p_\lambda & \longrightarrow & x_\lambda \\ \downarrow & & \downarrow \\ x'_\lambda & \longrightarrow & y_\lambda \end{array}$$

then the coproduct

$$\begin{array}{ccc} \coprod_{\lambda\in\Lambda} p_\lambda & \longrightarrow & \coprod_{\lambda\in\Lambda} x_\lambda \\ \downarrow & & \downarrow \\ \coprod_{\lambda\in\Lambda} x'_\lambda & \longrightarrow & \coprod_{\lambda\in\Lambda} y_\lambda \end{array}$$

is also a homotopy pullback square.

EXAMPLE 6.1.2. The example of most interest is when $\mathcal{S}$ is an essentially small triangulated category. Being essentially small, it satisfies 6.1.1.1. If it is closed under the formation of coproducts of fewer than α objects, then it satisfies also 6.1.1.2. The existence of homotopy pullback squares is automatic; see Definition 1.4.1 and Notation 1.4.2. Hence we automatically have 6.1.1.3 whenever $\mathcal{S}$ is triangulated. The fact that the coproducts of fewer than α homotopy pullback squares is a homotopy pullback square is also automatic for triangulated categories $\mathcal{S}$, following from Proposition 1.2.1. Hence 6.1.1.4 also comes for free, when $\mathcal{S}$ is triangulated.

As I said, this is the important example for us. We treat the slightly more general case only for the purpose of constructing certain counterexamples; see Section A.5. The reader not interested in counterexamples may safely restrict his attention to only essentially small triangulated categories $\mathcal{S}$, satisfying [TR5(α)]. Recall that $\mathcal{S}$ satisfies [TR5(α)] if it is closed under the formation of coproducts of fewer than α objects. The categories satisfying Hypothesis 6.1.1 are more general, but mostly this will play no role.

We remind the reader of Definition 5.1.1. Given an additive category $\mathcal{S}$, the category $\mathcal{C}at(\mathcal{S}^{op}, \mathcal{A}b)$ is the category whose objects are all additive functors $\mathcal{S}^{op} \longrightarrow \mathcal{A}b$, where $\mathcal{A}b$ is the category of all abelian groups. The morphisms in $\mathcal{C}at(\mathcal{S}^{op}, \mathcal{A}b)$ are the natural transformations.

The category $\mathcal{C}at(\mathcal{S}^{op}, \mathcal{A}b)$, being a functor category into the abelian category $\mathcal{A}b$, inherits the abelian structure of $\mathcal{A}b$. A short exact sequence in $\mathcal{C}at(\mathcal{S}^{op}, \mathcal{A}b)$ is three additive functors F', F and F'' from $\mathcal{S}^{op}$ to $\mathcal{A}b$, and two natural transformations

$$F' \Longrightarrow F, \qquad\qquad F \Longrightarrow F''$$

so that for every object $x \in \mathcal{S}$, the sequence

$$0 \longrightarrow F'(x) \longrightarrow F(x) \longrightarrow F''(x) \longrightarrow 0$$

is exact in $\mathcal{A}b$.

In this Chapter, we will be assuming that $\mathcal{S}$ satisfies Hypothesis 6.1.1, in particular is essentially small. It follows that the category $\mathcal{C}at(\mathcal{S}^{op}, \mathcal{A}b)$ has small Hom–sets.

DEFINITION 6.1.3. *Suppose $\mathcal{S}$ satisfies Hypothesis 6.1.1. The category $\mathcal{E}x(\mathcal{S}^{op}, \mathcal{A}b)$ is defined to be the full subcategory of $\mathcal{C}at(\mathcal{S}^{op}, \mathcal{A}b)$, whose objects are the functors $\mathcal{S}^{op} \longrightarrow \mathcal{A}b$ which take coproducts of fewer than α objects in $\mathcal{S}$ to products in $\mathcal{A}b$. In other words, let F be an additive functor $\mathcal{S}^{op} \longrightarrow \mathcal{A}b$. Then F will lie in the subcategory $\mathcal{E}x(\mathcal{S}^{op}, \mathcal{A}b) \subset \mathcal{C}at(\mathcal{S}^{op}, \mathcal{A}b)$ if, for every family $\{s_\lambda, \lambda \in \Lambda\}$ of fewer than α objects of $\mathcal{S}$, the natural map*

$$F\left(\coprod_{\lambda\in\Lambda} s_\lambda\right) \longrightarrow \prod_{\lambda\in\Lambda} F(s_\lambda)$$

is an isomorphism.

LEMMA 6.1.4. *Suppose $\mathcal{S}$ satisfies Hypothesis 6.1.1. Then the category $\mathcal{E}x(\mathcal{S}^{op}, \mathcal{A}b)$ is an abelian subcategory of $\mathcal{C}at(\mathcal{S}^{op}, \mathcal{A}b)$. That is, $\mathcal{E}x(\mathcal{S}^{op}, \mathcal{A}b)$ is an abelian category, and the inclusion*

$$\mathcal{E}x(\mathcal{S}^{op}, \mathcal{A}b) \qquad \subset \qquad \mathcal{C}at(\mathcal{S}^{op}, \mathcal{A}b)$$

is an exact functor.

Proof: Suppose $F \longrightarrow F'$ is a morphism in $\mathcal{E}x(\mathcal{S}^{op}, \mathcal{A}b)$. That is, F and F' are functors $\mathcal{S}^{op} \longrightarrow \mathcal{A}b$ taking coproducts of fewer than α objects to products, and $F \longrightarrow F'$ is a natural transformation. We need to show that the kernel and cokernel of the natural transformation, which are clearly objects of the big category $\mathcal{C}at(\mathcal{S}^{op}, \mathcal{A}b)$, actually lie in the subcategory $\mathcal{E}x(\mathcal{S}^{op}, \mathcal{A}b)$.

Complete the map $F \longrightarrow F'$ to an exact sequence in $\mathcal{C}at(\mathcal{S}^{op}, \mathcal{A}b)$

$$0 \longrightarrow K \longrightarrow F \longrightarrow F' \longrightarrow Q \longrightarrow 0.$$

Let $\{s_\lambda, \lambda \in \Lambda\}$ be a set of fewer than α objects in $\mathcal{S}$. Because F and F' lie in $\mathcal{E}x(\mathcal{S}^{op}, \mathcal{A}b)$, the natural maps

$$F\left(\coprod_{\lambda\in\Lambda} s_\lambda\right) \longrightarrow \prod_{\lambda\in\Lambda} F(s_\lambda)$$

$$F'\left(\coprod_{\lambda\in\Lambda} s_\lambda\right) \longrightarrow \prod_{\lambda\in\Lambda} F'(s_\lambda)$$

are both isomorphisms. We deduce that in the commutative square

$$\begin{array}{ccc} F\left(\coprod_{\lambda\in\Lambda} s_\lambda\right) & \longrightarrow & F'\left(\coprod_{\lambda\in\Lambda} s_\lambda\right) \\ \downarrow\wr & & \downarrow\wr \\ \prod_{\lambda\in\Lambda} F(s_\lambda) & \longrightarrow & \prod_{\lambda\in\Lambda} F'(s_\lambda) \end{array}$$

the vertical maps are both isomorphisms. But $\mathcal{A}b$ satisfies [AB4*]. Hence the product of the exact sequences

$$0 \longrightarrow K(s_\lambda) \longrightarrow F(s_\lambda) \longrightarrow F'(s_\lambda) \longrightarrow Q(s_\lambda) \longrightarrow 0$$

over $\lambda \in \Lambda$ is an exact sequence. In the comparison map

$$\begin{array}{ccccccc} K\left(\coprod_{\lambda\in\Lambda} s_\lambda\right) & \longrightarrow & F\left(\coprod_{\lambda\in\Lambda} s_\lambda\right) & \longrightarrow & F'\left(\coprod_{\lambda\in\Lambda} s_\lambda\right) & \longrightarrow & Q\left(\coprod_{\lambda\in\Lambda} s_\lambda\right) \\ \downarrow & & \downarrow\wr & & \downarrow\wr & & \downarrow \\ \prod_{\lambda\in\Lambda} K(s_\lambda) & \longrightarrow & \prod_{\lambda\in\Lambda} F(s_\lambda) & \longrightarrow & \prod_{\lambda\in\Lambda} F'(s_\lambda) & \longrightarrow & \prod_{\lambda\in\Lambda} Q(s_\lambda) \end{array}$$

both the top and bottom rows are exact. It easily follows that the natural maps

$$K\left(\coprod_{\lambda\in\Lambda} s_\lambda\right) \longrightarrow \prod_{\lambda\in\Lambda} K(s_\lambda)$$

$$Q\left(\coprod_{\lambda\in\Lambda} s_\lambda\right) \longrightarrow \prod_{\lambda\in\Lambda} Q(s_\lambda)$$

are both isomorphisms. □

LEMMA 6.1.5. *Suppose $\mathcal{S}$ satisfies Hypothesis 6.1.1. Then the category $\mathcal{E}x(\mathcal{S}^{op}, \mathcal{A}b)$ satisfies [AB3*]; it contains the product of any small set of its objects. Furthermore, the inclusion functor $\mathcal{E}x(\mathcal{S}^{op}, \mathcal{A}b) \subset \mathcal{C}at(\mathcal{S}^{op}, \mathcal{A}b)$ respects products.*

Proof: Let $\{F_\mu, \mu \in M\}$ be a set of objects in $\mathcal{E}x\big(\mathcal{S}^{op}, \mathcal{A}b\big)$. We can form the product in $\mathcal{C}at\big(\mathcal{S}^{op}, \mathcal{A}b\big)$ by the usual definition; for any $s \in \mathcal{S}$,

$$\left\{\prod_{\mu\in M} F_\mu\right\}(s) \quad = \quad \prod_{\mu\in M}\left\{F_\mu(s)\right\}.$$

What we need to show is that $\prod_{\mu\in M} F_\mu$ is an object of the category $\mathcal{E}x\big(\mathcal{S}^{op}, \mathcal{A}b\big)$.

Let therefore $\{s_\lambda, \lambda \in \Lambda\}$ be a family of fewer than α objects of $\mathcal{S}$. For each μ, the natural map

$$F_\mu\left(\coprod_{\lambda\in\Lambda} s_\lambda\right) \longrightarrow \prod_{\lambda\in\Lambda} F_\mu(s_\lambda)$$

is an isomorphism, because $F_\mu \in \mathcal{E}x\big(\mathcal{S}^{op}, \mathcal{A}b\big)$. Taking the product of these isomorphisms over all μ, we have that

$$\prod_{\mu\in M} F_\mu\left(\coprod_{\lambda\in\Lambda} s_\lambda\right) \longrightarrow \prod_{\lambda\in\Lambda}\left\{\prod_{\mu\in M} F_\mu(s_\lambda)\right\}$$

is an isomorphism, proving that $\prod_{\mu\in M} F_\mu$ is indeed an object of the category $\mathcal{E}x\big(\mathcal{S}^{op}, \mathcal{A}b\big)$. □

So far everything was very painless. Next we want to prove that the category $\mathcal{E}x\big(\mathcal{S}^{op}, \mathcal{A}b\big)$ has coproducts, and want to have a very explicit description of coproducts in $\mathcal{E}x\big(\mathcal{S}^{op}, \mathcal{A}b\big)$.

REMARK 6.1.6. The existence of coproducts in $\mathcal{E}x\big(\mathcal{S}^{op}, \mathcal{A}b\big)$ can be proved purely formally. One notes that the inclusion of $\mathcal{E}x\big(\mathcal{S}^{op}, \mathcal{A}b\big)$ into $\mathcal{C}at\big(\mathcal{S}^{op}, \mathcal{A}b\big)$ is exact and preserves products, hence preserves all inverse limits. Then one can show it has a left adjoint L; see for example Gabriel–Ulmer [**16**]. The coproduct of a family of objects in $\mathcal{E}x\big(\mathcal{S}^{op}, \mathcal{A}b\big)$ is the functor L applied to their coproduct in $\mathcal{C}at\big(\mathcal{S}^{op}, \mathcal{A}b\big)$. But as I said, we want an explicit description of the coproduct.

DEFINITION 6.1.7. *Suppose $\mathcal{S}$ satisfies Hypothesis 6.1.1. Let s be an object of $\mathcal{S}$. Let $\{F_\mu, \mu \in M\}$ be a set of objects in $\mathcal{E}x\big(\mathcal{S}^{op}, \mathcal{A}b\big)$. Define the set*

$$\left\{\bigvee_{\mu\in M} F_\mu\right\}(s) \quad = \quad \left\{\begin{array}{c}\textit{isomorphism classes of pairs}\\ s \longrightarrow \coprod_{\lambda\in\Lambda} s_\lambda, \qquad \beta \in \prod_{\lambda\in\Lambda} F_\lambda(s_\lambda)\\ \textit{with } \Lambda \subset M \textit{ of cardinality} < \alpha\end{array}\right\}$$

Note that since $\mathcal{S}$ is essentially small, $\left\{\bigvee_{\mu\in M} F_\mu\right\}(s)$ is indeed a small set. The idea is to construct $\left\{\coprod_{\mu\in M} F_\mu\right\}(s)$ by dividing $\left\{\bigvee_{\mu\in M} F_\mu\right\}(s)$ by a suitable equivalence relation.

DEFINITION 6.1.8. *Suppose $\mathcal{S}$ satisfies Hypothesis 6.1.1. Let s be an object of $\mathcal{S}$, and let $\{F_\mu, \mu \in M\}$ be a set of objects in $\mathcal{E}x(\mathcal{S}^{op}, \mathcal{A}b)$. Let us be given two elements of $\left\{\bigvee_{\mu\in M} F_\mu\right\}(s)$, that is pairs*

$$\begin{array}{ll} s \longrightarrow \coprod_{\lambda\in\Lambda} s_\lambda, & \beta \in \prod_{\lambda\in\Lambda} F_\lambda(s_\lambda) \\ s \longrightarrow \coprod_{\lambda\in\Lambda'} t_\lambda, & \beta' \in \prod_{\lambda\in\Lambda'} F_\lambda(t_\lambda) \end{array}$$

If $\lambda \notin \Lambda$, we adopt the convention that s_λ is defined to be zero, and similarly if $\lambda \notin \Lambda'$ define $t_\lambda = 0$. Recall that $\mathcal{S}$ is an additive category by Hypothesis 6.1.1.1, hence contains a zero object. The two maps

$$s \longrightarrow \coprod_{\lambda\in\Lambda} s_\lambda, \qquad s \longrightarrow \coprod_{\lambda\in\Lambda'} t_\lambda$$

can be combined to a single map

$$s \longrightarrow \coprod_{\lambda\in\Lambda\cup\Lambda'} s_\lambda \oplus t_\lambda$$

which exists because $\mathcal{S}$ is assumed an additive category, so the coproduct

$$\coprod_{\lambda\in\Lambda\cup\Lambda'} s_\lambda \oplus t_\lambda = \left\{\coprod_{\lambda\in\Lambda\cup\Lambda'} s_\lambda\right\} \oplus \left\{\coprod_{\lambda\in\Lambda\cup\Lambda'} t_\lambda\right\}$$

is a biproduct, in particular a product of the first and second sum. The two pairs

$$\begin{array}{ll} s \longrightarrow \coprod_{\lambda\in\Lambda} s_\lambda, & \beta \in \prod_{\lambda\in\Lambda} F_\lambda(s_\lambda) \\ s \longrightarrow \coprod_{\lambda\in\Lambda'} t_\lambda, & \beta' \in \prod_{\lambda\in\Lambda'} F_\lambda(t_\lambda) \end{array}$$

are defined to be equivalent if the map

$$s \longrightarrow \coprod_{\lambda\in\Lambda\cup\Lambda'} s_\lambda \oplus t_\lambda$$

factors as

$$s \longrightarrow \coprod_{\lambda\in\Lambda\cup\Lambda'} k_\lambda \xrightarrow{\coprod_{\lambda\in\Lambda\cup\Lambda'} f_\lambda} \coprod_{\lambda\in\Lambda\cup\Lambda'} s_\lambda \oplus t_\lambda$$

so that the images of β and β' in

$$\prod_{\lambda\in\Lambda\cup\Lambda'} F_\lambda(k_\lambda)$$

agree.

LEMMA 6.1.9. *Suppose $\mathcal{S}$ satisfies Hypothesis 6.1.1. Let s be an object of $\mathcal{S}$, and let $\{F_\mu, \mu \in M\}$ be a set of objects in $\mathcal{E}x\big(\mathcal{S}^{op}, \mathcal{A}b\big)$. Then the equivalence defined in Definition 6.1.8 is an equivalence relation.*

Proof: The relation is clearly reflexive and symmetric. We need to show it transitive. Suppose therefore that we have pairs of equivalent elements in $\left\{\bigvee_{\mu\in M} F_\mu\right\}(s)$

$$\begin{array}{ll} s \longrightarrow \coprod_{\lambda\in\Lambda} r_\lambda, & \beta \in \prod_{\lambda\in\Lambda} F_\lambda(r_\lambda) \\ s \longrightarrow \coprod_{\lambda\in\Lambda'} s_\lambda, & \beta' \in \prod_{\lambda\in\Lambda'} F_\lambda(s_\lambda) \end{array}$$

and

$$\begin{array}{ll} s \longrightarrow \coprod_{\lambda\in\Lambda'} s_\lambda, & \beta' \in \prod_{\lambda\in\Lambda'} F_\lambda(s_\lambda) \\ s \longrightarrow \coprod_{\lambda\in\Lambda''} t_\lambda, & \beta'' \in \prod_{\lambda\in\Lambda''} F_\lambda(t_\lambda). \end{array}$$

We need to show the equivalence of

$$\begin{array}{ll} s \longrightarrow \coprod_{\lambda\in\Lambda} r_\lambda, & \beta \in \prod_{\lambda\in\Lambda} F_\lambda(r_\lambda) \\ s \longrightarrow \coprod_{\lambda\in\Lambda''} t_\lambda, & \beta'' \in \prod_{\lambda\in\Lambda''} F_\lambda(t_\lambda). \end{array}$$

The equivalence of the pairs

$$\begin{array}{ll} s \longrightarrow \coprod_{\lambda\in\Lambda} r_\lambda, & \beta \in \prod_{\lambda\in\Lambda} F_\lambda(r_\lambda) \\ s \longrightarrow \coprod_{\lambda\in\Lambda'} s_\lambda, & \beta' \in \prod_{\lambda\in\Lambda'} F_\lambda(s_\lambda) \end{array}$$

means that the map

$$s \longrightarrow \coprod_{\lambda\in\Lambda\cup\Lambda'} r_\lambda \oplus s_\lambda$$

factors as

$$s \longrightarrow \coprod_{\lambda\in\Lambda\cup\Lambda'} k_\lambda \xrightarrow{\coprod_{\lambda\in\Lambda\cup\Lambda'} f_\lambda} \coprod_{\lambda\in\Lambda\cup\Lambda'} r_\lambda \oplus s_\lambda$$

so that the images of β and β' in

$$\prod_{\lambda\in\Lambda\cup\Lambda'} F_\lambda(k_\lambda)$$

agree. The equivalence of the pair

$$\begin{array}{ll} s \longrightarrow \coprod_{\lambda\in\Lambda'} s_\lambda, & \beta' \in \prod_{\lambda\in\Lambda'} F_\lambda(s_\lambda) \\ s \longrightarrow \coprod_{\lambda\in\Lambda''} t_\lambda, & \beta'' \in \prod_{\lambda\in\Lambda''} F_\lambda(t_\lambda) \end{array}$$

means that the map

$$s \longrightarrow \coprod_{\lambda\in\Lambda'\cup\Lambda''} s_\lambda \oplus t_\lambda$$

factors as

$$s \longrightarrow \coprod_{\lambda\in\Lambda'\cup\Lambda''} l_\lambda \xrightarrow{\coprod_{\lambda\in\Lambda'\cup\Lambda''} g_\lambda} \coprod_{\lambda\in\Lambda'\cup\Lambda''} s_\lambda \oplus t_\lambda$$

so that the images of β' and β'' in

$$\prod_{\lambda\in\Lambda'\cup\Lambda''} F_\lambda(l_\lambda)$$

agree. For each $\lambda \in \Lambda \cup \Lambda' \cup \Lambda''$ we have arrows

$$\begin{array}{ccc} & & r_\lambda \oplus l_\lambda \\ & & \Big\downarrow{\scriptstyle 1\oplus g_\lambda} \\ k_\lambda \oplus t_\lambda & \xrightarrow{f_\lambda\oplus 1} & r_\lambda \oplus s_\lambda \oplus t_\lambda \end{array}$$

and by Hypothesis 6.1.1.3 these may be completed to homotopy pullback squares

$$\begin{array}{ccc} m_\lambda & \longrightarrow & r_\lambda \oplus l_\lambda \\ \Big\downarrow & & \Big\downarrow{\scriptstyle 1\oplus g_\lambda} \\ k_\lambda \oplus t_\lambda & \xrightarrow{f_\lambda\oplus 1} & r_\lambda \oplus s_\lambda \oplus t_\lambda \end{array}$$

By Hypothesis 6.1.1.4, the coproduct of these is a homotopy pullback square

$$\begin{array}{ccc} \coprod_{\lambda\in\Lambda\cup\Lambda'\cup\Lambda''} m_\lambda & \longrightarrow & \coprod_{\lambda\in\Lambda\cup\Lambda'\cup\Lambda''} r_\lambda \oplus l_\lambda \\ \Big\downarrow & & \Big\downarrow \\ \coprod_{\lambda\in\Lambda\cup\Lambda'\cup\Lambda''} k_\lambda \oplus t_\lambda & \longrightarrow & \coprod_{\lambda\in\Lambda\cup\Lambda'\cup\Lambda''} r_\lambda \oplus s_\lambda \oplus t_\lambda \end{array}$$

On the other hand, we have an obvious commutative square

$$\begin{array}{ccc} s & \longrightarrow & \coprod_{\lambda\in\Lambda\cup\Lambda'\cup\Lambda''} r_\lambda \oplus l_\lambda \\ \Big\downarrow & & \Big\downarrow \\ \coprod_{\lambda\in\Lambda\cup\Lambda'\cup\Lambda''} k_\lambda \oplus t_\lambda & \longrightarrow & \coprod_{\lambda\in\Lambda\cup\Lambda'\cup\Lambda''} r_\lambda \oplus s_\lambda \oplus t_\lambda \end{array}$$

and the defining property of homotopy pullback squares tells us there is a map

$$s \longrightarrow \coprod_{\lambda\in\Lambda\cup\Lambda'\cup\Lambda''} m_\lambda$$

giving a map of the squares. But β and β' have the same image in $\prod_{\lambda\in\Lambda\cup\Lambda'} F_\lambda(k_\lambda)$, and β' and β'' have the same image in $\prod_{\lambda\in\Lambda'\cup\Lambda''} F_\lambda(l_\lambda)$. Hence β, β' and β'' all have the same image in $\prod_{\lambda\in\Lambda\cup\Lambda'\cup\Lambda''} F_\lambda(m_\lambda)$. □

DEFINITION 6.1.10. *Suppose $\mathcal{S}$ satisfies Hypothesis 6.1.1. Let s be an object of $\mathcal{S}$, and let $\{F_\mu, \mu \in M\}$ be a set of objects in $\mathcal{E}x(\mathcal{S}^{op}, \mathcal{A}b)$. The set*

$$\left\{\coprod_{\mu\in M} F_\mu\right\}(s)$$

is defined to be the quotient of $\left\{\bigvee_{\mu\in M} F_\mu\right\}(s)$ by the equivalence relation of Definition 6.1.8.

LEMMA 6.1.11. *Suppose $\mathcal{S}$ satisfies Hypothesis 6.1.1. Let s be an object of $\mathcal{S}$, and let $\{F_\mu, \mu \in M\}$ be a set of objects in $\mathcal{E}x(\mathcal{S}^{op}, \mathcal{A}b)$. The set $\left\{\coprod_{\mu\in M} F_\mu\right\}(s)$ has a natural structure of an abelian group.*

Proof: Given two elements

$$s \longrightarrow \coprod_{\lambda\in\Lambda} s_\lambda, \qquad \beta \in \prod_{\lambda\in\Lambda} F_\lambda(s_\lambda)$$
$$s \longrightarrow \coprod_{\lambda\in\Lambda'} t_\lambda, \qquad \beta' \in \prod_{\lambda\in\Lambda'} F_\lambda(t_\lambda)$$

we need to define their sum. It is the pair

$$s \longrightarrow \coprod_{\lambda\in\Lambda\cup\Lambda'} s_\lambda \oplus t_\lambda, \qquad \beta+\beta' \in \prod_{\lambda\in\Lambda\cup\Lambda'} F_\lambda(s_\lambda\oplus t_\lambda).$$

We leave it to the reader to check that this addition is well–defined; equivalent elements have equivalent sums. We also leave it to the reader to check that the associative and commutative law hold for this addition. The zero element of this group action is a pair

$$s \longrightarrow \coprod_{\lambda\in\Lambda} s_\lambda, \qquad 0 \in \prod_{\lambda\in\Lambda} F_\lambda(s_\lambda);$$

all such pairs are equivalent. We leave it to the reader to check that this is a neutral element for the addition, and that every element of $\left\{\coprod_{\mu\in M} F_\mu\right\}(s)$ has an additive inverse. □

LEMMA 6.1.12. *Suppose $\mathcal{S}$ satisfies Hypothesis 6.1.1. Let $\{F_\mu, \mu \in M\}$ be a set of objects in $\mathcal{E}x(\mathcal{S}^{op}, \mathcal{A}b)$. The assignment which sends $s \in \mathcal{S}$ to the abelian group $\left\{\coprod_{\mu\in M} F_\mu\right\}(s)$ can be naturally extended to an additive contravariant functor.*

Proof: We have to define this functor on morphisms. Given any morphism $f : s \longrightarrow t$ in $\mathcal{S}$, we need a map

$$\left\{\coprod_{\mu\in M} F_\mu\right\}(t) \longrightarrow \left\{\coprod_{\mu\in M} F_\mu\right\}(s).$$

Suppose therefore that we are given an element of $\left\{\coprod_{\mu\in M} F_\mu\right\}(t)$, that is a pair

$$t \longrightarrow \coprod_{\lambda\in\Lambda} t_\lambda, \qquad\qquad \beta \in \prod_{\lambda\in\Lambda} F_\lambda(t_\lambda).$$

Composition with $s \longrightarrow t$ gives the pair

$$s \longrightarrow t \longrightarrow \coprod_{\lambda\in\Lambda} t_\lambda, \qquad\qquad \beta \in \prod_{\lambda\in\Lambda} F_\lambda(t_\lambda)$$

which we may view as representing an element of $\left\{\coprod_{\mu\in M} F_\mu\right\}(s)$. This is our map

$$\left\{\coprod_{\mu\in M} F_\mu\right\}(t) \xrightarrow{\left\{\coprod_{\mu\in M} F_\mu\right\}(f)} \left\{\coprod_{\mu\in M} F_\mu\right\}(s).$$

We leave it to the reader to show that the map

$$\left\{\coprod_{\mu\in M} F_\mu\right\}(t) \xrightarrow{\left\{\coprod_{\mu\in M} F_\mu\right\}(f)} \left\{\coprod_{\mu\in M} F_\mu\right\}(s)$$

is a group homomorphism with the group structure as in Lemma 6.1.11. The reader will also easily check that the assignment sending $f : s \longrightarrow t$ to

$$\left\{\coprod_{\mu\in M} F_\mu\right\}(t) \xrightarrow{\left\{\coprod_{\mu\in M} F_\mu\right\}(f)} \left\{\coprod_{\mu\in M} F_\mu\right\}(s)$$

respects composition and identities; it defines a functor. Finally, it needs to be checked that the functor $\coprod_{\mu\in M} F_\mu$ is additive. This is also immediate from the definition; the functor clearly respects finite biproducts, since each F_μ does. In fact, we have more, as the next Lemma shows. □

LEMMA 6.1.13. *Suppose $\mathcal{S}$ satisfies Hypothesis 6.1.1. Let $\{F_\mu, \mu \in M\}$ be a set of objects in $\mathcal{E}x(\mathcal{S}^{op}, \mathcal{A}b)$. Then the functor $\coprod_{\mu\in M} F_\mu$ lies in the category $\mathcal{E}x(\mathcal{S}^{op}, \mathcal{A}b)$; it sends coproducts of fewer than α objects to products.*

Proof: We need to check that $\coprod_{\mu\in M} F_\mu$ is an object in the subcategory $\mathcal{E}x(\mathcal{S}^{op}, \mathcal{A}b)$ of the large category $\mathcal{C}at(\mathcal{S}^{op}, \mathcal{A}b)$. In other words, let $\{s_\gamma, \gamma \in \Gamma\}$ be a family of fewer than α objects in $\mathcal{S}$. We need to show that

$$\left\{\coprod_{\mu\in M} F_\mu\right\}\left[\coprod_{\gamma\in\Gamma} s_\gamma\right] \quad = \quad \prod_{\gamma\in\Gamma}\left\{\coprod_{\mu\in M} F_\mu\right\}(s_\gamma).$$

In any case, there is a natural map

$$\left\{\coprod_{\mu\in M} F_\mu\right\}\left[\coprod_{\gamma\in\Gamma} s_\gamma\right] \xrightarrow{\ \phi\ } \prod_{\gamma\in\Gamma}\left\{\coprod_{\mu\in M} F_\mu\right\}(s_\gamma);$$

we need to prove that ϕ is injective and surjective. Let us prove surjectivity first.

An element of $\prod_{\gamma\in\Gamma}\left\{\coprod_{\mu\in M} F_\mu\right\}(s_\gamma)$ is, for every $\gamma \in \Gamma$ a pair

$$s_\gamma \longrightarrow \coprod_{\lambda\in\Lambda_\gamma} t_\lambda^\gamma, \qquad\qquad \beta_\gamma \in \prod_{\lambda\in\Lambda_\gamma} F_\lambda(t_\lambda^\gamma)$$

where $\Lambda_\gamma \subset M$ is a subset of cardinality $< \alpha$. Let us put

$$\Lambda = \bigcup_{\gamma\in\Gamma} \Lambda_\gamma,$$

and adopt the notation that if $\lambda \notin \Lambda_\gamma$ we define t_λ^γ to be 0. Recall that, since each Λ_γ is of cardinality $< \alpha$ and the index set Γ is also of cardinality $< \alpha$, the union has cardinality bounded by the sum of fewer than α cardinals, each smaller than α. Since α is a regular cardinal, the cardinality of Λ is $< \alpha$.

The product

$$\prod_{\gamma\in\Gamma} \beta_\gamma \quad \in \quad \prod_{\gamma\in\Gamma}\prod_{\lambda\in\Lambda} F_\lambda(t_\lambda^\gamma)$$

can be viewed, by reversing the order of the product, as lying in the product

$$\prod_{\gamma\in\Gamma} \beta_\gamma \quad \in \quad \prod_{\lambda\in\Lambda}\prod_{\gamma\in\Gamma} F_\lambda(t_\lambda^\gamma)$$

and since each F_λ is in $\mathcal{E}x(\mathcal{S}^{op}, \mathcal{A}b)$, this is also

$$\prod_{\gamma\in\Gamma} \beta_\gamma \quad \in \quad \prod_{\lambda\in\Lambda} F_\lambda\left(\coprod_{\gamma\in\Gamma} t_\lambda^\gamma\right).$$

We have therefore produced a map

$$\coprod_{\gamma\in\Gamma} s_\gamma \longrightarrow \coprod_{\lambda\in\Lambda}\left\{\coprod_{\gamma\in\Gamma} t_\lambda^\gamma\right\}$$

and an element

$$\prod_{\gamma\in\Gamma} \beta_\gamma \quad \in \quad \prod_{\lambda\in\Lambda} F_\lambda\left(\coprod_{\gamma\in\Gamma} t_\lambda^\gamma\right)$$

and this pair is an element of $\left\{\coprod_{\mu\in M} F_\mu\right\}\left[\coprod_{\gamma\in\Gamma} s_\gamma\right]$. It is easy to check that it maps via ϕ to an element of $\prod_{\gamma\in\Gamma}\left\{\coprod_{\mu\in M} F_\mu\right\}(s_\gamma)$ equivalent to what we started with. Therefore ϕ is surjective. Next we must prove the injectivity of ϕ.

Suppose therefore we are given an element of the kernel of ϕ. That is, we have an element of $\left\{\coprod_{\mu\in M} F_\mu\right\}\left[\coprod_{\gamma\in\Gamma} s_\gamma\right]$ mapping to zero under ϕ. This element may be represented by a pair

$$\coprod_{\gamma\in\Gamma} s_\gamma \longrightarrow \coprod_{\lambda\in\Lambda} t_\lambda, \qquad \beta\in\prod_{\lambda\in\Lambda} F_\lambda(t_\lambda).$$

To say that the map ϕ takes this to zero is to assert that, for each of the natural inclusions $s_\gamma \longrightarrow \coprod_{\gamma\in\Gamma} s_\gamma$, the induced element

$$s_\gamma \longrightarrow \coprod_{\gamma\in\Gamma} s_\gamma \longrightarrow \coprod_{\lambda\in\Lambda} t_\lambda, \qquad \beta\in\prod_{\lambda\in\Lambda} F_\lambda(t_\lambda)$$

is equivalent to zero. The definition of the equivalence relation says that there must be objects k_λ^γ so that the map

$$s_\gamma \longrightarrow \coprod_{\lambda\in\Lambda} t_\lambda$$

factors as

$$s_\gamma \longrightarrow \coprod_{\lambda\in\Lambda} k_\lambda^\gamma \longrightarrow \coprod_{\lambda\in\Lambda} t_\lambda$$

and the image of β vanishes under the map

$$\prod_{\lambda\in\Lambda} F(t_\lambda) \longrightarrow \prod_{\lambda\in\Lambda} F(k_\lambda^\gamma).$$

But then, taking the product over all γ, the image of β by the map

$$\prod_{\lambda\in\Lambda} F(t_\lambda) \longrightarrow \prod_{\lambda\in\Lambda}\prod_{\gamma\in\Gamma} F(k_\lambda^\gamma) \quad = \quad \prod_{\lambda\in\Lambda} F\left(\coprod_{\gamma\in\Gamma} k_\lambda^\gamma\right)$$

also vanishes. We have factored

$$\coprod_{\gamma\in\Gamma} s_\gamma \longrightarrow \coprod_{\lambda\in\Lambda} t_\lambda$$

as

$$\coprod_{\gamma\in\Gamma} s_\gamma \longrightarrow \coprod_{\lambda\in\Lambda}\coprod_{\gamma\in\Gamma} k_\lambda^\gamma \longrightarrow \coprod_{\lambda\in\Lambda} t_\lambda$$

and β maps to zero in $\prod_{\lambda\in\Lambda} F\left(\coprod_{\gamma\in\Gamma} k_\lambda^\gamma\right)$. This establishes that the class in the kernel of ϕ is equivalent to zero. The kernel is trivial, and ϕ is injective. □

LEMMA 6.1.14. *Suppose $\mathcal{S}$ satisfies Hypothesis 6.1.1. Let $\{F_\mu, \mu \in M\}$ be a set of objects in $\mathcal{E}x(\mathcal{S}^{op}, \mathcal{A}b)$. Then there are natural transformations*

$$F_\mu \longrightarrow \coprod_{\mu\in M} F_\mu .$$

Proof: For s an object of $\mathcal{S}$, we define the map

$$F_\mu(s) \xrightarrow{\phi(s)} \left\{\coprod_{\mu\in M} F_\mu\right\}(s)$$

to take $\beta \in F_\mu(s)$ to the pair

$$1 : s \longrightarrow s \qquad\qquad \beta \in F_\mu(s).$$

We leave it to the reader to check that $\phi(s)$ is a group homomorphism. It remains to check that for any morphism $f : s \longrightarrow t$ in $\mathcal{S}$, the square

$$\begin{array}{ccc} F_\mu(t) & \xrightarrow{\phi(t)} & \left\{\coprod_{\mu\in M} F_\mu\right\}(t) \\ \downarrow & & \downarrow \\ F_\mu(s) & \xrightarrow{\phi(s)} & \left\{\coprod_{\mu\in M} F_\mu\right\}(s) \end{array}$$

commutes.

Pick $\beta \in F_\mu(t)$. The map $\phi(t)$ sends it to the pair

$$1 : t \longrightarrow t, \qquad\qquad \beta \in F_\mu(t).$$

Applying $\left\{\coprod_{\mu\in M} F_\mu\right\}(f)$ to the result we get

$$s \xrightarrow{f} t \xrightarrow{1} t, \qquad\qquad \beta \in F_\mu(t);$$

we remind the reader that the definition of the group homomorphism $\left\{\coprod_{\mu\in M} F_\mu\right\}(f)$ is given in Lemma 6.1.12.

On the other hand, under $F_\mu(t) \longrightarrow F_\mu(s)$ the element β maps to $F_\mu(f)(\beta)$. And $\phi(s)$ takes this to

$$1 : s \longrightarrow s, \qquad\qquad F_\mu(f)(\beta) \in F_\mu(s).$$

We must show the two pairs are equivalent.

For this, factor

$$s \xrightarrow{\begin{pmatrix} 1 \\ f \end{pmatrix}} s \oplus t$$

as

$$s \xrightarrow{1} s \xrightarrow{\begin{pmatrix} 1 \\ f \end{pmatrix}} s \oplus t,$$

and note that the images of $\beta \in F_\mu(t) \subset F_\mu(s \oplus t)$ and $F_\mu(f)(\beta) \in F_\mu(s) \subset F_\mu(s \oplus t)$ clearly agree. □

PROPOSITION 6.1.15. *Suppose $\mathcal{S}$ satisfies Hypothesis 6.1.1. Suppose $\{F_\mu, \mu \in M\}$ is a set of objects in $\mathcal{E}x(\mathcal{S}^{op}, \mathcal{A}b)$. The natural transformations*

$$F_\mu \longrightarrow \coprod_{\mu\in M} F_\mu$$

of Lemma 6.1.14 give $\coprod_{\mu\in M} F_\mu$ the structure of the coproduct of F_μ in the category $\mathcal{E}x(\mathcal{S}^{op}, \mathcal{A}b)$.

Proof: It needs to be checked that $\coprod_{\mu\in M} F_\mu$ has the universal property of a coproduct. Let G be any object in $\mathcal{E}x(\mathcal{S}^{op}, \mathcal{A}b)$. Suppose we are given maps $F_\mu \longrightarrow G$ for every μ. We need to factor them (uniquely) through

$$\coprod_{\mu\in M} F_\mu \xrightarrow{\phi} G.$$

Where must this map send an element of $\left\{\coprod_{\mu\in M} F_\mu\right\}(s)$? Recall that such an element is represented by a pair

$$s \longrightarrow \coprod_{\lambda\in\Lambda} s_\lambda, \qquad\qquad \beta \in \prod_{\lambda\in\Lambda} F_\lambda(s_\lambda).$$

To compute what its image by ϕ must be, note the following. If ϕ is to be a natural transformation, the square

$$\begin{array}{ccc}
\left\{\coprod_{\mu\in M} F_\mu\right\}\left(\coprod_{\lambda\in\Lambda} s_\lambda\right) & \longrightarrow & \left\{\coprod_{\mu\in M} F_\mu\right\}(s) \\
\downarrow & & \phi\downarrow \\
G\left(\coprod_{\lambda\in\Lambda} s_\lambda\right) & \longrightarrow & G(s)
\end{array}$$

must commute. We are now going to chase the image of

$$1 : \coprod_{\lambda\in\Lambda} s_\lambda \longrightarrow \coprod_{\lambda\in\Lambda} s_\lambda, \qquad \beta \in \prod_{\lambda\in\Lambda} F_\lambda(s_\lambda)$$

around this commutative square.

The map

$$\left\{\coprod_{\mu\in M} F_\mu\right\}\left(\coprod_{\lambda\in\Lambda} s_\lambda\right) \longrightarrow \left\{\coprod_{\mu\in M} F_\mu\right\}(s)$$

takes the pair

$$1 : \coprod_{\lambda\in\Lambda} s_\lambda \longrightarrow \coprod_{\lambda\in\Lambda} s_\lambda, \qquad \beta \in \prod_{\lambda\in\Lambda} F_\lambda(s_\lambda)$$

to the pair

$$s \longrightarrow \coprod_{\lambda\in\Lambda} s_\lambda \xrightarrow{1} \coprod_{\lambda\in\Lambda} s_\lambda, \qquad \beta \in \prod_{\lambda\in\Lambda} F_\lambda(s_\lambda),$$

by the definition of $\coprod_{\mu\in M} F_\mu$ applied to the morphism $s \longrightarrow \coprod_{\lambda\in\Lambda} s_\lambda$; see Lemma 6.1.12. The commutativity of

$$\begin{array}{ccc}
\left\{\coprod_{\mu\in M} F_\mu\right\}\left(\coprod_{\lambda\in\Lambda} s_\lambda\right) & \longrightarrow & \left\{\coprod_{\mu\in M} F_\mu\right\}(s) \\
\downarrow & & \phi\downarrow \\
G\left(\coprod_{\lambda\in\Lambda} s_\lambda\right) & \longrightarrow & G(s)
\end{array}$$

tells us that in order to compute the image of

$$s \longrightarrow \coprod_{\lambda\in\Lambda} s_\lambda, \qquad \beta \in \prod_{\lambda\in\Lambda} F_\lambda(s_\lambda)$$

in $G(s)$, it suffices to figure out what the composite

$$\begin{array}{ccc} \left\{\coprod_{\mu\in M} F_\mu\right\}\left(\coprod_{\lambda\in\Lambda} s_\lambda\right) & & \\ \downarrow & & \\ G\left(\coprod_{\lambda\in\Lambda} s_\lambda\right) & \longrightarrow & G(s) \end{array}$$

does to

$$1 : \coprod_{\lambda\in\Lambda} s_\lambda \longrightarrow \coprod_{\lambda\in\Lambda} s_\lambda, \qquad \beta \in \prod_{\lambda\in\Lambda} F_\lambda(s_\lambda).$$

Any element of the product $\prod_{\lambda\in\Lambda} F_\lambda(s_\lambda)$ can be written as a product

$$\beta = \prod_{\lambda\in\Lambda} \beta_\lambda,$$

with $\beta_\lambda \in F_\lambda(s_\lambda)$. Write β this way. Since we are assuming that the composite

$$F_\mu \longrightarrow \coprod_{\mu\in M} F_\mu \xrightarrow{\phi} G$$

is the given map $F_\mu \longrightarrow G$, the pair

$$1 : s_\lambda \longrightarrow s_\lambda, \qquad \beta_\lambda \in F_\lambda(s_\lambda)$$

must map to the image of $\beta_\lambda \in F_\lambda(s_\lambda)$ by $\phi_\lambda : F_\lambda(s_\lambda) \longrightarrow G(s_\lambda)$. The product of these elements, that is the pair

$$1 : \coprod_{\lambda\in\Lambda} s_\lambda \longrightarrow \coprod_{\lambda\in\Lambda} s_\lambda, \qquad \prod_{\lambda\in\Lambda} \beta_\lambda \in \prod_{\lambda\in\Lambda} F_\lambda(s_\lambda)$$

must map to

$$\prod_{\lambda\in\Lambda} \phi_\lambda(\beta_\lambda) \quad \in \quad \prod_{\lambda\in\Lambda} G(s_\lambda) \quad = \quad G\left(\coprod_{\lambda\in\Lambda} s_\lambda\right).$$

This computes the image of

$$1 : \coprod_{\lambda\in\Lambda} s_\lambda \longrightarrow \coprod_{\lambda\in\Lambda} s_\lambda, \qquad \prod_{\lambda\in\Lambda} \beta_\lambda \in \prod_{\lambda\in\Lambda} F_\lambda(s_\lambda)$$

under

$$\left\{\coprod_{\mu\in M} F_\mu\right\}\left(\coprod_{\lambda\in\Lambda} s_\lambda\right) \longrightarrow G\left(\coprod_{\lambda\in\Lambda} s_\lambda\right).$$

But then under the longer composite

$$\begin{array}{ccc} \left\{\coprod_{\mu\in M} F_\mu\right\}\left(\coprod_{\lambda\in\Lambda} s_\lambda\right) & & \\ \downarrow & & \\ G\left(\coprod_{\lambda\in\Lambda} s_\lambda\right) & \longrightarrow & G(s) \end{array}$$

the element

$$1 : \coprod_{\lambda\in\Lambda} s_\lambda \longrightarrow \coprod_{\lambda\in\Lambda} s_\lambda, \qquad \prod_{\lambda\in\Lambda} \beta_\lambda \in \prod_{\lambda\in\Lambda} F_\lambda(s_\lambda)$$

must map to the image of

$$\prod_{\lambda\in\Lambda} \phi_\lambda(\beta_\lambda) \quad \in \quad \prod_{\lambda\in\Lambda} G(s_\lambda) \quad = \quad G\left(\coprod_{\lambda\in\Lambda} s_\lambda\right)$$

under

$$G\left(\coprod_{\lambda\in\Lambda} s_\lambda\right) \longrightarrow G(s).$$

The commutativity of

$$\begin{array}{ccc} \left\{\coprod_{\mu\in M} F_\mu\right\}\left(\coprod_{\lambda\in\Lambda} s_\lambda\right) & \longrightarrow & \left\{\coprod_{\mu\in M} F_\mu\right\}(s) \\ \downarrow & & \downarrow{\scriptstyle\phi} \\ G\left(\coprod_{\lambda\in\Lambda} s_\lambda\right) & \longrightarrow & G(s) \end{array}$$

allows us to deduce that the image of

$$s \longrightarrow \coprod_{\lambda\in\Lambda} s_\lambda, \qquad \beta \in \prod_{\lambda\in\Lambda} F_\lambda(s_\lambda)$$

via ϕ must be the image of

$$\prod_{\lambda\in\Lambda} \phi_\lambda(\beta_\lambda) \quad \in \quad \prod_{\lambda\in\Lambda} G(s_\lambda) \quad = \quad G\left(\coprod_{\lambda\in\Lambda} s_\lambda\right)$$

via

$$G\left(\coprod_{\lambda\in\Lambda} s_\lambda\right) \longrightarrow G(s).$$

Thus the map ϕ is uniquely determined by the above. The reader is left to verify that this map is well–defined (takes equivalent pairs to the same element of $G(s)$), is a group homomorphism, is natural in s and that the composites

$$F_\mu \longrightarrow \coprod_{\mu \in M} F_\mu \xrightarrow{\ \phi\ } G$$

are the given maps $F_\mu \longrightarrow G$. □

REMARK 6.1.16. We have proved in this Section that the category $\mathcal{E}x(\mathcal{S}^{op}, \mathcal{A}b)$ is an abelian category satisfying [AB3] and [AB3*]; it is closed with respect to products and coproducts of its objects. The inclusion functor $\mathcal{E}x(\mathcal{S}^{op}, \mathcal{A}b) \subset \mathcal{C}at(\mathcal{S}^{op}, \mathcal{A}b)$ is exact and respects products, but decidedly does not respect coproducts.

LEMMA 6.1.17. *Suppose $\mathcal{S}$ satisfies Hypothesis 6.1.1. The inclusion functor $\mathcal{S} \longrightarrow \mathcal{E}x(\mathcal{S}^{op}, \mathcal{A}b)$ respects coproducts of $< \alpha$ objects*

Proof: There is an obvious, inclusion functor $\mathcal{S} \longrightarrow \mathcal{E}x(\mathcal{S}^{op}, \mathcal{A}b)$, which takes an object $s \in \mathcal{S}$ to the representable functor $\mathcal{S}(-, s)$. We wish to show that this functor preserves coproducts of fewer than α objects. Therefore let $\{t_\lambda, \lambda \in \Lambda\}$ be a set of objects in $\mathcal{S}$, with Λ of cardinality $< \alpha$. By 6.1.1.2, the coproduct of these objects exists in $\mathcal{E}x(\mathcal{S}^{op}, \mathcal{A}b)$.

We have a natural map

$$\coprod_{\lambda \in \Lambda} \mathcal{S}(-, t_\lambda) \xrightarrow{\ \phi\ } \mathcal{S}\left(-, \coprod_{\lambda \in \Lambda} t_\lambda\right)$$

and we wish to show the map an isomorphism. But this is an essentially immediate consequence of Definition 6.1.10. The point is that the map has an obvious inverse. The inverse

$$\coprod_{\lambda \in \Lambda} \mathcal{S}(-, t_\lambda) \xleftarrow{\ \psi\ } \mathcal{S}\left(-, \coprod_{\lambda \in \Lambda} t_\lambda\right)$$

takes an element of

$$\mathcal{S}\left(s, \coprod_{\lambda \in \Lambda} t_\lambda\right),$$

that is a map $s \longrightarrow \coprod_{\lambda \in \Lambda} t_\lambda$, to the pair

$$s \longrightarrow \coprod_{\lambda \in \Lambda} t_\lambda, \qquad\qquad 1 \in \prod_{\lambda \in \Lambda} \mathcal{S}(t_\lambda, t_\lambda)$$

which is an element of $\coprod_{\lambda\in\Lambda} \mathcal{S}(-,t_\lambda)$. It is obvious that $\phi\psi$ is the identity. But $\psi\phi$ takes an element of

$$\coprod_{\lambda\in\Lambda} \mathcal{S}(s,t_\lambda),$$

that is an equivaelnce class of pairs

$$s \longrightarrow \coprod_{\lambda\in\Lambda} s_\lambda, \qquad\qquad \beta \in \prod_{\lambda\in\Lambda} \mathcal{S}(s_\lambda,t_\lambda),$$

to

$$s \longrightarrow \coprod_{\lambda\in\Lambda} s_\lambda \longrightarrow \coprod_{\lambda\in\Lambda} t_\lambda, \qquad\qquad 1 \in \prod_{\lambda\in\Lambda} \mathcal{S}(t_\lambda,t_\lambda)$$

which is equivalent to it. □

6.2. The case of $\mathcal{S} = \mathcal{T}^\alpha$

Let $\mathcal{T}$ be a triangulated category satisfying [TR5]. Let α be the same regular cardinal that we have fixed throughout this Chapter.

LEMMA 6.2.1. *Suppose that $\mathcal{S}$ is an essentially small category, and is an α–localising subcategory of $\mathcal{T}$. Then $\mathcal{S}$ satisfies Hyposthesis 6.1.1.*

Proof: We need check that $\mathcal{S}$ satisfies all four parts of Hypothesis 6.1.1. We are supposing $\mathcal{S}$ is essentially small, and hence satisfies Hypothesis 6.1.1.1. We assume also that $\mathcal{S}$ is α–localising. This means $\mathcal{S}$ is triangulated and closed under coproducts of fewer than α of its objects. The closure under coproducts is Hypothesis 6.1.1.2. As in Example 6.1.2, Hypotheses 6.1.1.3 and 6.1.1.4 are automatic for a triangulated category; therefore $\mathcal{T}^\alpha$ satisfies all of Hypothesis 6.1.1. □

REMARK 6.2.2. In particular, the entire discussion of Section 6.1 applies, and we understand coproducts in $\mathcal{E}x(\mathcal{S}^{op},\mathcal{A}b)$ quite explicitly. If we do not insist that $\mathcal{S}$ be essentially small, it is not in general true that $\mathcal{C}at(\mathcal{S}^{op},\mathcal{A}b)$ will have small Hom–sets. Here we are primarily interested in categories with small Hom–sets.

In the situation $\mathcal{S} \subset \mathcal{T}$ as above, we have

LEMMA 6.2.3. *Let $\mathcal{T}$ be a triangulated category satisfying [TR5], and assume $\mathcal{S} \subset \mathcal{T}$ is an essentially small α–localising subcategory. There is a natural functor $\mathcal{T} \longrightarrow \mathcal{E}x(\mathcal{S}^{op},\mathcal{A}b)$ sending $t \in \mathcal{T}$ to the representable functor $\mathcal{T}(-,t)$, or more precisely to its restriction to $\mathcal{S}$. We will denote this restriction $\mathcal{T}(-,t)|_\mathcal{S}$.*

Proof: Clearly the restriction gives a functor to $\mathcal{C}at(\mathcal{S}^{op},\mathcal{A}b)$. We need to show that for $t \in \mathcal{T}$, the functor $\mathcal{T}(-,t)|_\mathcal{S}$ lies in $\mathcal{E}x(\mathcal{S}^{op},\mathcal{A}b)$, that is the functor $\mathcal{T}(-,t)|_{\mathcal{S}}$ sends coproducts in $\mathcal{S}$ of fewer than α objects to products in $\mathcal{A}b$. This is true simply because the coproducts in $\mathcal{S}$ agree with those in $\mathcal{T}$. The subcategory $\mathcal{S} \subset \mathcal{T}$ is α–localising. □

LEMMA 6.2.4. *Let $\mathcal{T}$ be a triangulated category satisfying [TR5], and assume $\mathcal{S} \subset \mathcal{T}$ is an essentially small α–localising subcategory. Then the functor $\mathcal{T} \longrightarrow \mathcal{E}x\big(\mathcal{S}^{op}, \mathcal{A}b\big)$ respects products.*

Proof: Let $\{t_\lambda, \lambda \in \Lambda\}$ be a set of objects in $\mathcal{T}$ whose product exists in $\mathcal{T}$. Then, for each $s \in \mathcal{S} \subset \mathcal{T}$,

$$\mathcal{T}\left(s, \prod_{\lambda\in\Lambda} t_\lambda\right) \quad = \quad \prod_{\lambda\in\Lambda} \mathcal{T}(s, t_\lambda)$$

and the right is just the product of $\mathcal{T}(-,t_\lambda)|_{\mathcal{S}}$ applied to s, as defined in Lemma 6.1.5. □

LEMMA 6.2.5. *Let $\mathcal{T}$ be a triangulated category satisfying [TR5], and assume $\mathcal{S} \subset \mathcal{T}$ is an essentially small α–localising subcategory. If $\mathcal{S}$ is α–perfect, then the functor $\mathcal{T} \longrightarrow \mathcal{E}x\big(\mathcal{S}^{op}, \mathcal{A}b\big)$ respects coproducts of fewer than α objects. If $\mathcal{S}$ is not only α–perfect, but every object of $\mathcal{S}$ is also α–small, then $\mathcal{T} \longrightarrow \mathcal{E}x\big(\mathcal{S}^{op}, \mathcal{A}b\big)$ respects all coproducts.*

Proof: Let $\{t_\lambda, \lambda \in \Lambda\}$ be a set of objects in $\mathcal{T}$. Since $\mathcal{T}$ satisfies [TR5], the coproduct exists in $\mathcal{T}$. We need to show that if $\mathcal{S}$ is α–perfect and the cardinality of Λ is $< \alpha$, or if $\mathcal{S}$ is not only α–perfect but also $\mathcal{S} \subset \mathcal{T}^{(\alpha)}$ and the cardinality of Λ is unrestricted, then

$$\mathcal{T}\left(-, \coprod_{\lambda\in\Lambda} t_\lambda\right)\Bigg|_{\mathcal{S}}$$

is the coproduct in $\mathcal{E}x\big(\mathcal{S}^{op}, \mathcal{A}b\big)$ of the functors $\mathcal{T}(-,t_\lambda)|_{\mathcal{S}}$. By the universal property of the coproduct in $\mathcal{E}x\big(\mathcal{S}^{op}, \mathcal{A}b\big)$, there is a natural map

$$\coprod_{\lambda\in\Lambda} \mathcal{T}(-,t_\lambda)|_{\mathcal{S}} \xrightarrow{\ \phi\ } \mathcal{T}\left(-, \coprod_{\lambda\in\Lambda} t_\lambda\right)\Bigg|_{\mathcal{S}}.$$

We need to check that ϕ is an isomorphism. We begin by showing it surjective.

Let us therefore begin with an element of

$$\mathcal{T}\left(s, \coprod_{\lambda\in\Lambda} t_\lambda\right)\Bigg|_{\mathcal{S}},$$

that is a map in $\mathcal{T}$ from an object s of $\mathcal{S}$ to the coproduct. If $\mathcal{S} \subset \mathcal{T}^{(\alpha)}$ then s is α–small, and any map

$$s \longrightarrow \coprod_{\lambda\in\Lambda} t_\lambda$$

factors as a map

$$s \longrightarrow \coprod_{\lambda\in\Lambda'} t_\lambda \qquad \subset \qquad \coprod_{\lambda\in\Lambda} t_\lambda$$

where $\Lambda' \subset \Lambda$ is a subset of cardinality $< \alpha$. If we are not assuming $\mathcal{S} \subset \mathcal{T}^{(\alpha)}$, then we assume anyway that the cardinality of Λ is $< \alpha$.

Because $\mathcal{S}$ is also α–perfect, the map factors further as

$$s \longrightarrow \coprod_{\lambda\in\Lambda'} s_\lambda \xrightarrow{\coprod_{\lambda\in\Lambda'} f_\lambda} \coprod_{\lambda\in\Lambda'} t_\lambda \qquad \subset \qquad \coprod_{\lambda\in\Lambda} t_\lambda$$

with $s_\lambda \in \mathcal{S}$. In other words, we have found a map

$$s \longrightarrow \coprod_{\lambda\in\Lambda'} s_\lambda$$

and an element

$$\coprod_{\lambda\in\Lambda'} f_\lambda \qquad \in \qquad \coprod_{\lambda\in\Lambda'} \mathcal{T}(s_\lambda, t_\lambda)|_{\mathcal{S}}$$

which map to our given

$$s \xrightarrow{f} \coprod_{\lambda\in\Lambda} t_\lambda;$$

we have found an element of $\coprod_{\lambda\in\Lambda} \mathcal{T}(s, t_\lambda)|_{\mathcal{S}}$ mapping to f. This establishes that ϕ is surjective.

Now choose an element of the kernel of ϕ. That is, a pair

$$s \longrightarrow \coprod_{\lambda\in\Lambda'} s_\lambda, \qquad\qquad \coprod_{\lambda\in\Lambda'} f_\lambda \quad \in \quad \coprod_{\lambda\in\Lambda'} \mathcal{T}(s_\lambda, t_\lambda)|_{\mathcal{S}}$$

mapping to zero. That is, the composite

$$s \longrightarrow \coprod_{\lambda\in\Lambda'} s_\lambda \xrightarrow{\coprod_{\lambda\in\Lambda'} f_\lambda} \coprod_{\lambda\in\Lambda'} t_\lambda \qquad \subset \qquad \coprod_{\lambda\in\Lambda} t_\lambda$$

vanishes. Because $\mathcal{S}$ is α–perfect, it is possible to factor each $f_\lambda : s_\lambda \longrightarrow t_\lambda$ as

$$s_\lambda \longrightarrow q_\lambda \longrightarrow t_\lambda$$

so that the q_λ are in $\mathcal{S}$, and the composite

$$s \longrightarrow \coprod_{\lambda\in\Lambda'} s_\lambda \longrightarrow \coprod_{\lambda\in\Lambda'} q_\lambda$$

already vanishes. Complete each $s_\lambda \longrightarrow q_\lambda$ to a triangle

$$k_\lambda \longrightarrow s_\lambda \longrightarrow q_\lambda \longrightarrow \Sigma k_\lambda$$

with $k_\lambda \in \mathcal{S}$. By Proposition 1.2.1 the direct sum is a triangle

$$\coprod_{\lambda\in\Lambda'} k_\lambda \longrightarrow \coprod_{\lambda\in\Lambda'} s_\lambda \longrightarrow \coprod_{\lambda\in\Lambda'} q_\lambda \longrightarrow \Sigma \coprod_{\lambda\in\Lambda'} k_\lambda$$

and the vanishing of the composite

$$s \longrightarrow \coprod_{\lambda\in\Lambda'} s_\lambda \longrightarrow \coprod_{\lambda\in\Lambda'} q_\lambda$$

means the map

$$s \longrightarrow \coprod_{\lambda\in\Lambda'} s_\lambda$$

factors as

$$s \longrightarrow \coprod_{\lambda\in\Lambda'} k_\lambda \longrightarrow \coprod_{\lambda\in\Lambda'} s_\lambda.$$

On the other hand, the composites

$$k_\lambda \longrightarrow s_\lambda \longrightarrow q_\lambda \longrightarrow t_\lambda$$

all vanish, since the first couple of maps are two maps of a triangle. We deduce that the pair

$$s \longrightarrow \coprod_{\lambda\in\Lambda'} s_\lambda, \qquad \prod_{\lambda\in\Lambda'} f_\lambda \quad\in\quad \prod_{\lambda\in\Lambda'} \mathcal{T}(s_\lambda, t_\lambda)|_{\mathcal{S}}$$

is equivalent to the pair

$$s \longrightarrow \coprod_{\lambda\in\Lambda'} s_\lambda, \qquad \prod_{\lambda\in\Lambda'} 0 \quad\in\quad \prod_{\lambda\in\Lambda'} \mathcal{T}(s_\lambda, t_\lambda)|_{\mathcal{S}},$$

in other words to the zero map. The kernel of ϕ is trivial, and ϕ is an isomorphism. □

PROPOSITION 6.2.6. *Let $\mathcal{T}$ be a triangulated category satisfying [TR5], and assume $\mathcal{S} \subset \mathcal{T}$ is an essentially small α–localising subcategory. The natural functor $\mathcal{T} \longrightarrow \mathcal{E}x(\mathcal{S}^{op}, \mathcal{A}b)$ is homological and respects products. If $\mathcal{S}$ is α–perfect, then the functor $\mathcal{T} \longrightarrow \mathcal{E}x(\mathcal{S}^{op}, \mathcal{A}b)$ respects coproducts of fewer than α objects. If $\mathcal{S}$ is not only α–perfect, but every object of $\mathcal{S}$ is also α–small, then $\mathcal{T} \longrightarrow \mathcal{E}x(\mathcal{S}^{op}, \mathcal{A}b)$ respects all coproducts.*

Proof: The fact that the functor respects products is Lemma 6.2.4, and the statements about coproducts are Lemma 6.2.5. The fact that $\mathcal{T} \longrightarrow \mathcal{E}x(\mathcal{S}^{op}, \mathcal{A}b)$ is homological is easy. Given a triangle in $\mathcal{T}$ of the form

$$r \longrightarrow s \longrightarrow t \longrightarrow \Sigma r$$

we need to check that

$$\mathcal{T}(-,r)|_{\mathcal{S}} \longrightarrow \mathcal{T}(-,s)|_{\mathcal{S}} \longrightarrow \mathcal{T}(-,t)|_{\mathcal{S}}$$

is exact in $\mathcal{E}x(\mathcal{S}^{op}, \mathcal{A}b)$, in other words gives an exact sequence when we evaluate it on any $k \in \mathcal{S}$. But this is the exactness of

$$\mathcal{T}(k,r) \longrightarrow \mathcal{T}(k,s) \longrightarrow \mathcal{T}(k,t)$$

which we know from Lemma 1.1.10. □

REMARK 6.2.7. The most interesting case of the above is $\mathcal{S} = \mathcal{T}^\alpha$, as in Definition 4.2.2. Suppose $\mathcal{T}^\alpha$ is essentially small. Lemma 4.2.5 asserts that $\mathcal{T}^\alpha$ is α–localising. By its definition, $\mathcal{T}^\alpha$ is contained in $\mathcal{T}^{(\alpha)}$, that is consists only of α–small objects. Furthermore, $\mathcal{T}^\alpha$ is α–perfect; it is in fact the largest α–perfect class in $\mathcal{T}^{(\alpha)}$. It follows that $\mathcal{S} = \mathcal{T}^\alpha \subset \mathcal{T}$ satisfies all the hypotheses of Proposition 6.2.6. We deduce that the natural functor

$$\mathcal{T} \longrightarrow \mathcal{E}x\Big(\{\mathcal{T}^\alpha\}^{op}, \mathcal{A}b\Big)$$

is a homological functor respecting all products and coproducts.

There is one more useful fact about the functors $\mathcal{T} \longrightarrow \mathcal{E}x(\mathcal{S}^{op}, \mathcal{A}b)$. To state it, we need to introduce one important definition.

DEFINITION 6.2.8. *Let $\mathcal{T}$ be a triangulated category satisfying [TR5], and assume $\mathcal{S} \subset \mathcal{T}$ is a triangulated subcategory. We say that $\mathcal{S}$* generates *$\mathcal{T}$ if*

$$\mathrm{Hom}(\mathcal{S}, x) = 0 \qquad \Longrightarrow \qquad x = 0.$$

That is, if x is an object of $\mathcal{T}$ and for all $s \in \mathcal{S}$ we have $\mathcal{T}(s,x) = 0$, then x is isomorphic to zero in $\mathcal{T}$.

LEMMA 6.2.9. *Let $\mathcal{T}$ be a triangulated category satisfying [TR5], and assume $\mathcal{S} \subset \mathcal{T}$ is an essentially small α–localising subcategory. Suppose that $\mathcal{S}$ generates $\mathcal{T}$, as in Definition 6.2.8. A morphism $x \longrightarrow y$ in $\mathcal{T}$ is an isomorphism if and only if its image by the functor*

$$\mathcal{T} \longrightarrow \mathcal{E}x(\mathcal{S}^{op}, \mathcal{A}b)$$

is an isomorphism in $\mathcal{E}x(\mathcal{S}^{op}, \mathcal{A}b)$.

Proof: One direction is obvious. If $x \longrightarrow y$ is an isomorphism in $\mathcal{T}$, then so is its image by

$$\mathcal{T} \longrightarrow \mathcal{E}x\Big(\{\mathcal{T}^\alpha\}^{op}, \mathcal{A}b\Big),$$

just because functors take isomorphisms to isomorphisms. We need to prove the converse.

Suppose therefore that $x \longrightarrow y$ is a morphism in $\mathcal{T}$, and that its image in $\mathcal{E}x(\mathcal{S}^{op}, \mathcal{A}b)$ is an isomorphism. We need to show that $x \longrightarrow y$ is an isomorphism in $\mathcal{T}$. In any case, we may complete to a triangle in $\mathcal{T}$

$$x \longrightarrow y \longrightarrow z \longrightarrow \Sigma x \longrightarrow \Sigma y.$$

Since the functor $\mathcal{T} \longrightarrow \mathcal{E}x(\mathcal{S}^{op}, \mathcal{A}b)$ is homological (see Proposition 6.2.6), this maps to an exact sequence in $\mathcal{E}x(\mathcal{S}^{op}, \mathcal{A}b)$. We are assuming that the functor $\mathcal{T} \longrightarrow \mathcal{E}x(\mathcal{S}^{op}, \mathcal{A}b)$ takes $x \longrightarrow y$ to an isomorphism. But then it also takes $\Sigma x \longrightarrow \Sigma y$ to an isomorphism. After all,

$$\mathcal{T}(-, \Sigma x)|_{\mathcal{S}} \longrightarrow \mathcal{T}(-, \Sigma y)|_{\mathcal{S}}$$

can also be written as

$$\mathcal{T}(\Sigma^{-1}-, x)|_{\mathcal{S}} \longrightarrow \mathcal{T}(\Sigma^{-1}-, y)|_{\mathcal{S}}$$

and $\Sigma^{-1} : \mathcal{S} \longrightarrow \mathcal{S}$ is an equivalence. From the exact sequence it follows that $\mathcal{T}(-, z)|_{\mathcal{S}}$ must vanish. That is, for all $s \in \mathcal{S}$, $\mathcal{T}(s, z) = 0$. But $\mathcal{S}$ generates; this means that z is isomorphic to zero, and by Corollary 1.2.6, $x \longrightarrow y$ is an isomorphism in $\mathcal{T}$. □

6.3. $\mathcal{E}x(\mathcal{S}^{op}, \mathcal{A}b)$ satisfies [AB4] and [AB4*], but not [AB5] or [AB5*]

We return now to considering the general case of

$$\mathcal{E}x(\mathcal{S}^{op}, \mathcal{A}b) \quad \subset \quad \mathcal{C}at(\mathcal{S}^{op}, \mathcal{A}b),$$

that is for this Section, $\mathcal{S}$ is an arbitrary category satisfying Hypothesis 6.1.1. The case $\mathcal{S} = \mathcal{T}^\alpha$ which we considered in Section 6.2 is a special case. Naturally, it is the case we are most interested in.

LEMMA 6.3.1. *Let $\mathcal{S}$ be a category satisfying Hypothesis 6.1.1. Then the abelian category $\mathcal{E}x(\mathcal{S}^{op}, \mathcal{A}b)$ satisfies [AB4*]; that is, products of exact sequences in $\mathcal{E}x(\mathcal{S}^{op}, \mathcal{A}b)$ are exact.*

Proof: The inclusion functor

$$\mathcal{E}x(\mathcal{S}^{op}, \mathcal{A}b) \longrightarrow \mathcal{C}at(\mathcal{S}^{op}, \mathcal{A}b)$$

respects exact sequences and products; see Lemma 6.1.4 for exact sequences, Lemma 6.1.5 for products. Take a family of exact sequences in $\mathcal{E}x(\mathcal{S}^{op}, \mathcal{A}b)$. Because the inclusion is exact, the sequence is exact in $\mathcal{C}at(\mathcal{S}^{op}, \mathcal{A}b)$. Because $\mathcal{A}b$, and hence also $\mathcal{C}at(\mathcal{S}^{op}, \mathcal{A}b)$, satisfy [AB4*], the product of the sequences in $\mathcal{C}at(\mathcal{S}^{op}, \mathcal{A}b)$ is also exact. But the product agrees in $\mathcal{E}x(\mathcal{S}^{op}, \mathcal{A}b)$ and $\mathcal{C}at(\mathcal{S}^{op}, \mathcal{A}b)$, and exactness agrees in the two categories. Hence the product sequence is exact in $\mathcal{E}x(\mathcal{S}^{op}, \mathcal{A}b)$. □

LEMMA 6.3.2. *Let $\mathcal{S}$ be a category satisfying Hypothesis 6.1.1. Then the abelian category $\mathcal{E}x(\mathcal{S}^{op}, \mathcal{A}b)$ satisfies [AB4]; that is, coproducts of exact sequences in $\mathcal{E}x(\mathcal{S}^{op}, \mathcal{A}b)$ are exact.*

Proof: Let M be a set, and suppose for each $\lambda \in M$ we have an exact sequence in $\mathcal{E}x(\mathcal{S}^{op}, \mathcal{A}b)$

$$0 \longrightarrow F'_\lambda \longrightarrow F_\lambda \longrightarrow F''_\lambda \longrightarrow 0.$$

We need to show that the coproduct of these sequences is exact. Since right exactness is clear, we need to show that

$$0 \longrightarrow \coprod_{\lambda\in M} F'_\lambda \xrightarrow{\ \phi\ } \coprod_{\lambda\in M} F_\lambda$$

is exact. For s an object in $\mathcal{S}$, pick an element in the kernel of

$$\left\{\coprod_{\lambda\in M} F'_\lambda\right\}(s) \xrightarrow{\ \phi\ } \left\{\coprod_{\lambda\in M} F_\lambda\right\}(s).$$

It is given by a subset $\Lambda \subset M$, where the cardinality of Λ is $< \alpha$, and a pair

$$s \longrightarrow \coprod_{\lambda\in\Lambda} s_\lambda, \qquad\qquad \beta \in \prod_{\lambda\in\Lambda} F'_\lambda(s_\lambda)$$

and the fact that the pair lies in the kernel means that under the map

$$\prod_{\lambda\in\Lambda} F'_\lambda(s) \xrightarrow{\ \phi\ } \prod_{\lambda\in\Lambda} F_\lambda(s)$$

$\beta \in \prod_{\lambda\in\Lambda} F'_\lambda(s)$ goes to an element $\phi(\beta)$, so that the pair

$$s \longrightarrow \coprod_{\lambda\in\Lambda} s_\lambda, \qquad\qquad \phi(\beta) \in \prod_{\lambda\in\Lambda} F_\lambda(s_\lambda)$$

is equivalent to zero. But by the definition of the equivalence relation (see Definition 6.1.8) this means that

$$s \longrightarrow \coprod_{\lambda\in\Lambda} s_\lambda$$

must factor as

$$s \longrightarrow \coprod_{\lambda\in\Lambda} k_\lambda \xrightarrow{\ \coprod_{\lambda\in\Lambda} f_\lambda\ } \coprod_{\lambda\in\Lambda} s_\lambda$$

so that under the map

$$\prod_{\lambda\in\Lambda} F_\lambda(s_\lambda) \longrightarrow \prod_{\lambda\in\Lambda} F_\lambda(k_\lambda)$$

the element $\phi(\beta)$ maps to zero. In other words, we have a commutative square of abelian groups

$$\begin{array}{ccc}
\prod_{\lambda\in\Lambda} F'_\lambda(s_\lambda) & \longrightarrow & \prod_{\lambda\in\Lambda} F'_\lambda(k_\lambda) \\
\downarrow & & \downarrow \\
\prod_{\lambda\in\Lambda} F_\lambda(s_\lambda) & \longrightarrow & \prod_{\lambda\in\Lambda} F_\lambda(k_\lambda)
\end{array}$$

and we have figured out that under the composite

$$\begin{array}{ccc}
\prod_{\lambda\in\Lambda} F'_\lambda(s_\lambda) & & \\
\downarrow & & \\
\prod_{\lambda\in\Lambda} F_\lambda(s_\lambda) & \longrightarrow & \prod_{\lambda\in\Lambda} F_\lambda(k_\lambda)
\end{array}$$

the element $\beta \in \prod_{\lambda\in\Lambda} F'_\lambda(s_\lambda)$ maps to zero. But by the commutativity it also maps to zero under the composite

$$\begin{array}{ccc}
\prod_{\lambda\in\Lambda} F'_\lambda(s_\lambda) & \longrightarrow & \prod_{\lambda\in\Lambda} F'_\lambda(k_\lambda) \\
& & {\scriptstyle\rho}\downarrow \\
& & \prod_{\lambda\in\Lambda} F_\lambda(k_\lambda).
\end{array}$$

Since the map ρ is injective (each map $F'_\lambda(k_\lambda) \longrightarrow F_\lambda(k_\lambda)$ is assumed injective), we deduce that the image of β via the map

$$\prod_{\lambda\in\Lambda} F'_\lambda(s_\lambda) \longrightarrow \prod_{\lambda\in\Lambda} F'_\lambda(k_\lambda)$$

already vanishes.

But then our pair

$$s \longrightarrow \coprod_{\lambda\in\Lambda} s_\lambda, \qquad\qquad \beta \in \prod_{\lambda\in\Lambda} F'_\lambda(s_\lambda)$$

is such that

$$s \longrightarrow \coprod_{\lambda\in\Lambda} s_\lambda$$

factors as

$$s \longrightarrow \coprod_{\lambda\in\Lambda} k_\lambda \xrightarrow{\coprod_{\lambda\in\Lambda} f_\lambda} \coprod_{\lambda\in\Lambda} s_\lambda$$

and β vanishes under the map

$$\prod_{\lambda\in\Lambda} F'_\lambda(s_\lambda) \longrightarrow \prod_{\lambda\in\Lambda} F'_\lambda(k_\lambda).$$

This means the pair

$$s \longrightarrow \coprod_{\lambda\in\Lambda} s_\lambda, \qquad \beta \in \prod_{\lambda\in\Lambda} F'_\lambda(s_\lambda)$$

is equivalent to zero. The kernel of

$$\left\{\coprod_{\lambda\in M} F'_\lambda\right\}(s) \xrightarrow{\ \phi\ } \left\{\coprod_{\lambda\in M} F_\lambda\right\}(s)$$

vanishes, and we have proved the left exactness. □

REMARK 6.3.3. Before we end this Section, we should warn the reader that in general, the category $\mathcal{E}x\big(\mathcal{S}^{op},\mathcal{A}b\big)$ satisfies neither [AB5] nor [AB5*]. We remind the reader that an abelian category is said to satisfy [AB5] if filtered direct limits of exact sequences are exact. It is said to satisfy [AB5*] if the dual category satisfies [AB5].

For [AB5*], this is clear; even the category $\mathcal{A}b$ does not satisfy [AB5*], and we can hardly expect a category of functors into $\mathcal{A}b$ to satisfy the condition.

More surprising is the fact that, in general, the category $\mathcal{E}x\big(\mathcal{S}^{op},\mathcal{A}b\big)$ need not satisfy [AB5]. In fact, let $\mathcal{T}$ be a triangulated category satisfying [TR5]. Let $\mathcal{S}\subset\mathcal{T}$ be an $\aleph_1$–localising subcategory. That is, $\mathcal{S}$ is closed under the formation of countable coproducts (in $\mathcal{T}$) of its objects. Suppose furthermore that $\mathcal{S}$ is $\aleph_1$–perfect. Let α be the cardinal $\aleph_1$.

By Proposition 6.2.6, the Yoneda map

$$\mathcal{T} \longrightarrow \mathcal{E}x\big(\mathcal{S}^{op},\mathcal{A}b\big),$$

that is the map sending an object $t\in\mathcal{T}$ to the functor $\mathcal{T}(-,t)|_{\mathcal{S}}$, is a homological functor respecting coproducts of fewer than $\alpha=\aleph_1$ objects. Let

$$X_0 \longrightarrow X_1 \longrightarrow X_2 \longrightarrow \cdots$$

be a sequence of objects and morphisms in $\mathcal{S}$. As in Definition 1.6.4, we can form the homotopy colimit, which is given by the triangle

$$\coprod_{i=0}^{\infty} X_i \xrightarrow{1-shift} \coprod_{i=0}^{\infty} X_i \longrightarrow \underrightarrow{\mathrm{Hocolim}}\, X_i \longrightarrow \Sigma\left\{\coprod_{i=0}^{\infty} X_i\right\}.$$

This sequence only involves countable coproducts, hence lies in $\mathcal{S}$. The functor

$$\mathcal{T} \longrightarrow \mathcal{E}x\big(\mathcal{S}^{op},\mathcal{A}b\big),$$

which respects countable coproducts, takes the map

$$\coprod_{i=0}^{\infty} X_i \xrightarrow{1 - shift} \coprod_{i=0}^{\infty} X_i$$

to the map

$$\coprod_{i=0}^{\infty} \mathcal{T}(-, X_i)|_{\mathcal{S}} \xrightarrow{1 - shift} \coprod_{i=0}^{\infty} \mathcal{T}(-, X_i)|_{\mathcal{S}}.$$

Since the abelian category $\mathcal{E}x(\mathcal{S}^{op}, \mathcal{A}b)$ satisfies [AB4], the kernel and cokernel compute colim^1 and colim terms, respectively. For a discussion of derived functors of limits see Section A.3, more particularly Remark A.3.6. In our case here, for the sequence

$$\mathcal{T}(-, X_0)|_{\mathcal{S}} \longrightarrow \mathcal{T}(-, X_1)|_{\mathcal{S}} \longrightarrow \mathcal{T}(-, X_2)|_{\mathcal{S}} \longrightarrow \cdots$$

in the abelian category $\mathcal{E}x(\mathcal{S}^{op}, \mathcal{A}b)$, we have that colim^1 is precisely the kernel of

$$\coprod_{i=0}^{\infty} \mathcal{T}(-, X_i)|_{\mathcal{S}} \xrightarrow{1 - shift} \coprod_{i=0}^{\infty} \mathcal{T}(-, X_i)|_{\mathcal{S}}.$$

But we have a vanishing composite

$$\Sigma^{-1} \underrightarrow{\operatorname{Hocolim}} X_i \longrightarrow \coprod_{i=0}^{\infty} X_i \xrightarrow{1 - shift} \coprod_{i=0}^{\infty} X_i$$

with $\Sigma^{-1} \underrightarrow{\operatorname{Hocolim}} X_i \in \mathcal{S}$. In other words, the map

$$\Sigma^{-1} \underrightarrow{\operatorname{Hocolim}} X_i \longrightarrow \coprod_{i=0}^{\infty} X_i$$

is an element of $\mathcal{T}\left(\Sigma^{-1} \underrightarrow{\operatorname{Hocolim}} X_i, \coprod_{i=0}^{\infty} X_i\right)\Big|_{\mathcal{S}}$, and lies in the kernel of

$$\begin{array}{ccc}
\coprod_{i=0}^{\infty} \mathcal{T}\left(\Sigma^{-1}\underrightarrow{\operatorname{Hocolim}} X_i, X_i\right)\Big|_{\mathcal{S}} & \xrightarrow{1 - shift} & \coprod_{i=0}^{\infty} \mathcal{T}\left(\Sigma^{-1}\underrightarrow{\operatorname{Hocolim}} X_i, X_i\right)\Big|_{\mathcal{S}} \\
\Big\downarrow \wr & & \Big\downarrow \wr \\
\mathcal{T}\left(\Sigma^{-1}\underrightarrow{\operatorname{Hocolim}} X_i, \coprod_{i=0}^{\infty} X_i\right)\Bigg|_{\mathcal{S}} & \xrightarrow{1 - shift} & \mathcal{T}\left(\Sigma^{-1}\underrightarrow{\operatorname{Hocolim}} X_i, \coprod_{i=0}^{\infty} X_i\right)\Bigg|_{\mathcal{S}}
\end{array}.$$

Since it is easy to find examples where this map fails to vanish, we see that in general, colim^1 can fail to vanish.

6.4. Projectives and injectives in the category $\mathcal{E}x(\mathcal{S}^{op}, \mathcal{A}b)$

Lemma 5.1.2 taught us that in the category $\mathcal{C}at(\mathcal{S}^{op}, \mathcal{A}b)$, the representable objects $\mathcal{S}(-, s)$ are projective. For this we needed nothing; $\mathcal{S}$ was only an additive category, not necessarily essentially small.

Now we leave the realm of the very general, and return to essentially small $\mathcal{S}$'s. The regular cardinal α is the one fixed throughout the Chapter, and the category $\mathcal{S}$ satisfies Hypothesis 6.1.1. It is an immediate consequence of Lemma 5.1.2, that the category $\mathcal{E}x(\mathcal{S}^{op}, \mathcal{A}b)$ has enough projectives. The category turns out in general *not* to have enough injectives. In this section, we will give a general discussion of the consequences of the existence of enough injectives; after all, some categories $\mathcal{E}x(\mathcal{S}^{op}, \mathcal{A}b)$ do have them. We refer the reader to an Apendix (see Section C.4) for a counterexample, showing that $\mathcal{E}x(\mathcal{S}^{op}, \mathcal{A}b)$ can fail to have enough injectives.

LEMMA 6.4.1. *Let $\mathcal{S}$ be a category satisfying Hypothesis 6.1.1. Let s be an object of the category $\mathcal{S}$. Then the representable functor $\mathcal{S}(-, s)$ is a projective object in the category $\mathcal{E}x(\mathcal{S}^{op}, \mathcal{A}b)$.*

Proof: Observe that the functor $Y_s(-) = \mathcal{S}(-, s)$ is an object of the category $\mathcal{E}x(\mathcal{S}^{op}, \mathcal{A}b)$. This is true because it clearly carries any coproduct of objects of $\mathcal{S}$ to a product of abelian groups. But by Lemma 5.1.2, the functor $\mathcal{S}(-, s)$ is projective as an object of $\mathcal{C}at(\mathcal{S}^{op}, \mathcal{A}b)$, hence also as an object of the exact subcategory $\mathcal{E}x(\mathcal{S}^{op}, \mathcal{A}b)$. □

LEMMA 6.4.2. *The projectives $\{\mathcal{S}(-, s), s \in \mathcal{S}\}$ give a generating set of projectives.*

Proof: We need to show that every non–zero object $F \in \mathcal{E}x(\mathcal{S}^{op}, \mathcal{A}b)$ admits a non–zero map

$$\mathcal{S}(-, s) \longrightarrow F(-)$$

for some $s \in \mathcal{S}$. But this is clear; if F in non–zero, then for some $s \in \mathcal{S}$, $F(s) \neq 0$. Yoneda's lemma says that $F(s)$ is in one–to–one correspondence with maps

$$\mathcal{S}(-, s) \longrightarrow F(-).$$

Hence there is a non–zero map $\mathcal{S}(-, s) \longrightarrow F(-)$. □

REMARK 6.4.3. Let $P = \coprod_{s \in \mathcal{S}} \mathcal{S}(-, s)$. Then P is a projective generator in the category $\mathcal{E}x(\mathcal{S}^{op}, \mathcal{A}b)$. From standard arguments, it follows formally that the category $\mathcal{E}x(\mathcal{S}^{op}, \mathcal{A}b)$ has enough projectives, and that any projective object is a direct summand of a coproduct of P's. We remind the reader how this goes.

LEMMA 6.4.4. *The category $\mathcal{E}x(\mathcal{S}^{op}, \mathcal{A}b)$ has enough projectives, and any projective object is a direct summand of a coproduct of P's.*

Proof: Let F be an object of $\mathcal{E}x(\mathcal{S}^{op}, \mathcal{A}b)$. Consider the set of all maps $P \longrightarrow F$, and take the coproduct. Let Q be the cokernel; that is we have an exact sequence

$$\coprod P \longrightarrow F \longrightarrow Q \longrightarrow 0.$$

Take any map $P \longrightarrow Q$. Since P is projective, the map factors through the epimorphism $F \longrightarrow Q$; it may be written as a composite $P \longrightarrow F \longrightarrow Q$. But any map $P \longrightarrow F$ factors thorough $\coprod P \longrightarrow F$, the coproduct of them all. Hence $P \longrightarrow F \longrightarrow Q$ composes to zero. This being true for all $P \longrightarrow Q$, it follows that Q vanishes. After all, P is a generator; any non–zero Q admits a non–zero map $P \longrightarrow Q$. Thus $Q = 0$ and $\coprod P \longrightarrow F$ is surjective.

This proves that $\mathcal{E}x(\mathcal{S}^{op}, \mathcal{A}b)$ has enough projectives. Now suppose F is projective. By the above, there is a surjective map

$$\coprod P \longrightarrow F.$$

But F is projective. Hence the identity $1 : F \longrightarrow F$ must factor through the surjective map $\coprod P \longrightarrow F$. We conclude that F is a direct summand of $\coprod P$. □

REMARK 6.4.5. The dual statements are far more subtle. It turns out that the category $\mathcal{E}x(\mathcal{S}^{op}, \mathcal{A}b)$ does not, in general, have enough injectives. For now, let us observe the trivial case. If $\alpha = \aleph_0$, then $\mathcal{E}x(\mathcal{S}^{op}, \mathcal{A}b) = \mathcal{C}at(\mathcal{S}^{op}, \mathcal{A}b)$ is a Grothendieck abelian category, and thus the existence of enough injectives is classical. But if $\alpha > \aleph_0$, we remind the reader that the category $\mathcal{E}x(\mathcal{S}^{op}, \mathcal{A}b)$ does not in general satisfy [AB5]. See Remark 6.3.3.

Even though enough injectives need not always exist, let us remind ourselves, briefly, what happens when they do.

LEMMA 6.4.6. *Suppose the category $\mathcal{E}x(\mathcal{S}^{op}, \mathcal{A}b)$ has enough injectives. Then it has an injective cogenerator. If I is an injective cogenerator, then any object F in $\mathcal{E}x(\mathcal{S}^{op}, \mathcal{A}b)$ admits an injection $F \longrightarrow \prod I$ into a product of I's. Any injective object is a direct summand of a product of I's.*

Proof: By Lemma 6.4.4, the category $\mathcal{E}x(\mathcal{S}^{op}, \mathcal{A}b)$ has a projective generator, which we will call P. Then every non–zero object k of $\mathcal{E}x(\mathcal{S}^{op}, \mathcal{A}b)$ admits a non–zero map $P \longrightarrow k$. That is, k contains the image of some non–zero map $P \longrightarrow k$, in other words, k contains a non–zero subobject, isomorphic to a quotient object of P.

Suppose that $\mathcal{E}x(\mathcal{S}^{op}, \mathcal{A}b)$ has enough injectives. For every quotient object of P, that is for every isomorphism class of exact sequences

$$P \longrightarrow q \longrightarrow 0,$$

choose an embedding $q \longrightarrow I_q$, with I_q an injective object in $\mathcal{E}x\big(\mathcal{S}^{op}, \mathcal{A}b\big)$. Let

$$I \quad = \quad \prod_{P \longrightarrow q \longrightarrow 0} I_q.$$

I assert that I is an injective cogenerator of $\mathcal{E}x\big(\mathcal{S}^{op}, \mathcal{A}b\big)$.

For let F be any object in $\mathcal{E}x\big(\mathcal{S}^{op}, \mathcal{A}b\big)$. Consider the set of all maps $F \longrightarrow I$, and take the product map,

$$F \longrightarrow \prod_{F \longrightarrow I} I.$$

Let k be the kernel of the map. I assert $k = 0$. For otherwise, k would have a non–zero subobject q, which is a quotient of P. We have a short exact

$$0 \longrightarrow q \longrightarrow k$$

and hence the embedding $q \longrightarrow I_q$ must factor as a map $k \longrightarrow I_q$. But k is a subobject of F, hence the map factors further as $F \longrightarrow I_q$, and finally

$$I_q \quad \subset \quad \prod_{P \longrightarrow q \longrightarrow 0} I_q \quad = \quad I$$

and we have a map $F \longrightarrow I$. By construction, this map fails to vanish on $q \subset k \subset F$. We have a map $F \longrightarrow I$, which fails to vanish on $k \subset F$, and k was defined as the kernel of

$$F \longrightarrow \prod_{F \longrightarrow I} I.$$

This is a contradiction, proving $k = 0$.

For every F, we have shown there is a monomorphism

$$F \longrightarrow \prod_{F \longrightarrow I} I.$$

If F is injective, this monomorphism must split, and F is a direct summand of a product of I's. $\square$

REMARK 6.4.7. The most interesting counterexample we have shows that, even when $\mathcal{S} = \mathcal{T}^\alpha$ for $\mathcal{T}$ a triangulated category satisfying [TR5], it may happen that $\mathcal{E}x\big(\mathcal{S}^{op}, \mathcal{A}b\big)$ does not have a cogenerator. If there is no cogenerator, then by Lemma 6.4.6, there cannot possibly be enough injectives. In the counterexample, which may be found in Section C.4, $\mathcal{T} = D(R)$, where R is any discrete valuation ring, and α is any regular cardinal $\geq \aleph_1$. But a slight modification of the argument allows us to find the same counterexample in $\mathcal{T} = D(\mathbb{Z})$, the derived category of $\mathbb{Z}$, or in $\mathcal{T}$ the homotopy category of spectra.

There is one more well–known fact about projective generators and injective cogenerators of which we want to remind the reader.

LEMMA 6.4.8. *Suppose I is an injective cogenerator for the category $\mathcal{E}x\big(\mathcal{S}^{op}, \mathcal{A}b\big)$. Then the sequence*

$$F' \longrightarrow F \longrightarrow F''$$

is exact in $\mathcal{E}x\big(\mathcal{S}^{op}, \mathcal{A}b\big)$ is and only if the sequence of abelian groups

$$\mathcal{E}x\big(\mathcal{S}^{op}, \mathcal{A}b\big)\Big\{F'', I\Big\} \longrightarrow \mathcal{E}x\big(\mathcal{S}^{op}, \mathcal{A}b\big)\Big\{F, I\Big\} \longrightarrow \mathcal{E}x\big(\mathcal{S}^{op}, \mathcal{A}b\big)\Big\{F', I\Big\}$$

is exact.

Proof: Let the homology to the sequence

$$F' \xrightarrow{\ \alpha\ } F \xrightarrow{\ \beta\ } F''$$

be H; that is,

$$H \quad = \quad \frac{\mathrm{Ker}(\beta)}{\mathrm{Im}(\alpha)}.$$

Because I is injective, we easily show that the homology of

$$\mathcal{E}x\big(\mathcal{S}^{op}, \mathcal{A}b\big)\Big\{F'', I\Big\} \longrightarrow \mathcal{E}x\big(\mathcal{S}^{op}, \mathcal{A}b\big)\Big\{F, I\Big\} \longrightarrow \mathcal{E}x\big(\mathcal{S}^{op}, \mathcal{A}b\big)\Big\{F', I\Big\}$$

is precisely $\mathcal{E}x\big(\mathcal{S}^{op}, \mathcal{A}b\big)\Big\{H, I\Big\}$. This will vanish precisely when H does. □

6.5. The relation between $A(\mathcal{T})$ and $\mathcal{E}x\Big(\{\mathcal{T}^\alpha\}^{op}, \mathcal{A}b\Big)$

As in the rest of this Chapter, let α be a fixed, regular cardinal.

In Chapter 5, more precisely Theorem 5.1.18, we learned about the universal homological functor. We remind the reader: given a triangulated category $\mathcal{T}$, there is a universal homological functor

$$\mathcal{T} \longrightarrow A(\mathcal{T}).$$

If $\mathcal{T}$ has small *Hom*–sets, and furthermore satisfies [TR5], there is a natural homological functor

$$\mathcal{T} \longrightarrow \mathcal{E}x\Big(\{\mathcal{T}^\alpha\}^{op}, \mathcal{A}b\Big).$$

This homological functor must factor through the universal homological functor. We have

$$\mathcal{T} \longrightarrow A(\mathcal{T}) \xrightarrow{\ \exists!\ } \mathcal{E}x\Big(\{\mathcal{T}^\alpha\}^{op}, \mathcal{A}b\Big).$$

We now propose to begin studying the functor $A(\mathcal{T}) \longrightarrow \mathcal{E}x\Big(\{\mathcal{T}^\alpha\}^{op}, \mathcal{A}b\Big)$. To simplify the notation, we will once again put $\mathcal{S} = \mathcal{T}^\alpha$, so that

$$\mathcal{E}x\Big(\{\mathcal{T}^\alpha\}^{op}, \mathcal{A}b\Big) \quad = \quad \mathcal{E}x\big(\mathcal{S}^{op}, \mathcal{A}b\big).$$

In this Section, we will prove that $\mathcal{E}x\big(\mathcal{S}^{op}, \mathcal{A}b\big)$ is a quotient of the category $A(\mathcal{T})$, in the sense of Gabriel. The reader is assumed to have some familiarity with Gabriel's [**15**]. For the reader's convenience, there is a condensed summary of the results we need in Appendix A, more specifically in Section A.2. In fact, this might be a good time for the reader to skim through Appendix A. In this Section, we use the results of Section A.2. In Section 7.1 we appeal to the work of Sections A.1. In the Sections 7.3 and 7.4, we depend mostly on the theory of Sections A.3 and A.4, although Sections A.1 also plays a rôle. Thus, for the next few Sections, we will be making heavy use of the theory of abelian categories, summarised for the reader's convenience in Appendix A.

LEMMA 6.5.1. *Let $\mathcal{T}$ be a triangulated category with small* Hom–*sets, satisfying [TR5]. The natural functor $\pi : A(\mathcal{T}) \longrightarrow \mathcal{E}x\big(\mathcal{S}^{op}, \mathcal{A}b\big)$ above is exact and respects coproducts.*

Proof: By Proposition 6.2.6 the functor $\mathcal{T} \longrightarrow \mathcal{E}x\big(\mathcal{S}^{op}, \mathcal{A}b\big)$ is homological and respects coproducts. From the fact that it is homological and from Theorem 5.1.18, we have that it factors

$$\mathcal{T} \longrightarrow A(\mathcal{T}) \stackrel{\pi}{\longrightarrow} \mathcal{E}x\big(\mathcal{S}^{op}, \mathcal{A}b\big)$$

with π an exact functor of abelian categories. Lemma 6.2.5 asserts that the functor $\mathcal{T} \longrightarrow \mathcal{E}x\big(\mathcal{S}^{op}, \mathcal{A}b\big)$ preserves coproducts. Lemma 5.1.24 tells us that, since $\mathcal{T}$ satisfies [TR5] and $\mathcal{T} \longrightarrow \mathcal{E}x\big(\mathcal{S}^{op}, \mathcal{A}b\big)$ preserves coproducts, it follows that

$$A(\mathcal{T}) \stackrel{\pi}{\longrightarrow} \mathcal{E}x\big(\mathcal{S}^{op}, \mathcal{A}b\big)$$

preserves coproducts. □

LEMMA 6.5.2. *The functor $\pi : A(\mathcal{T}) \longrightarrow \mathcal{E}x\big(\mathcal{S}^{op}, \mathcal{A}b\big)$ above is just the functor taking $F : \mathcal{T}^{op} \longrightarrow \mathcal{A}b$ to its restriction to $\mathcal{S} \subset \mathcal{T}$.*

Proof: The homological functor $H : \mathcal{T} \longrightarrow \mathcal{E}x\big(\mathcal{S}^{op}, \mathcal{A}b\big)$ is the functor taking an object $t \in \mathcal{T}$ to $\mathcal{T}(-,t)|_{\mathcal{S}}$. The functor $A(\mathcal{T}) \longrightarrow \mathcal{E}x\big(\mathcal{S}^{op}, \mathcal{A}b\big)$ is obtained as the universal factorisation

$$\mathcal{T} \longrightarrow A(\mathcal{T}) \stackrel{\pi}{\longrightarrow} \mathcal{E}x\big(\mathcal{S}^{op}, \mathcal{A}b\big)$$

of Theorem 5.1.18. By the proof of Theorem 5.1.18, the functor π is given as follows. An object $F \in A(\mathcal{T})$ is a functor $F : \mathcal{T}^{op} \longrightarrow \mathcal{A}b$, admitting a presentation

$$\mathcal{T}(-,s) \longrightarrow \mathcal{T}(-,t) \longrightarrow F(-) \longrightarrow 0$$

with $s, t \in \mathcal{T}$. The object $\pi(F)$ was obtained as the third term in the exact sequence

$$H(s) \longrightarrow H(t) \longrightarrow \pi(F) \longrightarrow 0.$$

In our case, we have an exact sequence of functors on $\mathcal{T}$

$$\mathcal{T}(-,s) \longrightarrow \mathcal{T}(-,t) \longrightarrow F(-) \longrightarrow 0.$$

If we restrict this sequence to $\mathcal{S} = \mathcal{T}^\alpha \subset \mathcal{T}$, we get an exact sequence functors on $\mathcal{S}$

$$\mathcal{T}(-,s)|_{\mathcal{S}} \longrightarrow \mathcal{T}(-,t)|_{\mathcal{S}} \longrightarrow F(-)|_{\mathcal{S}} \longrightarrow 0.$$

Since we know that $\mathcal{T}(-,s)|_{\mathcal{S}} = H(s)$ and $\mathcal{T}(-,t)|_{\mathcal{S}} = H(t)$, we deduce $\pi(F) = F(-)|_{\mathcal{S}}$. □

PROPOSITION 6.5.3. *Let $\mathcal{T}$ be a triangulated category with small* Hom*-sets, satisfying [TR5]. Put $\mathcal{S} = \mathcal{T}^\alpha$. The category $\mathcal{E}x(\mathcal{S}^{op}, \mathcal{A}b)$ is the quotient, in the sense of Gabriel, of the category $A(\mathcal{T})$ by a colocalizant subcategory we will denote $\mathcal{B}$, or $\mathcal{B}(\alpha)$ when we wish to remind ourselves of the dependence on the regular cardinal α. Precisely, $\pi : A(\mathcal{T}) \longrightarrow \mathcal{E}x(\mathcal{S}^{op}, \mathcal{A}b)$ is the quotient map, and it has a left adjoint L.*

Proof: We want to prove that the functor $\pi : A(\mathcal{T}) \longrightarrow \mathcal{E}x(\mathcal{S}^{op}, \mathcal{A}b)$ identifies $\mathcal{E}x(\mathcal{S}^{op}, \mathcal{A}b)$ as the Gabriel quotient $A(\mathcal{T})/\mathcal{B}$, with $\mathcal{B}$ a colocalizant subcategory. It suffices, by the dual of Proposition A.2.12, to produce a left adjoint $L : \mathcal{E}x(\mathcal{S}^{op}, \mathcal{A}b) \longrightarrow A(\mathcal{T})$ to the functor π, so that the unit of adjunction (dual to counit) is an isomorphism $\eta : 1 \longrightarrow \pi L$.

By Lemma 6.4.2, the representable functors on $\mathcal{S}$, that is the functors $\mathcal{S}(-,s)$, form a set of projective generators for $\mathcal{E}x(\mathcal{S}^{op}, \mathcal{A}b)$. Every object $F \in \mathcal{E}x(\mathcal{S}^{op}, \mathcal{A}b)$ admits a presentation

$$\coprod_{\lambda\in\Lambda} \mathcal{S}(-,s_\lambda) \longrightarrow \coprod_{\mu\in M} \mathcal{S}(-,t_\mu) \longrightarrow F(-) \longrightarrow 0.$$

Put Y_{s_λ} for the representable functor; that is

$$Y_{s_\lambda} \quad = \quad \mathcal{S}(-,s_\lambda).$$

Lemmas A.2.13 and A.2.15 give us that we need only prove the existence of the adjoints and the fact that η is an isomorphism on objects $\coprod_{\lambda\in\Lambda} Y_{s_\lambda}$ of $\mathcal{E}x(\mathcal{S}^{op}, \mathcal{A}b)$. More precisely, to prove the existence of the left adjoint L it suffices to show that the functor

$$\mathcal{E}x(\mathcal{S}^{op}, \mathcal{A}b)\left\{\coprod_{\lambda\in\Lambda} Y_{s_\lambda}, \pi(-)\right\}$$

is representable in $A(\mathcal{T})$. To show that the unit of adjunction is an isomorphism, observe that the functor π is exact by Lemma 6.5.1. Therefore Lemma A.2.15 applies, and we need only check that the natural transformation $\eta : 1 \longrightarrow \pi L$ is an isomorphism on objects $\coprod_{\lambda\in\Lambda} Y_{s_\lambda}$.

Let G be an arbitrary object in $A(\mathcal{T})$; that is, G is a functor $G : \mathcal{T}^{op} \longrightarrow \mathcal{A}b$, and by Lemma 5.1.5, G takes coproducts in $\mathcal{T}$ to products of abelian groups. By Lemma 6.5.2, $\pi(G)$ is just the restriction of G to $\mathcal{S} \subset \mathcal{T}$. To give a natural transformation

$$\coprod_{\lambda\in\Lambda} \mathcal{S}(-, s_\lambda) \longrightarrow G(-)|_{\mathcal{S}}$$

is to give, for each $\lambda \in \Lambda$, a natural transformation

$$\mathcal{S}(-, s_\lambda) \longrightarrow G(-)|_{\mathcal{S}}.$$

By Yoneda, this is the same as giving, for each $\lambda \in \Lambda$, an element of $G(s_\lambda)$. There is a 1–to–1 correspondence between natural transformations

$$\coprod_{\lambda\in\Lambda} \mathcal{S}(-, s_\lambda) \longrightarrow G(-)|_{\mathcal{S}}$$

and elements of

$$\prod_{\lambda\in\Lambda} G(s_\lambda).$$

Now by Lemma 5.1.5, G takes coproducts in $\mathcal{T}$ to products of abelian groups. That is,

$$G\left(\coprod_{\lambda\in\Lambda} s_\lambda\right) \quad = \quad \prod_{\lambda\in\Lambda} G(s_\lambda).$$

By Yoneda, elements of $G\left(\coprod_{\lambda\in\Lambda} s_\lambda\right)$ correspond 1–to–1 with natural tranformations

$$\mathcal{T}\left(-, \coprod_{\lambda\in\Lambda} s_\lambda\right) \longrightarrow G(-).$$

Summarising, there is a 1–to–1 natural correspondence between natural transformations

$$\coprod_{\lambda\in\Lambda} \mathcal{S}(-, s_\lambda) \longrightarrow G(-)|_{\mathcal{S}}$$

$$\mathcal{T}\left(-, \coprod_{\lambda\in\Lambda} s_\lambda\right) \longrightarrow G(-).$$

This exactly asserts that the functor

$$\mathcal{E}x\left(\mathcal{S}^{op}, \mathcal{A}b\right)\left\{\coprod_{\lambda\in\Lambda} Y_{s_\lambda}, \pi(-)\right\}$$

is representable in $A(\mathcal{T})$; in fact, it is represented by the object

$$\mathcal{T}\left(-, \coprod_{\lambda\in\Lambda} s_\lambda\right).$$

Thus the left adjoint L of π exists, and more concretely L takes

$$\coprod_{\lambda\in\Lambda} \mathcal{S}(-, s_\lambda) \qquad \text{to} \qquad \mathcal{T}\left(-, \coprod_{\lambda\in\Lambda} s_\lambda\right).$$

But then π takes this to its restriction to $\mathcal{S} \subset \mathcal{T}$, again by Lemma 6.5.2. But by Lemma 6.2.5 the restriction functor $\mathcal{T} \longrightarrow \mathcal{E}x(\mathcal{S}^{op}, \mathcal{A}b)$ respects coproducts. That is,

$$\mathcal{T}\left(-, \coprod_{\lambda\in\Lambda} s_\lambda\right)\Bigg|_{\mathcal{S}} \quad = \quad \coprod_{\lambda\in\Lambda} \mathcal{T}(-, s_\lambda)|_{\mathcal{S}} \quad = \quad \coprod_{\lambda\in\Lambda} \mathcal{S}(-, s_\lambda).$$

Thus πL takes $\coprod_{\lambda\in\Lambda} Y_{s_\lambda}$ to itself, proving that η is an isomorphism. □

COROLLARY 6.5.4. *Let $\mathcal{T}$ be a triangulated category with small* Hom*–sets, satisfying [TR5]. The natural functor $\pi : A(\mathcal{T}) \longrightarrow \mathcal{E}x(\mathcal{S}^{op}, \mathcal{A}b)$ respects products.*

Proof: The functor π has a left adjoint, and therefore must be left exact. □

REMARK 6.5.5. As a formal consequence of Proposition 6.5.3, (see also Proposition A.2.12) we have that the category $\mathcal{E}x(\mathcal{S}^{op}, \mathcal{A}b)$ is the quotient of $A(\mathcal{T})$ by a full subcategory $\mathcal{B} \subset A(\mathcal{T})$. The objects of $\mathcal{B} \subset A(\mathcal{T})$ are the objects F so that $\pi(F) = 0$. By Lemma 6.5.2, $\pi(F)$ is the restriction of F to $\mathcal{S} \subset \mathcal{T}$. Then $\mathcal{B}$ is the full subcategory of $A(\mathcal{T})$ consisting of the functors vanishing on $\mathcal{S}$. The functor

$$A(\mathcal{T}) \xrightarrow{\ \pi\ } \mathcal{E}x(\mathcal{S}^{op}, \mathcal{A}b)$$

respects products and coproducts. It respects coproducts by Lemma 6.5.1, products by Corollary 6.5.4. It follows that products and coproducts of objects in $\mathcal{B}$ lie in $\mathcal{B}$. The subcategory $\mathcal{B}$ is closed under subquotients, extensions, limits and colimits.

LEMMA 6.5.6. *We have an equivalence of categories $D(\mathcal{T}) \simeq A(\mathcal{T})$; see Definition 5.2.1 and Lemma 5.2.2. An object of $D(\mathcal{T})$ is a map $\{x \to y\} \in \mathcal{T}$. The object $\{x \to y\} \in D(\mathcal{T})$ lies in the kernel of*

$$D(\mathcal{T}) = A(\mathcal{T}) \longrightarrow \mathcal{E}x(\mathcal{S}^{op}, \mathcal{A}b)$$

if and only if the image of $\{x \to y\}$ via the map

$$\mathcal{T} \longrightarrow \mathcal{E}x(\mathcal{S}^{op}, \mathcal{A}b)$$

is zero.

Proof: By Remark 5.2.3, the object $\{x \to y\} \in D(\mathcal{T})$ may be thought of as a pair of maps in $A(\mathcal{T})$

$$\mathcal{T}(-,x) \xrightarrow{\alpha} F(-) \xrightarrow{\beta} \mathcal{T}(-,y),$$

with α epi and β mono. The equivalence $D(\mathcal{T}) \longrightarrow A(\mathcal{T})$ is simply the map that takes $\{x \to y\} \in D(\mathcal{T})$ to $F(-)$, that is to the image of the induced map in $A(\mathcal{T})$.

By Lemma 6.5.2, the map $A(\mathcal{T}) \longrightarrow \mathcal{E}x\big(\mathcal{S}^{op}, \mathcal{A}b\big)$ is just the restriction to $\mathcal{S} \subset \mathcal{T}$. It takes $F(-)$ to $F(-)|_{\mathcal{S}}$. That is, it takes $\{x \to y\} \in D(\mathcal{T})$ to the image of

$$\mathcal{T}(-,x)|_{\mathcal{S}} \longrightarrow \mathcal{T}(-,y)|_{\mathcal{S}}.$$

This image vanishes precisely when the image of $\{x \to y\}$ under the functor

$$\mathcal{T} \longrightarrow \mathcal{E}x\big(\mathcal{S}^{op}, \mathcal{A}b\big)$$

vanishes. □

This makes the next definition very natural.

DEFINITION 6.5.7. *A morphism $f : x \to y$ in $\mathcal{T}$ is called* α–phantom *if, under the natural map*

$$\mathcal{T} \longrightarrow \mathcal{E}x\Big(\{\mathcal{T}^\alpha\}^{op}, \mathcal{A}b\Big),$$

it maps to zero.

REMARK 6.5.8. With Definition 6.5.7, we can restate Lemma 6.5.6, to say that the kernel of the functor

$$D(\mathcal{T}) = A(\mathcal{T}) \longrightarrow \mathcal{E}x\Big(\{\mathcal{T}^\alpha\}^{op}, \mathcal{A}b\Big)$$

consists precisely of the α–phantom maps in $\mathcal{T}$, viewed as objects of $D(\mathcal{T})$. As above, we denote this kernel by $\mathcal{B} \subset D(\mathcal{T})$. When we want to emphasize the dependence on α, we will denote the kernel by $\mathcal{B}(\alpha)$.

We know, from Proposition 6.5.3, that $\mathcal{E}x\Big(\{\mathcal{T}^\alpha\}^{op}, \mathcal{A}b\Big)$ is the Gabriel quotient of $D(\mathcal{T})$ by $\mathcal{B}(\alpha)$ (see Remark 6.5.5). For many $\mathcal{T}$'s, we know that $\mathcal{T}^\alpha$ is essentially small for all α; for example, this is true when $\mathcal{T} = D(\mathbb{Z})$, the derived category of $\mathbb{Z}$. When this is the case, the categories $\mathcal{E}x\Big(\{\mathcal{T}^\alpha\}^{op}, \mathcal{A}b\Big)$ are well–powered for every α. This means every object of $\mathcal{E}x\Big(\{\mathcal{T}^\alpha\}^{op}, \mathcal{A}b\Big)$ has only a small set of quotients. We see in Corollary C.3.3 and Remark C.3.4 that the category $D(\mathcal{T})$ is not in general well–powered. Specifically, it is not well–powered for $\mathcal{T} = D(\mathbb{Z})$. It follows that, when $\mathcal{T} = D(\mathbb{Z})$, the category $D(\mathcal{T})$ can never agree with

$$\frac{D(\mathcal{T})}{B(\alpha)} = \mathcal{E}x\Big(\{\mathcal{T}^\alpha\}^{op}, \mathcal{A}b\Big).$$

We conclude that, for every regular cardinal α, $B(\alpha) \neq 0$. There are non-zero α-phantom maps for every α.

6.6. History of the results of Chapter 6

The results of this Chapter are essentially all new. Only the trivial case, that is $\alpha = \aleph_0$, has been studied at all.

CHAPTER 7

Homological properties of $\mathcal{E}x(\mathcal{S}^{op}, \mathcal{A}b)$

We have learned, in the previous Chapter, some of the basic properties of the categories $\mathcal{E}x(\mathcal{S}^{op}, \mathcal{A}b)$. In Appendix C, more specifically in Section C.4, we can see that in general the categories $\mathcal{E}x(\mathcal{S}^{op}, \mathcal{A}b)$ need not have enough injectives; in fact, they can fail to have cogenerators. See also Lemma 6.4.6 for the fact that, if $\mathcal{E}x(\mathcal{S}^{op}, \mathcal{A}b)$ fails to have a cogenerator, it certainly cannot have enough injectives.

But nevertheless, something positive is true. This Chapter will be devoted to proving the positive results we have. These positive results are fragmented and inconclusive. They are included for the benefit of future workers on the subject, who will hopefully be able to push them further. The casual reader is advised to skip this Chapter; the results do not affect the development later in the book.

One of the first questions to ask, is whether the categories $\mathcal{E}x(\mathcal{S}^{op}, \mathcal{A}b)$ satisfy [AB4.5]. In Proposition A.5.12, we saw that this need not be the case, for a general $\mathcal{S}$. In order to have a hope, we must at the very least assume $\mathcal{S}$ to be triangulated. Even if $\mathcal{S}$ is triangulated, I have not been able to prove that $\mathcal{E}x(\mathcal{S}^{op}, \mathcal{A}b)$ satisfies [AB4.5]. In this Chapter, I include what I could prove in that direction.

Also included are some other amusing homological facts about the categories $\mathcal{E}x(\mathcal{S}^{op}, \mathcal{A}b)$. Although for these it is not essential to assume that $\mathcal{S}$ is triangulated, we will make our life simpler by treating the restricted case. In other words, as was the case in the previous Chapter, α is a fixed regular cardinal. But from now on, $\mathcal{S}$ will be an essentially small triangulated category satisfying [TR5(α)].

7.1. $\mathcal{E}x(\mathcal{S}^{op}, \mathcal{A}b)$ as a locally presentable category

In this Section, we will be appealing to the material of Section A.1. Recall that Section A.1 deals with locally presentable categories. Lemma A.1.3 tells us that α–filtered colimits agree in $\mathcal{E}x(\mathcal{S}^{op}, \mathcal{A}b)$ and $\mathcal{C}at(\mathcal{S}^{op}, \mathcal{A}b)$, and as an immediate consequence α–filtered colimits in $\mathcal{E}x(\mathcal{S}^{op}, \mathcal{A}b)$ are exact. Proposition A.1.9 tells us that the category $\mathcal{E}x(\mathcal{S}^{op}, \mathcal{A}b)$ is locally presentable. That is, for every object $a \in \mathcal{E}x(\mathcal{S}^{op}, \mathcal{A}b)$, there exists a cardinal

β, in general depending on a, so that $\text{Hom}(a, -)$ commutes with β–filtered colimits. As a very special case, we have

LEMMA 7.1.1. *Let $\mathcal{S} \longrightarrow \mathcal{E}x(\mathcal{S}^{op}, \mathcal{A}b)$ be the Yoneda inclusion; it is the functor sending $s \in \mathcal{S}$ to $\mathcal{S}(-, s)$. For every $s \in \mathcal{S}$, its image in $\mathcal{E}x(\mathcal{S}^{op}, \mathcal{A}b)$ is an α–presentable object. That is, $Hom(s, -)$ commutes with α–filtered colimits in $\mathcal{E}x(\mathcal{S}^{op}, \mathcal{A}b)$.*

Proof: By Yoneda's Lemma, for $s \in \mathcal{S}$ and $F \in \mathcal{E}x(\mathcal{S}^{op}, \mathcal{A}b)$,

$$\text{Hom}(s, F) \quad = \quad \mathcal{E}x(\mathcal{S}^{op}, \mathcal{A}b)\Big\{\mathcal{S}(-, s), F(-)\Big\} \quad = \quad F(s).$$

By Lemma A.1.3, α–filtered colimits agree in $\mathcal{E}x(\mathcal{S}^{op}, \mathcal{A}b)$ and $\mathcal{C}at(\mathcal{S}^{op}, \mathcal{A}b)$. For an α–filtered colimit,

$$\Big\{\underrightarrow{\text{colim}} F\Big\}(s) \quad = \quad \underrightarrow{\text{colim}}\,\{F(s)\}.$$

But Yoneda's Lemma identifies this as

$$\text{Hom}\left(s, \underrightarrow{\text{colim}} F\right) \quad = \quad \underrightarrow{\text{colim}}\Big\{\text{Hom}(s, F)\Big\}.$$

□

COROLLARY 7.1.2. *In particular, any map*

$$\mathcal{S}(-, s) \longrightarrow \coprod_{\lambda \in \Lambda} F_\lambda(-)$$

factors as

$$\mathcal{S}(-, s) \longrightarrow \coprod_{\lambda \in M} F_\lambda(-) \qquad \subset \qquad \coprod_{\lambda \in \Lambda} F_\lambda(-)$$

where $M \subset \Lambda$, and the cardinality of M is $< \alpha$.

Proof: The object $\coprod_{\lambda \in \Lambda} F_\lambda(-)$ is the direct limit of all the coproducts over subsets of cardinality $< \alpha$. This is an α–filtered colimit. Hence by Lemma 7.1.1, a map from $\mathcal{S}(-, s)$ into this α–filtered colimit must factor through one of the terms. □

So far, we have not used the fact that the category $\mathcal{S}$ is triangulated. Now we will start.

LEMMA 7.1.3. *Let $\mathcal{S}$ be an essentially small triangulated category, satisfying [TR5(α)]. Then the abelian category $A(\mathcal{S})$ satisfies [AB4.5(α)], and there is a natural full embedding*

$$A(\mathcal{S}) \longrightarrow \mathcal{E}x(\mathcal{S}^{op}, \mathcal{A}b)$$

which is exact, and respects coproducts of fewer than α objects.

Proof: By Lemma 5.1.24, if $\mathcal{S}$ is a triangulated category satisfying [TR5(α)], then $A(\mathcal{S})$ is an abelian category which satisfies [AB3(α)]; coproducts of fewer than α objects exist in $\mathcal{S}$. From Corollary 5.1.23, we know that the category $A(\mathcal{S})$ has enough injectives. It therefore follows that $A(\mathcal{S})$ satisfies [AB4.5(α)]; see Lemma A.3.15, Definition A.3.16 and Remark A.3.17.

The other statements of the Lemma, which still remain to be proved, are that the natural functor

$$A(\mathcal{S}) \longrightarrow \mathcal{E}x\big(\mathcal{S}^{op}, \mathcal{A}b\big)$$

is a full embedding, is exact, and preserves coproducts of fewer than α objects. First we should remind the reader what the natural functor is. Recall that we always have a map

$$\mathcal{S} \longrightarrow \mathcal{E}x\big(\mathcal{S}^{op}, \mathcal{A}b\big)$$

taking $s \in \mathcal{S}$ to the representable functor $\mathcal{S}(-, s)$. This functor is homological, and therefore factors uniquely (up to canonical equivalence) through the universal homological functor of Theorem 5.1.18. It factors as

$$\mathcal{S} \longrightarrow A(\mathcal{S}) \xrightarrow{\exists!} \mathcal{E}x\big(\mathcal{S}^{op}, \mathcal{A}b\big)$$

where the canonical functor $A(\mathcal{S}) \longrightarrow \mathcal{E}x\big(\mathcal{S}^{op}, \mathcal{A}b\big)$ is exact. We need to show that this unique functor is a full embedding(=fully faithful), and respects coproducts of fewer than α objects.

Observe that by Lemma 6.1.17, the composite functor

$$\mathcal{S} \longrightarrow A(\mathcal{S}) \longrightarrow \mathcal{E}x\big(\mathcal{S}^{op}, \mathcal{A}b\big)$$

respects coproducts of fewer than α objects. From the fact that $\mathcal{S}$ satisfies [TR5(α)], coupled with Lemma 5.1.24, we deduce that the functor $A(\mathcal{S}) \longrightarrow \mathcal{E}x\big(\mathcal{S}^{op}, \mathcal{A}b\big)$ also respects coproducts of fewer than α objects. It remains only to show the functor fully faithful.

Consider now the longer composite

$$\mathcal{S} \longrightarrow A(\mathcal{S}) \longrightarrow \mathcal{E}x\big(\mathcal{S}^{op}, \mathcal{A}b\big) \qquad \subset \qquad \mathcal{C}at\big(\mathcal{S}^{op}, \mathcal{A}b\big).$$

It is clear that the canonical functor $A(\mathcal{S}) \longrightarrow \mathcal{C}at\big(\mathcal{S}^{op}, \mathcal{A}b\big)$ is a full embedding. The category $A(\mathcal{S})$ is, by definition, a full subcategory of $\mathcal{C}at\big(\mathcal{S}^{op}, \mathcal{A}b\big)$; see Definition 5.1.3. But then the inclusion of $A(\mathcal{S})$ in $\mathcal{E}x\big(\mathcal{S}^{op}, \mathcal{A}b\big)$ must also be fully faithful. □

REMARK 7.1.4. In Lemma 7.1.1, we proved that every object in $\mathcal{S} \subset \mathcal{E}x\big(\mathcal{S}^{op}, \mathcal{A}b\big)$ is α–presentable. In Lemma 7.1.3, we factored the inclusion as $\mathcal{S} \subset A(\mathcal{S}) \subset \mathcal{E}x\big(\mathcal{S}^{op}, \mathcal{A}b\big)$. It is natural to ask if the objects in the larger $A(\mathcal{S}) \supset \mathcal{S}$ are also α–presentable. The answer is yes, and now we wish to prove it.

LEMMA 7.1.5. *Every object $F \in A(\mathcal{S}) \subset \mathcal{E}x\big(\mathcal{S}^{op}, \mathcal{A}b\big)$ is α–presentable.*

Proof: Let F be an object in $A(\mathcal{S})$. By Definition 5.1.3, it admits a presentation

$$\mathcal{S}(-,s) \longrightarrow \mathcal{S}(-,t) \longrightarrow F(-) \longrightarrow 0,$$

and $\mathcal{S}(-,s)$, $\mathcal{S}(-,t)$ are object in $\mathcal{S} \subset \mathcal{E}x\big(\mathcal{S}^{op}, \mathcal{A}b\big)$, hence α–presentable by Lemma 7.1.1. Taking maps in $\mathcal{E}x\big(\mathcal{S}^{op}, \mathcal{A}b\big)$ into an α–filtered colimit, we deduce an exact sequence

$$\begin{array}{ccccc} 0 & \longrightarrow & \mathrm{Hom}(F, \varinjlim\phi) & \longrightarrow & \mathrm{Hom}(t, \varinjlim\phi) \\ & & & & \downarrow \\ & & & & \mathrm{Hom}(s, \varinjlim\phi) \end{array}$$

Since s and t are α–presentable, this identifies as

$$\begin{array}{ccccc} 0 & \longrightarrow & \mathrm{Hom}(F, \varinjlim\phi) & \longrightarrow & \varinjlim\Big\{\mathrm{Hom}(t, \phi)\Big\} \\ & & & & \downarrow \\ & & & & \varinjlim\Big\{\mathrm{Hom}(s, \phi)\Big\}. \end{array}$$

But the category $\mathcal{A}b$ of abelian groups satisfies [AB5]; filtered colimits are exact. Hence this identifies as

$$\begin{array}{ccccc} 0 & \longrightarrow & \varinjlim\Big\{\mathrm{Hom}(F, \phi)\Big\} & \longrightarrow & \varinjlim\Big\{\mathrm{Hom}(t, \phi)\Big\} \\ & & & & \downarrow \\ & & & & \varinjlim\Big\{\mathrm{Hom}(s, \phi)\Big\}, \end{array}$$

which allows us to deduce that

$$\varinjlim\Big\{\mathrm{Hom}(F, \phi)\Big\} \quad = \quad \mathrm{Hom}(F, \varinjlim\phi).$$

□

7.2. Homological objects in $\mathcal{E}x\big(\mathcal{S}^{op}, \mathcal{A}b\big)$

In keeping with our conventions for this Chapter, the regular cardinal α is fixed, and $\mathcal{S}$ is an essentially small triangulated category satisfying [TR5(α)], that is closed under the formation of coproducts of $< \alpha$ of its objects.

REMARK 7.2.1. The next few lemmas concern homological functors. Recall: an object $F \in \mathcal{E}x\big(\mathcal{S}^{op}, \mathcal{A}b\big)$, that is a functor

$$F : \mathcal{S}^{op} \longrightarrow \mathcal{A}b$$

is called *homological* if it takes triangles to long exact sequences.

In the remainder of the Chapter, we will be considering many functors into the abelian category $\mathcal{E}x\big(\mathcal{S}^{op},\mathcal{A}b\big)$, and some of these will be homological. There exist interesting homological functors into the abelian category $\mathcal{E}x\big(\mathcal{S}^{op},\mathcal{A}b\big)$. But the abelian category is itself a category of functors, whose objects may be homological. Because this could lead to nightmarish confusions, an object of $\mathcal{E}x\big(\mathcal{S}^{op},\mathcal{A}b\big)$ which is homological, as a functor

$$F:\mathcal{S}^{op}\longrightarrow\mathcal{A}b,$$

will be called a *homological object of* $\mathcal{E}x\big(\mathcal{S}^{op},\mathcal{A}b\big)$.

LEMMA 7.2.2. *Let $\mathcal{S}$ be an essentially small triangulated category, satisfying [TR5(α)]. An α–filtered colimit of homological objects in $\mathcal{E}x\big(\mathcal{S}^{op},\mathcal{A}b\big)$ is a homological object in $\mathcal{E}x\big(\mathcal{S}^{op},\mathcal{A}b\big)$.*

Proof: By Lemma A.1.3, α–filtered colimits agree in $\mathcal{E}x\big(\mathcal{S}^{op},\mathcal{A}b\big)$ and $\mathcal{C}at\big(\mathcal{S}^{op},\mathcal{A}b\big)$. Suppose $\{F_i, i\in\mathcal{I}\}$ is an α–directed family of homological objects. Let

$$x\longrightarrow y\longrightarrow z\longrightarrow\Sigma x$$

be a triangle in $\mathcal{S}$. For each $i\in\mathcal{I}$, the sequence

$$F_i(z)\longrightarrow F_i(y)\longrightarrow F_i(x)$$

is an exact sequence of abelian groups. The category $\mathcal{A}b$ of abelian groups satisfies [AB5], and hence the sequence of abelian groups

$$\operatorname*{colim}_{i\in\mathcal{I}} F_i(z)\longrightarrow\operatorname*{colim}_{i\in\mathcal{I}} F_i(y)\longrightarrow\operatorname*{colim}_{i\in\mathcal{I}} F_i(x)$$

is exact. But since these colimits agree in $\mathcal{E}x\big(\mathcal{S}^{op},\mathcal{A}b\big)$ and $\mathcal{C}at\big(\mathcal{S}^{op},\mathcal{A}b\big)$, this is precisely the colimit of the functors F_i, taken in $\mathcal{E}x\big(\mathcal{S}^{op},\mathcal{A}b\big)$, applied to

$$x\longrightarrow y\longrightarrow z\longrightarrow\Sigma x.$$

Since the sequence is exact, it follows that $\operatorname{colim}_{i\in\mathcal{I}} F_i$ is homological. □

EXAMPLE 7.2.3. Any coproduct in $\mathcal{E}x\big(\mathcal{S}^{op},\mathcal{A}b\big)$ of representable functors is homological. For let $\{s_\lambda,\lambda\in\Lambda\}$ be a set of objects in $\mathcal{S}$. The object of $\mathcal{E}x\big(\mathcal{S}^{op},\mathcal{A}b\big)$

$$\coprod_{\lambda\in\Lambda}\mathcal{S}(-,s_\lambda)$$

is, as in Corollary 7.1.2, the colimit of all coproducts over subsets of Λ of cardinality $<\alpha$. This is an α–filtered colimit, and hence by Lemma 7.2.2, it suffices to prove it an α–filtered colimit of homological functors. In other words, we are reduced to showing that

$$\coprod_{\lambda\in\Lambda}\mathcal{S}(-,s_\lambda)$$

is homological when the cardinality of Λ is $< \alpha$. But from Lemma 6.1.17, we learn that

$$\coprod_{\lambda\in\Lambda} \mathcal{S}(-, s_\lambda) \quad = \quad \mathcal{S}\left(-, \coprod_{\lambda\in\Lambda} s_\lambda\right).$$

Being representable, it is homological.

It turns out that there is a converse to Lemma 7.2.2. Lemma 7.2.2 asserts that every α–filtered colimit of homological objects is homological. But it turns out that every homological object is an α–filtered colimit of representables.

LEMMA 7.2.4. *Let $\mathcal{S}$ be an essentially small triangulated category, satisfying [TR5(α)]. Any homological object $F \in \mathcal{E}x(\mathcal{S}^{op}, \mathcal{A}b)$ is an α–filtered colimit of representable objects.*

Proof: Since α–filtered colimits are the same in the categories $\mathcal{E}x(\mathcal{S}^{op}, \mathcal{A}b)$ and $\mathcal{C}at(\mathcal{S}^{op}, \mathcal{A}b)$, it suffices to prove that F is an α–filtered colimit, in $\mathcal{C}at(\mathcal{S}^{op}, \mathcal{A}b)$, of representables. Clearly, F is the colimit of all the representables mapping to it. It needs to be shown that this colimit is α–filtered.

Suppose we are given any collection of $< \alpha$ representables mapping to F. I assert that, in the category of representables mapping to F, there is a coproduct. We have

$$\mathcal{S}(-, s_\lambda) \longrightarrow F(-)$$

for $\lambda \in \Lambda$, with the cardinality of Λ being $< \alpha$. By Yoneda, this gives, for every $\lambda \in \Lambda$, an element of $f_\lambda \in F(s_\lambda)$. That is, we have an element $\prod_{\lambda\in\Lambda} f_\lambda \in \prod_{\lambda\in\Lambda} F(s_\lambda)$. But F lies in $\mathcal{E}x(\mathcal{S}^{op}, \mathcal{A}b)$, and therefore

$$\prod_{\lambda\in\Lambda} F(s_\lambda) \quad = \quad F\left(\coprod_{\lambda\in\Lambda} s_\lambda\right).$$

We have constructed an element $\prod_{\lambda\in\Lambda} f_\lambda \in F\left(\coprod_{\lambda\in\Lambda} s_\lambda\right)$. By Yoneda, it corresponds to a morphism

$$\mathcal{S}\left(-, \coprod_{\lambda\in\Lambda} s_\lambda\right) \longrightarrow F(-).$$

What we have proved is that the morphisms

$$\mathcal{S}(-, s_\lambda) \longrightarrow F(-)$$

all factor through

$$\mathcal{S}\left(-, \coprod_{\lambda\in\Lambda} s_\lambda\right) \longrightarrow F(-).$$

We next want to show that

$$\mathcal{S}\left(-, \coprod_{\lambda \in \Lambda} s_\lambda\right) \longrightarrow F(-)$$

is the coproduct, in the category of representables mapping to F, of the objects

$$\mathcal{S}(-, s_\lambda) \longrightarrow F(-).$$

Suppose therefore that we are given a $\mathcal{S}(-,t) \longrightarrow F(-)$, and for each $\lambda \in \Lambda$ a commutative diagram

$$\begin{array}{ccc} \mathcal{S}(-, s_\lambda) & \longrightarrow & \mathcal{S}(-,t) \\ \downarrow & & \downarrow \\ F(-) & \xrightarrow{1} & F(-). \end{array}$$

To prove that

$$\mathcal{S}\left(-, \coprod_{\lambda \in \Lambda} s_\lambda\right) \longrightarrow F(-)$$

is the coproduct of

$$\mathcal{S}(-, s_\lambda) \longrightarrow F(-),$$

we must produce a unique factorisation through

$$\begin{array}{ccccc} \mathcal{S}(-, s_\lambda) & \longrightarrow & \mathcal{S}\left(-, \coprod_{\lambda \in \Lambda} s_\lambda\right) & \longrightarrow & \mathcal{S}(-,t) \\ \downarrow & & \downarrow & & \downarrow \\ F(-) & \xrightarrow{1} & F(-) & \xrightarrow{1} & F(-). \end{array}$$

But our morphism $\mathcal{S}(-,t) \longrightarrow F(-)$ corresponds to $f \in F(t)$. The given commutative diagrams

$$\begin{array}{ccc} \mathcal{S}(-, s_\lambda) & \longrightarrow & \mathcal{S}(-,t) \\ \downarrow & & \downarrow \\ F(-) & \xrightarrow{1} & F(-) \end{array}$$

tells us that, for every $\lambda \in \Lambda$, the the natural map $F(t) \longrightarrow F(s_\lambda)$ takes $f \in F(t)$ to $f_\lambda \in F(s_\lambda)$. But then the natural map

$$\coprod_{\lambda \in \Lambda} s_\lambda \longrightarrow t$$

induces a map $F(t) \longrightarrow F(\coprod_{\lambda\in\Lambda} s_\lambda)$, which takes $f \in F(t)$ to $\prod_{\lambda\in\Lambda} f_\lambda \in F(\coprod_{\lambda\in\Lambda} s_\lambda)$. This establishes the factorisation

$$\begin{array}{ccccc} \mathcal{S}(-,s_\lambda) & \longrightarrow & \mathcal{S}\left(-,\coprod_{\lambda\in\Lambda} s_\lambda\right) & \longrightarrow & \mathcal{S}(-,t) \\ \downarrow & & \downarrow & & \downarrow \\ F(-) & \xrightarrow{1} & F(-) & \xrightarrow{1} & F(-). \end{array}$$

The uniqueness is obvious.

To prove that the category of natural transformations

$$\mathcal{S}(-,s) \longrightarrow F(-)$$

is α–filtered, it remains to show that any two morphisms are coequalised. Given two objects

$$\mathcal{S}(-,s) \longrightarrow F(-) \qquad \text{and} \qquad \mathcal{S}(-,t) \longrightarrow F(-)$$

and two morphisms between them, we need to show they can be coequalised.

Two objects as above, and two morphisms between them, amount to a diagram

$$\mathcal{S}(-,s) \underset{g}{\overset{f}{\rightrightarrows}} \mathcal{S}(-,t) \xrightarrow{\ \theta\ } F(-)$$

with $\theta f = \theta g$. By Yoneda's Lemma, the natural transformation

$$\mathcal{S}(-,t) \xrightarrow{\ \theta\ } F(-)$$

corresponds to an element $x \in F(t)$. The fact that $\theta f = \theta g$ corresponds, via Yoneda, to the statement that the two maps

$$F(t) \underset{F(g)}{\overset{F(f)}{\rightrightarrows}} F(s)$$

take $x \in F(t)$ to the same element of $F(s)$. In other words, $f - g : s \longrightarrow t$ induces a map $F(t) \longrightarrow F(s)$, taking $x \in F(t)$ to zero. Now complete $f - g : s \longrightarrow t$ to a triangle

$$r \longrightarrow s \xrightarrow{f-g} t \longrightarrow \Sigma r.$$

Because F is homological, the sequence

$$F(\Sigma r) \longrightarrow F(t) \xrightarrow{F(f-g)} F(s)$$

is exact. Now $x \in F(t)$ lies in the kernel of the map $F(f-g)$, and hence there is a $y \in F(\Sigma r)$ mapping to $x \in F(t)$. By Yoneda, this means the natural transformation

$$\mathcal{S}(-,t) \xrightarrow{\ \theta\ } F(-)$$

factors as

$$\mathcal{S}(-,t) \longrightarrow \mathcal{S}(-,\Sigma r) \longrightarrow F(-)$$

and the two composites

$$\mathcal{S}(-,s) \underset{g}{\overset{f}{\rightrightarrows}} \mathcal{S}(-,t) \longrightarrow \mathcal{S}(-,\Sigma r)$$

are equal. The given two maps

$$\mathcal{S}(-,s) \underset{g}{\overset{f}{\rightrightarrows}} \mathcal{S}(-,t) \xrightarrow{\ \theta\ } F(-)$$

are coequalised by a map into

$$\mathcal{S}(-,\Sigma r) \longrightarrow F(-).$$

□

LEMMA 7.2.5. *Let $\mathcal{S}$ be an essentially small triangulated category, satisfying [TR5(α)]. Suppose*

$$0 \longrightarrow F \longrightarrow G \longrightarrow H \longrightarrow 0$$

is a short exact sequence in $\mathcal{E}x(\mathcal{S}^{op}, \mathcal{A}b)$. If any two of F, G or H are homological objects, then so is the third.

Proof: Let

$$x \longrightarrow y \longrightarrow z \longrightarrow \Sigma x$$

be a triangle in $\mathcal{S}$. Consider the short exact sequence of chain complexes

$$\begin{array}{ccccccccc}
0 & & 0 & & 0 & & 0 & & 0 \\
\downarrow & & \downarrow & & \downarrow & & \downarrow & & \downarrow \\
F\{\Sigma x\} & \longrightarrow & Fz & \longrightarrow & Fy & \longrightarrow & Fx & \longrightarrow & F\{\Sigma^{-1}z\} \\
\downarrow & & \downarrow & & \downarrow & & \downarrow & & \downarrow \\
G\{\Sigma x\} & \longrightarrow & Gz & \longrightarrow & Gy & \longrightarrow & Gx & \longrightarrow & G\{\Sigma^{-1}z\} \\
\downarrow & & \downarrow & & \downarrow & & \downarrow & & \downarrow \\
H\{\Sigma x\} & \longrightarrow & Hz & \longrightarrow & Hy & \longrightarrow & Hx & \longrightarrow & H\{\Sigma^{-1}z\} \\
\downarrow & & \downarrow & & \downarrow & & \downarrow & & \downarrow \\
0 & & 0 & & 0 & & 0 & & 0
\end{array}$$

From the long exact sequence of the homology of these chain complexes we learn that, if any two of the rows are exact, then so is the third. □

7.3. A technical lemma and some consequences

The regular cardinal α is fixed, as was assumed throughout the Chapter. The category $\mathcal{S}$ is an essentially small triangulated category satisfying [TR5(α)], that is closed under the formation of coproducts of $< \alpha$ of its objects.

We warn the reader that this section and the next rely heavily on the material developed in Appendix A. We will make frequent use of the notions of Sections A.3 and A.4. The reader should be familiar with the definition of a sequence of length γ (Definition A.3.5), a Mittag–Leffler sequence (Definition A.3.10), and the lemma about the vanishing of $\underleftarrow{\lim}^n$ of Mittag–Leffler sequences in the presence of enough projectives (Lemma A.3.15). Definition A.3.16 encapsulates it concisely, if somewhat mysteriously: an abelian category satisfies [AB4.5*] if it satisfies the vanishing of $\underleftarrow{\lim}^n$ for Mittag–Leffler sequences. This summarises the main facts we appeal to in Section A.3.

From Section A.4, we need the fact that the derived functors of the limit of a sequence agree with those of a cofinal subsequence (Proposition A.4.8), and the fact that we can compute the derived functor of the limit of a Mittag–Leffler sequence very concretely by injective resolutions, as in Construction A.4.10.

REMARK 7.3.1. Actually, we will be applying mostly the duals of the results in Appendix A. We will be studying sequences of length γ which are covariant; that is $F : \mathcal{I}(\gamma) \longrightarrow \mathcal{E}x(\mathcal{S}^{op}, \mathcal{A}b)$ (see Definition A.3.5). We consider, in this Section, sequences $\mathcal{I} \longrightarrow \mathcal{A}$, not sequences $\mathcal{I}^{op} \longrightarrow \mathcal{A}$. We will be considering their colimits rather than limits, and the interesting sequences will be co–Mittag–Leffler, which is the dual of Mittag–Leffler. We will freely apply the duals of facts proved in Appendix A.

Suppose γ is an ordinal, and $\gamma < \alpha$. Suppose we are given a sequence $F : \mathcal{I}(\gamma) \longrightarrow \mathcal{E}x(\mathcal{S}^{op}, \mathcal{A}b)$. If it so happens that F is co–Mittag–Leffler and factors through

$$\mathcal{I}(\gamma) \xrightarrow{\quad G \quad} A(\mathcal{S}) \xrightarrow{\quad \phi \quad} \mathcal{E}x(\mathcal{S}^{op}, \mathcal{A}b),$$

then the functor $G : \mathcal{I}(\gamma) \longrightarrow A(\mathcal{S})$ must also be co–Mittag–Leffler, since $\phi : A(\mathcal{S}) \longrightarrow \mathcal{E}x(\mathcal{S}^{op}, \mathcal{A}b)$ is an exact embedding respecting coproducts of $< \alpha$ objects; see Lemma 7.1.3. Because $A(\mathcal{S})$ satisfies [AB4.5(α)], it follows that $\underrightarrow{\text{colim}}^n G = 0$ if $n \geq 1$.

Now $A(\mathcal{S})$ and $\mathcal{E}x(\mathcal{S}^{op}, \mathcal{A}b)$ are abelian categories satisfying [AB4(α)]; for $A(\mathcal{S})$ we see this because, by Lemma 7.1.3, $A(\mathcal{S})$ even satisfies the stronger [AB4.5(α)]. For $\mathcal{E}x(\mathcal{S}^{op}, \mathcal{A}b)$ we have that, by Lemma 6.3.2, it satisfies [AB4], and hence the weaker [AB4(α)]. The inclusion $A(\mathcal{S}) \longrightarrow \mathcal{E}x(\mathcal{S}^{op}, \mathcal{A}b)$ is exact and respects coproducts of $< \alpha$ objects. Lemma A.3.2

now tells us that, for colimits over partially ordered sets of cardinality $< \alpha$,

$$\underrightarrow{\operatorname{colim}}^n \{\phi G\} \quad = \quad \phi\left\{\underrightarrow{\operatorname{colim}}^n G\right\}.$$

Our index set $\mathcal{I}(\gamma)$ has cardinality $< \alpha$, and since $\underrightarrow{\operatorname{colim}}^n G = 0$, so is

$$\underrightarrow{\operatorname{colim}}^n \{\phi G\} \quad = \quad \phi\left\{\underrightarrow{\operatorname{colim}}^n G\right\} \quad = \quad \phi\{0\}.$$

That is, for $n \geq 1$, $F = \phi G$ has vanishing $\underrightarrow{\operatorname{colim}}^n$.

The essence of this Section is to reduce the case of a general co–Mittag–Leffler sequence in $\mathcal{E}x(\mathcal{S}^{op}, \mathcal{A}b)$ to the above. The unfortunate technical detail is that this does not quite work. We end up proving the following. Suppose for all co–Mittag–Leffler sequence F of length $< \gamma$ and all $n \geq 1$, $\underrightarrow{\operatorname{colim}}^n F = 0$. Then any co–Mittag–Leffler sequence F of length γ satisfies $\underrightarrow{\operatorname{colim}}^n F = 0$ for $n \geq 2$. If we wish to do a transfinite induction, this is not quite enough. In Section 7.4, we will see how the author tried (unsuccessfully so far) to worm his way around this difficulty. Because we will need some variants of the statement given above, the main lemma of this Section is quite awkward to state.

REMARK 7.3.2. From now until the end of Section 7.4, we will assume that our fixed regular cardinal α is $> \aleph_0$. When $\alpha = \aleph_0$, the inclusion

$$\mathcal{E}x(\mathcal{S}^{op}, \mathcal{A}b) \qquad \subset \qquad \mathcal{C}at(\mathcal{S}^{op}, \mathcal{A}b)$$

is an equality. The functors in $\mathcal{C}at(\mathcal{S}^{op}, \mathcal{A}b)$ are the additive functors $\mathcal{S}^{op} \longrightarrow \mathcal{A}b$, and hence they must respect finite biproducts. Therefore they send coproducts of fewer than $\alpha = \aleph_0$ objects in $\mathcal{S}$ to products of abelian groups.

In the case $\alpha = \aleph_0$, the category $\mathcal{E}x(\mathcal{S}^{op}, \mathcal{A}b) = \mathcal{C}at(\mathcal{S}^{op}, \mathcal{A}b)$ is a Grothendieck abelian category. It satisfies [AB5], hence [AB4.5]. In this Section and the next, we are out to prove that [AB4.5] is true even for $\alpha > \aleph_0$. The case $\alpha = \aleph_0$ being trivial, in the remainder of the Section we assume $\alpha > \aleph_0$.

One consequence is that the category $\mathcal{S}$ is closed under countable coproducts. After all, $\mathcal{S}$ satisfies [TR5(α)] and $\alpha > \aleph_0$. By Proposition 1.6.8, the category $\mathcal{S}$ is closed under splitting idempotents. By Corollary 5.1.23, the projective–injective objects of $A(\mathcal{S})$ are precisely the representable functors. Recall that for a general $\mathcal{S}$, the projective–injectives are direct summands of representables. The assumption that $\alpha > \aleph_0$ guarantees that a direct summand of a representable functor is representable.

REMARK 7.3.3. We wish to remind ourselves about projective objects. The objects $\mathcal{S}(-, s)$ are projective in $\mathcal{E}x(\mathcal{S}^{op}, \mathcal{A}b)$ by Lemma 6.4.1. They are projective in $A(\mathcal{S})$ by Lemma 5.1.11. Any projective object in $\mathcal{E}x(\mathcal{S}^{op}, \mathcal{A}b)$ which happens to lie in $A(\mathcal{S}) \subset \mathcal{E}x(\mathcal{S}^{op}, \mathcal{A}b)$ is clearly projective in the exact subcategory $A(\mathcal{S})$. By the last paragraph of Remark 7.3.2, it must be a representable $\mathcal{S}(-,s)$. The representable functors $\mathcal{S}(-,s)$ are precisely the projective objects of $A(\mathcal{S})$, and they can also be characterised as the projective objects in $\mathcal{E}x(\mathcal{S}^{op}, \mathcal{A}b)$ which happen to lie in $A(\mathcal{S})$.

Recall that the category $\mathcal{E}x(\mathcal{S}^{op}, \mathcal{A}b)$ has enough projectives. The objects $\mathcal{S}(-,s)$ are projective, as are all their coproducts. Lemma 6.4.2 shows that the projectives $\mathcal{S}(-,s)$ generate. More precisely, in Lemma 6.4.4 we see that any object of $\mathcal{E}x(\mathcal{S}^{op}, \mathcal{A}b)$ is a quotient of a coproduct of $\mathcal{S}(-,s)$'s, and any projective object is a direct summand of such a coproduct.

Let P be the class of all coproducts of representable functors $\mathcal{S}(-,s)$ in $\mathcal{E}x(\mathcal{S}^{op}, \mathcal{A}b)$. By the above, every object of $\mathcal{E}x(\mathcal{S}^{op}, \mathcal{A}b)$ is a quotient of an object in P. By the dual of Lemma A.4.3, every object in the category of sequences

$$\mathcal{I}(\gamma) \longrightarrow \mathcal{E}x(\mathcal{S}^{op}, \mathcal{A}b)$$

is a quotient of a projective object. The projective objects may be chosen to be coproducts of the duals of $F^i_{a_i}$'s. That is, define

$$G^i_a : \mathcal{I}(\gamma) \longrightarrow \mathcal{E}x(\mathcal{S}^{op}, \mathcal{A}b)$$

by the formula

$$G^i_a(j) = \begin{cases} 0 & \text{if } i > j \\ a & \text{if } i \le j \end{cases}$$

Then for any (covariant) sequence F of length γ, there is a surjection

$$\coprod_{i \in \mathcal{I}(\gamma)} G^i_{a_i} \longrightarrow F,$$

where the a_i's are projective in $\mathcal{E}x(\mathcal{S}^{op}, \mathcal{A}b)$. By Remark A.4.4, we may even choose the a_i's to be objects in the class P. That is the a_i's are coproducts of representables.

In other words, given any functor

$$F : \mathcal{I}(\gamma) \longrightarrow \mathcal{E}x(\mathcal{S}^{op}, \mathcal{A}b)$$

we may find a surjection $G \longrightarrow F$, where G is a particularly nice projective object as above. In Construction A.4.10, we saw that concretely, such G's satisfy

$$\frac{Gi}{\operatorname{colim}_{j<i} Gj} = \coprod_{\lambda \in \Lambda} \mathcal{S}(-, s_\lambda).$$

We recall; if

$$G = \coprod_{i \in \mathcal{I}} G^i_{a_i},$$

then a_i is given as

$$\frac{Gi}{\operatorname*{colim}_{j<i} Gj} \quad = \quad a_i.$$

For our particularly nice projective G's, we assume that a_i is a coproduct of representables.

REMARK 7.3.4. The next Lemma is somewhat technical. What is really happening is that three facts we want, namely Corollaries 7.3.6, 7.3.7 and 7.3.8, have essentially the same proof. For the sake of efficiency, we state and prove a somewhat convoluted Lemma 7.3.5, and deduce Corollaries 7.3.6, 7.3.7 and 7.3.8 as consequences.

LEMMA 7.3.5. *Let $\alpha > \aleph_0$ be our fixed regular cardinal. Suppose $\mathcal{S}$ is a triangulated category satisfying [TR5(α)]. Let γ be an ordinal $< \alpha$. Suppose we are given a co–Mittag–Leffler sequence*

$$F : \mathcal{I}(\gamma) \longrightarrow \mathcal{E}x\big(\mathcal{S}^{op}, \mathcal{A}b\big).$$

Suppose furthermore that there is a short exact sequence of co–Mittag–Leffler objects in the category of sequences

$$0 \longrightarrow F \longrightarrow G \longrightarrow H \longrightarrow 0.$$

Suppose that for every ordinal $i < \gamma$, the sequence

$$0 \longrightarrow \operatorname*{colim}_{j<i} Fj \longrightarrow \operatorname*{colim}_{j<i} Gj \longrightarrow \operatorname*{colim}_{j<i} Hj \longrightarrow 0$$

is exact in $\mathcal{E}x\big(\mathcal{S}^{op}, \mathcal{A}b\big)$. Finally suppose that G is of the special form, so that

$$\frac{Gi}{\operatorname*{colim}_{j<i} Gj} \quad = \quad \coprod_{\lambda\in\Lambda_i} G^i_\lambda,$$

with $G^i_\lambda \in A(\mathcal{S}) \subset \mathcal{E}x\big(\mathcal{S}^{op}, \mathcal{A}b\big)$. Then for all $n \geq 1$

$$\operatorname*{colim}_{j<\gamma}{}^n F = 0.$$

If we furthermore happen to know that

7.3.5.1. *For all $i \in \mathcal{I}(\gamma)$,*

$$\frac{Gi}{\operatorname*{colim}_{j<i} Gj} \quad \text{and} \quad \frac{Hi}{\operatorname*{colim}_{j<i} Hj}$$

are homological objects in $\mathcal{E}x\big(\mathcal{S}^{op}, \mathcal{A}b\big)$

then $\underrightarrow{\operatorname{colim}} F$ is also a homological object in $\mathcal{E}x\big(\mathcal{S}^{op}, \mathcal{A}b\big)$.

Proof: It is only fair to warn the reader that the proof of this Lemma is long and technical. Before reading it, the reader might wish to glance ahead to Corollary 7.3.6 on page 249, Corollary 7.3.7 on page 251, and Corollary 7.3.8 on page 253, to see how the result is applied.

For the reader who did not follow the advice offered in the last paragraph, we should briefly note one interesting class of G's to which the hypothesis above applies. If G is a slightly special projective object as in Remark 7.3.3, then it has the desired form. In other words, the Lemma will prove, among other things, the vanishing of $\underrightarrow{\mathrm{colim}}^n F$ for all co–Mittag–Leffler subfunctors $F \subset G$, with co–Mittag–Leffler quotients G/F, for the slightly special projective G's of Remark 7.3.3.

Now we turn to the proof of the Lemma. By induction on the ordinal γ, we assume the Lemma true for all ordinals $i < \gamma$. In Construction A.4.10, we saw that F admits a resolution by projective objects

$$\cdots \longrightarrow G_2 \longrightarrow G_1 \longrightarrow G_0 \longrightarrow F \longrightarrow 0$$

We can combine this with the short exact sequence

$$0 \longrightarrow F \longrightarrow G \longrightarrow H \longrightarrow 0$$

to extend to a slightly longer exact sequence

$$\cdots \longrightarrow G_2 \longrightarrow G_1 \longrightarrow G_0 \longrightarrow G \longrightarrow H \longrightarrow 0.$$

The G_i may be chosen to be the special projectives as in Remark 7.3.3. Choose and fix such a projective resolution. Let i be an ordinal, $i < \gamma$. By our induction hypothesis, $\underrightarrow{\mathrm{colim}}^n F$ vanishes when F is restricted to $\mathcal{I}(i) \subset \mathcal{I}(\gamma)$, where $i < \gamma$. For $i < \gamma$, the sequence

$$\longrightarrow \operatorname*{colim}_{j<i} G_1(j) \longrightarrow \operatorname*{colim}_{j<i} G_0(j) \longrightarrow \operatorname*{colim}_{j<i} F(j) \longrightarrow 0$$

is exact. By the hypothesis of the Lemma, so is

$$0 \longrightarrow \operatorname*{colim}_{j<i} Fj \longrightarrow \operatorname*{colim}_{j<i} Gj \longrightarrow \operatorname*{colim}_{j<i} Hj \longrightarrow 0.$$

Splicing these sequences, we deduce the exactness of

$$\longrightarrow \operatorname*{colim}_{j<i} G_0(j) \longrightarrow \operatorname*{colim}_{j<i} G(j) \longrightarrow \operatorname*{colim}_{j<i} H(j) \longrightarrow 0$$

Hence the sequence below is also exact

$$\longrightarrow \frac{G_0(i)}{\operatorname*{colim}_{j<i} G_0(j)} \longrightarrow \frac{G(i)}{\operatorname*{colim}_{j<i} G(j)} \longrightarrow \frac{H(i)}{\operatorname*{colim}_{j<i} H(j)} \longrightarrow 0.$$

As in Construction A.4.10, we rewrite this. For each ordinal $i < \gamma$ there is an exact sequence

$$\cdots \longrightarrow g_i^2 \longrightarrow g_i^1 \longrightarrow g_i^0 \longrightarrow g_i \longrightarrow h_i \longrightarrow 0.$$

In other words,

$$\frac{G_n(i)}{\operatorname*{colim}_{j<i} G_n(j)} = g_i^n, \qquad \frac{G(i)}{\operatorname*{colim}_{j<i} G(j)} = g_i, \qquad \frac{H(i)}{\operatorname*{colim}_{j<i} H(j)} = h_i.$$

By construction, the projective objects G_n may be chosen so that g_i^n are objects in the class P, that is

$$g_i^n \quad = \quad \coprod \mathcal{S}(-, s_\lambda).$$

By the hypothesis of the Lemma, the object G is such that

$$\frac{G(i)}{\operatorname*{colim}_{j<i} G(j)} \quad = \quad g_i \quad = \quad \coprod G_\lambda^i,$$

with $G_\lambda^i \in A(\mathcal{S})$. For the projective objects G_n we know

$$G_n(i) \quad = \quad \coprod_{j \le i} g_i^n.$$

The objects G and H are not so simple. $G(i)$ and $H(i)$ are some extensions of the building blocks $g_j, j \le i$ and $h_j, j \le i$ respectively.

So much is true under the general hypothesis of the Lemma. If we furthermore assume 7.3.5.1, that is we assume that

$$\frac{G(i)}{\operatorname*{colim}_{j<i} G(j)} = g_i \qquad \text{and} \qquad \frac{H(i)}{\operatorname*{colim}_{j<i} H(j)} = h_i$$

are both homological objects in $\mathcal{E}x\big(\mathcal{S}^{op}, \mathcal{A}b\big)$, then in the exact sequence above

$$\cdots \longrightarrow g_i^2 \longrightarrow g_i^1 \longrightarrow g_i^0 \longrightarrow g_i \longrightarrow h_i \longrightarrow 0$$

the objects g_i and h_i are homological. We can break it into two exact sequences

$$\cdots \longrightarrow g_i^2 \longrightarrow g_i^1 \longrightarrow g_i^0 \longrightarrow f_i \longrightarrow 0$$

and

$$0 \longrightarrow f_i \longrightarrow g_i \longrightarrow h_i \longrightarrow 0.$$

In the second exact sequence, g_i and h_i are homological. Lemma 7.2.5 now tells us that

$$\frac{F(i)}{\operatorname*{colim}_{j<i} F(j)} \quad = \quad f_i$$

is also homological.

Return now to the general case; that is, we are not necessarily assuming 7.3.5.1. We want to prove that the derived functors of the colimit vanish on the sequence F. In other words, we want to prove the exactness of the colimit sequence

$$\cdots \longrightarrow \coprod_{i\in\mathfrak{I}(\gamma)} g_i^2 \longrightarrow \coprod_{i\in\mathfrak{I}(\gamma)} g_i^1 \longrightarrow \coprod_{i\in\mathfrak{I}(\gamma)} g_i^0.$$

Of course, to say this sequence of functors is exact is to say that, when we evaluate it at $s \in \mathcal{S}$, we obtain an exact sequence of abelian groups

$$\cdots \longrightarrow \left\{\coprod_{i\in\mathfrak{I}(\gamma)} g_i^2\right\}(s) \longrightarrow \left\{\coprod_{i\in\mathfrak{I}(\gamma)} g_i^1\right\}(s) \longrightarrow \left\{\coprod_{i\in\mathfrak{I}(\gamma)} g_i^0\right\}(s).$$

Take an element k in the kernel of one of the differentials of this sequence. For the sake of definiteness, k is an element

$$k \quad \in \quad \left\{\coprod_{i\in\mathfrak{I}(\gamma)} g_i^n\right\}(s).$$

But recall that each g_i^n is assumed to be a coproduct of representables. That is, $g_i^n = \coprod_{\lambda\in\Lambda_i} \mathcal{S}(-, s_\lambda)$. Therefore k lies in

$$k \quad \in \quad \left\{\coprod_{i\in\mathfrak{I}(\gamma)} \coprod_{\lambda\in\Lambda_i^n} \mathcal{S}(-, s_\lambda)\right\}(s).$$

And the key point is that, by the definition of the coproduct in $\mathcal{E}x(\mathcal{S}^{op}, \mathcal{A}b)$ (Definition 6.1.10; see also Lemma 6.1.17), an element in this coproduct is a morphism

$$s \longrightarrow \coprod_{\lambda\in\Lambda'} s_\lambda$$

where Λ' is a subset of $\cup_{i\in\mathfrak{I}(\gamma)}\Lambda_i^n$ *whose cardinality is* $< \alpha$.

In other words, although k lies in a gigantic coproduct, it really exists in a much smaller one. For each g_i^n we can choose a direct summand $u_i^n \subset g_i^n$. The object u_i^n is a coproduct of fewer than α representables, hence is representable. It is a projective object of $A(\mathcal{S}) \subset \mathcal{E}x(\mathcal{S}^{op}, \mathcal{A}b)$. And k lies in

$$k \quad \in \quad \left\{\coprod_{i\in\mathfrak{I}(\gamma)} u_i^n\right\}(s) \quad \subset \quad \left\{\coprod_{i\in\mathfrak{I}(\gamma)} g_i^n\right\}(s).$$

From now on, we will consider a great many direct summands $u_i^n \subset g_i^n$. Note that, in all of these summands, we assume we have a fixed decomposition of g_i^n as

$$g_i^n \quad = \quad \coprod_{\lambda\in\Lambda_i^n} \mathcal{S}(-, s_\lambda),$$

and that all the $u_i^n \subset g_i^n$ considered are direct sums over parts of the set Λ_i^n. This being assumed, if we construct an increasing family of u_i^ns, then its direct limit is also a direct summand of g_i^n.

The idea of the rest of the proof is to show that k lies in a subcomplex of

$$\cdots \longrightarrow \left\{\coprod_{i\in\mathcal{I}(\gamma)} g_i^2\right\}(s) \longrightarrow \left\{\coprod_{i\in\mathcal{I}(\gamma)} g_i^1\right\}(s) \longrightarrow \left\{\coprod_{i\in\mathcal{I}(\gamma)} g_i^0\right\}(s)$$

which is entirely contained in $A(\mathcal{S})$. We know that [AB4.5(α)] holds for $A(\mathcal{S})$, by Lemma 7.1.3. This will allow us to deduce the vanishing of k. We will proceed in two steps.

7.3.5.2. *Suppose the complex*

$$\cdots \longrightarrow \coprod_{i\in\mathcal{I}(\gamma)} g_i^2 \longrightarrow \coprod_{i\in\mathcal{I}(\gamma)} g_i^1 \longrightarrow \coprod_{i\in\mathcal{I}(\gamma)} g_i^0$$

is as above; that is, it comes from a slightly special projective resolution of a co–Mittag–Leffler sequence F of length γ in $\mathcal{E}x\big(\mathcal{S}^{op}, \mathcal{A}b\big)$. Suppose for each ordinal $i < \gamma$ and each $0 \leq n < \infty$, we are given some direct summand $u_i^n \subset g_i^n$, with $u_i^n \in A(\mathcal{S})$.

Then, by increasing u_i^n if necessary, one can find direct summands

$$u_i^n \quad \subset \quad h_i^n \quad \subset \quad g_i^n$$

so that

7.3.5.2.1. *The objects h_i^n are projective objects in*

$$A(\mathcal{S}) \subset \mathcal{E}x\big(\mathcal{S}^{op}, \mathcal{A}b\big).$$

7.3.5.2.2. *The objects h_i^n form a complex*

$$\cdots \longrightarrow \coprod_{i\in\mathcal{I}(\gamma)} h_i^2 \longrightarrow \coprod_{i\in\mathcal{I}(\gamma)} h_i^1 \longrightarrow \coprod_{i\in\mathcal{I}(\gamma)} h_i^0$$

which is a subcomplex of

$$\cdots \longrightarrow \coprod_{i\in\mathcal{I}(\gamma)} g_i^2 \longrightarrow \coprod_{i\in\mathcal{I}(\gamma)} g_i^1 \longrightarrow \coprod_{i\in\mathcal{I}(\gamma)} g_i^0.$$

Proof: Recall that g_i^n can be expressed

$$g_i^n \quad = \quad \coprod_{\lambda\in\Lambda_i^n} \mathcal{S}(-, s_\lambda)$$

and that u_i^n is assumed to be the coproduct over some subset of Λ_i^n, of cardinality $< \alpha$. Thus

$$u_i^n \quad = \quad \coprod_{\lambda\in M_i^n} \mathcal{S}(-, s_\lambda),$$

and $M_i^n \subset \Lambda_i^n$ is of cardinality $< \alpha$. Put $M_i^n = {}_0M_i^n$. We will now show how to construct, inductively, a sequence

$$ {}_0M_i^n \subset {}_1M_i^n \subset {}_2M_i^n \subset \cdots \subset \Lambda_i^n $$

so that all the ${}_mM_i^n$ have cardinality $< \alpha$.

Suppose we have constructed ${}_mM_i^n$. For each $\lambda \in {}_mM_i^n$, we can consider the composite map

$$ \mathcal{S}(-, s_\lambda) \quad \subset \quad g_i^n \quad \subset \quad \coprod_{i \in \mathcal{I}(\gamma)} g_i^n \xrightarrow{\ \partial\ } \coprod_{i \in \mathcal{I}(\gamma)} g_i^{n-1}. $$

Of course, each g_i^{n-1} is itself a coproduct. This composite is a map

$$ \mathcal{S}(-, s_\lambda) \xrightarrow{\ \partial\ } \coprod_{i \in \mathcal{I}(\gamma)} \coprod_{\mu \in \Lambda_i^{n-1}} \mathcal{S}(-, s_\mu). $$

By Corollary 7.1.2, this map factors through a coproduct of fewer than α terms.

For each $\lambda \in M_i^n$, choose a set M_λ, of cardinality $< \alpha$, contained in

$$ M_\lambda \quad \subset \quad \bigcup_{i \in \mathcal{I}(\gamma)} \Lambda_i^{n-1}, $$

so that the differential

$$ \mathcal{S}(-, s_\lambda) \xrightarrow{\ \partial\ } \coprod_{i \in \mathcal{I}(\gamma)} \coprod_{\mu \in \Lambda_i^{n-1}} \mathcal{S}(-, s_\mu) $$

factors through the coproduct over M_λ. Now the union

$$ M^{n-1} \quad = \quad \bigcup_{i \in \mathcal{I}(\gamma)} \bigcup_{\lambda \in {}_mM_i^n} M_\lambda $$

is a union of $< \alpha$ sets, each of cardinality $< \alpha$. Therefore its cardinality is $< \alpha$. Put

$$ {}_{m+1}M_i^{n-1} \quad = \quad {}_mM_i^{n-1} \cup \left\{ M^{n-1} \cap \Lambda_i^{n-1} \right\}. $$

This construction guarantees that, if $\lambda \in {}_mM_i^n$, the differential

$$ \mathcal{S}(-, s_\lambda) \xrightarrow{\ \partial\ } \coprod_{i \in \mathcal{I}(\gamma)} \coprod_{\mu \in \Lambda_i^{n-1}} \mathcal{S}(-, s_\mu) $$

factors through

$$ \coprod_{i \in \mathcal{I}, \mu \in {}_{m+1}M_i^{n-1}} \mathcal{S}(-, s_\mu). $$

Now define

$$ N_i^n \quad = \quad \bigcup_{m=0}^{\infty} {}_mM_i^n. $$

The set N_i^n is the union of countably many sets of cardinality $< \alpha$. We are assuming $\alpha > \aleph_0$; therefore N_i^n is the union of fewer than α sets, each of cardinality $< \alpha$. Since α is regular, the cardinality of N_i^n must be $< \alpha$. Put

$$h_i^n \quad = \quad \coprod_{\lambda \in N_i^n} \mathcal{S}(-, s_\lambda) \quad \subset \quad \coprod_{\lambda \in \Lambda_i^n} \mathcal{S}(-, s_\lambda).$$

This defines a summand $h_i^n \subset g_i^n$, containing u_i^n, so that

$$\cdots \longrightarrow \coprod_{i \in \mathcal{I}(\gamma)} h_i^2 \longrightarrow \coprod_{i \in \mathcal{I}(\gamma)} h_i^1 \longrightarrow \coprod_{i \in \mathcal{I}(\gamma)} h_i^0$$

is a subcomplex of

$$\cdots \longrightarrow \coprod_{i \in \mathcal{I}(\gamma)} g_i^2 \longrightarrow \coprod_{i \in \mathcal{I}(\gamma)} g_i^1 \longrightarrow \coprod_{i \in \mathcal{I}(\gamma)} g_i^0.$$

□

7.3.5.3. *Suppose we are in the situation of the proof of Lemma 7.3.5. To remind the reader: suppose the complex*

$$\cdots \longrightarrow \coprod_{i \in \mathcal{I}(\gamma)} g_i^2 \longrightarrow \coprod_{i \in \mathcal{I}(\gamma)} g_i^1 \longrightarrow \coprod_{i \in \mathcal{I}(\gamma)} g_i^0$$

is obtained from a slightly special projective resolution of a co–Mittag–Leffler sequence F in $\mathcal{E}x\big(\mathcal{S}^{op}, \mathcal{A}b\big)$. And suppose further that F admits a short exact sequence of co–Mittag–Lefflers

$$0 \longrightarrow F \longrightarrow G \longrightarrow H \longrightarrow 0$$

and that G is of the special form, so that

$$\frac{Gi}{\operatorname*{colim}_{j<i} Gj} \quad = \quad \coprod_{\lambda \in \Lambda_i} G_\lambda^i,$$

with $G_\lambda^i \in A(\mathcal{S}) \subset \mathcal{E}x\big(\mathcal{S}^{op}, \mathcal{A}b\big)$. We remind the reader further that, for each $i \in \mathcal{I}(\gamma)$, we then have an exact sequence

$$\cdots \longrightarrow g_i^2 \longrightarrow g_i^1 \longrightarrow g_i^0 \longrightarrow g_i \longrightarrow h_i \longrightarrow 0,$$

where

$$g_i \quad = \quad \coprod_{\lambda \in \Lambda_i} G_\lambda^i.$$

Suppose that we are given a subcomplex

$$\cdots \longrightarrow \coprod_{i \in \mathcal{I}(\gamma)} h_i^2 \longrightarrow \coprod_{i \in \mathcal{I}(\gamma)} h_i^1 \longrightarrow \coprod_{i \in \mathcal{I}(\gamma)} h_i^0$$

of the complex

$$\cdots \longrightarrow \coprod_{i\in\mathfrak{I}(\gamma)} g_i^2 \longrightarrow \coprod_{i\in\mathfrak{I}(\gamma)} g_i^1 \longrightarrow \coprod_{i\in\mathfrak{I}(\gamma)} g_i^0,$$

with $h_i^n \subset g_i^n$ a direct summand, and $h_i^n \in A(\mathcal{S})$.

Then the following is true:

7.3.5.3.1. *If $n \geq 1$, one may choose a direct summand $u_i^{n+1} \subset g_i^{n+1}$, $u_i^{n+1} \in A(\mathcal{S})$, so that in the commutative diagram*

$$\begin{array}{ccccc} & & h_i^n & \xrightarrow{\partial} & h_i^{n-1} \\ & & {\scriptstyle f}\downarrow & & \downarrow \\ u_i^{n+1} & \longrightarrow & g_i^n & \longrightarrow & g_i^{n-1} \end{array}$$

the map f takes the kernel of $\partial : h_i^n \longrightarrow h_i^{n-1}$ to the image of $u_i^{n+1} \longrightarrow g_i^n$.

7.3.5.3.2. *If $n = 0$, there exists a direct summand $u_i^1 \subset g_i^1$, $u_i^1 \in A(\mathcal{S})$, so that in the diagram*

$$\begin{array}{ccccc} & & h_i^0 & & \\ & & \downarrow & & \\ u_i^1 & \longrightarrow & g_i^0 & \longrightarrow & g_i \end{array}$$

the image of h_i^0 in the cokernel of $u_i^1 \longrightarrow g_i^0$ injects into g_i.

7.3.5.3.3. *On top of the standard hypotheses of the Lemma, now assume 7.3.5.1 as well. That is, in the exact sequences*

$$\cdots \longrightarrow g_i^2 \longrightarrow g_i^1 \longrightarrow g_i^0 \longrightarrow g_i \longrightarrow h_i \longrightarrow 0,$$

assume g_i and h_i are homological objects in $\mathcal{E}x\big(\mathcal{S}^{op}, \mathcal{A}b\big)$. Then we may also find a direct summand $u_i^0 \subset g_i^0$, $u_i^0 \in A(\mathcal{S})$, so that in the commutative square

$$\begin{array}{ccc} h_i^0 & \xrightarrow{\partial} & g_i \\ \downarrow & & {\scriptstyle 1}\downarrow \\ h_i^0 \oplus u_i^0 & \xrightarrow{\partial} & g_i \end{array}$$

the inclusion of the images of the two differentials ∂ factors through a representable $\mathcal{S}(-, s)$.

Proof: First we prove 7.3.5.3.1. Consider therefore the map $h_i^n \longrightarrow h_i^{n-1}$. It is a morphism in $A(\mathcal{S})$, hence its kernel lies in $A(\mathcal{S})$, hence we can find a projective in $A(\mathcal{S})$ surjecting to the kernel. There is an exact sequence

$$\mathcal{S}(-, p) \longrightarrow h_i^n \xrightarrow{\partial} h_i^{n-1}$$

in $A(\mathcal{S}) \subset \mathcal{E}x(\mathcal{S}^{op}, \mathcal{A}b)$. On the other hand we know that the sequence

$$g_i^{n+1} \longrightarrow g_i^n \longrightarrow g_i^{n-1}$$

is exact in $\mathcal{E}x(\mathcal{S}^{op}, \mathcal{A}b)$. We have a commutative diagram

$$\begin{array}{ccccc} \mathcal{S}(-,p) & \longrightarrow & h_i^n & \xrightarrow{\partial} & h_i^{n-1} \\ & & {\scriptstyle f}\downarrow & & \downarrow \\ g_i^{n+1} & \longrightarrow & g_i^n & \longrightarrow & g_i^{n-1} \end{array}$$

and the commutativity establishes the vanishing of the composite

$$\begin{array}{ccc} \mathcal{S}(-,p) & \longrightarrow & h_i^n \\ & & {\scriptstyle f}\downarrow \\ & & g_i^n & \longrightarrow & g_i^{n-1}. \end{array}$$

The map $\mathcal{S}(-,p) \longrightarrow g_i^n$ factors through the kernel of $g_i^n \longrightarrow g_i^{n-1}$. But $\mathcal{S}(-,p)$ is a projective object; hence the map factors through anything surjecting to the kernel. There is a map $\mathcal{S}(-,p) \longrightarrow g_i^{n+1}$ rendering commutative the diagram

$$\begin{array}{ccccc} \mathcal{S}(-,p) & \longrightarrow & h_i^n & \xrightarrow{\partial} & h_i^{n-1} \\ \downarrow & & {\scriptstyle f}\downarrow & & \downarrow \\ g_i^{n+1} & \longrightarrow & g_i^n & \longrightarrow & g_i^{n-1}. \end{array}$$

By Corollary 7.1.2, the map

$$\mathcal{S}(-,p) \longrightarrow g_i^{n+1} = \coprod_{\lambda \in \Lambda_i^{n+1}} \mathcal{S}(-,s_\lambda)$$

factors through a coproduct of fewer than α terms, which we call $u_i^{n+1} \subset g_i^{n+1}$. We deduce the commutative diagram

$$\begin{array}{ccccc} \mathcal{S}(-,p) & \longrightarrow & h_i^n & \xrightarrow{\partial} & h_i^{n-1} \\ \downarrow & & {\scriptstyle f}\downarrow & & \downarrow \\ u_i^{n+1} & \longrightarrow & g_i^n & \longrightarrow & g_i^{n-1} \end{array}$$

and this completes the proof of 7.3.5.3.1; the kernel of $h_i^n \longrightarrow h_i^{n-1}$ is the image of $\mathcal{S}(-,p)$, and it maps under $f : h_i^n \longrightarrow g_i^n$ into the image of $u_i^{n+1} \longrightarrow g_i^n$.

Now for the proof of 7.3.5.3.2. Consider the diagram

$$\begin{array}{ccccc} & & h_i^0 & & \\ & & \downarrow & & \\ g_i^1 & \longrightarrow & g_i^0 & \longrightarrow & g_i. \end{array}$$

Recall that we are assuming about g_i that it is a coproduct; more precisely

$$g_i = \coprod_{\lambda \in \Lambda} G_\lambda^i,$$

with $G_\lambda^i \in A(\mathcal{S})$. In any case, since h_i^0 is projective in $A(\mathcal{S})$, it is isomorphic to $\mathcal{S}(-,t)$ for some $t \in \mathcal{S}$. The map

$$\mathcal{S}(-,t) = h_i^0 \longrightarrow g_i = \coprod_{\lambda \in \Lambda} G_\lambda^i$$

must factor, by Corollary 7.1.2, through a coproduct of fewer than α objects. This coproduct lies in $A(\mathcal{S})$. Call it h_i^{-1}. We deduce a commutative diagram

$$\begin{array}{ccccc} & & h_i^0 & \longrightarrow & h_i^{-1} \\ & & \downarrow & & \downarrow \\ g_i^1 & \longrightarrow & g_i^0 & \longrightarrow & g_i \end{array}$$

where the bottom row is exact, and the top row lies in $A(\mathcal{S})$. As in the proof of 7.3.5.3.1, we may find a direct summand $u_i^1 \subset g_i^1$, $u_i^1 \in A(\mathcal{S})$, and a commutative diagram

$$\begin{array}{ccccc} \mathcal{S}(-,p) & \longrightarrow & h_i^0 & \longrightarrow & h_i^{-1} \\ \downarrow & & \downarrow & & \downarrow \\ u_i^1 & \longrightarrow & g_i^0 & \longrightarrow & g_i \end{array}$$

where the top row is exact. This says precisely that the image of $h_i^0 \longrightarrow h_i^{-1} \subset g_i$ agrees with the image of h_i^0 in the cokernel of $u_i^1 \longrightarrow g_i^0$.

It remains to prove 7.3.5.3.3. We assume therefore that we are in the situation of 7.3.5.1. We have exact sequences

$$\cdots \longrightarrow g_i^2 \longrightarrow g_i^1 \longrightarrow g_i^0 \longrightarrow f_i \longrightarrow 0$$

and

$$0 \longrightarrow f_i \longrightarrow g_i \longrightarrow h_i \longrightarrow 0,$$

and f_i, g_i and h_i are homological objects in $\mathcal{E}x\big(\mathcal{S}^{op}, \mathcal{A}b\big)$. We are given a direct summand $h_i^0 \subset g_i^0$. The kernel of the composite

$$h_i^0 \longrightarrow g_i^0 \longrightarrow f_i \longrightarrow g_i$$

is the same as the kernel of

$$h_i^0 \longrightarrow g_i^0 \longrightarrow f_i,$$

since the map $f_i \longrightarrow g_i$ is mono. On the other hand, in the proof of 7.3.5.3.2 above we saw that the image of the natural map $h_i^0 \longrightarrow g_i$ is the cokernel of $\mathcal{S}(-,p) \longrightarrow h_i^0$. We have a vanishing composite

$$\mathcal{S}(-,p) \longrightarrow h_i^0 \longrightarrow f_i.$$

Now recall that h_i^0 is projective; it must be $\mathcal{S}(-,q)$ for some $q \in \mathcal{S}$. Complete to a triangle

$$p \longrightarrow q \longrightarrow r \longrightarrow \Sigma p.$$

Our vanishing composite becomes

$$\mathcal{S}(-,p) \longrightarrow \mathcal{S}(-,q) \longrightarrow f_i.$$

By Yoneda, the map $\mathcal{S}(-,q) \longrightarrow f_i$ corresponds to an element $\theta \in f_i(q)$. The vanishing of the composite

$$\mathcal{S}(-,p) \longrightarrow \mathcal{S}(-,q) \longrightarrow f_i$$

means that the image of θ under

$$f_i(q) \longrightarrow f_i(p)$$

is zero. But f_i is a homological object of $\mathcal{E}x\big(\mathcal{S}^{op}, \mathcal{A}b\big)$. The sequence

$$f_i(r) \longrightarrow f_i(q) \longrightarrow f_i(p)$$

is exact. This means that θ lies in the image of

$$f_i(r) \longrightarrow f_i(q).$$

Again, by Yoneda's lemma, this means the map

$$\mathcal{S}(-,q) \longrightarrow f_i$$

must factor as

$$\mathcal{S}(-,q) \longrightarrow \mathcal{S}(-,r) \longrightarrow f_i.$$

Recalling that $h_i^0 = \mathcal{S}(-,q)$, we have a factoring

$$h_i^0 \longrightarrow \mathcal{S}(-,r) \longrightarrow f_i.$$

On the other hand, $\mathcal{S}(-,r)$ is projective, and the map $\mathcal{S}(-,r) \longrightarrow f_i$ must factor through the epimorphism $g_i^0 \longrightarrow f_i$. It factors as

$$\mathcal{S}(-,r) \longrightarrow g_i^0 \longrightarrow f_i.$$

But g_i^0 is a coproduct of representables, and by Corollary 7.1.2, a map from $\mathcal{S}(-,r)$ into a coproduct factors through a coproduct of fewer than α terms. There exists $u_i^0 \subset g_i^0$, $u_i^0 \in A(\mathcal{S})$, and a factorisation

$$h_i^0 \longrightarrow \mathcal{S}(-,r) \longrightarrow u_i^0 \subset g_i^0 \longrightarrow f_i.$$

Put another way, we have a commutative diagram where the top row is exact

$$\begin{array}{ccccc} \mathcal{S}(-,p) & \longrightarrow & h_i^0 & \xrightarrow{\partial} & g_i \\ & & \downarrow & & \downarrow{\scriptstyle 1} \\ & & u_i^0 & \xrightarrow{\partial} & g_i. \end{array}$$

The inclusion of the cokernel of $\mathcal{S}(-,p) \longrightarrow h_i^0$ into g_i factors through $h_i^0 \longrightarrow \mathcal{S}(-,r) \longrightarrow u_i^0$ as above. In other words, the assertion of 7.3.5.3.3 is true. There exists a commutative square

$$\begin{array}{ccc} h_i^0 & \xrightarrow{\partial} & g_i \\ \downarrow & & \downarrow{\scriptstyle 1} \\ h_i^0 \oplus u_i^0 & \xrightarrow{\partial} & g_i \end{array}$$

as asserted. In fact, for out choice of u_i^0, the h_i^0 in the bottom left corner of this square is superfluous. □

Next we use the above two steps, to conclude the proof of Lemma 7.3.5. We want to show the exactness of the sequence

$$\cdots \longrightarrow \coprod_{i\in\mathcal{I}(\gamma)} g_i^2 \longrightarrow \coprod_{i\in\mathcal{I}(\gamma)} g_i^1 \longrightarrow \coprod_{i\in\mathcal{I}(\gamma)} g_i^0.$$

We showed that any homology class is really contained in $\coprod_{i\in\mathcal{I}} u_i^n$, with $u_i^n \subset g_i^n$ a direct summand, and $u_i^n \in A(\mathcal{S})$. In 7.3.5.2 we showed that it is possible to find a subcomplex

$$\cdots \longrightarrow \coprod_{i\in\mathcal{I}(\gamma)} h_i^2 \longrightarrow \coprod_{i\in\mathcal{I}(\gamma)} h_i^1 \longrightarrow \coprod_{i\in\mathcal{I}(\gamma)} h_i^0$$

contained in $A(\mathcal{S})$, so that $u_i^n \subset h_i^n$. Call this complex ${}_0h$. We will now proceed to construct a sequence of subcomplexes of

$$\cdots \longrightarrow \coprod_{i\in\mathcal{I}(\gamma)} g_i^2 \longrightarrow \coprod_{i\in\mathcal{I}(\gamma)} g_i^1 \longrightarrow \coprod_{i\in\mathcal{I}(\gamma)} g_i^0.$$

Our first subcomplex is ${}_0h$, which we denote also

$$\cdots \longrightarrow \coprod_{i\in\mathcal{I}(\gamma)} {}_0h_i^2 \longrightarrow \coprod_{i\in\mathcal{I}(\gamma)} {}_0h_i^1 \longrightarrow \coprod_{i\in\mathcal{I}(\gamma)} {}_0h_i^0.$$

Suppose we have constructed ${}_mh$, that is a complex

$$\cdots \longrightarrow \coprod_{i\in\mathcal{I}(\gamma)} {}_mh_i^2 \longrightarrow \coprod_{i\in\mathcal{I}(\gamma)} {}_mh_i^1 \longrightarrow \coprod_{i\in\mathcal{I}(\gamma)} {}_mh_i^0.$$

Then, as in 7.3.5.3, we can construct objects u_i^n. Recall: these objects come with commutative diagrams

$$\begin{array}{ccccc}
\mathcal{S}(-,p) & \longrightarrow & {}_m h_i^n & \xrightarrow{\partial} & {}_m h_i^{n-1} \\
\downarrow & & \downarrow f & & \downarrow \\
u_i^{n+1} & \longrightarrow & g_i^n & \longrightarrow & g_i^{n-1}
\end{array}$$

where the top row is exact. Here, if $n = 0$, we interpret g_i^{-1} to be g_i. These diagrams define objects u_i^{n+1} for $n \geq 0$. In the situation where the hypothesis of 7.3.5.3.3 holds, that is g_i and h_i are homological, then the conclusion of 7.3.5.3.3 allows us to also choose $u_i^0 \subset g_i^0$.

By 7.3.5.2, there is a subcomplex

$$\cdots \longrightarrow \coprod_{i\in\mathcal{I}(\gamma)} {}_{m+1}h_i^2 \longrightarrow \coprod_{i\in\mathcal{I}(\gamma)} {}_{m+1}h_i^1 \longrightarrow \coprod_{i\in\mathcal{I}(\gamma)} {}_{m+1}h_i^0$$

of the complex

$$\cdots \longrightarrow \coprod_{i\in\mathcal{I}(\gamma)} g_i^2 \longrightarrow \coprod_{i\in\mathcal{I}(\gamma)} g_i^1 \longrightarrow \coprod_{i\in\mathcal{I}(\gamma)} g_i^0$$

so that ${}_{m+1}h_i^n \subset g_i^n$ contains ${}_m h_i^n$ and u_i^n. This defines the complex ${}_{m+1}h$.

What we have done is constructed a sequence of complexes ${}_m h$, and the inclusion ${}_m h \longrightarrow {}_{m+1}h$ is, for each $i \in \mathcal{I}$, a map of complexes

$$\begin{array}{ccccccc}
\cdots & \longrightarrow & {}_m h_i^2 & \longrightarrow & {}_m h_i^1 & \longrightarrow & {}_m h_i^0 \\
& & \downarrow & & \downarrow & & \downarrow \\
\cdots & \longrightarrow & {}_{m+1}h_i^2 & \longrightarrow & {}_{m+1}h_i^1 & \longrightarrow & {}_{m+1}h_i^0 .
\end{array}$$

The fact that $u_i^n \subset {}_{m+1}h_i^n$ means that we have commutative diagrams

$$\begin{array}{ccccc}
\mathcal{S}(-,p) & \longrightarrow & {}_m h_i^n & \xrightarrow{\partial} & {}_m h_i^{n-1} \\
\downarrow & & \downarrow & & \downarrow \\
u_i^{n+1} & \longrightarrow & {}_{m+1}h_i^n & \longrightarrow & {}_{m+1}h_i^{n-1} \\
\downarrow & & \downarrow 1 & & \downarrow 1 \\
{}_{m+1}h_i^{n+1} & \longrightarrow & {}_{m+1}h_i^n & \longrightarrow & {}_{m+1}h_i^{n-1}
\end{array}$$

The exactness of the top row means that the kernel of $\partial : {}_m h_i^n \longrightarrow {}_m h_i^{n-1}$ maps to zero in the homology of the bottom row. In other words, the map of chain complexes

$$\begin{array}{ccccccc}
\cdots \longrightarrow & {}_mh_i^2 & \longrightarrow & {}_mh_i^1 & \xrightarrow{{}_m\partial_0} & {}_mh_i^0 \\
& \downarrow & & \downarrow & & \downarrow \\
\cdots \longrightarrow & {}_{m+1}h_i^2 & \longrightarrow & {}_{m+1}h_i^1 & \xrightarrow{{}_{m+1}\partial_0} & {}_{m+1}h_i^0
\end{array}$$

induces the zero map in homology, except when $n = 0$. The statement for $n = 0$ is that the image of the cokernel of ${}_m\partial_0$ in the cokernel of ${}_{m+1}\partial_0$ injects into g_i. If we further assume the hypothesis of 7.3.5.3.3 holds, that is g_i and h_i are homological, then the construction of

$${}_mh_i^0 \oplus u_i^0 \subset {}_{m+1}h_i^0$$

guarantees that the map of images

$$\text{Im}\left\{{}_mh_i^0 \longrightarrow g_i\right\} \longrightarrow \text{Im}\left\{{}_{m+1}h_i^0 \longrightarrow g_i\right\}$$

factors through some representable $\mathcal{S}(-, r_m)$.

Now let the subcomplex h be the colimit of the ${}_mh$. That is, we have a complex

$$\cdots \longrightarrow \coprod_{i\in\mathcal{I}(\gamma)} h_i^2 \longrightarrow \coprod_{i\in\mathcal{I}(\gamma)} h_i^1 \longrightarrow \coprod_{i\in\mathcal{I}(\gamma)} h_i^0.$$

It clearly lies in $A(\mathcal{S})$, since $A(\mathcal{S})$ is closed under colimits of fewer than α objects, and $\aleph_0 < \alpha$. I assert that this complex has the property that for each $i \in \mathcal{I}$, the sequence

$$\cdots \longrightarrow h_i^2 \longrightarrow h_i^1 \longrightarrow h_i^0$$

is exact. The sequence is the colimit of sequences

$$\cdots \longrightarrow {}_mh_i^2 \longrightarrow {}_mh_i^1 \longrightarrow {}_mh_i^0$$

and there is a spectral sequence for computing its homology. The easiest way to see the existence of this spectral sequence is to consider the double complex given by the map of complexes

$$\coprod_{m=0}^{\infty} {}_mh_i \xrightarrow{1-shift} \coprod_{m=0}^{\infty} {}_mh_i.$$

There are two spectral sequences that compute the homology of this double complex. One quickly degenerates to the colimit sequence

$$\cdots \longrightarrow h_i^2 \longrightarrow h_i^1 \longrightarrow h_i^0.$$

The other can be used to compute its homology.

The terms in the spectral sequence involve

$$\underrightarrow{\text{colim}} H^{-n}({}_mh_i) \qquad \text{and} \qquad \underrightarrow{\text{colim}}^1 H^{-n}({}_mh_i).$$

If $n \geq 1$, the map

$$H^{-n}({}_m h_i) \longrightarrow H^{-n}({}_{m+1} h_i)$$

vanishes. In other words, the sequence of maps

$$H^{-n}({}_0 h_i) \longrightarrow H^{-n}({}_1 h_i) \longrightarrow H^{-n}({}_2 h_i) \longrightarrow \cdots$$

is a cofinal subsequence of

$$H^{-n}({}_0 h_i) \longrightarrow 0 \longrightarrow H^{-n}({}_1 h_i) \longrightarrow 0 \longrightarrow \cdots$$

as is the sequence

$$0 \longrightarrow 0 \longrightarrow 0 \longrightarrow \cdots$$

By Proposition A.4.8, all three sequences have the same $\underrightarrow{\mathrm{colim}}$ and $\underrightarrow{\mathrm{colim}}^1$; but for the zero sequence this clearly vanishes.

If $n = 0$, we have that the image of

$$H^0({}_m h_i) \longrightarrow H^0({}_{m+1} h_i)$$

injects into g_i. In other words, it stabilises. The sequence

$$H^0({}_0 h_i) \xrightarrow{\phi_0} H^0({}_1 h_i) \xrightarrow{\phi_1} H^0({}_2 h_i) \longrightarrow \cdots$$

is a cofinal subsequence of

$$H^0({}_0 h_i) \longrightarrow \mathrm{Im}(\phi_0) \longrightarrow H^0({}_1 h_i) \longrightarrow \mathrm{Im}(\phi_1) \longrightarrow \cdots$$

as is

$$\mathrm{Im}(\phi_0) \longrightarrow \mathrm{Im}(\phi_1) \longrightarrow \mathrm{Im}(\phi_2) \longrightarrow \cdots$$

By Proposition A.4.8, all three sequences have the same $\underrightarrow{\mathrm{colim}}$ and $\underrightarrow{\mathrm{colim}}^1$. On the other hand, the sequence

$$\mathrm{Im}(\phi_0) \longrightarrow \mathrm{Im}(\phi_1) \longrightarrow \mathrm{Im}(\phi_2) \longrightarrow \cdots$$

is co–Mittag–Leffler. Being a co–Mittag–Leffler sequence in $A(\mathcal{S})$, it must have vanishing $\underrightarrow{\mathrm{colim}}^1$.

If we furthermore assume the hypothesis of 7.3.5.1, that is that g_i and h_i are homological, then we know a little more. We know that in the sequence

$$\mathrm{Im}(\phi_0) \longrightarrow \mathrm{Im}(\phi_1) \longrightarrow \mathrm{Im}(\phi_2) \longrightarrow \cdots$$

above, the maps $\mathrm{Im}(\phi_n) \longrightarrow \mathrm{Im}(\phi_{n+1})$ factor through representable objects $\mathcal{S}(-, r_n)$. The sequence above is cofinal in

$$\mathrm{Im}(\phi_0) \longrightarrow \mathcal{S}(-, r_0) \longrightarrow \mathrm{Im}(\phi_1) \longrightarrow \mathcal{S}(-, r_1) \longrightarrow \cdots$$

as is

$$\mathcal{S}(-, r_0) \longrightarrow \mathcal{S}(-, r_1) \longrightarrow \mathcal{S}(-, r_2) \longrightarrow \cdots$$

Once again, Proposition A.4.8 allows us to conclude that all three sequences have the same $\underrightarrow{\operatorname{colim}}$ and $\underrightarrow{\operatorname{colim}}^1$. We already know that $\underrightarrow{\operatorname{colim}}^1$ vanishes. It follows that we have an exact sequence in $A(\mathcal{S}) \subset \mathcal{E}x\big(\mathcal{S}^{op}, \mathcal{A}b\big)$

$$0 \longrightarrow \coprod_{n=0}^{\infty} \mathcal{S}(-, r_0) \xrightarrow{1-shift} \coprod_{n=0}^{\infty} \mathcal{S}(-, r_0) \longrightarrow \underrightarrow{\operatorname{colim}} \longrightarrow 0.$$

Representables are projective–injective in the Frobenius abelian category $A(\mathcal{S})$. Their coproduct is projective, hence injective. And since the first two terms in this exact sequence are injective, the sequence splits and the third term is also projective–injective, hence representable. Under the hypothesis of 7.3.5.1, the only non–vanishing term in the spectral sequence is a representable object in $A(\mathcal{S})$.

In the spectral sequence, there is only one non–zero term. And that term is the homology of the complex

$$\cdots \longrightarrow h_i^2 \longrightarrow h_i^1 \longrightarrow h_i^0.$$

The homology is all concentrated in degree 0. If we furthermore assume the hypothesis of 7.3.5.1, that is that g_i and h_i are homological, then we know further that this unique non–zero term is a representable $\mathcal{S}(-, s)$.

But now it follows that the complex

$$\cdots \longrightarrow \coprod_{i \in \mathcal{I}(\gamma)} h_i^2 \longrightarrow \coprod_{i \in \mathcal{I}(\gamma)} h_i^1 \longrightarrow \coprod_{i \in \mathcal{I}(\gamma)} h_i^0$$

is induced by a projective resolution of a co–Mittag–Leffler sequence in $A(\mathcal{S})$. Define $F'(j)$ to be the cokernel of

$$\coprod_{i \leq j} h_i^1 \longrightarrow \coprod_{i \leq j} h_i^0.$$

This gives a co–Mittag–Leffler sequence

$$F' : \mathcal{I}(\gamma) \longrightarrow A(\mathcal{S}).$$

This sequence comes with a projective resolution, and h is a chain complex computing its $\underrightarrow{\operatorname{colim}}^n$. Since co–Mittag–Leffler sequences in $A(\mathcal{S})$ have vanishing $\underrightarrow{\operatorname{colim}}^n$, the chain complex

$$\cdots \longrightarrow \coprod_{i \in \mathcal{I}(\gamma)} h_i^2 \longrightarrow \coprod_{i \in \mathcal{I}(\gamma)} h_i^1 \longrightarrow \coprod_{i \in \mathcal{I}(\gamma)} h_i^0$$

is acyclic except at $n = 0$. Since any homology class in the complex

$$\cdots \longrightarrow \coprod_{i \in \mathcal{I}(\gamma)} g_i^2 \longrightarrow \coprod_{i \in \mathcal{I}(\gamma)} g_i^1 \longrightarrow \coprod_{i \in \mathcal{I}(\gamma)} g_i^0$$

is supported on some subcomplex h as above, it follows that the complex g is acyclic.

Let the co–Mittag–Leffler sequence $F' : \mathcal{I}(\gamma) \longrightarrow A(\mathcal{S})$ be as above. We have an exact sequence

$$\cdots \longrightarrow h_i^2 \longrightarrow h_i^1 \longrightarrow h_i^0 \longrightarrow \frac{F'i}{\operatorname{colim}_{j<i} F'j} \longrightarrow 0$$

From now until the end of the proof, assume furthermore that the hypothesis of 7.3.5.1 holds, that is that g_i and h_i are homological. Under this assumption we have that the unique non–zero homology of

$$\cdots \longrightarrow h_i^2 \longrightarrow h_i^1 \longrightarrow h_i^0$$

is a representable in $A(\mathcal{S})$. That is, the object

$$\frac{F'i}{\operatorname{colim}_{j<i} F'j}$$

is projective–injective. But then an easy induction shows that for all $i \leq \gamma$,

$$\operatorname{colim}_{j<i} F'j \quad = \quad \coprod_{j<i} \frac{F'j}{\operatorname{colim}_{k<j} F'k}$$

is projective–injective. All the extensions must split. Taking the colimit over all ordinals $< \gamma$, we have that $\operatorname{colim}_{j<\gamma} F'j$ is representable, hence a homological object in $\mathcal{E}x(\mathcal{S}^{op}, \mathcal{A}b)$.

This is true for any subcomplex

$$\cdots \longrightarrow \coprod_{i\in\mathcal{I}(\gamma)} h_i^2 \longrightarrow \coprod_{i\in\mathcal{I}(\gamma)} h_i^1 \longrightarrow \coprod_{i\in\mathcal{I}(\gamma)} h_i^0$$

of the complex

$$\cdots \longrightarrow \coprod_{i\in\mathcal{I}(\gamma)} g_i^2 \longrightarrow \coprod_{i\in\mathcal{I}(\gamma)} g_i^1 \longrightarrow \coprod_{i\in\mathcal{I}(\gamma)} g_i^0$$

as constructed in the proof. But these good subcomplexes contain any subcomplex of cardinality $< \alpha$. In other words, $\underrightarrow{\operatorname{colim}} F$ is the α–filtered colimit of $\underrightarrow{\operatorname{colim}} F'$, with F' as constructed in the proof. This makes $\underrightarrow{\operatorname{colim}} F$ an α–filtered colimit of homological objects in $\mathcal{E}x(\mathcal{S}^{op}, \mathcal{A}b)$, and by Lemma 7.2.2, we deduce that $\underrightarrow{\operatorname{colim}} F$ is homological. □

COROLLARY 7.3.6. *Let $\gamma < \alpha$ be an ordinal. Suppose for any ordinal $i < \gamma$, for any co–Mittag–Leffler sequence H' in $\mathcal{E}x(\mathcal{S}^{op}, \mathcal{A}b)$ of length i, and for any integer $n \geq 1$, $\underrightarrow{\operatorname{colim}}^n H' = 0$.*

Let H be a co–Mittag–Leffler sequence of length γ. Then for all $n \geq 2$,

$$\underrightarrow{\operatorname{colim}}^n H = 0.$$

Proof: Let H be a co–Mittag–Leffler sequence of length γ. We can always find a slightly special projective object G in the category of sequences, and an epimorphism $G \longrightarrow H$, as in Remark 7.3.3. We deduce an exact sequence

$$0 \longrightarrow F \longrightarrow G \longrightarrow H \longrightarrow 0.$$

The functor H is co–Mittag–Leffler by hypothesis, the functor G because it is a projective object. We wish to apply Lemma 7.3.5. To this end, we must prove that the functor F is co–Mittag–Leffler, and that for any $i < \gamma$,

$$0 \longrightarrow \operatorname*{colim}_{j<i} Fj \longrightarrow \operatorname*{colim}_{j<i} Gj \longrightarrow \operatorname*{colim}_{j<i} Hj \longrightarrow 0$$

is exact.

Let $i < \gamma$ be an ordinal. Now H is a co–Mittag–Leffler sequence on $\mathfrak{I}(\gamma)$, and hence the restriction of H to $\mathfrak{I}(i) \subset \mathfrak{I}(\gamma)$ is also co–Mittag–Leffler. The long exact sequence for the left derived functors of colimit gives

$$\operatorname*{colim}^1_{h<i} H \longrightarrow \operatorname*{colim}_{h<i} Fh \longrightarrow \operatorname*{colim}_{h<i} Gh \longrightarrow \operatorname*{colim}_{h<i} Hh \longrightarrow 0.$$

But by the hypothesis of the Corollary, $\underrightarrow{\operatorname{colim}}^1$ vanishes on co–Mittag–Leffler sequences of length $i < \gamma$, in other words

$$\operatorname*{colim}^1_{h<i} H \quad = \quad 0.$$

Hence we have, for $i < \gamma$, an exact sequence

$$0 \longrightarrow \operatorname*{colim}_{h<i} Fh \longrightarrow \operatorname*{colim}_{h<i} Gh \longrightarrow \operatorname*{colim}_{h<i} Hh \longrightarrow 0.$$

Now let $i \le j < \gamma$ be ordinals. We deduce a commutative diagram with exact rows

$$\begin{array}{ccccccccc} 0 & \longrightarrow & \operatorname*{colim}_{h<i} Fh & \longrightarrow & \operatorname*{colim}_{h<i} Gh & \longrightarrow & \operatorname*{colim}_{h<i} Hh & \longrightarrow & 0 \\ & & \downarrow & & \downarrow & & \downarrow & & \\ 0 & \longrightarrow & Fj & \longrightarrow & Gj & \longrightarrow & Hj & \longrightarrow & 0 \end{array}$$

Since G is co–Mittag–Leffler, the map

$$\operatorname*{colim}_{h<i} Gh \longrightarrow Gj$$

is mono. The composite

$$\begin{array}{ccc} \operatorname*{colim}_{h<i} Fh & \longrightarrow & \operatorname*{colim}_{h<i} Gh \\ & & \downarrow \\ & & Gj \end{array}$$

is a composite of two monomorphisms, hence a monomorphism. But the commutativity of

$$\begin{array}{ccc} \operatorname*{colim}_{h<i} Fh & \longrightarrow & \operatorname*{colim}_{h<i} Gh \\ \downarrow & & \downarrow \\ Fj & \longrightarrow & Gj \end{array}$$

means that it is equal to

$$\begin{array}{ccc} \operatorname*{colim}_{h<i} Fh & & \\ \downarrow & & \\ Fj & \longrightarrow & Gj \end{array}$$

and hence

$$\operatorname*{colim}_{h<i} Fh \longrightarrow Fj$$

is monomorphic. By Remark A.3.11, this says that F is co–Mittag–Leffler.

Lemma 7.3.5 now applies, and we deduce that for $n \geq 1$, $\underrightarrow{\operatorname{colim}}^n F = 0$. The exact sequence for the left derived functor of the colimit gives

$$\underrightarrow{\operatorname{colim}}^{n+1} G \longrightarrow \underrightarrow{\operatorname{colim}}^{n+1} H \longrightarrow \underrightarrow{\operatorname{colim}}^n F.$$

If $n \geq 1$, then $\underrightarrow{\operatorname{colim}}^n F$ vanishes by the above, while $\underrightarrow{\operatorname{colim}}^{n+1} G$ vanishes because G is projective. Hence $\underrightarrow{\operatorname{colim}}^{n+1} H$ vanishes if $n \geq 1$, in other words $\underrightarrow{\operatorname{colim}}^n H$ vanishes if $n \geq 2$. □

COROLLARY 7.3.7. *Let the hypotheses be as in Corollary 7.3.6. That is, let $\gamma < \alpha$ be an ordinal. Suppose for any ordinal $i < \gamma$, for any co–Mittag–Leffler sequence H' in $\mathcal{E}x(\mathcal{S}^{op}, \mathcal{A}b)$ of length i, and for any integer $n \geq 1$, $\underrightarrow{\operatorname{colim}}^n H' = 0$.*

Let H be a co–Mittag–Leffler sequence of length γ. Suppose that for each $i \in \mathcal{I}(\gamma)$, Hi is a homological object of $\mathcal{E}x(\mathcal{S}^{op}, \mathcal{A}b)$. Then for all $i < \gamma$, the objects

$$\frac{Hi}{\operatorname*{colim}_{j<i} Hj} \qquad \textit{and} \qquad \operatorname*{colim}_{j<i} Hj$$

are homological.

Proof: We will prove, by induction on the ordinal $i < \gamma$, that

7.3.7.1. *For all $j \leq i$, $\operatorname{colim}_{k<j} Hk$ is homological.*

7.3.7.2. *For all $j \leq i$, the object*

$$\frac{Hj}{\operatorname{colim}_{k<j} Hk}$$

is homological.

The strategy of the proof will be to show that

7.3.7.3. If 7.3.7.1 holds for i, then 7.3.7.2 holds for i.

7.3.7.4. If 7.3.7.2 holds for all $j < i$, then 7.3.7.1 holds for i.

The proof of 7.3.7.3 is trivial. If 7.3.7.1 holds for i, that means that, for all $j \leq i$, $\operatorname{colim}_{k<j} Hj$ is homological. In the exact sequence

$$0 \longrightarrow \operatorname{colim}_{k<j} Hk \longrightarrow Hj \longrightarrow \frac{Hj}{\operatorname{colim}_{k<j} Hk} \longrightarrow 0,$$

the first two terms are homological. By Lemma 7.2.5, so is the third; we deduce that $\frac{Hj}{\operatorname{colim}_{k<j} Hk}$ is homological for $j \leq i$, that is 7.3.7.2.

Now for the proof of 7.3.7.4. Assume therefore that i is an ordinal $< \gamma$, and for all $j < i$ we have 7.3.7.2. That is, we assume that for all $j \in \mathcal{I}(i)$, the quotient

$$\frac{Hj}{\operatorname{colim}_{k<j} Hk}$$

is homological. We want to show that for $j \leq i$, the object $\operatorname{colim}_{k<j} Hk$ is homological. Reducing i if necessary, we may assume $j = i$.

As in the proof of Corollary 7.3.6, choose an exact sequence of co–Mittag–Lefflers

$$0 \longrightarrow F \longrightarrow G \longrightarrow H \longrightarrow 0,$$

with G a special projective. For any $j < i$,

$$\frac{Gj}{\operatorname{colim}_{k<j} Gk} \quad = \quad \coprod \mathcal{S}(-, s_\lambda).$$

By Example 7.2.3, it is homological.

We therefore find ourselves in the situation of Lemma 7.3.5, but more precisely we are in the situation where the hypothesis of 7.3.5.1 holds as well. For all $j < i$, the functors

$$\frac{Gj}{\operatorname{colim}_{k<j} Gk} \qquad \text{and} \qquad \frac{Hj}{\operatorname{colim}_{k<j} Hk}$$

are homological. The conclusion of 7.3.5.1 tells us that $\operatorname{colim}_{j<i} Fj$ is also homological.

Now we have an exact sequence

$$\operatorname*{colim}_{j<i}{}^1 H \longrightarrow \operatorname*{colim}_{j<i} F \longrightarrow \operatorname*{colim}_{j<i} G \longrightarrow \operatorname*{colim}_{j<i} H \longrightarrow 0.$$

By the hypothesis of the Corollary, $\underrightarrow{\operatorname{colim}}^1$ vanishes on all co–Mittag–Leffler sequences of length $i < \gamma$. Hence $\operatorname*{colim}_{j<i}{}^1 H = 0$, and we have an exact

$$0 \longrightarrow \operatorname*{colim}_{j<i} F \longrightarrow \operatorname*{colim}_{j<i} G \longrightarrow \operatorname*{colim}_{j<i} H \longrightarrow 0.$$

We have just proved that $\operatorname{colim}_{j<i} F$ is homological. The object $\operatorname{colim}_{j<i} G$ is a coproduct of representables, hence homological by Example 7.2.3. From Lemma 7.2.5, we can now deduce that $\operatorname{colim}_{j<i} H$ is also homological. □

COROLLARY 7.3.8. *If G is a co–Mittag–Leffler sequence in $\mathcal{E}x(\mathcal{S}^{op}, \mathcal{A}b)$ of length $< \alpha$, and G satisfies the hypothesis*

7.3.8.1. *For all i*

$$\frac{Gi}{\operatorname*{colim}_{j<i} Gj} = \coprod_{\lambda\in\Lambda_i} G^i_\lambda$$

with $G^i_\lambda \in A(\mathcal{S})$,

then for $n \geq 1$, $\underrightarrow{\operatorname{colim}}^n G = 0$.

Proof: Apply Lemma 7.3.5, letting $F = G$ and $H = 0$ in the exact sequence

$$0 \longrightarrow F \longrightarrow G \longrightarrow H \longrightarrow 0.$$

That is, we consider the short exact sequence of co–Mittag–Lefflers

$$0 \longrightarrow G \stackrel{1}{\longrightarrow} G \longrightarrow 0 \longrightarrow 0.$$

Clearly, any colimit of this exact sequence is exact. Lemma 7.3.5 applies, and tells us that for $n \geq 1$, $\underrightarrow{\operatorname{colim}}^n G = 0$. □

7.4. The derived functors of colimits in $\mathcal{E}x(\mathcal{S}^{op}, \mathcal{A}b)$

Section 7.3 proved a number of technical lemmas about the derived functor of the colimit, applied to certain co–Mittag–Leffler sequences in $\mathcal{E}x(\mathcal{S}^{op}, \mathcal{A}b)$. In this Section, we propose to outline how the author hoped to deduce that $\underrightarrow{\operatorname{colim}}^n$ vanishes on co–Mittag–Leffler sequences. We must somehow apply our lemmas.

REMARK 7.4.1. In Remark A.3.11, we learned that a sequence F of length γ is co–Mittag–Leffler if and only if, for every $j \leq k < \gamma$, the map

$$\operatorname*{colim}_{i<j} Gi \longrightarrow Gk$$

is mono. Of course, it would be much more pleasant to only have to check the case $j = k$. Under suitable vanishing for $\underrightarrow{\text{colim}}^n$, this works.

LEMMA 7.4.2. *Let $\mathcal{A}$ be an abelian category. Let Q be a class of objects in $\mathcal{A}$. Suppose γ is an ordinal, and further*

> 7.4.2.1. *If β is an ordinal $\beta < \gamma$, and $H : \mathcal{I}(\beta) \longrightarrow \mathcal{A}$ a co–Mittag–Leffler sequence, so that for all $j \in \mathcal{I}(\beta)$*
>
> $$\frac{Hj}{\operatorname*{colim}_{i<j} Hi} \qquad \in \qquad Q,$$
>
> *then for $n \geq 1$, $\underrightarrow{\text{colim}}^n H = 0$.*

Suppose we are given a sequence $G : \mathcal{I}(\gamma) \longrightarrow \mathcal{A}$ so that

> 7.4.2.2. *For $0 < j \in \mathcal{I}(\gamma)$ any ordinal, the map*
>
> $$\operatorname*{colim}_{i<j} Gi \longrightarrow Gj$$
>
> *is mono, and*
>
> $$\frac{Gj}{\operatorname*{colim}_{i<j} Gi} \qquad \in \qquad Q.$$

Then the sequence G is automatically co–Mittag–Leffler.

Proof: Note first that Condition 7.4.2.2 places no hypothesis on $G0$. The hypothesis is on $j > 0$. We have that $G1/G0$, $G2/G1$, etc. are in Q. But no hypothesis is made on $G0$.

Now for the proof. If j is a limit ordinal, then 7.4.2.2 asserts that the map

$$\operatorname*{colim}_{i<j} Gi \longrightarrow Gj$$

is mono. This is exactly A.3.10.2 of Definition A.3.10. It remains to check that G satisfies A.3.10.1. That is, for every $i < j < \gamma$, we must show the map

$$Gi \longrightarrow Gj$$

is mono.

By induction, we may assume this is true for all $i < j < \beta$, with $\beta < \gamma$. We want to show that it remains true for all $i < j \leq \beta$. The interesting case, which is not included in the induction hypothesis, is where $j = \beta$. We are given by induction that the map

$$Gi \longrightarrow Gj$$

is mono, if $i < j < \beta$, and want to deduce that

$$Gi \longrightarrow G\beta$$

is mono for $i < \beta$.

If β is a successor ordinal, then $\beta = j + 1$. The induction hypothesis asserts that

$$Gi \longrightarrow Gj$$

is mono. From 7.4.2.2, we know that so is

$$\operatorname*{colim}_{i<j+1} Gi \longrightarrow G(j+1);$$

but this is just the map

$$Gj \longrightarrow G(j+1) \quad = \quad G\beta.$$

The composite

$$Gi \longrightarrow Gj \longrightarrow G(j+1) \quad = \quad G\beta$$

is the composite of two monomorphisms, hence mono.

The remaining case is when β is a limit ordinal. By 7.4.2.2, the map

$$\operatorname*{colim}_{j<\beta} Gj \longrightarrow G\beta$$

is mono. We want to show that for $i < \beta$, the composite

$$Gi \longrightarrow \operatorname*{colim}_{j<\beta} Gj \longrightarrow G\beta$$

is mono. It clearly sufices to prove that the natural map

$$Gi \longrightarrow \operatorname*{colim}_{j<\beta} Gj$$

is mono.

To do this, it is convenient to consider the following short exact sequence of objects in $\mathcal{C}at\big(\mathfrak{I}^{op}(\beta), \mathcal{A}\big)$

$$0 \longrightarrow F \longrightarrow G \longrightarrow H \longrightarrow 0.$$

G is our given functor, restricted from $\mathfrak{I}(\gamma)$ to the subset $\mathfrak{I}(\beta)$. The sequence F is given by the rule

$$Fj \quad = \quad \begin{cases} Gj & \text{if } j \leq i \\ Gi & \text{if } j > i \end{cases}$$

while H is given by the rule

$$Hj \quad = \quad \begin{cases} 0 & \text{if } j \leq i \\ \frac{Gj}{Gi} & \text{if } j > i \end{cases}$$

In other words, the short exact sequence

$$0 \longrightarrow F \longrightarrow G \longrightarrow H \longrightarrow 0$$

is just

$$\begin{array}{ccccccccc}
0 & & 0 & & 0 & & 0 & & 0 \\
\downarrow & & \downarrow & & \downarrow & & \downarrow & & \downarrow \\
G(i-2) & \longrightarrow & G(i-1) & \longrightarrow & G(i) & \longrightarrow & G(i) & \longrightarrow & G(i) \\
\downarrow & & \downarrow & & \downarrow & & \downarrow & & \downarrow \\
G(i-2) & \longrightarrow & G(i-1) & \longrightarrow & G(i) & \longrightarrow & G(i+1) & \longrightarrow & G(i+2) \\
\downarrow & & \downarrow & & \downarrow & & \downarrow & & \downarrow \\
0 & \longrightarrow & 0 & \longrightarrow & 0 & \longrightarrow & \frac{G(i+1)}{G(i)} & \longrightarrow & \frac{G(i+2)}{G(i)} \\
\downarrow & & \downarrow & & \downarrow & & \downarrow & & \downarrow \\
0 & & 0 & & 0 & & 0 & & 0
\end{array}$$

Our induction hypothesis is that, restricted to $\mathcal{I}(\beta)$, the sequence G is co–Mittag–Leffler. Hence so is the sequence H. But the sequence H now satisfies the hypothesis of 7.4.2.1; it is a co–Mittag–Leffler sequence of length $< \gamma$, with

$$\frac{Hj}{\underset{i<j}{\operatorname{colim}} Hi} \quad \in \quad Q.$$

By the conclusion of 7.4.2.1, $\underrightarrow{\operatorname{colim}}^1 H = 0$. Now the long exact sequence for the colim gives an exact

$$\underrightarrow{\operatorname{colim}}^1 H \longrightarrow \underrightarrow{\operatorname{colim}} F \longrightarrow \underrightarrow{\operatorname{colim}} G$$

and since $\underrightarrow{\operatorname{colim}}^1 H = 0$, we have an injection

$$\underrightarrow{\operatorname{colim}} F \longrightarrow \underrightarrow{\operatorname{colim}} G.$$

But $\underrightarrow{\operatorname{colim}} F = Gi$, and hence the map

$$Gi \longrightarrow \underset{j<\beta}{\operatorname{colim}}\, Gj$$

is mono. □

EXAMPLE 7.4.3. Let G be a sequence

$$G : \mathcal{I}(\alpha+1) \longrightarrow \mathcal{E}x(\mathcal{S}^{op}, \mathcal{A}b).$$

Suppose it satisfies

7.4.3.1. For $j \geq 0$, the map

$$Gj \longrightarrow G(j+1)$$

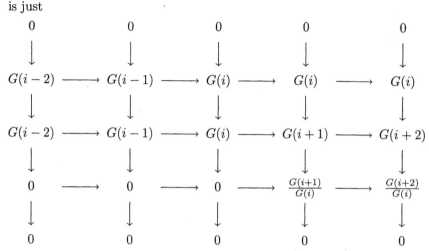

is mono, and the quotient $G(j+1)/G(j)$ is a coproduct of objects in $A(\mathcal{S}) \subset \mathcal{E}x\big(\mathcal{S}^{op}, \mathcal{A}b\big)$.

7.4.3.2. For j a limit ordinal,

$$Gj \quad = \quad \operatorname*{colim}_{i<j} Gi.$$

Then the sequence G is automatically co–Mittag–Leffler.

The point is that Lemma 7.4.2 applies. Let $\mathcal{A}$ be the abelian category $\mathcal{E}x\big(\mathcal{S}^{op}, \mathcal{A}b\big)$, and let Q be the class of objects which are coproducts of objects in $A(\mathcal{S})$. Then 7.4.3.1 and 7.4.3.2 are 7.4.2.2 for, respectively, the case of a successor and a limit ordinal. It remains to show that 7.4.2.1 holds.

That means, we need to know that, for any ordinal $\beta < \alpha + 1$, for any co–Mittag–Leffler

$$H : \mathcal{I}(\beta) \longrightarrow \mathcal{E}x\big(\mathcal{S}^{op}, \mathcal{A}b\big)$$

with

$$\frac{Hj}{\operatorname*{colim}_{i<j} Hi} \quad \in \quad Q,$$

and for any $n \geq 1$, $\underrightarrow{\operatorname{colim}}^n H = 0$. If $\beta < \alpha$, this is just exactly Corollary 7.3.8. If $\beta = \alpha$, then $\mathcal{I}(\alpha)$ is an α–filtered category. By Lemma A.1.3, α–filtered colimits agree in $\mathcal{E}x\big(\mathcal{S}^{op}, \mathcal{A}b\big)$ and $\mathcal{C}at\big(\mathcal{S}^{op}, \mathcal{A}b\big)$. In particular, colimits over the category $\mathcal{I}(\alpha)$ are the same in $\mathcal{E}x\big(\mathcal{S}^{op}, \mathcal{A}b\big)$ as in the larger category $\mathcal{C}at\big(\mathcal{S}^{op}, \mathcal{A}b\big)$, hence are exact. Therefore if $n \geq 1$, then $\underrightarrow{\operatorname{colim}}^n$ vanishes for sequences

$$H : \mathcal{I}(\alpha) \longrightarrow \mathcal{E}x\big(\mathcal{S}^{op}, \mathcal{A}b\big).$$

REMARK 7.4.4. Suppose $\mathcal{T}$ is a triangulated category statisfying [TR5], and $\mathcal{S} = \mathcal{T}^\alpha$. Any object F of $\mathcal{E}x\big(\mathcal{S}^{op}, \mathcal{A}b\big)$ has a projective presentation

$$\coprod_{\lambda\in\Lambda} \mathcal{S}(-, s_\lambda) \longrightarrow \coprod_{\mu\in M} \mathcal{S}(-, t_\mu) \longrightarrow F(-) \longrightarrow 0.$$

By Lemma 6.2.5, the homological functor $\mathcal{T} \longrightarrow \mathcal{E}x\big(\mathcal{S}^{op}, \mathcal{A}b\big)$ preserves coproducts. Hence the presentation above may be rewritten

$$\mathcal{T}\left(-, \coprod_{\lambda\in\Lambda} s_\lambda\right)\Bigg|_{\mathcal{S}} \longrightarrow \mathcal{T}\left(-, \coprod_{\mu\in M} t_\mu\right)\Bigg|_{\mathcal{S}} \longrightarrow F(-) \longrightarrow 0.$$

In the triangulated category $\mathcal{T}$, there is a triangle

$$\coprod_{\lambda\in\Lambda} s_\lambda \longrightarrow \coprod_{\mu\in M} t_\mu \longrightarrow y \longrightarrow \Sigma\left\{\coprod_{\lambda\in\Lambda} s_\lambda\right\}.$$

Hence we have an exact sequence in $\mathcal{E}x(\mathcal{S}^{op}, \mathcal{A}b)$

$$\mathcal{T}\left(-, \coprod_{\lambda\in\Lambda} s_\lambda\right)\bigg|_{\mathcal{S}} \longrightarrow \mathcal{T}\left(-, \coprod_{\mu\in M} t_\mu\right)\bigg|_{\mathcal{S}} \longrightarrow \mathcal{T}(-,y)|_{\mathcal{S}},$$

and it follows that there is a monomorphism

$$F(-) \longrightarrow \mathcal{T}(-,y)|_{\mathcal{S}}.$$

In other words, $F \in \mathcal{E}x(\mathcal{S}^{op}, \mathcal{A}b)$ may be embedded in $\mathcal{T}(-,y)|_{\mathcal{S}}$ for some $y \in \mathcal{T}$. Thus every F may be embedded in a functor $\mathcal{S}^{op} \longrightarrow \mathcal{A}b$ which is *homological;* it takes triangles to exact sequences. The next two Lemmas prove that this is true even if we do not assume $\mathcal{S} = \mathcal{T}^\alpha$.

LEMMA 7.4.5. *Let $\mathcal{S}$ be an essentially small triangulated category, satisfying [TR5(α)]. An object G of $\mathcal{E}x(\mathcal{S}^{op}, \mathcal{A}b)$ is a homological functor*

$$G : \mathcal{S}^{op} \longrightarrow \mathcal{A}b$$

if and only if, for every object

$$x \in A(\mathcal{S}) \subset \mathcal{E}x(\mathcal{S}^{op}, \mathcal{A}b),$$

the following Ext*–group vanishes*

$$\mathrm{Ext}^1(x, G) = 0.$$

Here, the Ext *groups are computed in $\mathcal{E}x(\mathcal{S}^{op}, \mathcal{A}b)$.*

Proof: Suppose that, for all $x \in A(\mathcal{S})$,

$$\mathrm{Ext}^1(x, G) = 0.$$

We need to show that G is homological. That is, for any triangle in $\mathcal{S}$

$$r \longrightarrow s \longrightarrow t \longrightarrow \Sigma s,$$

we must show the sequence

$$Gt \longrightarrow Gs \longrightarrow Gr$$

is exact.

Now the triangle

$$r \longrightarrow s \longrightarrow t \longrightarrow \Sigma s$$

gives an exact sequence in $A(\mathcal{S})$

$$\mathcal{S}(-,r) \longrightarrow \mathcal{S}(-,s) \longrightarrow \mathcal{S}(-,t)$$

which can be broken into two exact sequences in $A(\mathcal{S})$

$$\mathcal{S}(-,r) \longrightarrow \mathcal{S}(-,s) \longrightarrow P(-) \longrightarrow 0$$

$$0 \longrightarrow P(-) \longrightarrow \mathcal{S}(-,t) \longrightarrow Q(-) \longrightarrow 0.$$

Applying the functor $\mathcal{E}x(\mathcal{S}^{op}, \mathcal{A}b)\Big\{-, G\Big\}$ to these exact sequences, we deduce exact sequences

$$0 \longrightarrow \mathrm{Hom}(P, G) \longrightarrow Gs \longrightarrow Gr$$

$$Gt \longrightarrow \mathrm{Hom}(P, G) \longrightarrow \mathrm{Ext}^1(Q, G)$$

The vanishing of $\mathrm{Ext}^1(Q, G)$ means these two exact sequences can be pieced together to an exact sequence

$$Gt \longrightarrow Gs \longrightarrow Gr.$$

This proves that G is homological.

Now suppose that G is homological; we wish to prove the vanishing of Ext^1. Let x be any object of $A(\mathcal{S})$. As in Definition 5.1.3, x admits a presentation

$$\mathcal{S}(-, s) \longrightarrow \mathcal{S}(-, t) \longrightarrow x(-) \longrightarrow 0.$$

The map $s \longrightarrow t$ may be completed to a triangle

$$r \longrightarrow s \longrightarrow t \longrightarrow \Sigma r;$$

this allows us to extend the resolution of $x(-)$ to

$$\mathcal{S}(-, r) \longrightarrow \mathcal{S}(-, s) \longrightarrow \mathcal{S}(-, t) \longrightarrow x(-) \longrightarrow 0.$$

The above is the beginning of a projective resolution for x, in either the category $A(\mathcal{S})$ or the category $\mathcal{E}x(\mathcal{S}^{op}, \mathcal{A}b)$. Viewing it as a projective resolution in the category $\mathcal{E}x(\mathcal{S}^{op}, \mathcal{A}b)$ permits us to compute Ext^1; the group $\mathrm{Ext}^1(x, G)$ is the homology of the sequence

$$\begin{array}{ccc} \mathcal{E}x(\mathcal{S}^{op}, \mathcal{A}b)\Big\{\mathcal{S}(-, t), G(-)\Big\} & \longrightarrow & \mathcal{E}x(\mathcal{S}^{op}, \mathcal{A}b)\Big\{\mathcal{S}(-, s), G(-)\Big\} \\ & & \downarrow \\ & & \mathcal{E}x(\mathcal{S}^{op}, \mathcal{A}b)\Big\{\mathcal{S}(-, r), G(-)\Big\}. \end{array}$$

Yoneda's Lemma identifies this as

$$Gt \longrightarrow Gs \longrightarrow Gr,$$

which is exact since G is homological. Hence $\mathrm{Ext}^1(x, G) = 0$. □

LEMMA 7.4.6. *Let $\mathcal{S}$ be an essentially small triangulated category, satisfying [TR5(α)]. Let F be an object of $\mathcal{E}x(\mathcal{S}^{op}, \mathcal{A}b)$; that is,*

$$F : \mathcal{S}^{op} \longrightarrow \mathcal{A}b$$

is a functor taking coproducts of fewer than α objects to products of abelian groups.

Then there exists a monomorphism $F \longrightarrow G$ in $\mathcal{E}x(\mathcal{S}^{op}, \mathcal{A}b)$, where

$$G : \mathcal{S}^{op} \longrightarrow \mathcal{A}b$$

is homological.

Proof: By Lemma 7.4.5, we are reduced to showing that F may be embedded in an object G, so that $\mathrm{Ext}^1(x, G) = 0$ for all $x \in A(\mathcal{S})$. Recall that the category $\mathcal{S}$ is essentially small, and hence so is the category $A(\mathcal{S})$. Let Λ be a set of representatives for the isomorphism classes of objects in $A(\mathcal{S})$. For each $\lambda \in \Lambda$, we denote the object in $A(\mathcal{S})$ that corresponds to it by t_λ. For each $\lambda \in \Lambda$, choose an exact sequence

$$0 \longrightarrow r_\lambda \longrightarrow \mathcal{S}(-, s_\lambda) \longrightarrow t_\lambda \longrightarrow 0$$

where $\mathcal{S}(-, s_\lambda)$ is a projective object.

Now we define a map which sends an object x of $\mathcal{E}x(\mathcal{S}^{op}, \mathcal{A}b)$ to a morphism $x \longrightarrow \overline{x}$. The construction is as follows. Let M be the set of all maps $r_\lambda \longrightarrow x$; that is

$$M = \{r_\lambda \to x \mid \lambda \in \Lambda\}.$$

We have maps

$$\begin{array}{ccc} \coprod_{\{r_\lambda \to x\} \in M} r_\lambda & \longrightarrow & \coprod_{\{r_\lambda \to x\} \in M} \mathcal{S}(-, s_\lambda) \\ \big\downarrow & & \\ x & & \end{array}$$

and we define $x \longrightarrow \overline{x}$ to be given by the pushout

$$\begin{array}{ccc} \coprod_{\{r_\lambda \to x\} \in M} r_\lambda & \longrightarrow & \coprod_{\{r_\lambda \to x\} \in M} \mathcal{S}(-, s_\lambda) \\ \big\downarrow & & \big\downarrow \\ x & \longrightarrow & \overline{x} \end{array}$$

Now recall that the sequences

$$0 \longrightarrow r_\lambda \longrightarrow \mathcal{S}(-, s_\lambda) \longrightarrow t_\lambda \longrightarrow 0$$

are exact, and that $\mathcal{E}x(\mathcal{S}^{op}, \mathcal{A}b)$ satisfies [AB4] by Lemma 6.3.2. The sequence

$$0 \longrightarrow \coprod r_\lambda \longrightarrow \coprod \mathcal{S}(-, s_\lambda) \longrightarrow \coprod t_\lambda \longrightarrow 0$$

is therefore also exact, and the pushout gives a commutative diagram with exact rows

$$\begin{array}{ccccccccc} 0 & \longrightarrow & \coprod r_\lambda & \longrightarrow & \coprod \mathcal{S}(-, s_\lambda) & \longrightarrow & \coprod t_\lambda & \longrightarrow & 0 \\ & & \big\downarrow & & \big\downarrow & & {\scriptstyle 1}\big\downarrow & & \\ 0 & \longrightarrow & x & \longrightarrow & \overline{x} & \longrightarrow & \coprod t_\lambda & \longrightarrow & 0. \end{array}$$

We deduce that $x \longrightarrow \overline{x}$ is mono, and $\overline{x}/x$ is a coproduct of objects in $A(\mathcal{S})$.

Now recall; we are given an object $F \in \mathcal{E}x(\mathcal{S}^{op}, \mathcal{A}b)$, and want to embed it in a homological object G. We propose to define a sequence of length $\alpha+1$ in $\mathcal{E}x(\mathcal{S}^{op}, \mathcal{A}b)$. We define this sequence,

$$S : \mathcal{I}(\alpha+1) \longrightarrow \mathcal{E}x(\mathcal{S}^{op}, \mathcal{A}b)$$

inductively, as follows.

7.4.6.1. $S(0) \in \mathcal{E}x(\mathcal{S}^{op}, \mathcal{A}b)$ is defined to be F.

7.4.6.2. The map $S(i) \longrightarrow S(i+1)$ is defined to be

$$S(i) \longrightarrow \overline{S(i)}$$

as above.

7.4.6.3. If i is a limit ordinal, then

$$S(i) \quad = \quad \operatorname*{colim}_{j<i} S(j).$$

By Example 7.4.3.2, we know that the sequence is co–Mittag–Leffler. In particular, it follows that the map

$$F \quad = \quad S(0) \longrightarrow S(\alpha)$$

is a monomorphism. Define $G = S(\alpha)$. It remains to prove that G is homological.

We know that

$$G \quad = \quad S(\alpha) \quad = \quad \operatorname*{colim}_{j<\alpha} S(j).$$

This is an α–filtered colimit. Let $\lambda \in \Lambda$, and suppose we are given a map

$$r_\lambda \longrightarrow G \quad = \quad \operatorname*{colim}_{j<\alpha} S(j).$$

By Lemma 7.1.5, the object $r_\lambda \in A(\mathcal{S})$ is α–presentable, and therefore its map into the α–filtered colimit must factor as

$$r_\lambda \longrightarrow S(j) \longrightarrow \operatorname*{colim}_{j<\alpha} S(j).$$

By the construction of $S(j+1) = \overline{S(j)}$, we have a commutative square

$$\begin{array}{ccc} r_\lambda & \longrightarrow & \mathcal{S}(-, s_\lambda) \\ \downarrow & & \downarrow \\ S(j) & \longrightarrow & S(j+1). \end{array}$$

This says that the map $r_\lambda \longrightarrow G$ must factor as

$$r_\lambda \longrightarrow \mathcal{S}(-, s_\lambda) \longrightarrow G.$$

But now we have the exact sequence

$$0 \longrightarrow r_\lambda \longrightarrow \mathcal{S}(-, s_\lambda) \longrightarrow t_\lambda \longrightarrow 0.$$

Applying $\text{Hom}(-, G)$ to it, we deduce an exact sequence

$$\begin{array}{ccccc} \text{Hom}\Big\{\mathcal{S}(-, s_\lambda), G\Big\} & \longrightarrow & \text{Hom}(r_\lambda, G) & \longrightarrow & \text{Ext}^1(t_\lambda, G) \\ & & & & \downarrow \\ & & & & \text{Ext}^1\Big\{\mathcal{S}(-, s_\lambda), G\Big\}. \end{array}$$

We have just proved that any map $r_\lambda \longrightarrow G$ factors as

$$r_\lambda \longrightarrow \mathcal{S}(-, s_\lambda) \longrightarrow G,$$

that is the map

$$\text{Hom}\Big\{\mathcal{S}(-, s_\lambda), G\Big\} \longrightarrow \text{Hom}(r_\lambda, G)$$

is surjective. Since $\mathcal{S}(-, s_\lambda)$ is projective in the category $\mathcal{E}x\big(\mathcal{S}^{op}, \mathcal{A}b\big)$, we must have

$$\text{Ext}^1\Big\{\mathcal{S}(-, s_\lambda), G\Big\} = 0.$$

The exact sequence now permits us to conclude that

$$\text{Ext}^1(t_\lambda, G) = 0;$$

this is true for all $\lambda \in \Lambda$, meaning for all isomorphism classes of $t_\lambda \in A(\mathcal{S})$. By Lemma 7.4.5, G must be homological. □

LEMMA 7.4.7. *Let $\mathcal{S}$ be an essentially small triangulated category, satisfying [TR5(α)]. Let γ be an ordinal, $\gamma < \alpha$. Suppose that for all ordinals $i < \gamma$, any co–Mittag–Leffler sequence*

$$H : \mathcal{I}(i) \longrightarrow \mathcal{E}x\big(\mathcal{S}^{op}, \mathcal{A}b\big),$$

and any integer $n \geq 1$, $\underrightarrow{\text{colim}}^n H = 0$.

Let F be a co–Mittag–Leffler sequence in $\mathcal{E}x\big(\mathcal{S}^{op}, \mathcal{A}b\big)$ of length γ. There exists an exact sequence of co–Mittag–Lefflers

$$0 \longrightarrow F \longrightarrow G \longrightarrow H \longrightarrow 0,$$

where for all $i \in \mathcal{I}(\gamma)$, $Gi \in \mathcal{E}x\big(\mathcal{S}^{op}, \mathcal{A}b\big)$ is homological.

Proof: We will define the short exact

$$0 \longrightarrow F \longrightarrow G \longrightarrow H \longrightarrow 0$$

by induction. If j is a limit ordinal, and the sequence of co–Mittag–Lefflers

$$0 \longrightarrow F \longrightarrow G \longrightarrow H \longrightarrow 0$$

has been defined on $\mathcal{I}(i)$ for all $i < j$, then one defines the short exact sequence on $\mathcal{I}(j)$ to be the union. It remains to show how to extend from $\mathcal{I}(j) \subset \mathcal{I}(\gamma)$ to $\mathcal{I}(j+1)$. Suppose therefore that $j < \gamma$, and on $\mathcal{I}(j)$ we have a short exact sequence co–Mittag–Lefflers

$$0 \longrightarrow F \longrightarrow G \longrightarrow H \longrightarrow 0$$

Suppose for $i \in \mathcal{I}(j)$, $Gi \in \mathcal{E}x\big(\mathcal{S}^{op}, \mathcal{A}b\big)$ is homological. We need to define the extension to $\mathcal{I}(j+1)$.

In any case, by the hypothesis of the Lemma, the co–Mittag–Leffler sequence H on $\mathcal{I}(j)$ has vanishing $\underrightarrow{\mathrm{colim}}^1$. The exact sequence

$$\operatorname*{colim}_{i<j}{}^1 H \longrightarrow \operatorname*{colim}_{i<j} F \longrightarrow \operatorname*{colim}_{i<j} G$$

means that the map

$$\operatorname*{colim}_{i<j} F \longrightarrow \operatorname*{colim}_{i<j} G$$

is mono. We can form the pushout diagram

$$\begin{array}{ccc} \operatorname*{colim}_{i<j} F & \longrightarrow & \operatorname*{colim}_{i<j} G \\ \downarrow & & \downarrow \\ Fj & \longrightarrow & X \end{array}$$

Of course, this means we have a map of short exact sequences

$$\begin{array}{ccccccccc} 0 & \longrightarrow & \operatorname*{colim}_{i<j} Fj & \longrightarrow & \operatorname*{colim}_{i<j} Gj & \longrightarrow & \operatorname*{colim}_{i<j} Hj & \longrightarrow & 0 \\ & & \downarrow & & \downarrow & & {\scriptstyle 1}\downarrow & & \\ 0 & \longrightarrow & Fj & \longrightarrow & X & \longrightarrow & \operatorname*{colim}_{i<j} Hj & \longrightarrow & 0 \end{array}$$

By Lemma 7.4.6, we can choose an embedding of X into a homological object. Define Gj to be a homological object, for which there is an embedding $X \longrightarrow Gj$. Define Hj by the pushout square

$$\begin{array}{ccc} X & \longrightarrow & \operatorname*{colim}_{i<j} Hi \\ \downarrow & & \downarrow \\ Gj & \longrightarrow & Hj \end{array}$$

Now the sequence F is co–Mittag–Leffler, which means the map

$$\operatorname*{colim}_{i<j} F \longrightarrow Fj$$

is mono. The pushout square

$$\begin{array}{ccc} \operatorname*{colim}_{i<j} F & \longrightarrow & \operatorname*{colim}_{i<j} G \\ \downarrow & & \downarrow \\ Fj & \longrightarrow & X \end{array}$$

tells us that the map $\operatorname{colim}_{i<j} Gi \longrightarrow X$ is also mono. By construction, the map $X \longrightarrow Gj$ is mono. Hence the composite $\operatorname{colim}_{i<j} Gi \longrightarrow Gj$ is mono.

Finally, the map $X \longrightarrow Gj$ is mono, and we have a pushout square

$$\begin{array}{ccc} X & \longrightarrow & \operatorname*{colim}_{i<j} Hj \\ \downarrow & & \downarrow \\ Gj & \longrightarrow & Hj. \end{array}$$

We immediately learn that $\operatorname{colim}_{i<j} Hi \longrightarrow Hj$ is also mono.

The functors G and H on $\mathcal{I}(j+1) \subset \mathcal{I}(\gamma)$ satisfy the hypothesis that, for all $i < j+1$, the maps

$$\operatorname*{colim}_{h<i} Gh \longrightarrow Gi, \qquad \operatorname*{colim}_{h<i} Hh \longrightarrow Hi$$

are both mono. For $i < j$ this is true by induction, for $i = j$ by the above. By the hypothesis of the Lemma, all $\underrightarrow{\operatorname{colim}}^n$ vanish for co–Mittag–Leffler sequences of length $< \gamma$. Lemma 7.4.2 now applies. In 7.4.2.1 and 7.4.2.2, we may take Q to be the class of all objects of $\mathcal{E}x\big(\mathcal{S}^{op}, \mathcal{A}b\big)$. We deduce that both G and H are co–Mittag–Leffler on $\mathcal{I}(j+1)$.

Finally, the exactness of

$$0 \longrightarrow Fj \longrightarrow Gj \longrightarrow Hj \longrightarrow 0$$

extends the short exact sequence

$$0 \longrightarrow F \longrightarrow G \longrightarrow H \longrightarrow 0$$

from $\mathcal{I}(j)$ to $\mathcal{I}(j+1)$. □

This ends what I can prove. For a while, I thought I could deduce [AB4.5]. But the argument I had for the next Lemma seems to have a gap. Let me give the Lemma as a problem, and show how, if it is true, [AB4.5] follows immediately.

PROBLEM 7.4.8. Let $\mathcal{S}$ be a triangulated category satisfying [TR5(α)]. Suppose γ is an ordinal $< \alpha$, and suppose

7.4.8.1. *For all $i < \gamma$, all co–Mittag–Leffler sequences H of length i, and all integers $n \geq 1$, $\underrightarrow{\operatorname{colim}}^n H = 0$.*

Let G be a sequence of length γ, that is a functor

$$G : \mathcal{I}(\gamma) \longrightarrow \mathcal{E}x(\mathcal{S}^{op}, \mathcal{A}b).$$

Suppose G is co–Mittag–Leffler, and for all $i \in \mathcal{I}(\gamma)$, $Gi \in \mathcal{E}x(\mathcal{S}^{op}, \mathcal{A}b)$ is a homological object. Our problem is the following. Is it true that, for all integers $n \geq 1$, $\underrightarrow{\text{colim}}^n G = 0$?

Assuming Problem 7.4.8 has a positive answer, life is easy. We have

LEMMA 7.4.9. *Let $\mathcal{S}$ be a triangulated category satisfying [TR5(α)]. Suppose Problem 7.4.8 has a positive answer. Let γ be an ordinal, $\gamma < \alpha$. Let F be a co–Mittag–Leffler sequence of length γ in $\mathcal{E}x(\mathcal{S}^{op}, \mathcal{A}b)$. Then for all $n \geq 1$, $\underrightarrow{\text{colim}}^n F = 0$.*

Proof: We prove this by induction on the ordinal γ, the case of finite ordinals being trivial. Suppose therefore that the vanishing of $\underrightarrow{\text{colim}}^n$, $n \geq 1$ is true for all co–Mittag–Leffler sequences of length $i < \gamma < \alpha$. We want to extend to sequences of length γ.

Let F be a co–Mittag–Leffler sequence of length γ. By Lemma 7.4.7, there exists an exact sequence of co–Mittag–Lefflers of length γ

$$0 \longrightarrow F \longrightarrow G \longrightarrow H \longrightarrow 0,$$

and furthermore, Gi is homological for all $i \in \mathcal{I}(\gamma)$. By the positive answer to Problem 7.4.8, we have that for all $n \geq 1$, $\underrightarrow{\text{colim}}^n G = 0$. But from Corollary 7.3.6, $\underrightarrow{\text{colim}}^n H = 0$ whenever $n \geq 2$. The long exact homology sequence gives an exact

$$\underrightarrow{\text{colim}}^{n+1} H \longrightarrow \underrightarrow{\text{colim}}^n F \longrightarrow \underrightarrow{\text{colim}}^n G,$$

and if $n \geq 1$, we have $\underrightarrow{\text{colim}}^n G = 0 = \underrightarrow{\text{colim}}^{n+1} H$, and we deduce that $\underrightarrow{\text{colim}}^n F = 0$. □

PROPOSITION 7.4.10. *Let $\mathcal{S}$ be a triangulated category, and assume $\mathcal{S}$ satisfies [TR5(α)]. Suppose Problem 7.4.8 has a positive answer. Then the category $\mathcal{E}x(\mathcal{S}^{op}, \mathcal{A}b)$ satisfies [AB4.5]; co–Mittag–Leffler sequences have vanishing $\underrightarrow{\text{colim}}^n$, $n \geq 1$.*

Proof: Let γ be an ordinal, F a co–Mittag–Leffler sequence of length γ. Let β be the cofinality of γ. That is, β is the least cardinal with a cofinal map

$$\mathcal{I}(\beta) \longrightarrow \mathcal{I}(\gamma).$$

The cardinal β is clearly regular, and by Proposition A.4.8, $\underrightarrow{\text{colim}}^n F$ agrees with $\underrightarrow{\text{colim}}^n$ of the restriction of F to $\mathcal{I}(\beta)$. And the restriction of F to $\mathcal{I}(\beta)$ is also co–Mittag–Leffler.

If $\beta < \alpha$, then by Lemma 7.4.9, $\underrightarrow{\operatorname{colim}}^n F = 0$ for $n \geq 1$. If $\beta \geq \alpha$, then the regular cardinal β is α–filtered. The ordered set $\mathcal{I}(\beta)$ is α–filtered, and by Lemma A.1.3, colimits of functors

$$F : \mathcal{I}(\beta) \longrightarrow \mathcal{E}x(\mathcal{S}^{op}, \mathcal{A}b)$$

agree in $\mathcal{E}x(\mathcal{S}^{op}, \mathcal{A}b)$ and $\mathcal{C}at(\mathcal{S}^{op}, \mathcal{A}b)$. Hence these colimits are exact, and $\underrightarrow{\operatorname{colim}}^n F = 0$ for $n \geq 1$. □

7.5. The adjoint to the inclusion of $\mathcal{E}x(\mathcal{S}^{op}, \mathcal{A}b)$

Once again, the regular cardinal α is the one fixed throughout the Chapter. The category $\mathcal{S}$ is assumed essentially small and triangulated, and satisfies [TR5(α)].

The natural embedding

$$i : \mathcal{E}x(\mathcal{S}^{op}, \mathcal{A}b) \qquad \subset \qquad \mathcal{C}at(\mathcal{S}^{op}, \mathcal{A}b)$$

is an exact functor preserving products. It is natural to ask whether it has a left adjoint. In this Section we will see that it does, and learn an explicit description of the left adjoint. The reader is referred to Gabriel and Ulmer's [**16**] for a more general discussion. It turns out that this adjoint exists in very great generality, for functor categories of limit–preserving functors into sets.

As we said in the previous paragraph, we prove that the functor i has a left adjoint. This much generalises infinitely, as the reader can see in Gabriel and Ulmer's [**16**]. But then we go further. The left adjoint

$$j : \mathcal{C}at(\mathcal{S}^{op}, \mathcal{A}b) \longrightarrow \mathcal{E}x(\mathcal{S}^{op}, \mathcal{A}b)$$

had left derived functors, denoted $L^n j$. We prove that, for any $n \geq 1$ and $F \in \mathcal{E}x(\mathcal{S}^{op}, \mathcal{A}b)$,

$$L^n j\{iF\} = 0.$$

One application of these facts is the statement that, for any F and G objects of $\mathcal{E}x(\mathcal{S}^{op}, \mathcal{A}b)$, the extension groups $\operatorname{Ext}^n(F, G)$ are the same, whether computed in the category $\mathcal{E}x(\mathcal{S}^{op}, \mathcal{A}b)$ or $\mathcal{C}at(\mathcal{S}^{op}, \mathcal{A}b)$.

PROPOSITION 7.5.1. *Let $\mathcal{S}$ be a category satisfying Hypothesis 6.1.1. Then the natural functor*

$$\mathcal{E}x(\mathcal{S}^{op}, \mathcal{A}b) \xrightarrow{\ i\ } \mathcal{C}at(\mathcal{S}^{op}, \mathcal{A}b)$$

has a left adjoint j.

Proof: First, by Lemma 5.1.2, the category $\mathcal{C}at(\mathcal{S}^{op}, \mathcal{A}b)$ has enough projectives. More explicitly, the representable (Yoneda) functors are projective. We denote them $Y_s(-) = \mathcal{S}(-, s)$. And the set of all Yoneda projectives as above is a generating set. Every object $F \in \mathcal{C}at(\mathcal{S}^{op}, \mathcal{A}b)$ admits a presentation

$$\bigoplus_{\lambda\in\Lambda} \mathcal{S}(-, s_\lambda) \longrightarrow \bigoplus_{\mu\in M} \mathcal{S}(-, t_\mu) \longrightarrow F(-) \longrightarrow 0,$$

where $\oplus$ stands for the coproduct in $\mathcal{C}at(\mathcal{S}^{op}, \mathcal{A}b)$. By the dual of Proposition A.2.13, to prove the existence of a left adjoint to the functor

$$\mathcal{E}x(\mathcal{S}^{op}, \mathcal{A}b) \xrightarrow{\ i\ } \mathcal{C}at(\mathcal{S}^{op}, \mathcal{A}b)$$

it suffices to prove the representability of the functors

$$\mathcal{C}at(\mathcal{S}^{op}, \mathcal{A}b)\left[\bigoplus_{\lambda\in\Lambda} Y_s,\ i(-)\right]$$

Explicitly, for each projective object in $\mathcal{C}at(\mathcal{S}^{op}, \mathcal{A}b)$ of the form

$$\bigoplus_{\lambda\in\Lambda} \mathcal{S}(-, s_\lambda) \quad = \quad \bigoplus_{\lambda\in\Lambda} Y_s(-),$$

the functor sending $G \in \mathcal{E}x(\mathcal{S}^{op}, \mathcal{A}b)$ to maps in $\mathcal{C}at(\mathcal{S}^{op}, \mathcal{A}b)$

$$\bigoplus_{\lambda\in\Lambda} \mathcal{S}(-, s_\lambda) \longrightarrow iG(-)$$

needs to be proved representable in $\mathcal{E}x(\mathcal{S}^{op}, \mathcal{A}b)$. But to give a map in $\mathcal{C}at(\mathcal{S}^{op}, \mathcal{A}b)$

$$\bigoplus_{\lambda\in\Lambda} \mathcal{S}(-, s_\lambda) \longrightarrow iG(-)$$

is to give, by the universal property of the coproduct, for each $\lambda \in \Lambda$ a map

$$\mathcal{S}(-, s_\lambda) \longrightarrow iG(-).$$

The representable functor $\mathcal{S}(-, s_\lambda)$ lies in $\mathcal{E}x(\mathcal{S}^{op}, \mathcal{A}b)$; it takes coproducts to products. The above maps are therefore maps in $\mathcal{E}x(\mathcal{S}^{op}, \mathcal{A}b)$, and by the universal property of the coproduct in $\mathcal{E}x(\mathcal{S}^{op}, \mathcal{A}b)$, we deduce a map

$$\coprod_{\lambda\in\Lambda} \mathcal{S}(-, s_\lambda) \longrightarrow G(-).$$

In other words,

$$\mathcal{C}at(\mathcal{S}^{op}, \mathcal{A}b)\left[\bigoplus_{\lambda\in\Lambda} Y_s(-), iG(-)\right] \quad = \quad \mathcal{E}x(\mathcal{S}^{op}, \mathcal{A}b)\left[\coprod_{\lambda\in\Lambda} Y_s(-), G(-)\right].$$

$\square$

REMARK 7.5.2. Proposition 7.5.1 proves more then just the existence of the adjoint. It gives the adjoint very explicitly. Given an object $F \in \mathcal{C}at(\mathcal{S}^{op}, \mathcal{A}b)$, to find what the left adjoint to

$$\mathcal{E}x(\mathcal{S}^{op}, \mathcal{A}b) \xrightarrow{\ i\ } \mathcal{C}at(\mathcal{S}^{op}, \mathcal{A}b)$$

takes it to, consider any presentation

$$\bigoplus_{\lambda \in \Lambda} \mathcal{S}(-, s_\lambda) \longrightarrow \bigoplus_{\mu \in M} \mathcal{S}(-, t_\mu) \longrightarrow F(-) \longrightarrow 0.$$

The functor j, being a left adjoint, is right exact. It follows that j takes the above exact sequence in $\mathcal{C}at(\mathcal{S}^{op}, \mathcal{A}b)$ to an exact sequence in $\mathcal{E}x(\mathcal{S}^{op}, \mathcal{A}b)$. In other words, the sequence below is exact and computes jF

$$\coprod_{\lambda \in \Lambda} \mathcal{S}(-, s_\lambda) \longrightarrow \coprod_{\mu \in M} \mathcal{S}(-, t_\mu) \longrightarrow jF(-) \longrightarrow 0.$$

REMARK 7.5.3. We have proved the existence of a left adjoint j to the natural inclusion functor

$$\mathcal{E}x(\mathcal{S}^{op}, \mathcal{A}b) \xrightarrow{\ i\ } \mathcal{C}at(\mathcal{S}^{op}, \mathcal{A}b).$$

It is clear that, being a left adjoint, j is right exact. Since the category $\mathcal{C}at(\mathcal{S}^{op}, \mathcal{A}b)$ has enough projectives, the functor j has left derived functors. Denote them

$$L^n j : \mathcal{C}at(\mathcal{S}^{op}, \mathcal{A}b) \longrightarrow \mathcal{E}x(\mathcal{S}^{op}, \mathcal{A}b).$$

As usual, $L^n j$ may be computed using projective resolutions. Let F be an object of $\mathcal{C}at(\mathcal{S}^{op}, \mathcal{A}b)$. Form a projective resolution of F in $\mathcal{C}at(\mathcal{S}^{op}, \mathcal{A}b)$

$$\cdots \longrightarrow \bigoplus_{\lambda \in \Lambda_2} Y_{s_\lambda} \longrightarrow \bigoplus_{\lambda \in \Lambda_1} Y_{s_\lambda} \longrightarrow \bigoplus_{\lambda \in \Lambda_0} Y_{s_\lambda} \longrightarrow F \longrightarrow 0$$

and then $L^n jF$ is the nth homology of the complex

$$\cdots \longrightarrow \coprod_{\lambda \in \Lambda_2} Y_{s_\lambda} \longrightarrow \coprod_{\lambda \in \Lambda_1} Y_{s_\lambda} \longrightarrow \coprod_{\lambda \in \Lambda_0} Y_{s_\lambda} \longrightarrow 0.$$

We next propose to study this for the special case of functors of the form iF, with $F \in \mathcal{E}x(\mathcal{S}^{op}, \mathcal{A}b)$.

LEMMA 7.5.4. *Let F be a homological object of $\mathcal{E}x(\mathcal{S}^{op}, \mathcal{A}b)$. Then for all $n \geq 1$, we have*

$$L^n j\{iF\} = 0.$$

Proof: We are given that F is homological, and hence by Lemma 7.2.4, it is an α–filtered colimit of representable functors. But α–filtered colimits are the same in $\mathcal{E}x(\mathcal{S}^{op}, \mathcal{A}b)$ and $\mathcal{C}at(\mathcal{S}^{op}, \mathcal{A}b)$, and in particular are exact in $\mathcal{E}x(\mathcal{S}^{op}, \mathcal{A}b)$. The higher derived functors of this colimit vanish.

Suppose F is written as a colimit of a functor

$$f : \mathcal{I} \longrightarrow \mathcal{S},$$

where $\mathcal{I}$ is α–filtered. In Lemma A.3.2, we saw that there are canonical resolutions computing the derived functor of the colimit. In both $\mathcal{E}x(\mathcal{S}^{op}, \mathcal{A}b)$ and $\mathcal{C}at(\mathcal{S}^{op}, \mathcal{A}b)$, the derived functors of this α–filtered colimit vanish, in other words the sequences below are exact

$$\longrightarrow \bigoplus_{i_0 \rightarrow i_1} f(i_0) \longrightarrow \bigoplus_{i_0} f(i_0) \longrightarrow iF \longrightarrow 0$$

and

$$\longrightarrow \coprod_{i_0 \rightarrow i_1} f(i_0) \longrightarrow \coprod_{i_0} f(i_0) \longrightarrow F \longrightarrow 0.$$

But the first can be viewed as a projective resolution for iF in $\mathcal{E}x(\mathcal{S}^{op}, \mathcal{A}b)$, while the second is j applied to it. By Remark 7.5.3, the homology of the second complex is $L^n j\{iF\}$; since the complex is exact, we deduce that $L^n j\{iF\} = 0$ if $n \geq 1$. □

PROPOSITION 7.5.5. *Let F be any object of $\mathcal{E}x(\mathcal{S}^{op}, \mathcal{A}b)$. Then for all $n \geq 1$, $L^n j\{iF\} = 0$.*

Proof: The category $\mathcal{E}x(\mathcal{S}^{op}, \mathcal{A}b)$ has enough projectives, more specifically any object F admits a surjection $P \longrightarrow F$ with P a special projective of the form

$$P \quad = \quad \coprod_{\lambda \in \Lambda} \mathcal{S}(-, s_\lambda).$$

P is a coproduct of representables, hence homological by Example 7.2.3. Now Lemma 7.5.4 tells us that, for all $n \geq 1$, $L^n j\{iP\} = 0$.

Complete the surjection $P \longrightarrow F$ to an exact sequence

$$0 \longrightarrow G \longrightarrow P \longrightarrow F \longrightarrow 0.$$

Since the inclusion

$$i : \mathcal{E}x(\mathcal{S}^{op}, \mathcal{A}b) \longrightarrow \mathcal{C}at(\mathcal{S}^{op}, \mathcal{A}b)$$

is exact, there is an exact sequence in $\mathcal{C}at(\mathcal{S}^{op}, \mathcal{A}b)$

$$0 \longrightarrow iG \longrightarrow iP \longrightarrow iF \longrightarrow 0.$$

We deduce a long exact sequence for the δ–functor $L^n j$. For $n \geq 1$ we have exact sequences

$$L^{n+1} j\{iP\} \longrightarrow L^{n+1} j\{iF\} \longrightarrow L^n j\{iG\} \longrightarrow L^n j\{iP\}.$$

By Lemma 7.5.4 we know that

$$L^{n+1} j\{iP\} \quad = \quad 0 \quad = \quad L^n j\{iP\},$$

and hence $L^{n+1}j\{iF\} \simeq L^n j\{iG\}$.

For $n = 0$, we have an exact sequence

$$L^1 j\{iP\} \longrightarrow L^1 j\{iF\} \longrightarrow j\{iG\} \longrightarrow j\{iP\}.$$

On the other hand, the counit of adjunction gives a commutative square

$$\begin{array}{ccc} j\{iG\} & \longrightarrow & j\{iP\} \\ {\scriptstyle \varepsilon_G}\downarrow & & {\scriptstyle \varepsilon_P}\downarrow \\ G & \longrightarrow & P. \end{array}$$

We know that the inclusion i is fully faithful. Lemma A.2.9 allows us to deduce that both ε_G and ε_P above are isomorphisms. Since the bottom row is injective, so is the top. Hence the map

$$L^1 j\{iP\} \longrightarrow L^1 j\{iF\}$$

must be surjective. But by Lemma 7.5.4, we know that $L^1 j\{iP\} = 0$. Hence $L^1 j\{iF\} = 0$.

We have proved that, for every object $F \in \mathcal{E}x(\mathcal{S}^{op},\mathcal{A}b)$, $L^1 j\{iF\} = 0$. Furthermore, for any $F \in \mathcal{E}x(\mathcal{S}^{op},\mathcal{A}b)$ there exists a $G \in \mathcal{E}x(\mathcal{S}^{op},\mathcal{A}b)$ so that, for all $n \geq 1$, $L^{n+1}j\{iF\} \simeq L^n j\{iG\}$. By dimension shifting, $L^n j\{iF\} = 0$ for all $n \geq 1$. □

One application we have for Proposition 7.5.5 is the following

PROPOSITION 7.5.6. *If F and G are any two objects of $\mathcal{E}x(\mathcal{S}^{op},\mathcal{A}b)$ and $n \geq 0$, then* $\mathrm{Ext}^n(F,G)$ *is the same, whether computed in $\mathcal{E}x(\mathcal{S}^{op},\mathcal{A}b)$ or $\mathcal{C}at(\mathcal{S}^{op},\mathcal{A}b)$. In symbols, we have*

$$\mathrm{Ext}^n_{\mathcal{E}x(\mathcal{S}^{op},\mathcal{A}b)}(F,G) \quad = \quad \mathrm{Ext}^n_{\mathcal{C}at(\mathcal{S}^{op},\mathcal{A}b)}(iF,iG).$$

Proof: Take a projective resolution of iF in $\mathcal{C}at(\mathcal{S}^{op},\mathcal{A}b)$

$$\cdots \longrightarrow \bigoplus_{\lambda\in\Lambda_2} Y_{s_\lambda} \longrightarrow \bigoplus_{\lambda\in\Lambda_1} Y_{s_\lambda} \longrightarrow \bigoplus_{\lambda\in\Lambda_0} Y_{s_\lambda} \longrightarrow iF \longrightarrow 0.$$

Then $L^n j\{iF\}$ is the nth homology of the complex

$$\cdots \longrightarrow \coprod_{\lambda\in\Lambda_2} Y_{s_\lambda} \longrightarrow \coprod_{\lambda\in\Lambda_1} Y_{s_\lambda} \longrightarrow \coprod_{\lambda\in\Lambda_0} Y_{s_\lambda} \longrightarrow 0.$$

By Proposition 7.5.5, for $n \geq 1$ we have $L^n j\{iF\} = 0$. For $n = 0$, observe $L^0 j\{iF\} = jiF$, and since i is fully faithful with a left adjoint j, Lemma A.2.9 tells us that the counit of adjunction $jiF \longrightarrow F$ is an isomorphism. We deduce an exact sequence in $\mathcal{E}x(\mathcal{S}^{op},\mathcal{A}b)$

$$\cdots \longrightarrow \coprod_{\lambda\in\Lambda_2} Y_{s_\lambda} \longrightarrow \coprod_{\lambda\in\Lambda_1} Y_{s_\lambda} \longrightarrow \coprod_{\lambda\in\Lambda_0} Y_{s_\lambda} \longrightarrow F \longrightarrow 0.$$

This is, of course, nothing more nor less than a projective resolution, in $\mathcal{E}x\big(\mathcal{S}^{op}, \mathcal{A}b\big)$, for the functor F. Mapping the complex

$$\cdots \longrightarrow \coprod_{\lambda\in\Lambda_2} Y_{s_\lambda} \longrightarrow \coprod_{\lambda\in\Lambda_1} Y_{s_\lambda} \longrightarrow \coprod_{\lambda\in\Lambda_0} Y_{s_\lambda} \longrightarrow 0$$

into G and taking homology, we obtain the groups

$$\mathrm{Ext}^n_{\mathcal{E}x\left(\mathcal{S}^{op},\mathcal{A}b\right)}(F,G).$$

Mapping the complex

$$\cdots \longrightarrow \bigoplus_{\lambda\in\Lambda_2} Y_{s_\lambda} \longrightarrow \bigoplus_{\lambda\in\Lambda_1} Y_{s_\lambda} \longrightarrow \bigoplus_{\lambda\in\Lambda_0} Y_{s_\lambda} \longrightarrow 0$$

into iG and taking homology, we obtain the groups

$$\mathrm{Ext}^n_{\mathcal{C}at\left(\mathcal{S}^{op},\mathcal{A}b\right)}(iF,iG).$$

But

$$\mathcal{E}x(\mathcal{S}^{op},\mathcal{A}b)\left(\coprod_{\lambda\in\Lambda} Y_{s_\lambda}, G\right) \quad = \quad \mathcal{C}at(\mathcal{S}^{op},\mathcal{A}b)\left(\bigoplus_{\lambda\in\Lambda} Y_{s_\lambda}, iG\right).$$

It immediately follows that the complexes of abelian groups whose homology is

$$\mathrm{Ext}^n_{\mathcal{E}x\left(\mathcal{S}^{op},\mathcal{A}b\right)}(F,G) \qquad \text{respectively} \qquad \mathrm{Ext}^n_{\mathcal{C}at\left(\mathcal{S}^{op},\mathcal{A}b\right)}(iF,iG)$$

coincide. □

REMARK 7.5.7. In this Section, we assumed that the category $\mathcal{S}$ is essentially small, triangulated and satisfies[TR5(α)]. However, the statements remain true even if it only satisfies Hypothesis 6.1.1. The reader will note that in the proof of Proposition 7.5.5, we only apply Lemma 7.5.4 to an object P which is a coporduct of representables. Such objects are α–filtered colimits of representables even when $\mathcal{S}$ is not triangulated.

7.6. History of the results in Chapter 7

The results in this Chapter are all new.

CHAPTER 8

Brown representability

8.1. Preliminaries

In this Chapter, all categories are assumed to have small Hom sets. Sometimes we will explicitly remind the reader of this; even when we do not, it is assumed. Let us make some definitions about possible sets of generators for $\mathcal{T}$.

DEFINITION 8.1.1. *(cf. Definition 6.2.8). Let $\mathcal{T}$ be a triangulated category satisfying [TR5]. A set T of objects of $\mathcal{T}$ is called a* generating set *if*

8.1.1.1. $\{\text{Hom}(T, x) = 0\} \Longrightarrow \{x = 0\}$*; that is, if $x \in \mathcal{T}$ satisfies*

$$\forall t \in T, \text{Hom}(t, x) = 0$$

then x is isomorphic in $\mathcal{T}$ to 0.

8.1.1.2. *Up to isomorphisms, T is closed under suspension and desuspension; that means that given an object $t \in T$ and an integer n, there is an object in T isomorphic to $\Sigma^n t$.*

DEFINITION 8.1.2. *Let $\mathcal{T}$ be a triangulated category satisfying [TR5]. A set T of objects of $\mathcal{T}$ is called a β–*perfect generating set *for $\mathcal{T}$ if it is a generating set as in Definition 8.1.1, and $T \cup \{0\}$ is β–perfect as in Definition 3.3.1.*

Note that we do not insist that $0 \in T$; this is mostly for convenience of notation.

REMARK 8.1.3. Let α and β be any infinite cardinals. Let $\mathcal{T}$ be a triangulated category satisfying [TR5]. Let T be a β–perfect generating set for $\mathcal{T}$. By Proposition 3.2.5 $\langle T\rangle^\alpha$ is an essentially small category. Let S be a set of objects of $\langle T\rangle^\alpha$ containing T, and containing at least one representative in each isomorphism class of objects. We assert that S is also a β–perfect generating set for $\mathcal{T}$. After all, it is a set closed under suspension (up to ismorphism). Since it contains T, it generates. The fact that it is β–perfect is Theorem 3.3.9 and Lemma 3.3.2.

DEFINITION 8.1.4. *Let β be an infinite cardinal. A triangulated category $\mathcal{T}$ satisfying [TR5] is called* β–perfectly generated *if it contains some β–perfect generating set T.*

REMARK 8.1.5. Replacing T by a set S equivalent to $\langle T\rangle^{\alpha}$, we may assume that S is, up to equivalence of subcategories, the set of objects of an α–localising triangulated subcategory.

An even better situation is when the perfect generating set may be chosen so that its objects are also β–small. This gives

DEFINITION 8.1.6. *Let $\mathcal{T}$ be a triangulated category satisfying [TR5]. Let β be a regular cardinal. A set T of objects of $\mathcal{T}$ is called a* β–compact generating set for $\mathcal{T}$ *if it is a β–perfect generating set as in Definition 8.1.2, and all the objects of T are β–small. That is, $T \subset \mathcal{T}^{(\beta)}$.*

REMARK 8.1.7. Let T be a β–compact generating set for $\mathcal{T}$. Then T is a β–perfect subset of $\mathcal{T}^{(\beta)}$, hence contained in the largest such, which is $\mathcal{T}^{\beta}$. That is, T is contained in the subcategory $\mathcal{T}^{\beta}$ of β–compact objects in $\mathcal{T}$. Hence the name β–compact generating set.

The categories possessing β–compact generating sets for some regular cardinal β are particularly nice. We will call them *well generated.* We cannot call them compactly generated, since the term already exists in the literature, referring to categories which possess an $\aleph_0$–compact generating set.

Before we leave this introductory section, which focused mostly on the definitions of various types of generating sets, let us remind the reader of the (unrelated) definition of homotopy colimits of countable sequences.

DEFINITION 1.6.4 *Let $\mathcal{T}$ be a triangulated category satisfying [TR5]. (This means that $\mathcal{T}$ is closed under small coproducts). Let*

$$X_0 \xrightarrow{j_1} X_1 \xrightarrow{j_2} X_2 \xrightarrow{j_3} \cdots$$

be a sequence of objects and morphisms in $\mathcal{T}$. The homotopy colimit *of the sequence, denoted $\underrightarrow{\mathrm{Hocolim}}\, X_i$, is by definition given, up to non–canonical isomorphism, by the triangle*

$$\coprod_{i=0}^{\infty} X_i \xrightarrow{1-shift} \coprod_{i=0}^{\infty} X_i \longrightarrow \underrightarrow{\mathrm{Hocolim}}\, X_i \longrightarrow \Sigma\left\{\coprod_{i=0}^{\infty} X_i\right\}$$

where the shift map

$$\coprod_{i=0}^{\infty} X_i \xrightarrow{shift} \coprod_{i=0}^{\infty} X_i$$

is the direct sum of $j_{i+1} : X_i \to X_{i+1}$. *In other words, the map* $\{1 - shift\}$ *is the infinite matrix*

$$\begin{pmatrix} 1_{X_0} & 0 & 0 & 0 & \cdots \\ -j_1 & 1_{X_1} & 0 & 0 & \cdots \\ 0 & -j_2 & 1_{X_2} & 0 & \cdots \\ 0 & 0 & -j_3 & 1_{X_3} & \cdots \\ \vdots & \vdots & \vdots & \vdots & \end{pmatrix}$$

8.2. Brown representability

We will be studying representable functors from triangulated categories $\mathcal{T}^{op}$ to the category $\mathcal{A}b$ of abelian groups. Clearly, the functor $\mathcal{T}(-, h)$ is homological and takes coproducts to products. We define

DEFINITION 8.2.1. *A triangulated category* $\mathcal{T}$ *is said to* satisfy the representability theorem *if*

8.2.1.1. *The category* $\mathcal{T}$ *satisfies [TR5].*

8.2.1.2. *Any functor* $H : \mathcal{T}^{op} \longrightarrow \mathcal{A}b$, *which is homological, and sends coproducts in* $\mathcal{T}$ *to products in* $\mathcal{A}b$, *is representable; it is naturally isomorphic to* $\mathcal{T}(-, h)$ *for some* $h \in \mathcal{T}$.

REMARK 8.2.2. The main theorems of the Chapter will show that, if $\mathcal{T}$ or $\mathcal{T}^{op}$ have sufficiently nice generating sets (see Section 8.1), then the representability theorem holds in $\mathcal{T}$.

The first theorem of this type was proved by Brown [**7**]. For this reason, theorems of this type are usually referred to as Brown representability.

The key to all the representability theorems we will prove is the following Lemma.

LEMMA 8.2.3. *Let* $\mathcal{T}$ *be a triangulated category with small* Hom*–sets satisfying [TR5],* T *a set of objects in* $\mathcal{T}$. *Suppose* T *is essentially closed under suspension; that means it contains objects isomorphic to all suspensions of its objects. Let* $H : \mathcal{T}^{op} \longrightarrow \mathcal{A}b$ *be a homological functor. That is,* H *is contravariant and takes triangles to long exact sequences. Suppose the natural map*

$$H\left(\coprod_{\lambda \in \Lambda} t_\lambda\right) \longrightarrow \prod_{\lambda \in \Lambda} H(t_\lambda)$$

is an isomorphism for all small coproducts in $\mathcal{T}$.

Then it is possible to construct a sequence of objects and morphisms in $\mathcal{T}$

$$X_0 \xrightarrow{\ j_1\ } X_1 \xrightarrow{\ j_2\ } X_2 \xrightarrow{\ j_3\ } \cdots$$

so that

8.2.3.1. *For every i, the objects X_i lie in $\langle T \rangle$, the smallest localising subcategory of $\mathcal{T}$ containing T.*

8.2.3.2. *For each i, there is a natural transformation of functors on $\mathcal{T}$*

$$\mathcal{T}(-, X_i) \longrightarrow H(-).$$

These are compatible in that the diagram

$$\begin{array}{ccccc} & & \mathcal{T}(-, X_i) & & \\ & \swarrow & & \searrow & \\ \mathcal{T}(-, X_{i+1}) & & \longrightarrow & & H(-) \end{array}$$

commutes for every i.

8.2.3.3. *Let $X = \underrightarrow{\text{Hocolim}}\, X_i$. There is a natural transformation $\mathcal{T}(-, X) \longrightarrow H(-)$ rendering commutative the triangles*

$$\begin{array}{ccccc} & & \mathcal{T}(-, X_i) & & \\ & \swarrow & & \searrow & \\ \mathcal{T}(-, X) & & \longrightarrow & & H(-) \end{array}$$

for every i.

8.2.3.4. *For every object $t \in T$, the image of the map*

$$\mathcal{T}(t, X_i) \longrightarrow \mathcal{T}(t, X_{i+1})$$

maps isomorphically to $H(t)$ via

$$\mathcal{T}(t, X_{i+1}) \longrightarrow H(t).$$

Proof: The rest of this Section will be devoted to the proof of this Lemma. Since the proof is somewhat technical, on first reading the reader might do well to skip this proof and pass on to the applications.

We have fixed a set T of objects in $\mathcal{T}$, where up to isomorphism $T = \Sigma T$. Let U_0 be defined as

$$U_0 = \bigcup_{t \in T} H(t).$$

Elements of U_0 can be thought of as pairs (α, t) with $\alpha \in H(t)$. Put

$$X_0 = \coprod_{(\alpha,t)\in U_0} t.$$

Clearly, X_0 is an object of $\langle T\rangle$. Also, by the hypothesis that H takes coproducts to products,

$$H(X_0) = \prod_{(\alpha,t)\in U_0} H(t),$$

and there is an obvious element in $H(X_0)$, namely the element which is $\alpha \in H(t)$ for $(\alpha, t) \in U_0$. Call this element $\alpha_0 \in H(X_0)$. The construction is such that if $t \longrightarrow X_0$ is the inclusion of t into $X_0 = \coprod_{(\alpha,t)\in U_0} t$ corresponding to $(\alpha, t) \in U_0$, then the induced map $H(X_0) \longrightarrow H(t)$ takes $\alpha_0 \in H(X_0)$ to $\alpha \in H(t)$.

To give an object X_0 and an element $\alpha_0 \in H(X_0)$ is by Yoneda's lemma the same as giving a natural transformation

$$\phi_0 : \mathfrak{T}(-, X_0) \longrightarrow H(-),$$

and what we have seen is precisely that

$$\phi_0(t) : \mathfrak{T}(t, X_0) \longrightarrow H(t)$$

is surjective, for all $t \in T$.

Suppose that for some $i \geq 0$ we have defined an object X_i of $\langle T\rangle$, and a natural transformation

$$\mathfrak{T}(-, X_i) \longrightarrow H(-).$$

Suppose further that for $t \in T$, the map

$$\mathfrak{T}(t, X_i) \longrightarrow H(t)$$

is surjective. We want to continue by induction to define $X_{i+1} \in \langle T\rangle$, and a map $X_i \longrightarrow X_{i+1}$ so that

$$\mathfrak{T}(-, X_i) \longrightarrow H(-)$$

factors as

$$\mathfrak{T}(-, X_i) \longrightarrow \mathfrak{T}(-, X_{i+1}) \longrightarrow H(-).$$

Define U_{i+1} by

$$U_{i+1} = \bigcup_{t\in T} \ker\left\{\mathfrak{T}(t, X_i) \longrightarrow H(t)\right\}.$$

An element of U_{i+1} can be thought of as a pair (f, t), where $t \in T$ and $f : t \longrightarrow X_i$ is a morphism. Put

$$K_{i+1} = \coprod_{(f,t)\in U_{i+1}} t,$$

and let $K_{i+1} \longrightarrow X_i$ be the map which is f on the factor t corresponding to the pair (f,t). Let X_{i+1} be given by the triangle

$$K_{i+1} \longrightarrow X_i \longrightarrow X_{i+1} \longrightarrow \Sigma K_{i+1}.$$

Since $X_i \in \langle T \rangle$ by induction, while $K_{i+1} \in \langle T \rangle$ because it is a coproduct of $t \in T$, we deduce from the triangle that $X_{i+1} \in \langle T \rangle$. This constructs $X_i \longrightarrow X_{i+1}$; next we show that

$$\mathcal{T}(-, X_i) \longrightarrow H(-)$$

factors as

$$\mathcal{T}(-, X_i) \longrightarrow \mathcal{T}(-, X_{i+1}) \longrightarrow H(-).$$

We have a map $\mathcal{T}(-, X_i) \longrightarrow H(-)$, which by Yoneda's lemma corresponds to an element $\alpha_i \in H(X_i)$. Under the map

$$\begin{array}{ccccc} H(X_i) & & & & \\ \downarrow & & & & \\ H(K_{i+1}) & = & H\left(\coprod\limits_{(f,t)\in U_{i+1}} t\right) & = & \prod\limits_{(f,t)\in U_{i+1}} H(t) \end{array}$$

the element $\alpha_i \in H(X_i)$ maps to zero; the $f : t \longrightarrow X_i$ were chosen so that the induced map $\mathcal{T}(t, X_i) \longrightarrow H(t)$ vanishes. But H is a homological functor; the exact sequence

$$H(X_{i+1}) \xrightarrow{\ k\ } H(X_i) \xrightarrow{\ j\ } H(K_{i+1})$$

coupled with the fact that $j(\alpha_i) = 0$, guarantees that there exists $\alpha_{i+1} \in H(X_{i+1})$ with $k(\alpha_{i+1}) = \alpha_i$. Choose such an α_{i+1}. There is a corresponding natural transformation

$$\mathcal{T}(-, X_{i+1}) \longrightarrow H(-)$$

rendering commutative the triangle

$$\begin{array}{ccc} & \mathcal{T}(-, X_i) & \\ \swarrow & & \searrow \\ \mathcal{T}(-, X_{i+1}) & \longrightarrow & H(-) \end{array}$$

For $t \in T$, the fact that the composite

$$\mathcal{T}(t, X_i) \longrightarrow \mathcal{T}(t, X_{i+1}) \longrightarrow H(t)$$

is surjective implies that so is the second map $\mathcal{T}(t, X_{i+1}) \longrightarrow H(t)$.

The other thing we should observe about this construction is the definition of K_{i+1}. The set U_{i+1} is the set of pairs

$$t \in T, \qquad f : t \longrightarrow X_i$$

which map to zero under $\mathcal{T}(-, X_i) \longrightarrow H(-)$. The map $K_{i+1} \longrightarrow X_i$ is the coproduct of all such maps $t \longrightarrow X_i$. Any map $f : t \longrightarrow X_i$ in the kernel of $\mathcal{T}(t, X_i) \longrightarrow H(t)$ is in the image of $\mathcal{T}(t, K_{i+1}) \longrightarrow \mathcal{T}(t, X_i)$. Since the composite

$$\mathcal{T}(t, K_{i+1}) \longrightarrow \mathcal{T}(t, X_i) \longrightarrow \mathcal{T}(t, X_{i+1})$$

vanishes, we deduce that the kernel of

$$\mathcal{T}(t, X_i) \longrightarrow \mathcal{T}(t, X_{i+1})$$

contains the kernel of the longer map

$$\mathcal{T}(t, X_i) \longrightarrow \mathcal{T}(t, X_{i+1}) \longrightarrow H(t).$$

The kernels must therefore be equal. The image of

$$\mathcal{T}(t, X_i) \longrightarrow \mathcal{T}(t, X_{i+1})$$

is isomorphic to the image of

$$\mathcal{T}(t, X_i) \longrightarrow H(t)$$

which is all of $H(t)$, since $\mathcal{T}(t, X_i) \longrightarrow H(t)$ is surjective.

By induction, this establishes 8.2.3.1, 8.2.3.2 and 8.2.3.4. The only remaining statement to prove is 8.2.3.3, which allows us to factor the whole sequence of maps through the homotopy colimit. Recall that 8.2.3.3 asserts

8.2.3.3: Let $X = \underrightarrow{\mathrm{Hocolim}}\, X_i$. That is, X is given by a triangle

$$\coprod_i X_i \xrightarrow{1 - shift} \coprod_i X_i \longrightarrow \underrightarrow{\mathrm{Hocolim}}\, X_i \longrightarrow \Sigma\left\{\coprod_i X_i\right\}.$$

There is a natural transformation $\mathcal{T}(-, X) \longrightarrow H(-)$ so that all the triangles

$$\begin{array}{ccc} & \mathcal{T}(-, X_i) & \\ \swarrow & & \searrow \\ \mathcal{T}(-, X) & \longrightarrow & H(-) \end{array}$$

commute. That is, for every i the triangle commutes.

Now we want to prove 8.2.3.3.

Consider the triangle

$$\coprod_i X_i \xrightarrow{1-shift} \coprod_i X_i \longrightarrow \underrightarrow{\mathrm{Hocolim}}\, X_i \longrightarrow \Sigma\left\{\coprod_i X_i\right\}$$

Applying the cohomological functor H, we get an exact sequence

$$\begin{array}{ccccc} H(X) & \longrightarrow & H\left(\coprod_i X_i\right) & \stackrel{\text{1-shift}}{\longrightarrow} & H\left(\coprod_i X_i\right) \\ & & \| & & \| \\ & & \prod_i H(X_i) & \stackrel{\text{1-shift}}{\longrightarrow} & \prod_i H(X_i). \end{array}$$

The element $\prod_i \alpha_i \in \prod_i H(X_i)$ is in the kernel of (1-shift), and hence there is an $\alpha \in H(X)$ mapping to it. By Yoneda, α corresponds to a natural transformation

$$\mathcal{T}(-,X) \longrightarrow H(-),$$

and the fact that α maps to $\prod \alpha_i \in H(\coprod X_i)$ means that the diagram

$$\begin{array}{ccccc} & & \mathrm{Hom}(-,X_i) & & \\ & \swarrow & & \searrow & \\ \mathrm{Hom}(-,X) & & \longrightarrow & & H(-) \end{array}$$

commutes for all i. □

8.3. The first representability theorem

In this Section we will use Lemma 8.2.3 to prove a representability result. First another helpful lemma.

LEMMA 8.3.1. *Let the hypotheses be as in Lemma 8.2.3, except that we insist now that T be a generating set for $\mathcal{T}$, as in Definition 8.1.1. We remind the reader: $\mathcal{T}$ is a triangulated category with small* Hom*-sets satisfying [TR5], T* a generating set *of objects in $\mathcal{T}$. For every homological functor $H : \mathcal{T}^{op} \longrightarrow \mathcal{A}b$ taking coproducts in $\mathcal{T}$ to products in $\mathcal{A}b$, there is (at least one) sequence X_i and homotopy colimit X as in the conclusion of*

Lemma 8.2.3. We have $X = \underrightarrow{\text{Hocolim}}\, X_i$*, and a map* $\mathcal{T}(-,X) \longrightarrow H(-)$ *as in 8.2.3.3. If, for every* H *and every* $t \in T$*, the map*

$$\mathcal{T}(t,X) \longrightarrow H(t)$$

is injective, then

8.3.1.1. *The category* $\mathcal{T}$ *is the smallest localising subcategory containing* T*; that is* $\mathcal{T} = \langle T \rangle$.

8.3.1.2. *For every* H *the map*

$$\mathcal{T}(-,X) \xrightarrow{\ \phi\ } H(-)$$

is an isomorphism, in particular H *is representable.*

Proof: By 8.2.3.4 the map $\mathcal{T}(t,X_i) \longrightarrow H(t)$ is surjective for all i. But it factors as

$$\mathcal{T}(t,X_i) \longrightarrow \mathcal{T}(t,X) \longrightarrow H(t)$$

and hence $\mathcal{T}(t,X) \longrightarrow H(t)$ is also surjective. We are given, by hypothesis, that the map $\mathcal{T}(t,X) \longrightarrow H(t)$ is injective. Hence, for every $t \in T$, the map $\mathcal{T}(t,X) \longrightarrow H(t)$ is an isomorphism.

Let $\mathcal{S} \subset \mathcal{T}$ be the full subcategory of objects $y \in \mathcal{T}$ such that, for all $n \in \mathbb{Z}$, the map $\phi(\Sigma^n y) : \mathcal{T}(\Sigma^n y, X) \longrightarrow H(\Sigma^n y)$ is an isomorphism. This category contains T, is triangulated, and is closed under coproducts. Thus $\phi : \mathcal{T}(-,X) \longrightarrow H(-)$ is an isomorphism on $\langle T \rangle$. If we prove 8.3.1.1, that is $\mathcal{T} = \langle T \rangle$, then 8.3.1.2 follows.

Let Y be an object of $\mathcal{T}$; we would like to prove $Y \in \langle T \rangle$. Consider the functor

$$H(-) \quad = \quad \mathcal{T}(-,Y).$$

By Lemma 8.2.3 we have a sequence of X_i's with homotopy colimit X, and a map $\mathcal{T}(-,X) \longrightarrow \mathcal{T}(-,Y)$. The X_i's all lie in $\langle T \rangle$, and hence so does $X = \underrightarrow{\text{Hocolim}}\, X_i$. But under the hypothesis of the present Lemma, for each $t \in T$ the map

$$\mathcal{T}(t,X) \longrightarrow \mathcal{T}(t,Y)$$

is an isomorphism.

The natural transformation

$$\mathcal{T}(-,X) \longrightarrow \mathcal{T}(-,Y)$$

comes from a map $X \longrightarrow Y$. Complete this to a triangle in $\mathcal{T}$

$$X \longrightarrow Y \longrightarrow Z \longrightarrow \Sigma X.$$

The long exact sequence we get by applying the functor $\mathcal{T}(t,-)$, coupled with the fact that $\mathcal{T}(t,X) \longrightarrow \mathcal{T}(t,Y)$ is an isomorphism, implies that $\mathcal{T}(t,Z) = 0$. For each $t \in T$, there are no non–zero map $t \longrightarrow Z$. But T

is a generating set; this implies, by Definition 8.1.1, that $Z = 0$. Hence $X \longrightarrow Y$ is an isomorphism. But X is in $\langle T \rangle$, hence so is Y. □

REMARK 8.3.2. Concretely, this means if we wish to prove that $\mathcal{T}$ satisfies the representability theorem, we can proceed as follows. Choose a generating set T. If we can prove that for any homological H sending coproducts to products, and any sequence X_i as in the conclusion of Lemma 8.2.3, the maps $\mathcal{T}(t, X) \longrightarrow H(t)$ are injective for $t \in T$, then in fact $\mathcal{T}$ satisfies the representability theorem. We also get, as an extra bonus, that $\mathcal{T} = \langle T \rangle$. The proofs we will give of representability theorems will amount to choosing T carefully and proving that the maps $\mathcal{T}(t, X) \longrightarrow H(t)$ of Lemma 8.2.3 are injective.

THEOREM 8.3.3. *Suppose $\mathcal{T}$ is a triangulated category with small* Hom*–sets, satisfying [TR5]. Suppose T is an $\aleph_1$–perfect generating set for $\mathcal{T}$ (see Definition 8.1.2). Then $\mathcal{T}$ satisfies the representability theorem, and furthermore $\langle T \rangle = \mathcal{T}$. We remind the reader of the statement of the representability theorem. For every homological functor $H : \mathcal{T}^{op} \longrightarrow \mathcal{A}b$ taking coproducts in $\mathcal{T}$ to products in $\mathcal{A}b$, there is an object $h \in \mathcal{T}$ and a natural isomorphism*

$$\mathcal{T}(-, h) \quad = \quad H(-).$$

Proof: Since T is $\aleph_1$–perfect, by Lemma 3.3.7 so is $\mathcal{S} = \langle T \rangle^{\aleph_1}$. Since T is a set, by Proposition 3.2.5 the category $\mathcal{S} = \langle T \rangle^{\aleph_1}$ is essentially small. Replace T by a set of objects of $\mathcal{S}$ containing at least one representative in each isomorphism class. Let $\alpha = \aleph_1$ in the work of Chapter 6. Let $\mathcal{E}x(\mathcal{S}^{op}, \mathcal{A}b)$ be as in Definition 6.1.3. That is, the objects of $\mathcal{E}x(\mathcal{S}^{op}, \mathcal{A}b)$ are the functors $\mathcal{S} \longrightarrow \mathcal{A}b$ which take coproducts in $\mathcal{S}$ of fewer than $\aleph_1$ objects to products in $\mathcal{A}b$.

Lemma 8.2.3 showed the existence of a sequence X_i in $\mathcal{T}$ satisfying 8.2.3.1–8.2.3.4. This sequence gives a sequence of functors on $\mathcal{T}$

$$\mathcal{T}(-, X_0) \longrightarrow \mathcal{T}(-, X_1) \longrightarrow \mathcal{T}(-, X_2) \longrightarrow \cdots$$

and a map from the sequence to $H(-)$. If we restrict the sequence to $\mathcal{S} \subset \mathcal{T}$, we have a sequence in $\mathcal{E}x(\mathcal{S}^{op}, \mathcal{A}b)$. The functors $\mathcal{T}(-, X_i)|_{\mathcal{S}}$ and $H(-)|_{\mathcal{S}}$ all lie in $\mathcal{E}x(\mathcal{S}^{op}, \mathcal{A}b)$ because they take coproducts in $\mathcal{T}$ to products, hence certainly take coproducts of fewer than $\aleph_1$ objects in $\mathcal{S}$ to products of abelian groups.

But up to isomorphism, the objects of $\mathcal{S}$ are the objects of T. By 8.2.3.4 we know that for $t \in T$, the maps

$$\mathcal{T}(t, X_i) \longrightarrow H(t)$$

are surjective, and the kernel is precisely the kernel of

$$\mathcal{T}(t, X_i) \longrightarrow \mathcal{T}(t, X_{i+1}).$$

In other words, in the sequence

$$\mathcal{T}(-,X_0)|_{\mathcal{S}} \longrightarrow \mathcal{T}(-,X_1)|_{\mathcal{S}} \longrightarrow \mathcal{T}(-,X_2)|_{\mathcal{S}} \longrightarrow \cdots$$

we have that for $i \geq 1$, there is a natural isomorphism

$$\mathcal{T}(-,X_i)|_{\mathcal{S}} \quad = \quad H(-)|_{\mathcal{S}} \oplus K_i(-)$$

where $H(-)|_{\mathcal{S}}$ can be thought of as the image of

$$\mathcal{T}(-,X_{i-1})|_{\mathcal{S}} \longrightarrow \mathcal{T}(-,X_i)|_{\mathcal{S}}$$

while $K_i(-)$ is the kernel of

$$\mathcal{T}(-,X_i)|_{\mathcal{S}} \longrightarrow \mathcal{T}(-,X_{i+1})|_{\mathcal{S}}.$$

Rephrasing this still differently, it follows from 8.2.3.4 that the sequence

$$\mathcal{T}(-,X_1)|_{\mathcal{S}} \longrightarrow \mathcal{T}(-,X_2)|_{\mathcal{S}} \longrightarrow \mathcal{T}(-,X_3)|_{\mathcal{S}} \longrightarrow \cdots$$

is the direct sum of the two sequences

$$\begin{array}{ccccccc} H(-)|_{\mathcal{S}} & \xrightarrow{1} & H(-)|_{\mathcal{S}} & \xrightarrow{1} & H(-)|_{\mathcal{S}} & \xrightarrow{1} & \cdots \\ K_1(-) & \xrightarrow{0} & K_2(-) & \xrightarrow{0} & K_3(-) & \xrightarrow{0} & \cdots \end{array}$$

It is trivial that we get two short exact sequences

$$\begin{array}{ccccccccc} 0 & \longrightarrow & \coprod_{i=1}^{\infty} H(-)|_{\mathcal{S}} & \xrightarrow{1-shift} & \coprod_{i=1}^{\infty} H(-)|_{\mathcal{S}} & \longrightarrow & H(-)|_{\mathcal{S}} & \longrightarrow & 0 \\ 0 & \longrightarrow & \coprod_{i=1}^{\infty} K_i & \xrightarrow{1-shift} & \coprod_{i=1}^{\infty} K_i & \longrightarrow & 0 & \longrightarrow & 0 \end{array}$$

The first is split exact, while in the second the map $\{shift\} = 0$, making $1 - shift = 1$ an isomorphism. Adding up the sequences, we get a short exact sequence in $\mathcal{E}x\big(\mathcal{S}^{op}, \mathcal{A}b\big)$

$$0 \to \coprod_{i=1}^{\infty} \mathcal{T}(-,X_i)|_{\mathcal{S}} \xrightarrow{1-shift} \coprod_{i=1}^{\infty} \mathcal{T}(-,X_i)|_{\mathcal{S}} \longrightarrow H(-)|_{\mathcal{S}} \to 0.$$

It does no harm to add $\mathcal{T}(-,X_0)|_{\mathcal{S}}$ to the first and second term of the sequence; the sequence

$$0 \to \coprod_{i=0}^{\infty} \mathcal{T}(-,X_i)|_{\mathcal{S}} \xrightarrow{1-shift} \coprod_{i=0}^{\infty} \mathcal{T}(-,X_i)|_{\mathcal{S}} \longrightarrow H(-)|_{\mathcal{S}} \to 0$$

is also exact.

By Proposition 6.2.6, the functor $\mathcal{T} \longrightarrow \mathcal{E}x\big(\mathcal{S}^{op}, \mathcal{A}b\big)$ preserves coproducts. Hence we can identify

$$\coprod_{i=0}^{\infty} \mathcal{T}(-,X_i)|_{\mathcal{S}} \quad = \quad \mathcal{T}\left(-, \coprod_{i=0}^{\infty} X_i\right)\Bigg|_{\mathcal{S}}.$$

We therefore have a short exact sequence

$$0 \longrightarrow \mathcal{T}\left(-, \coprod_{i=0}^{\infty} X_i\right)\Big|_{\mathcal{S}} \xrightarrow{1 - shift} \mathcal{T}\left(-, \coprod_{i=0}^{\infty} X_i\right)\Big|_{\mathcal{S}} \longrightarrow H(-)|_{\mathcal{S}} \longrightarrow 0.$$

On the other hand, the functor $\mathcal{T} \longrightarrow \mathcal{E}x(\mathcal{S}^{op}, \mathcal{A}b)$ is homological by Proposition 6.2.6. It takes the triangle

$$\coprod_{i=0}^{\infty} X_i \xrightarrow{1 - shift} \coprod_{i=0}^{\infty} X_i \longrightarrow X \longrightarrow \Sigma\left\{\coprod_{i=0}^{\infty} X_i\right\}$$

to the long exact sequence

$$\begin{array}{ccccc} \mathcal{T}\left(-, \coprod_{i=0}^{\infty} X_i\right)\Big|_{\mathcal{S}} & \xrightarrow{1 - shift} & \mathcal{T}\left(-, \coprod_{i=0}^{\infty} X_i\right)\Big|_{\mathcal{S}} & \longrightarrow & \mathcal{T}(-, X)|_{\mathcal{S}} \\ & & & & \downarrow \\ & & & & \mathcal{T}\left(-, \Sigma\left\{\coprod_{i=0}^{\infty} X_i\right\}\right)\Big|_{\mathcal{S}}. \end{array}$$

The injectivity of

$$\mathcal{T}\left(-, \Sigma\left\{\coprod_{i=0}^{\infty} X_i\right\}\right)\Big|_{\mathcal{S}} \xrightarrow{1 - shift} \mathcal{T}\left(-, \Sigma\left\{\coprod_{i=0}^{\infty} X_i\right\}\right)\Big|_{\mathcal{S}}$$

guarantees the surjectivity of

$$\mathcal{T}\left(-, \coprod_{i=0}^{\infty} X_i\right)\Big|_{\mathcal{S}} \longrightarrow \mathcal{T}(-, X)|_{\mathcal{S}}$$

so we really have a short exact sequence

$$0 \longrightarrow \mathcal{T}\left(-, \coprod_{i=0}^{\infty} X_i\right)\Big|_{\mathcal{S}} \xrightarrow{1 - shift} \mathcal{T}\left(-, \coprod_{i=0}^{\infty} X_i\right)\Big|_{\mathcal{S}} \longrightarrow \mathcal{T}(-, X)|_{\mathcal{S}} \longrightarrow 0.$$

Now the commutativity of the diagram with exact rows

$$\begin{array}{ccccccc} 0 \longrightarrow \mathcal{T}\left(-, \coprod_{i=0}^{\infty} X_i\right)\Big|_{\mathcal{S}} & \xrightarrow{1 - shift} & \mathcal{T}\left(-, \coprod_{i=0}^{\infty} X_i\right)\Big|_{\mathcal{S}} & \longrightarrow & \mathcal{T}(-, X)|_{\mathcal{S}} \longrightarrow 0 \\ \| \downarrow \wr & & \| \downarrow \wr & & \phi \downarrow \\ 0 \longrightarrow \mathcal{T}\left(-, \coprod_{i=0}^{\infty} X_i\right)\Big|_{\mathcal{S}} & \xrightarrow{1 - shift} & \mathcal{T}\left(-, \coprod_{i=0}^{\infty} X_i\right)\Big|_{\mathcal{S}} & \longrightarrow & H(-)|_{\mathcal{S}} \longrightarrow 0 \end{array}$$

tells us that $\phi : \mathcal{T}(-, X)|_{\mathcal{S}} \longrightarrow H(-)|_{\mathcal{S}}$ is an isomorphism. □

8.4. Corollaries of Brown representability

Let us recall first conclusion 8.3.1.1 of Lemma 8.3.1. We deduced in Theorem 8.3.3 not only that $\mathcal{T}$ satisfies the representability theorem, but also that $\mathcal{T} = \langle T \rangle$. Let us reiterate this.

PROPOSITION 8.4.1. *Let $\mathcal{T}$ be a triangulated category with small* Hom*–sets, satisfying [TR5]. Let T be a set of objects of $\mathcal{T}$ closed under suspension; $\Sigma T = T$. If $\mathcal{T} = \langle T \rangle$, then T is a generating set. If we assume further that T is an $\aleph_1$–perfect set of objects in $\mathcal{T}$, then the converse also holds. If T is a generating set, then $\mathcal{T} = \langle T \rangle$. We remind the reader: T is a generating set if and only if*

$$\{\mathrm{Hom}(T, x) = 0\} \qquad \Longrightarrow \qquad \{x = 0\}.$$

Proof: It is an immediate consequence of Theorem 8.3.3 that if T is an $\aleph_1$–perfect generating set, then $\mathcal{T} = \langle T \rangle$. We want to prove the converse, for which the hypothesis on $\aleph_1$–perfection is, as we said, superfluous. Let T be any set of objects in $\mathcal{T}$, with $\Sigma T = T$. Suppose that $\mathcal{T} = \langle T \rangle$. We want to show that T generates.

Let $x \in \mathcal{T}$ be an object so that $\mathrm{Hom}(T, x) = 0$. Consider the full subcategory $\mathcal{S} \subset \mathcal{T}$ defined by

$$Ob(\mathcal{S}) = \Big\{ s \in \mathcal{T} \,|\, \forall n \in \mathbb{Z},\ \mathrm{Hom}(\Sigma^n s, x) = 0 \Big\}.$$

Now $\mathcal{S}$ contains T by hypothesis, and is a triangulated subcategory of $\mathcal{T}$ closed under coproducts. That is, $\mathcal{S}$ is a localising subcategory of $\mathcal{T}$ containing T. Since $\mathcal{T} = \langle T \rangle$ is the minimal such, we must have $\mathcal{S} = \mathcal{T}$. That is, for all $s \in \mathcal{T}$, $\mathrm{Hom}(s, x) = 0$. Hence $x = 0$, and T generates. □

PROPOSITION 8.4.2. *Suppose $\mathcal{T}$ is a triangulated category with small* Hom*–sets, satisfying [TR5]. Suppose $\mathcal{T}$ is well–generated (for the difinitions, see Definition 8.1.6 and Remark 8.1.7). Then*

8.4.2.1. *The representability theorem holds for $\mathcal{T}$.*

8.4.2.2. *For every cardinal β the category $\mathcal{T}^\beta$ is essentially small.*

8.4.2.3.

$$\mathcal{T} = \bigcup_\beta \mathcal{T}^\beta.$$

Proof: The category $\mathcal{T}$ is well–generated. This means that, for some regular cardinal α, there exists a α–perfect generating set $T \subset \mathcal{T}^{(\alpha)}$. Since the objects of T are all α–small ($T \subset \mathcal{T}^{(\alpha)}$), Lemma 4.2.1 tells us that for any infinite cardinal β, T is also β–perfect. In particular, T is $\aleph_1$–perfect. From Theorem 8.3.3 we learn that the representability theorem holds for $\mathcal{T}$, and that $\mathcal{T} = \langle T \rangle$.

But then Lemma 4.4.5 tells us that for any regular cardinal $\beta \geq \alpha$, $\mathcal{T}^\beta = \langle T\rangle^\beta$. Since T is a set, Proposition 3.2.5 tells us that $\langle T\rangle^\beta$ is essentially small. This is true for every regular $\beta \geq \alpha$. But Lemma 4.2.3 says that if $\gamma < \beta$, then $\mathcal{T}^\gamma \subset \mathcal{T}^\beta$. Since for any γ we can choose a regular cardinal β bigger than both α and γ, it follows that $\mathcal{T}^\gamma$ is essentially small for all infinite γ.

But now we have

$$\begin{aligned} \mathcal{T} &= \langle T\rangle \\ &= \bigcup_\beta \langle T\rangle^\beta \\ &= \bigcup_\beta \mathcal{T}^\beta. \end{aligned}$$

□

REMARK 8.4.3. Proposition 8.4.2 really means the following. Let $\mathcal{T}$ be a triangulated category with small *Hom*–sets, satisfying [TR5]. The triangulated category $\mathcal{T}$ is well–generated if and only if for all β, $\mathcal{T}^\beta$ is essentially small, and for β large enough $\mathcal{T}^\beta$ generates.

If $\mathcal{T}^\beta$ is essentially small and generates, choose a set T of objects of $\mathcal{T}^\beta$ containing representatives from each isomorphism class. Clearly, T is a β–compact generating set. Hence $\mathcal{T}$ is well–generated.

Suppose $\mathcal{T}$ is well–generated. That is, for some infinite cardinal α, $\mathcal{T}$ admits an α–compact generating set T. By Proposition 8.4.2, for every regular cardinal $\beta \geq \alpha$ we have that $T \subset \langle T\rangle^\beta = \mathcal{T}^\beta$; that means that $\mathcal{T}^\beta$ is essentially small, and since it contains the generating set T it also generates.

In practice, the only time one finds it useful to consider other generating sets T is when proving the category well–generated. Given a triangulated category $\mathcal{T}$, it is often awkward to compute $\mathcal{T}^\alpha$ directly, and even worse to try to show that for all α, $\mathcal{T}^\alpha$ is essentially small, and for all α sufficiently large it generates. It is much easier to produce one α–perfect generating set $T \subset \mathcal{T}^{(\alpha)}$.

Another important consequence of the representability theorem is

THEOREM 8.4.4. *Let $\mathcal{S}$ and $\mathcal{T}$ be triangulated categories with small* Hom*–sets. Suppose that*

> 8.4.4.1. *$\mathcal{S}$ satisfies the representability theorem; this will happen, for example, if $\mathcal{S}$ is $\aleph_1$–perfectly generated.*
>
> 8.4.4.2. *$F : \mathcal{S} \longrightarrow \mathcal{T}$ be a triangulated functor.*
>
> 8.4.4.3. *F respects coproducts. Recall that we are not assuming coproducts exist in $\mathcal{T}$; only that the images of coproducts in $\mathcal{S}$ are*

coproducts in $\mathcal{T}$. That is, let $\{s_\lambda, \lambda \in \Lambda\}$ be a set of objects of $\mathcal{S}$. We know that the representability theorem holds in $\mathcal{S}$, and in particular, by 8.2.1.1, $\mathcal{S}$ satisfies [TR5]. The coproduct of these objects exists in $\mathcal{S}$. But then we have, in $\mathcal{T}$, maps

$$F(s_\lambda) \longrightarrow F\left(\coprod_{\lambda\in\Lambda} s_\lambda\right).$$

To say that F respects coproducts is to say that these maps give the object on the right the structure of a coproduct in $\mathcal{T}$ of the objects on the left.

We assert that under the hypotheses above, F has a right adjoint $G : \mathcal{T} \longrightarrow \mathcal{S}$. That is there is, for every $s \in \mathcal{S}$ and $t \in \mathcal{T}$, an isomorphism natural in s and t

$$\mathcal{T}(Fs, t) \quad = \quad \mathcal{S}(s, Gt).$$

Proof: Let t be an object in $\mathcal{T}$. Consider the functor

$$H(-) \quad = \quad \mathcal{T}\Big(F(-), t\Big).$$

The functor F takes triangles to triangles and coproducts to coproducts. The functor $\mathcal{T}(-, t)$ takes triangles to long exact sequences and coproducts to products. The composite functor $\mathcal{T}\Big(F(-), t\Big)$ is a cohomological functor on $\mathcal{S}$ taking coproducts to products. By 8.4.4.1, it is representable. For each t, choose an object $Gt \in \mathcal{S}$ representing it; that is

$$\mathcal{T}\Big(F(-), t\Big) \quad = \quad \mathcal{S}(-, Gt).$$

Given a map $t \to t'$ in $\mathcal{T}$, we have a natural transformation

$$\mathcal{T}\Big(F(-), t\Big) \longrightarrow \mathcal{T}\Big(F(-), t'\Big),$$

and hence a natural transformation

$$\mathcal{S}(-, Gt) \longrightarrow \mathcal{S}(-, Gt').$$

By Yoneda's lemma, this must arise from a unique map $Gt \longrightarrow Gt'$, and one proves easily that this gives G the structure of a functor and makes the isomorphism

$$\mathcal{T}(Fs, t) \quad = \quad \mathcal{S}(s, Gt)$$

natural in s and t. □

Recall that in Lemma 5.3.6 we established that the adjoint G of a triangulated functor F is triangulated.

EXAMPLE 8.4.5. Let $\mathcal{T}$ be a triangulated category with small Hom–sets. Suppose the representability theorem holds for $\mathcal{T}$. Let $\mathcal{S}$ be a localising subcategory. By Corollary 3.2.11, the category $\mathcal{T}/\mathcal{S}$ also satisfies [TR5], and more importantly the natural functor $F : \mathcal{T} \longrightarrow \mathcal{T}/\mathcal{S}$ preserves coproducts.

If we know that the category $\mathcal{T}/\mathcal{S}$ has small Hom–sets, then it follows that the functor $F : \mathcal{T} \longrightarrow \mathcal{T}/\mathcal{S}$ has a right adjoint $G : \mathcal{T}/\mathcal{S} \longrightarrow \mathcal{T}$. This right adjoint is called the *Bousfield localisation functor* of $\mathcal{T}$ with respect to $\mathcal{S}$.

PROPOSITION 8.4.6. *Let $\mathcal{T}$ be a triangulated category with small* Hom–*sets. Suppose the representability theorem holds for $\mathcal{T}$. Then $\mathcal{T}$ satisfies [TR5*]; it contains a product of every small set of objects.*

Proof: Let $\{X_\lambda, \lambda \in \Lambda\}$ be a set of objects in $\mathcal{T}$. For each $\lambda \in \Lambda$, the functor $\mathcal{T}(-, X_\lambda)$ is a cohomological functor by Remark 1.1.11. Being representable, it sends coproducts in $\mathcal{T}$ to products of abelian groups. But then the functor

$$H(-) \quad = \quad \prod_{\lambda \in \Lambda} \mathcal{T}(-, X_\lambda)$$

is also a cohomological functor sending coproducts in $\mathcal{T}$ to products of abelian groups. Because the representability theorem holds for $\mathcal{T}$, the functor H is representable. There is an object $X \in \mathcal{T}$ so that

$$\mathcal{T}(-, X) \quad = \quad \prod_{\lambda \in \Lambda} \mathcal{T}(-, X_\lambda).$$

But then X satisfies the universal property of a product; it is the product of the X_λ's. □

8.5. Applications in the presence of injectives

In this Section, we look a little more closely at what happens, if the category $\mathcal{E}x\Big(\{\mathcal{T}^\alpha\}^{op}, \mathcal{A}b\Big)$ has enough injectives. We remind the reader that enough injectives need not exist. See the counterexample of Section C.4.

REMARK 8.5.1. Before proceeding to Proposition 8.5.2, let us remind the reader of Proposition 8.4.2 and Theorem 8.3.3. Let $\mathcal{T}$ be a well–generated triangulated category. By Proposition 8.4.2, more precisely by 8.4.2.2, we have that for every infinite α, $\mathcal{T}^\alpha$ is essentially small. In Theorem 8.3.3, we learned that $\mathcal{T}$ satisfies the representability theorem. The next Proposition therefore applies to well–generated categories.

PROPOSITION 8.5.2. *Let $\mathcal{T}$ be a triangulated category with small* Hom–*sets, satisfying [TR5]. Let α be a regular cardinal. Suppose the category $\mathcal{T}^\alpha$*

is essentially small. By the results of Chapter 6, more specifically Proposition 6.2.6 and Remark 6.2.7, the natural homological functor

$$\mathcal{T} \longrightarrow \mathcal{E}x\Big(\{\mathcal{T}^\alpha\}^{op}, \mathcal{A}b\Big)$$

preserves coproducts and products.

To simplify the notation, write $\mathcal{S} = \mathcal{T}^\alpha$. We remind the reader that $\mathcal{E}x\big(\mathcal{S}^{op}, \mathcal{A}b\big)$ is the category of functors $\mathcal{S}^{op} \longrightarrow \mathcal{A}b$, sending coproducts of fewer than α objects to products.

Suppose $\mathcal{T}$ satisfies the representability theorem. Let I be an injective object in $\mathcal{E}x\big(\mathcal{S}^{op}, \mathcal{A}b\big)$. There exists an object $GI \in \mathcal{T}$ so that:

8.5.2.1. *For any object $x \in \mathcal{T}$, there is a natural isomorphism*

$$\mathcal{T}(x, GI) \longrightarrow \mathcal{E}x\big(\mathcal{S}^{op}, \mathcal{A}b\big)\Big[\mathcal{T}(-,x)|_{\mathcal{S}}, I(-)\Big].$$

Proof: Let I be an injective object in the category $\mathcal{E}x\big(\mathcal{S}^{op}, \mathcal{A}b\big)$. Then we can construct a functor $\mathcal{T}^{op} \longrightarrow \mathcal{A}b$, denoted

$$\mathcal{E}x\big(\mathcal{S}^{op}, \mathcal{A}b\big)\Big[\mathcal{T}(-,x)|_{\mathcal{S}}, I(-)\Big]$$

which takes $x \in \mathcal{T}$ to the group of maps in $\mathcal{E}x\big(\mathcal{S}^{op}, \mathcal{A}b\big)$

$$\mathcal{T}(-,x)|_{\mathcal{S}} \longrightarrow I(-).$$

By Remark 6.2.7, the functor taking x to $\mathcal{T}(-,x)|_{\mathcal{S}}$ is a homological functor respecting coproducts

$$\mathcal{T} \longrightarrow \mathcal{E}x\big(\mathcal{S}^{op}, \mathcal{A}b\big).$$

In $\mathcal{E}x\big(\mathcal{S}^{op}, \mathcal{A}b\big)$, the object I is injective. Taking maps into I preserves exact sequences, and clearly takes coproducts to products. Thus the functor taking $x \in \mathcal{T}$ to the group of maps in $\mathcal{E}x\big(\mathcal{S}^{op}, \mathcal{A}b\big)$

$$\mathcal{E}x\big(\mathcal{S}^{op}, \mathcal{A}b\big)\Big[\mathcal{T}(-,x)|_{\mathcal{S}}, I(-)\Big]$$

is a homological functor $\mathcal{T}^{op} \longrightarrow \mathcal{A}b$, taking coproducts to products. We are assuming the representability theorem holds for $\mathcal{T}$. It follows that there exists an object $GI \in \mathcal{T}$, so that maps $x \longrightarrow GI$ are in one–to–one correspondence with maps

$$\mathcal{T}(-,x)|_{\mathcal{S}} \longrightarrow I(-).$$

□

COROLLARY 8.5.3. *Let $\mathcal{T}$ be a triangulated category with small* Hom*–sets, satisfying [TR5]. Let α be a regular cardinal. Suppose the category $\mathcal{T}^\alpha$ is essentially small. Suppose $\mathcal{T}$ satisfies the representability theorem. Finally, suppose that the category $\mathcal{E}x\Big(\{\mathcal{T}^\alpha\}^{op}, \mathcal{A}b\Big)$ has enough injectives.*

Then the natural functor of Section 6.5

$$\pi : A(\mathcal{T}) \longrightarrow \mathcal{E}x\big(\mathcal{S}^{op}, \mathcal{A}b\big)$$

has a right adjoint G.

Proof: In Proposition 8.5.2, we proved that for $x \in \mathcal{T}$ and I an injective object in $\mathcal{E}x\big(\mathcal{S}^{op}, \mathcal{A}b\big)$, the functor sending $x \in \mathcal{T}$ to

$$\mathcal{E}x\big(\mathcal{S}^{op}, \mathcal{A}b\big)\Big[\pi x, I\Big]$$

is representable. In other words, there is an isomorphism, natural in $x \in \mathcal{T}$,

$$\mathcal{T}(x, GI) \xrightarrow{\sim} \mathcal{E}x\big(\mathcal{S}^{op}, \mathcal{A}b\big)\Big[\pi x, I(-)\Big].$$

Both functors, viewed as functors in x, are homological functors $\mathcal{T}^{op} \longrightarrow \mathcal{A}b$. They are the restrictions to $x \in \mathcal{T} \subset A(\mathcal{T})$ of, respectively,

$$A(\mathcal{T})(x, GI) \qquad \text{and} \qquad \mathcal{E}x\big(\mathcal{S}^{op}, \mathcal{A}b\big)\Big[\pi x, I(-)\Big].$$

The restrictions of these functors to $\mathcal{T} \subset A(\mathcal{T})$ are isomorphic. From the canonical nature of the factorisation of a homological functor through the universal one (see Theorem 5.1.18), it follows that the functors are isomorphic on all of $A(\mathcal{T})$.

We want to show that the functor π has a right adjoint. We have already shown that the adjoint is well–defined on injective objects $I \in \mathcal{E}x\big(\mathcal{S}^{op}, \mathcal{A}b\big)$. But by hypothesis, the category $\mathcal{E}x\big(\mathcal{S}^{op}, \mathcal{A}b\big)$ has enough injectives. Every object has an injective copresentation. From Lemma A.2.13, it follows that the functor G extends to all of $\mathcal{E}x\big(\mathcal{S}^{op}, \mathcal{A}b\big)$, to define a right adjoint for π. □

REMARK 8.5.4. From Proposition 6.5.3, we already know that the category $\mathcal{E}x\big(\mathcal{S}^{op}, \mathcal{A}b\big)$ is a Gabriel quotient $A(\mathcal{T})/\mathcal{B}$. In Proposition 6.5.3, we proved that the quotient is colocalizant. In Corollary 8.5.3, we saw that if $\mathcal{T}$ satisfies the representability theorem and if $\mathcal{E}x\Big(\{\mathcal{T}^\alpha\}^{op}, \mathcal{A}b\Big)$ has enough injectives, then the quotient is also localizant. The map to the quotient has a right adjoint as well as a left adjoint.

From Proposition A.2.10, we know that the counit of adjuction $\varepsilon : \pi G \longrightarrow 1$ is an isomorphism. Concretely, for any $I \in \mathcal{E}x\big(\mathcal{S}^{op}, \mathcal{A}b\big)$,

$$\mathcal{T}(-, GI)|_{\mathcal{S}} \quad = \quad I(-).$$

LEMMA 8.5.5. *Let $\mathcal{T}$ be a triangulated category satisfying [TR5]. Let α be a regular cardinal. Suppose that $\mathcal{S} = \mathcal{T}^\alpha$ is essentially small.*

If furthermore the natural projection

$$\pi : A(\mathcal{T}) \longrightarrow \mathcal{E}x\big(\mathcal{S}^{op}, \mathcal{A}b\big)$$

admits a right adjoint $G : \mathcal{E}x\big(\mathcal{S}^{op}, \mathcal{A}b\big) \longrightarrow A(\mathcal{T})$, then the abelian category $\mathcal{E}x\big(\mathcal{S}^{op}, \mathcal{A}b\big)$ has a cogenerator.

Proof: By Lemma 6.4.4, the category $\mathcal{E}x\big(\mathcal{S}^{op}, \mathcal{A}b\big)$ has a projective generator P. Let $\{q_\lambda, \lambda \in \Lambda\}$ be a set of representatives for all the quotients of P. Every non–zero object of $\mathcal{E}x\big(\mathcal{S}^{op}, \mathcal{A}b\big)$ contains a a non–zero subobject of the form q_λ.

Let I be an injective object in $A(\mathcal{T})$, which admits an embedding

$$\prod_{\lambda \in \Lambda} Gq_\lambda \longrightarrow I.$$

We assert that πI is a cogenerator for the abelian category $\mathcal{E}x\big(\mathcal{S}^{op}, \mathcal{A}b\big)$. Take any $x \in \mathcal{E}x\big(\mathcal{S}^{op}, \mathcal{A}b\big)$, and consider the map

$$x \longrightarrow \prod_{x \longrightarrow \pi I} \pi I.$$

Let k be the kernel of the map. We wish to show that $k = 0$. Let q_λ be a quotient of P embedding in k. Then we have monomorphisms

$$q_\lambda \longrightarrow k \longrightarrow x.$$

But the functor G has a left adjoint, and therefore is left exact. We deduce monomorphisms in $A(\mathcal{T})$

$$Gq_\lambda \longrightarrow Gk \longrightarrow Gx.$$

Now I is an injective object of $A(\mathcal{T})$, containing Gq_λ. The monomorphism $Gq_\lambda \longrightarrow I$ extends to a map $Gx \longrightarrow I$, and applying the functor π we have a commutative diagram

$$\begin{array}{ccccccc}
0 & \longrightarrow & \pi Gq_\lambda & \longrightarrow & \pi Gx & \longrightarrow & \pi I \\
 & & \Big\downarrow{\scriptstyle \varepsilon_{q_\lambda}} & & \Big\downarrow{\scriptstyle \varepsilon_x} & & \\
0 & \longrightarrow & q_\lambda & \longrightarrow & x & &
\end{array}$$

The map $\pi Gq_\lambda \longrightarrow \pi I$ is a monomorphism since $Gq_\lambda \longrightarrow I$ is, and π is exact. By Proposition A.2.10, we know that the vertical maps $\varepsilon : \pi G \longrightarrow 1$ are isomorphisms. We have therefore found a map $x \longrightarrow \pi I$, which is a monomorphism on $q_\lambda \subset k$, and k is the kernel of all maps $x \longrightarrow \pi I$. It follows that $q_\lambda = 0$. The only subobject of k of the form q_λ is the zero object. Hence $k = 0$, and x embeds into a product of πI's. □

REMARK 8.5.6. We therefore know the implications

$$\left\{\begin{array}{c} \mathcal{E}x\big(\mathcal{S}^{op}, \mathcal{A}b\big) \\ \text{has enough} \\ \text{injectives} \end{array}\right\} \Longrightarrow \left\{\begin{array}{c} A(\mathcal{T}) \longrightarrow \mathcal{E}x\big(\mathcal{S}^{op}, \mathcal{A}b\big) \\ \text{has a right adjoint} \end{array}\right\} \Longrightarrow \left\{\begin{array}{c} \mathcal{E}x\big(\mathcal{S}^{op}, \mathcal{A}b\big) \\ \text{has a} \\ \text{cogenerator} \end{array}\right\}$$

In Section C.4, we see an example where $\mathcal{E}x(\mathcal{S}^{op}, \mathcal{A}b)$ does not have a cogenerator. This example therefore teaches us not only that there need not be enough injectives, but also that there need not be a right adjoint to $\pi : A(\mathcal{T}) \longrightarrow \mathcal{E}x(\mathcal{S}^{op}, \mathcal{A}b)$.

It might be instructive to reformulate some of the statements about injectives. Given an injective in $\mathcal{E}x(\mathcal{S}^{op}, \mathcal{A}b)$, we saw in Proposition 8.5.2 that there is an object GI in $\mathcal{T}$ with certain properties. We next want to identify the objects $GI \in \mathcal{T}$.

Recall Proposition 6.5.3: $\mathcal{E}x(\mathcal{S}^{op}, \mathcal{A}b)$ is the Gabriel quotient of $A(\mathcal{T})$, where we send to zero the class of objects vanishing under π. If we use the identification $A(\mathcal{T}) = D(\mathcal{T})$ of Definition 5.2.1 and Lemma 5.2.2, the class of objects mapping to zero under π is easier to understand; see Lemma 6.5.6 and Definition 6.5.7. Let us remind the reader.

The category $D(\mathcal{T})$ has for its objects the morphisms $\{x \longrightarrow y\} \in \mathcal{T}$. A morphism in $D(\mathcal{T})$ is an equivalence class of commutative squares

$$\begin{array}{ccc} x & \longrightarrow & y \\ \downarrow & & \downarrow \\ x' & \longrightarrow & y'. \end{array}$$

A square as above is equivalent to zero if the two equal composites

$$\begin{array}{ccc} x & & \\ \downarrow & & \\ x' & \longrightarrow & y' \end{array} \qquad\qquad \begin{array}{ccc} x & \longrightarrow & y \\ & & \downarrow \\ & & y' \end{array}$$

both vanish. The functor $\pi : D(\mathcal{T}) \longrightarrow \mathcal{E}x(\mathcal{S}^{op}, \mathcal{A}b)$ takes the object $\{x \longrightarrow y\} \in D(\mathcal{T})$ to the image of

$$\mathcal{T}(-, x)|_{\mathcal{S}} \longrightarrow \mathcal{T}(-, y)|_{\mathcal{S}}.$$

The objects $\{x \longrightarrow y\} \in D(\mathcal{T})$ map to zero under π precisely if the induced map

$$\mathcal{T}(-, x)|_{\mathcal{S}} \longrightarrow \mathcal{T}(-, y)|_{\mathcal{S}}$$

vanishes; that is, $x \longrightarrow y$ is an α–phantom map.

LEMMA 8.5.7. *Let $\mathcal{T}$ be a triangulated category satisfying the representability theorem, and let α be a regular cardinal. Put $\mathcal{S} = \mathcal{T}^{\alpha}$, and assume $\mathcal{S}$ is essentially small.*

Let I be an injective object in $\mathcal{E}x(\mathcal{S}^{op}, \mathcal{A}b)$. Then any α–phantom map $x \longrightarrow GI$ vanishes. Furthermore, πGI is naturally isomorphic to I.

Proof: By Proposition 8.5.2, there is a natural isomorphism

$$\mathcal{T}(x, GI) \longrightarrow \mathcal{E}x(\mathcal{S}^{op}, \mathcal{A}b)\Big[\mathcal{T}(-, x)|_{\mathcal{S}}, I(-)\Big].$$

Let $x \longrightarrow y$ be an α–phantom map; it induces the zero map

$$\mathcal{T}(-,x)|_{\mathcal{S}} \longrightarrow \mathcal{T}(-,y)|_{\mathcal{S}},$$

hence the zero map

$$\mathcal{E}x\big(\mathcal{S}^{op},\mathcal{A}b\big)\Big[\mathcal{T}(-,y)|_{\mathcal{S}},I(-)\Big] \longrightarrow \mathcal{E}x\big(\mathcal{S}^{op},\mathcal{A}b\big)\Big[\mathcal{T}(-,x)|_{\mathcal{S}},I(-)\Big].$$

By the naturality of the isomorphism of Proposition 8.5.2, this is also the zero map

$$\mathcal{T}(y,GI) \longrightarrow \mathcal{T}(x,GI).$$

Now if $y = GI$, we discover that any α–phantom map $f : x \longrightarrow GI$ induces the zero map

$$\mathcal{T}(GI,GI) \longrightarrow \mathcal{T}(x,GI).$$

In particular, $f = 1_{\{GI\}} \circ f = 0$. Thus any α–phantom map $f : x \longrightarrow GI$ must vanish.

It remains to show that πGI is naturally isomorphic to I. In Remark 8.5.4, we observed that when there are enough injectives and G extends to a right adjoint to π, then the counit of adjunction $\varepsilon_I : \pi GI \longrightarrow I$ is an isomorphism for all I, not only for I injective. But even when there are not enough injectives, for any injective I we have a natural isomorphism $\pi GI \simeq I$. This is what we now want to prove.

By the definition of GI, there is a natural isomorphism

$$\mathcal{T}[x,GI] \quad = \quad \mathcal{E}x\big(\mathcal{S}^{op},\mathcal{A}b\big)\Big\{\pi x, I\Big\}.$$

Now put $x = s \in \mathcal{S} \subset \mathcal{T}$. Then

$$\pi GI(s) \quad = \quad \mathcal{T}[s,GI] \quad = \quad \mathcal{E}x\big(\mathcal{S}^{op},\mathcal{A}b\big)\Big\{s,I\Big\} \quad = \quad I(s).$$

□

LEMMA 8.5.8. *Let $\mathcal{T}$ be a triangulated category satisfying the representability theorem, and let α be a regular cardinal. Put $\mathcal{S} = \mathcal{T}^{\alpha}$, and assume $\mathcal{S}$ is essentially small.*

Suppose t is an object of $\mathcal{T}$, and suppose further that any α–phantom map $x \longrightarrow t$ in $\mathcal{T}$ vanishes. Let

$$\pi : A(\mathcal{T}) \longrightarrow \mathcal{E}x\big(\mathcal{S}^{op},\mathcal{A}b\big)$$

be the usual projection.

If a is an object of $A(\mathcal{T})$ with $\pi a = 0$, then $A(\mathcal{T})[a,t] = 0$.

Proof: It is convenient to work with $D(\mathcal{T}) \simeq A(\mathcal{T})$. An object $a \in D(\mathcal{T})$ is a morphism $\{f : x \longrightarrow y\}$ in $\mathcal{T}$. The object $t \in \mathcal{T}$ maps to $\{1 : t \longrightarrow t\} \in$

$D(\mathcal{T})$, under the universal homological functor $\mathcal{T} \longrightarrow D(\mathcal{T})$. A morphism $a \longrightarrow t$ is an equivalence class of commutative squares

$$\begin{array}{ccc} x & \xrightarrow{f} & y \\ \downarrow & & \downarrow \\ t & \xrightarrow{1} & t. \end{array}$$

If $\pi a = 0$, then $\{f : x \longrightarrow y\}$ is an α–phantom map in $\mathcal{T}$. Hence so is the composite $x \xrightarrow{f} y \longrightarrow t$. By the hypothesis of the lemma, any α–phantom map into t is zero. In particular, the map $x \xrightarrow{f} y \longrightarrow t$ must vanish. In other words, the two equal composites

$$\begin{array}{ccc} x & & \\ \downarrow & & \\ t & \xrightarrow{1} & t \end{array} \qquad\qquad \begin{array}{ccc} x & \xrightarrow{f} & y \\ & & \downarrow \\ & & t \end{array}$$

vanish, making the commutative square

$$\begin{array}{ccc} x & \xrightarrow{f} & y \\ \downarrow & & \downarrow \\ t & \xrightarrow{1} & t \end{array}$$

equivalent to zero. That is, any map $a \longrightarrow t$, in $D(\mathcal{T})$, must vanish. □

LEMMA 8.5.9. *Let $\mathcal{T}$ be a triangulated category satisfying the representability theorem, and let α be a regular cardinal. Put $\mathcal{S} = \mathcal{T}^\alpha$, and assume $\mathcal{S}$ is essentially small.*

Suppose t is an object of $\mathcal{T}$, and suppose further that any α–phantom map $x \longrightarrow t$ in $\mathcal{T}$ vanishes. Let

$$\pi : A(\mathcal{T}) \longrightarrow \mathcal{E}x\big(\mathcal{S}^{op}, \mathcal{A}b\big)$$

be the usual projection.

Then πt is an injective object of $\mathcal{E}x\big(\mathcal{S}^{op}, \mathcal{A}b\big)$, and $t = G\pi t$.

Proof: Let $x \longrightarrow y$ be a monomorphism in $\mathcal{E}x\big(\mathcal{S}^{op}, \mathcal{A}b\big)$. The functor π has a left adjoint L, which is not left exact. But we can form an exact sequence

$$0 \longrightarrow k \longrightarrow Lx \longrightarrow Ly.$$

The functor π is exact; hence we have a commutative diagram, where the bottom row is an exact sequence

$$\begin{array}{ccccccc} & & & & x & \longrightarrow & y \\ & & & & {\scriptstyle\eta_x}\downarrow & & {\scriptstyle\eta_y}\downarrow \\ 0 & \longrightarrow & \pi k & \longrightarrow & \pi Lx & \longrightarrow & \pi Ly. \end{array}$$

By Proposition 6.5.3, the maps η_x, η_y are both isomorphisms. Since $x \longrightarrow y$ is a monomorphism in $\mathcal{E}x\big(\mathcal{S}^{op}, \mathcal{A}b\big)$, it follows that $\pi k = 0$. By Lemma 8.5.8 we conclude that $A(\mathcal{T})[k,t] = 0$.

But as $t \in \mathcal{T} \subset A(\mathcal{T})$, Corollary 5.1.23 tells us that t is injective, as an object of the abelian category $A(\mathcal{T})$. The exact sequence

$$0 \longrightarrow k \longrightarrow Lx \longrightarrow Ly$$

gives rise to an exact sequence

$$A(\mathcal{T})[Ly,t] \longrightarrow A(\mathcal{T})[Lx,t] \longrightarrow A(\mathcal{T})[k,t] \longrightarrow 0.$$

Since we know that $A(\mathcal{T})[k,t] = 0$, we deduce a surjective map

$$A(\mathcal{T})[Ly,t] \longrightarrow A(\mathcal{T})[Lx,t].$$

But L is left adjoint to π, hence this surjective map identifies as

$$\mathcal{E}x\big(\mathcal{S}^{op}, \mathcal{A}b\big)\Big\{y, \pi t\Big\} \longrightarrow \mathcal{E}x\big(\mathcal{S}^{op}, \mathcal{A}b\big)\Big\{x, \pi t\Big\}.$$

Since this is surjective for any monomorphism $x \longrightarrow y$ in $\mathcal{E}x\big(\mathcal{S}^{op}, \mathcal{A}b\big)$, it follows that πt is an injective object.

Proposition 8.5.2 now tells us that we can form an object $G\pi t \in \mathcal{T}$, and I assert that it is canonically isomorphic to t. Recall that for any injective object $I \in \mathcal{E}x\big(\mathcal{S}^{op}, \mathcal{A}b\big)$, $GI \in \mathcal{T}$ is defined by representing the functor

$$x \quad \mapsto \quad \mathcal{E}x\big(\mathcal{S}^{op}, \mathcal{A}b\big)\Big\{\pi x, I\Big\}.$$

In the case where $I = \pi t$, with $t \in \mathcal{T}$ as above, we have

$$\mathcal{E}x\big(\mathcal{S}^{op}, \mathcal{A}b\big)\Big\{\pi x, \pi t\Big\} \quad = \quad A(\mathcal{T})[L\pi x, t]$$

just because L is left adjoint to π. But we also have the counit of adjunction

$$\varepsilon_x : L\pi x \longrightarrow x.$$

Complete it to an exact sequence in $A(\mathcal{T})$

$$0 \longrightarrow k \longrightarrow L\pi x \xrightarrow{\varepsilon_x} x \longrightarrow q \longrightarrow 0.$$

Now $\pi : A(\mathcal{T}) \longrightarrow \mathcal{E}x(\mathcal{S}^{op}, \mathcal{A}b)$ is an exact functor; in the commutative diagram below, the bottom row is exact

$$\begin{array}{ccccccccc} & & \pi x & \xrightarrow{1} & \pi x & & & & \\ & & {\scriptstyle \eta_{\pi x}}\downarrow & & \downarrow{\scriptstyle 1} & & & & \\ 0 \longrightarrow & \pi k & \longrightarrow \pi L \pi x & \xrightarrow{\pi\varepsilon_x} & \pi x & \longrightarrow & \pi q & \longrightarrow & 0. \end{array}$$

Since $\eta : 1 \Longrightarrow \pi L$ is an isomorphism, so is $\pi\varepsilon_x$. The exact sequence tells us that

$$\pi k = 0 = \pi q.$$

From Lemma 8.5.8, it now follows that

$$A(\mathcal{T})[k, t] \quad = \quad 0 \quad = \quad A(\mathcal{T})[q, t].$$

But $t \in \mathcal{T} \subset A(\mathcal{T})$ is an injective object. The exact sequence

$$0 \longrightarrow k \longrightarrow L\pi x \xrightarrow{\varepsilon_x} x \longrightarrow q \longrightarrow 0$$

gives an exact sequence

$$0 = A(\mathcal{T})[q, t] \longrightarrow A(\mathcal{T})[x, t] \longrightarrow A(\mathcal{T})[L\pi x, t] \longrightarrow A(\mathcal{T})[k, t] = 0.$$

Hence $\mathcal{T}(x, t) = A(\mathcal{T})[x, t]$ is naturally isomorphic to $A(\mathcal{T})[L\pi x, t]$, which in turn is naturally isomorphic to

$$\mathcal{E}x(\mathcal{S}^{op}, \mathcal{A}b)\Big\{\pi x, \pi t\Big\}.$$

This means $G\pi t = t$. □

REMARK 8.5.10. Lemmas 8.5.7 and 8.5.9 allow us to identify injective objects $I \in \mathcal{E}x(\mathcal{S}^{op}, \mathcal{A}b)$ with objects $t \in \mathcal{T}$ so that all α–phantom maps $x \longrightarrow t$ vanish. If $t \in \mathcal{T}$ admits no non-zero α–phantom maps $x \longrightarrow t$, then πt is an injective object of $\mathcal{E}x(\mathcal{S}^{op}, \mathcal{A}b)$. If I is an injective object of $\mathcal{E}x(\mathcal{S}^{op}, \mathcal{A}b)$, then $GI \in \mathcal{T}$ admits no non-zero α–phantom maps $x \longrightarrow GI$. Furthermore, $\pi GI \simeq I$, and $G\pi t \simeq t$.

We will say that an object $t \in \mathcal{T}$ is *orthogonal to the α–phantom maps* if all α–phantom maps $x \longrightarrow t$ vanish. Thus injective objects in $\mathcal{E}x(\mathcal{S}^{op}, \mathcal{A}b)$ are in 1–to–1 correspondence with objects $t \in \mathcal{T}$, orthogonal to the α–phantom maps. It might be helpful to state this in more generality.

DEFINITION 8.5.11. *Let $\mathcal{T}$ be an additive category. An* ideal $\mathcal{I}$ *of morphisms in the category $\mathcal{T}$ is a class of morphisms, closed under addition, and so that if $g \in \mathcal{I}$ and f and h are arbitrary morphisms, then $fgh \in \mathcal{I}$, whenever the composite exists.*

EXAMPLE 8.5.12. If $\mathcal{T}$ is a triangulated category satisfying [TR5] and α is a regular cardinal, the class $\mathcal{I}$ of all α–phantom maps is an ideal.

DEFINITION 8.5.13. *Given an ideal $\mathcal{I}$ of morphisms, its orthogonal, denoted $\mathcal{I}^\perp$, is the collection of all objects $t \in \mathcal{T}$, so that if $\{x \longrightarrow t\} \in \mathcal{I}$, then $x \longrightarrow t$ is the zero map. If T is any class of objects in $\mathcal{T}$, the ideal $I\{T\}$ is the ideal of all morphisms $\{f : x \longrightarrow y\} \in \mathcal{T}$, so that for any $t \in T$, all composites $x \xrightarrow{f} y \longrightarrow t$ vanish.*

REMARK 8.5.14. Clearly for any ideal $\mathcal{I}$, $\mathcal{I} \subset I\{\mathcal{I}^\perp\}$. It may happen that equality holds. That is, under favorable conditions, one has

$$\mathcal{I} \quad = \quad I\{\mathcal{I}^\perp\}.$$

The next lemma treats the case where $\mathcal{I}$ is the ideal of α–phantom maps, in a triangulated category $\mathcal{T}$.

LEMMA 8.5.15. *Let $\mathcal{T}$ be a triangulated category satisfying the representability theorem, and let α be a regular cardinal. Put $\mathcal{S} = \mathcal{T}^\alpha$, and assume $\mathcal{S}$ is essentially small. Let $\mathcal{I}$ be the ideal of α–phantom maps.*

The category $\mathcal{E}x\big(\mathcal{S}^{op}, \mathcal{A}b\big)$ has enough injectives if and only if

$$\mathcal{I} \quad = \quad I\{\mathcal{I}^\perp\}.$$

Proof: Recall that

$$\pi : D(\mathcal{T}) = A(\mathcal{T}) \longrightarrow \mathcal{E}x\big(\mathcal{S}^{op}, \mathcal{A}b\big)$$

is a Gabriel quotient map. Every object in $\mathcal{E}x\big(\mathcal{S}^{op}, \mathcal{A}b\big)$ is of the form πa, with a an object of $D(\mathcal{T})$. The category $\mathcal{E}x\big(\mathcal{S}^{op}, \mathcal{A}b\big)$ will have enough injectives if and only if every non–zero object πa in $\mathcal{E}x\big(\mathcal{S}^{op}, \mathcal{A}b\big)$ admits a non–zero map to an injective object. By Remark 8.5.10, injective objects in $\mathcal{E}x\big(\mathcal{S}^{op}, \mathcal{A}b\big)$ are all of the form πt, with $t \in \mathcal{I}^\perp$. In other words, the category $\mathcal{E}x\big(\mathcal{S}^{op}, \mathcal{A}b\big)$ has enough injectives iff, for any object $a \in D(\mathcal{T})$ not in the kernel of π, there is a non–zero morphism $\pi a \longrightarrow \pi t$, with $t \in \mathcal{I}^\perp$. By Lemma 8.5.9, whenever $t \in \mathcal{I}^\perp$,

$$D(\mathcal{T})\big[a, t\big] \quad = \quad \mathcal{E}x\big(\mathcal{S}^{op}, \mathcal{A}b\big)\big\{\pi a, \pi t\big\}.$$

Therefore $\mathcal{E}x\big(\mathcal{S}^{op}, \mathcal{A}b\big)$ will have enough injectives if and only if, for any object $a \in D(\mathcal{T})$ with πa non–zero, there is a non–zero map (in $D(\mathcal{T})$) $a \longrightarrow t$, with $t \in \mathcal{I}^\perp$.

Now remember that an object $a \in D(\mathcal{T})$, with πa non–zero, is a morphism $\{x \longrightarrow y\} \in \mathcal{T}$ which is not α–phantom. The category $\mathcal{E}x\big(\mathcal{S}^{op}, \mathcal{A}b\big)$ has enough injectives precisely if, for every non–α–phantom $\{x \longrightarrow y\} \in \mathcal{T}$, there is a non–zero map, in $D(\mathcal{T})$, into $t \in \mathcal{I}^\perp$.

But a map from $\{f : x \longrightarrow y\}$ to t is an equivalence class of commutative squares

$$\begin{array}{ccc} x & \xrightarrow{f} & y \\ \downarrow & & \downarrow \\ t & \xrightarrow{1} & t. \end{array}$$

To say that the square is not equivalent to zero, is to say that the equal composites

$$\begin{array}{ccc} x & & \\ \downarrow & & \\ t & \xrightarrow{1} & t \end{array} \qquad \begin{array}{ccc} x & \xrightarrow{f} & y \\ & & \downarrow \\ & & t \end{array}$$

do not vanish. In other words, a non–zero map from $\{f : x \longrightarrow y\}$ to t exists if and only if there is a non–zero composite $x \xrightarrow{f} y \longrightarrow t$. There will be enough injectives, if and only if a non–zero composite $x \xrightarrow{f} y \longrightarrow t$ exists for every non–α–phantom $\{f : x \longrightarrow y\} \in \mathcal{T}$. Equivalently, if $x \xrightarrow{f} y \longrightarrow t$ vanishes for every $t \in \mathcal{I}^{\perp}$, then $f : x \longrightarrow y$ should be an α–phantom map. That is,

$$\mathcal{I} \quad = \quad I\{\mathcal{I}^{\perp}\}.$$

□

REMARK 8.5.16. From the counterexample of Section C.4, we learn that in general, the inclusion $\mathcal{I} \subset I\{\mathcal{I}^{\perp}\}$ can be proper.

Lemma 8.5.15 gives an equivalence between the existence of enough injectives in $\mathcal{E}x\big(\mathcal{S}^{op}, \mathcal{A}b\big)$, and a statement purely about α–phantom maps in $\mathcal{T}$. Let us find another, equivalent formulation.

LEMMA 8.5.17. *Let $\mathcal{T}$ be a triangulated category satisfying the representability theorem, and let α be a regular cardinal. Put $\mathcal{S} = \mathcal{T}^{\alpha}$, and assume $\mathcal{S}$ is essentially small. Let $\mathcal{I}$ be the ideal of α–phantom maps.*

The category $\mathcal{E}x\big(\mathcal{S}^{op}, \mathcal{A}b\big)$ has enough injectives if and only if, for every object $z \in \mathcal{T}$, there exists a triangle

$$y \xrightarrow{\ f\ } z \longrightarrow t \longrightarrow \Sigma y$$

where $t \in \mathcal{I}^{\perp}$ and $f : y \longrightarrow z$ is an α–phantom map (that is, $f \in \mathcal{I}$).

Proof: The category $\mathcal{E}x\big(\mathcal{S}^{op}, \mathcal{A}b\big)$ has enough injectives if and only if every object $a \in \mathcal{E}x\big(\mathcal{S}^{op}, \mathcal{A}b\big)$ has an embedding into an injective object. Now recall that the functor

$$\pi : A(\mathcal{T}) \longrightarrow \mathcal{E}x\big(\mathcal{S}^{op}, \mathcal{A}b\big)$$

has a left adjoint L. The category $A(\mathcal{T})$ always has enough injectives; they are the objects $z \in \mathcal{T}$. Given an object $a \in \mathcal{E}x(\mathcal{S}^{op}, \mathcal{A}b)$, there is an embedding $La \longrightarrow z, z \in \mathcal{T}$. Since π is exact, this gives an embedding

$$a = \pi La \longrightarrow \pi z.$$

Therefore, the category $\mathcal{E}x(\mathcal{S}^{op}, \mathcal{A}b)$ has enough injectives if and only if all objects of the form $\{\pi z, z \in \mathcal{T}\}$ can be embedded in injective objects $I \in \mathcal{E}x(\mathcal{S}^{op}, \mathcal{A}b)$.

By Remark 8.5.10, the injective objects in $\mathcal{E}x(\mathcal{S}^{op}, \mathcal{A}b)$ can be identified with objects $t \in \mathcal{I}^{\perp}$. The category $\mathcal{E}x(\mathcal{S}^{op}, \mathcal{A}b)$ has enough injectives if and only if, for every object $z \in \mathcal{T}$, there is an embedding in $\mathcal{E}x(\mathcal{S}^{op}, \mathcal{A}b)$

$$\pi z \longrightarrow \pi t,$$

with $t \in \mathcal{I}^{\perp}$. By Lemma 8.5.9, whenever $t \in \mathcal{I}^{\perp}$,

$$\mathcal{T}(z,t) \quad = \quad \mathcal{E}x(\mathcal{S}^{op}, \mathcal{A}b)\Big\{\pi z, \pi t\Big\}.$$

In other words, there are enough injectives in $\mathcal{E}x(\mathcal{S}^{op}, \mathcal{A}b)$ if and only if, for every object $z \in \mathcal{T}$, there is a morphism $\{z \longrightarrow t, t \in \mathcal{I}^{\perp}\}$ in $\mathcal{T}$, so that $\pi z \longrightarrow \pi t$ is a monomorphism.

Consider now the triangle

$$y \xrightarrow{\ f\ } z \longrightarrow t \longrightarrow \Sigma y.$$

The functor π takes it to an exact sequence in $\mathcal{E}x(\mathcal{S}^{op}, \mathcal{A}b)$. That is,

$$\pi y \xrightarrow{\ \pi f\ } \pi z \longrightarrow \pi t$$

is a exact in $\mathcal{E}x(\mathcal{S}^{op}, \mathcal{A}b)$. The map $\pi z \longrightarrow \pi t$ will be a monomorphism if and only if $\pi f = 0$, that is $f \in \mathcal{I}$. Summarising, the category $\mathcal{E}x(\mathcal{S}^{op}, \mathcal{A}b)$ has enough injectives if and only if, for every object $z \in \mathcal{T}$, there is a triangle

$$y \xrightarrow{\ f\ } z \longrightarrow t \longrightarrow \Sigma y$$

with $t \in \mathcal{I}^{\perp}$ and $f \in \mathcal{I}$. □

Combining Lemmas 8.5.15 and 8.5.17, we have

PROPOSITION 8.5.18. *Let $\mathcal{T}$ be a triangulated category satisfying the representability theorem, and let α be a regular cardinal. Put $\mathcal{S} = \mathcal{T}^{\alpha}$, and assume $\mathcal{S}$ is essentially small. Let $\mathcal{I}$ be the ideal of α–phantom maps.*

The following are equivalent:

8.5.18.1. *The category $\mathcal{E}x(\mathcal{S}^{op}, \mathcal{A}b)$ has enough injectives.*

8.5.18.2. *The ideal $\mathcal{I}$ satisfies $\mathcal{I} = I\{\mathcal{I}^{\perp}\}$*

8.5.18.3. *For every object $z \in \mathcal{T}$, there is a triangle*

$$y \xrightarrow{f} z \longrightarrow t \longrightarrow \Sigma y$$

with $t \in \mathcal{I}^{\perp}$ and $f \in \mathcal{I}$.

Proof: By Lemma 8.5.15, we have the equivalence

$$8.5.18.1 \Longleftrightarrow 8.5.18.2.$$

By Lemma 8.5.17, we also have the equivalence

$$8.5.18.1 \Longleftrightarrow 8.5.18.3.$$

□

COROLLARY 8.5.19. *Let $\mathcal{T}$ be a triangulated category satisfying the representability theorem, and let α be a regular cardinal. Put $\mathcal{S} = \mathcal{T}^{\alpha}$, and assume $\mathcal{S}$ is essentially small. Suppose $\mathcal{E}x\big(\mathcal{S}^{op}, \mathcal{A}b\big)$ has enough injectives.*

Then every object $z \in \mathcal{T}$ admits a maximal α–phantom map $y \longrightarrow z$. That is, $y \longrightarrow z$ is an α–phantom map, and every α–phantom map $x \longrightarrow z$ factors as

$$x \longrightarrow y \longrightarrow z.$$

Proof: Let $\mathcal{I}$ be the ideal of α–phantom maps. We are assuming that $\mathcal{E}x\big(\mathcal{S}^{op}, \mathcal{A}b\big)$ has enough injectives, and hence by Lemma 8.5.17, every object $z \in \mathcal{T}$ admits a triangle

$$y \xrightarrow{f} z \longrightarrow t \longrightarrow \Sigma y,$$

with $t \in \mathcal{I}^{\perp}$ and $f \in \mathcal{I}$. I assert that the map $f : y \longrightarrow z$ is a maximal α–phantom map.

The fact that $f : y \longrightarrow z$ is α–phantom is given; $f \in \mathcal{I}$ is granted to us. What remains to prove is its maximality. Suppose $g : x \longrightarrow z$ is any α–phantom map. The composite

$$x \xrightarrow{g} z \longrightarrow t$$

is an α–phantom map $x \longrightarrow t$, with $t \in \mathcal{I}^{\perp}$. All such maps vanish. From the triangle

$$y \xrightarrow{f} z \longrightarrow t \longrightarrow \Sigma y,$$

we have that $x \longrightarrow z$ must factor as $x \longrightarrow y \longrightarrow z$. □

Recall that in Remark 8.5.6 we saw the implications

$$\left\{\begin{array}{c}\mathcal{E}x\big(\mathcal{S}^{op}, \mathcal{A}b\big)\\ \text{has enough}\\ \text{injectives}\end{array}\right\} \Longrightarrow \left\{\begin{array}{c}A(\mathcal{T}) \longrightarrow \mathcal{E}x\big(\mathcal{S}^{op}, \mathcal{A}b\big)\\ \text{has a right adjoint}\end{array}\right\} \Longrightarrow \left\{\begin{array}{c}\mathcal{E}x\big(\mathcal{S}^{op}, \mathcal{A}b\big)\\ \text{has a}\\ \text{cogenerator}\end{array}\right\}$$

In the last few Lemmas, summarised in Proposition 8.5.18, we have explained concretely, in terms of the ideal of α–phantom maps and its orthogonal, what it means to have enough injectives in $\mathcal{E}x(\mathcal{S}^{op}, \mathcal{A}b)$. In other words, the reader can now express somewhat more concretely, what is implied by the fact that enough injectives need not exist, as in the counterexample of Section C.4.

But the counterexample of Section C.4 says more than just that there are not enough injectives. It says there is, in general, no cogenerator. In the spirit of reformulating everything in terms of α–phantom maps, let us see what it means to say that there is no right adjoint to $A(\mathcal{T}) \longrightarrow \mathcal{E}x(\mathcal{S}^{op}, \mathcal{A}b)$.

LEMMA 8.5.20. *Let $\mathcal{T}$ be a triangulated category satisfying [TR5]. Let α be a regular cardinal. Put $\mathcal{S} = \mathcal{T}^\alpha$, and assume $\mathcal{S}$ is essentially small.*

The functor

$$\pi : A(\mathcal{T}) \longrightarrow \mathcal{E}x(\mathcal{S}^{op}, \mathcal{A}b)$$

will have a right adjoint if and only if, for every object $z \in \mathcal{T}$, there is a maximal α–phantom map $y \longrightarrow z$. That is, there is an α–phantom map $y \longrightarrow z$, and given any other α–phantom map $x \longrightarrow z$, there is a (non–unique) factorisation

$$x \longrightarrow y \longrightarrow z.$$

Proof: The map

$$\pi : A(\mathcal{T}) \longrightarrow \mathcal{E}x(\mathcal{S}^{op}, \mathcal{A}b)$$

is a quotient map by Proposition 6.5.3. The abelian category $A(\mathcal{T})$ has enough injectives. Therefore Proposition A.2.20 applies: a right adjoint will exist if and only if every injective object $z \in \mathcal{T} \subset A(\mathcal{T})$ has a maximal subobject, among those in the kernel of π.

It is convenient to use the description $D(\mathcal{T}) \simeq A(\mathcal{T})$. Recall Proposition 5.2.6. The subobjects, in $D(\mathcal{T})$, of an object $z \in \mathcal{T}$, may be represented by morphisms in $\mathcal{T}$ of the form $y \longrightarrow z$. A subobject $y \longrightarrow z$ contains $x \longrightarrow z$ if the map $x \longrightarrow z$ factors as

$$x \longrightarrow y \longrightarrow z.$$

The kernel of π is identified with the α–phantom maps $x \longrightarrow z$. Therefore the existence of a maximal subobject of z, among those belonging to the kernel of π, is equivalent to the existence of a maximal α–phantom map $y \longrightarrow z$. □

REMARK 8.5.21. Proposition 8.5.18 rephrases for us the existence of enough injectives, in terms of α–phantom maps. Lemma 8.5.20 rephrases the existence of an adjoint $G : \mathcal{E}x(\mathcal{S}^{op}, \mathcal{A}b) \longrightarrow A(\mathcal{T})$, also in terms of α–phantom maps. From Proposition 8.5.2 we know, that the existence of enough injectives implies the existence of a right adjoint. Corollary 8.5.19 rephrases this in terms of α–phantom maps.

That is, in Proposition 8.5.18 we saw that $\mathcal{E}x\big(\mathcal{S}^{op}, \mathcal{A}b\big)$ has enough injectives if and only if, for every object $z \in \mathcal{T}$ there is a triangle

$$y \xrightarrow{\ f\ } z \longrightarrow t \longrightarrow \Sigma y,$$

with $t \in \mathcal{I}^\perp$ and $f \in \mathcal{I}$. In Corollary 8.5.19, we learned that in any such triangle, the map $y \longrightarrow z$ is a maximal α–phantom map into z. Finally, Lemma 8.5.20 teaches us that the existence of maximal α–phantom maps $y \longrightarrow z$ is equivalent to the existence of a right adjoint G to the natural functor

$$\pi : A(\mathcal{T}) \longrightarrow \mathcal{E}x\big(\mathcal{S}^{op}, \mathcal{A}b\big).$$

In other words, in terms of α–phantom maps, we recover Proposition 8.5.2. We learn the implication

$$\left\{\begin{array}{c}\mathcal{E}x\big(\mathcal{S}^{op}, \mathcal{A}b\big)\\ \text{has enough}\\ \text{injectives}\end{array}\right\} \Longrightarrow \left\{\begin{array}{c}A(\mathcal{T}) \longrightarrow \mathcal{E}x\big(\mathcal{S}^{op}, \mathcal{A}b\big)\\ \text{has a right adjoint}\end{array}\right\}.$$

From Lemma 8.5.20, we know that there is a right adjoint

$$G : \mathcal{E}x\big(\mathcal{S}^{op}, \mathcal{A}b\big) \longrightarrow A(\mathcal{T}),$$

if and only if every z in $\mathcal{T}$ admits a maximal α–phantom map $y \longrightarrow z$. If we furthermore know that there are enough injectives in $\mathcal{E}x\big(\mathcal{S}^{op}, \mathcal{A}b\big)$, then the maximal α–phantom map $y \longrightarrow z$ may be so chosen, that in the triangle

$$y \longrightarrow z \longrightarrow t \longrightarrow \Sigma y,$$

the object t is orthogonal to the α–phantom maps.

From Section C.4, we know by example that $\mathcal{E}x\big(\mathcal{S}^{op}, \mathcal{A}b\big)$ need have neither enough injectives, nor a right adjoint to $\pi : A(\mathcal{T}) \longrightarrow \mathcal{E}x\big(\mathcal{S}^{op}, \mathcal{A}b\big)$. There need not be enough objects $t \in \mathcal{I}^\perp$, but even worse than that, there may be objects $z \in \mathcal{T}$, admitting no maximal α–phantom map $y \longrightarrow z$.

REMARK 8.5.22. If I is an injective cogenerator of $\mathcal{E}x\big(\mathcal{S}^{op}, \mathcal{A}b\big)$, we will denote GI by $\mathbb{BC}$. The notation stands for *Brown–Comenetz* objects of $\mathcal{T}$. When we wish to remind ourselves of the dependence on the choice of α and on $\mathcal{T}$, we will write them $\mathbb{BC}(\alpha, \mathcal{T})$.

Of course, $\mathbb{BC}$ need not exist, as illustrated by Section C.4.

Let $\mathcal{T}$ be an α–compactly generated triangulated category. We remind the reader: this means that $\mathcal{T}^\alpha$ is essentially small, and that the homological functor

$$\mathcal{T} \longrightarrow \mathcal{E}x\Big(\{\mathcal{T}^\alpha\}^{op}, \mathcal{A}b\Big)$$

does not annihilate any object. Suppose furthermore that $\mathbb{BC}(\alpha, \mathcal{T})$ exists; there is an injective cogenerator for $\mathcal{E}x\Big(\{\mathcal{T}^\alpha\}^{op}, \mathcal{A}b\Big)$.

Because the functor

$$\mathcal{T} \longrightarrow \mathcal{E}x\Big(\{\mathcal{T}^\alpha\}^{op}, \mathcal{A}b\Big)$$

does not annihilate any object, for any x an object of $\mathcal{T}$, if $\mathcal{T}(-,x)|_{\mathcal{T}^\alpha} = 0$, then $x = 0$

It follows that the set of all suspensions of $\mathbb{BC}$ cogenerates $\mathcal{T}$. Given a non–zero object $x \in \mathcal{T}$, the functor $\mathcal{T}(-,x)|_{\mathcal{T}^\alpha}$ is non–zero, and hence has a non–zero map to the injective cogenerator

$$\mathcal{T}(-,x)|_{\mathcal{T}^\alpha} \longrightarrow \mathcal{T}(-,\mathbb{BC})|_{\mathcal{T}^\alpha}.$$

By 8.5.2.1 and Lemma 8.5.7, such a map is induced by a morphism $x \longrightarrow \mathbb{BC}$. We deduce there is a non–zero map $x \longrightarrow \mathbb{BC}$. This is true for every non–zero x, which means precisely that in the dual category, a set containing $\mathbb{BC}$ and closed under suspension is a generating set.

8.6. The second representability theorem: Brown representability for the dual

In this Section, we prove the theorem

THEOREM 8.6.1. *Let α be a regular cardinal. Suppose $\mathcal{T}$ is an α–compactly generated triangulated category. Suppose further that the category $\mathcal{E}x\Big(\{\mathcal{T}^\alpha\}^{op}, \mathcal{A}b\Big)$ has enough injectives. Then the representability theorem holds for the dual of $\mathcal{T}$. Every homological functor $\mathcal{T} \longrightarrow \mathcal{A}b$ which takes products to products is naturally isomorphic to $\mathcal{T}(h,-)$.*

For α as above, let $\mathbb{BC} = \mathbb{BC}(\alpha, \mathcal{T})$ be a corresponding Brown–Comenetz object. We remind the reader; this means that $\mathcal{T}(-,\mathbb{BC})|_{\mathcal{S}}$ is an injective cogenerator of $\mathcal{E}x\Big(\{\mathcal{T}^\alpha\}^{op}, \mathcal{A}b\Big)$. We assert further that, with $\mathbb{BC}$ as above, the dual category $\mathcal{T}^{op}$ satisfies $\langle \mathbb{BC} \rangle = \mathcal{T}^{op}$.

Proof: By Proposition 8.4.6, the category $\mathcal{T}$ satisfies [TR5*]. Put $\mathcal{S} = \mathcal{T}^\alpha$. Because $\mathcal{T}$ is α–compactly generated, $\mathcal{S}$ generates $\mathcal{T}$. We wish to choose a generating set for the dual; we choose the set $T = \{\Sigma^n \mathbb{BC}, n \in \mathbb{Z}\}$ of Remark 8.5.22. Lemmas 8.2.3 and 8.3.1 apply. For each $H : \mathcal{T} \longrightarrow \mathcal{A}b$ sending products to products, there exists a sequence of X_i's in the dual of $\mathcal{T}$ satisfying 8.2.3.1–8.2.3.4, and to prove the representability theorem as well as show that $\langle \mathbb{BC} \rangle = \mathcal{T}^{op}$, it suffices by Lemma 8.3.1 to prove the map $\mathcal{T}^{op}(\Sigma^n \mathbb{BC}, X) \longrightarrow H(\Sigma^n \mathbb{BC})$ injective.

Let us state concretely, in $\mathcal{T}$ rather than its dual, what we have in 8.2.3.1–8.2.3.4. For each H there is a sequence in $\mathcal{T}$

$$\cdots \longrightarrow X_2 \longrightarrow X_1 \longrightarrow X_0.$$

For each i there is a map $\mathcal{T}(X_i, -) \longrightarrow H(-)$, compatible with the maps of the sequence. The all important 8.2.3.4 tells us that for any object $t \in T$, the map $\mathcal{T}(X_i, t) \longrightarrow H(t)$ is surjective, and its kernel is the kernel of $\mathcal{T}(X_i, t) \longrightarrow \mathcal{T}(X_{i+1}, t)$.

The objects of $t \in T$ are of the form $t = \Sigma^n \mathbb{BC}$; without loss take $n = 0$, that is take $t = \mathbb{BC}$. We have that, for $i \geq 1$, the group $\mathcal{T}(X_i, \mathbb{BC})$ is the direct sum of the kernel of

$$\mathcal{T}(X_i, \mathbb{BC}) \longrightarrow \mathcal{T}(X_{i+1}, \mathbb{BC})$$

and the image of

$$\mathcal{T}(X_{i-1}, \mathbb{BC}) \longrightarrow \mathcal{T}(X_i, \mathbb{BC}),$$

and the sequence of images maps to each other by isomorphisms. But recall that for any $y \in \mathcal{T}$, the group of maps $y \longrightarrow \mathbb{BC}$ is canonically isomorphic to the group of maps

$$\mathcal{T}(-, y)|_{\mathcal{S}} \longrightarrow I(-)$$

with I a fixed injective cogenerator of $\mathcal{E}x\big(\mathcal{S}^{op}, \mathcal{A}b\big)$. (Here $\mathcal{S} = \mathcal{T}^\alpha$, to make the notation less crowded). What the above means is the following. We have natural maps

$$\mathcal{T}(-, X_i)|_{\mathcal{S}} \xleftarrow{\ \rho\ } \mathcal{T}(-, X_{i+1})|_{\mathcal{S}}, \qquad \mathcal{T}(-, X_{i-1})|_{\mathcal{S}} \xleftarrow{\ \theta\ } \mathcal{T}(-, X_i)|_{\mathcal{S}}$$

in $\mathcal{E}x\big(\mathcal{S}^{op}, \mathcal{A}b\big)$. When we take maps to the injective cogenerator, they induce the two maps

$$\mathcal{T}(X_i, \mathbb{BC}) \xrightarrow{\ \overline{\rho}\ } \mathcal{T}(X_{i+1}, \mathbb{BC}) \qquad \mathcal{T}(X_{i-1}, \mathbb{BC}) \xrightarrow{\ \overline{\theta}\ } \mathcal{T}(X_i, \mathbb{BC})$$

The fact that $\mathcal{T}(X_i, \mathbb{BC}) = \mathrm{Im}(\overline{\theta}) \oplus \mathrm{Ker}(\overline{\rho})$, coupled with the fact that taking Hom into an injective cogenerator reflects and preserves homology, allows us to deduce that $\mathcal{T}(-, X_i)|_{\mathcal{S}} = \mathrm{Im}(\theta) \oplus \mathrm{Coker}(\rho)$.

In other words, the sequence

$$\cdots \longrightarrow \mathcal{T}(-, X_3)|_{\mathcal{S}} \longrightarrow \mathcal{T}(-, X_2)|_{\mathcal{S}} \longrightarrow \mathcal{T}(-, X_1)|_{\mathcal{S}}$$

is the direct sum of two sequences

$$\cdots \xrightarrow{\ 1\ } F(-) \xrightarrow{\ 1\ } F(-) \xrightarrow{\ 1\ } F(-)$$
$$\cdots \xrightarrow{\ 0\ } Q_3(-) \xrightarrow{\ 0\ } Q_2(-) \xrightarrow{\ 0\ } Q_1(-)$$

and the group $H(\mathbb{BC})$ is naturally identified with the group of maps in $\mathcal{E}x\big(\mathcal{S}^{op}, \mathcal{A}b\big)$

$$F(-) \longrightarrow I(-).$$

The two sequences

$$\cdots \xrightarrow{\ 1\ } F(-) \xrightarrow{\ 1\ } F(-) \xrightarrow{\ 1\ } F(-)$$
$$\cdots \xrightarrow{\ 0\ } Q_3(-) \xrightarrow{\ 0\ } Q_2(-) \xrightarrow{\ 0\ } Q_1(-)$$

yield two short exact sequences in $\mathcal{E}x(\mathcal{S}^{op}, \mathcal{A}b)$

$$0 \to F(-) \longrightarrow \prod_{i=1}^{\infty} F(-) \xrightarrow{1-shift} \prod_{i=1}^{\infty} F(-) \to 0$$

$$0 \to 0 \longrightarrow \prod_{i=1}^{\infty} Q_i(-) \xrightarrow{1-shift} \prod_{i=1}^{\infty} Q_i(-) \to 0$$

which add up to the exact sequence

$$0 \to F(-) \longrightarrow \prod_{i=1}^{\infty} \mathcal{T}(-, X_i)|_{\mathcal{S}} \xrightarrow{1-shift} \prod_{i=1}^{\infty} \mathcal{T}(-, X_i)|_{\mathcal{S}} \to 0.$$

Adding the extra term $\mathcal{T}(-, X_0)|_{\mathcal{S}}$ is harmless; the sequence

$$0 \to F(-) \longrightarrow \prod_{i=0}^{\infty} \mathcal{T}(-, X_i)|_{\mathcal{S}} \xrightarrow{1-shift} \prod_{i=0}^{\infty} \mathcal{T}(-, X_i)|_{\mathcal{S}} \to 0$$

is also exact in $\mathcal{E}x(\mathcal{S}^{op}, \mathcal{A}b)$.

From Lemma 6.2.4 we know that the functor $\mathcal{T} \longrightarrow \mathcal{E}x(\mathcal{S}^{op}, \mathcal{A}b)$ which sends t to $\mathcal{T}(-, t)|_{\mathcal{S}}$ respects products. Therefore, in the exact sequence above we can identify

$$\begin{array}{ccccccc} 0 \to F(-) & \longrightarrow & \prod_{i=0}^{\infty} \mathcal{T}(-, X_i)|_{\mathcal{S}} & \xrightarrow{1-shift} & \prod_{i=0}^{\infty} \mathcal{T}(-, X_i)|_{\mathcal{S}} & \to 0 \\ & & \downarrow \wr & & \downarrow \wr & \\ & & \mathcal{T}\left(-, \prod_{i=0}^{\infty} X_i\right)\Big|_{\mathcal{S}} & \xrightarrow{1-shift} & \mathcal{T}\left(-, \prod_{i=0}^{\infty} X_i\right)\Big|_{\mathcal{S}} & \end{array}$$

Taking maps into I, we deduce an exact sequence

$$0 \leftarrow H(\mathbb{BC}) \longleftarrow \mathcal{T}\left(\prod_{i=0}^{\infty} X_i, \mathbb{BC}\right) \xleftarrow{1-shift} \mathcal{T}\left(\prod_{i=0}^{\infty} X_i, \mathbb{BC}\right) \leftarrow 0.$$

On the other hand, we have a triangle

$$X \longrightarrow \prod_{i=0}^{\infty} X_i \xrightarrow{1-shift} \prod_{i=0}^{\infty} X_i \longrightarrow \Sigma X$$

and hence a long exact sequence

$$\leftarrow \mathcal{T}(X, \mathbb{BC}) \longleftarrow \mathcal{T}\left(\prod_{i=0}^{\infty} X_i, \mathbb{BC}\right) \xleftarrow{1-shift} \mathcal{T}\left(\prod_{i=0}^{\infty} X_i, \mathbb{BC}\right) \leftarrow$$

The injectivity of

$$\mathcal{T}\left(\prod_{i=0}^{\infty} X_i, \mathbb{BC}\right) \xleftarrow{1-shift} \mathcal{T}\left(\prod_{i=0}^{\infty} X_i, \mathbb{BC}\right)$$

and all its suspensions means that we really get a short exact sequence

$$0 \longleftarrow \mathcal{T}(X, \mathbb{BC}) \longleftarrow \mathcal{T}\left(\prod_{i=0}^{\infty} X_i, \mathbb{BC}\right) \xleftarrow{1-shift} \mathcal{T}\left(\prod_{i=0}^{\infty} X_i, \mathbb{BC}\right) \longleftarrow 0.$$

There is a natural map of short exact sequences

$$\begin{array}{ccccccc} 0 \longleftarrow \mathcal{T}(X, \mathbb{BC}) & \longleftarrow & \mathcal{T}\left(\prod_{i=0}^{\infty} X_i, \mathbb{BC}\right) & \xleftarrow{1-shift} & \mathcal{T}\left(\prod_{i=0}^{\infty} X_i, \mathbb{BC}\right) & \longleftarrow 0 \\ \downarrow & & \| \wr & & \| \wr & \\ 0 \longleftarrow H(\mathbb{BC}) & \longleftarrow & \mathcal{T}\left(\prod_{i=0}^{\infty} X_i, \mathbb{BC}\right) & \xleftarrow{1-shift} & \mathcal{T}\left(\prod_{i=0}^{\infty} X_i, \mathbb{BC}\right) & \longleftarrow 0 \end{array}$$

from which it immediately follows that

$$\mathcal{T}(X, \mathbb{BC}) \longrightarrow H(\mathbb{BC})$$

is an isomorphism. Since this is also true for suspensions of $\mathbb{BC}$, it is true for all objects of $T = \{\Sigma^n \mathbb{BC}, n \in \mathbb{Z}\}$. The Theorem now follows from Lemma 8.3.1. □

8.7. History of the results in Chapter 8

Representability theorems and the existence of adjoints is perhaps the oldest and most venerable application of triangulated categories, but the theorems about them were always viewed as difficult. In constructing the derived category Verdier was motivated by the problem of finding a good formalism for Grothendieck's duality theorem. The duality theorem is an assertion that some triangulated functor between triangulated categories has a right adjoint. The literature is truly immense. Let me mention only Hartshorne's [**19**] and Deligne's [**10**] for the classical proof of the existence of the adjoint. Many people worked extensively on generalisations and analogues. Perhaps most notable is the work of Lipman.

All the results I mentioned look at specific functors $F : \mathcal{S} \longrightarrow \mathcal{T}$, and very concretely and explicitly, using the detailed structure of $\mathcal{S}$ and $\mathcal{T}$, construct the right adjoint.

The fact that the existence of such adjoints should be a formal consequence of the axioms of triangulated categories came from homotopy theory. Brown [**7**] proved a special case of Proposition 8.4.2. Precisely, he proved the case where $\mathcal{T}$ is not only well–generated, but more precisely $\aleph_0$–compactly generated. For a long time, nothing was known about the dual. The first representability theorem for duals of well–generated triangulated categories is the author's [**25**]. The result given there is narrower than the one here, but is also based on injective objects. In the present generality, Theorem 8.6.1 is new.

In homotopy theory, there are two types of applications for representability theorems. They can be used to construct adjoints, in particular Bousfield localisation functors. We will see much more about the significance of this. But they can also be used to construct objects in $\mathcal{T}$; we have already seen one example, the construction of the object $\mathbb{BC}$ in Proposition 8.5.2. The objects $\mathbb{BC}$ constructed here are generalisations to α–compactly generated categories of the Brown–Comenetz duals of finite spectra, constructed by Brown and Comenetz in [**8**]. The construction of Brown and Comenetz works only for $\aleph_0$–compactly generated categories.

I recently learned that Franke has, independently, also obtained a general Brown representability theorem. It is similar in spirit to Theorem 8.3.3. The similarity is spirit does not extend to a similarity in detail; neither theorem seems to easily imply the other, and the proofs are totally different. The reader is referred to Franke's [**11**].

CHAPTER 9

Bousfield localisation

9.1. Basic properties

Let $\mathcal{T}$ be a triangulated category, $\mathcal{S} \subset \mathcal{T}$ a triangulated subcategory. In Theorem 2.1.8 we learned how to construct the Verdier quotient $\mathcal{T}/\mathcal{S}$. There is a natural localisation map $F : \mathcal{T} \longrightarrow \mathcal{T}/\mathcal{S}$. In Example 8.4.5 we learned that under suitable hypotheses, the functor F has a right adjoint. We remind the reader of the hypoteses.

Suppose $\mathcal{T}$ is a triangulated category with small Hom–sets, satisfying [TR5]. Suppose further that the representability theorem holds for $\mathcal{T}$. Let $\mathcal{S}$ be a localising subcategory. Assume that the Verdier quotient $\mathcal{T}/\mathcal{S}$ is a category with small Hom–sets. In Example 8.4.5 we saw that the natural functor $F : \mathcal{T} \longrightarrow \mathcal{T}/\mathcal{S}$ has a right adjoint $G : \mathcal{T}/\mathcal{S} \longrightarrow \mathcal{T}$. This adjoint is called the Bousfield localisation functor.

Of course, this adjoint may exist even if $\mathcal{T}$ does not satisfy the hypotheses of Example 8.4.5. We say that a Bousfield localisation exists for the pair $\mathcal{S} \subset \mathcal{T}$ if the adjoint exists. More precisely

DEFINITION 9.1.1. *Let $\mathcal{T}$ be a triangulated category with small Hom–sets. Let $\mathcal{S}$ be a thick subcategory. We say that a* Bousfield localisation functor exists for the pair $\mathcal{S} \subset \mathcal{T}$ *when there is a right adjoint to the natural functor*

$$F : \mathcal{T} \longrightarrow \mathcal{T}/\mathcal{S}.$$

We will call the adjoint the Bousfield localisation functor, *and denote it* $G : \mathcal{T}/\mathcal{S} \longrightarrow \mathcal{T}$.

LEMMA 9.1.2. *Suppose a Bousfield localisation functor exists for the pair $\mathcal{S} \subset \mathcal{T}$. Let s be an object of $\mathcal{S}$, t an object of $\mathcal{T}/\mathcal{S}$, which is an object of $\mathcal{T}$; the objects in the two categories are the same. Any map $s \longrightarrow Gt$ vanishes.*

Proof: By the adjunction, we have

$$\mathcal{T}(s, Gt) \quad = \quad \mathcal{T}/\mathcal{S}\big(Fs, t\big),$$

but since $s \in \mathcal{S}$, we must have $Fs = 0$. □

DEFINITION 9.1.3. *Let $\mathcal{S}$ be a class of objects in a triangulated category $\mathcal{T}$. An object $t \in \mathcal{T}$ is called $\mathcal{S}$–*local *if, for any object $s \in \mathcal{S}$, the maps $s \longrightarrow t$ all vanish. If $\mathcal{S}$ is not just any class of objects but a thick subcategory, and if a Bousfield localisation functor exists for the pair $\mathcal{S} \subset \mathcal{T}$, then Lemma 9.1.2 proves that for any $t \in \mathcal{T}$, the object Gt is $\mathcal{S}$–local.*

DEFINITION 9.1.4. *Let $\mathcal{S}$ be a class of objects in a triangulated category $\mathcal{T}$. An object $t \in \mathcal{T}$ is called $\mathcal{S}$–colocal if it is local in the dual category, that is if any map $t \longrightarrow s$, $s \in \mathcal{S}$ vanishes.*

LEMMA 9.1.5. *Let $\mathcal{T}$ be a triangulated category, $\mathcal{S} \subset \mathcal{T}$ a triangulated subcategory. Let t be an $\mathcal{S}$–local object of $\mathcal{T}$. Then the natural map*

$$\phi : \mathcal{T}(x,t) \longrightarrow \mathcal{T}/\mathcal{S}\big(Fx, Ft\big)$$

is an isomorphism, for all $x \in \mathcal{T}$.

Proof: By the construction of $\mathcal{T}/\mathcal{S}$ and the functor F, a map in $\mathcal{T}/\mathcal{S}$ from Fx to Ft is an equivalence class of diagrams in $\mathcal{T}$

$$\begin{array}{ccc} x' & \longrightarrow & t \\ {\scriptstyle\alpha}\big\downarrow & & \\ x & & \end{array}$$

with $\alpha \in Mor_{\mathcal{S}}$ (see Definition 2.1.11). Pick a representative for such a class. To show ϕ surjective, we need to show that there is a map $x \longrightarrow t$ equivalent to it.

Choose therefore a diagram

$$\begin{array}{ccc} x' & \longrightarrow & t \\ {\scriptstyle\alpha}\big\downarrow & & \\ x & & \end{array}$$

as above. Since $\alpha \in Mor_{\mathcal{S}}$, in the triangle

$$s \longrightarrow x' \stackrel{\alpha}{\longrightarrow} x \longrightarrow \Sigma s$$

the object s lies in $\mathcal{S}$. Applying the cohomological functor $\mathcal{T}(-,t)$ to the triangle, we get an exact sequence of abelian groups

$$\mathcal{T}(s,t) \longleftarrow \mathcal{T}(x',t) \xleftarrow{\mathcal{T}(\alpha,t)} \mathcal{T}(x,t) \longleftarrow \mathcal{T}(\Sigma s,t).$$

Since s and Σs both lie in $\mathcal{S}$ and t is $\mathcal{S}$–local, the groups $\mathcal{T}(s,t)$ and $\mathcal{T}(\Sigma s,t)$ both vanish. The map $\mathcal{T}(\alpha,t)$ is an isomorphism, and in particular there is a morphism $x \longrightarrow t$ in $\mathcal{T}$ rendering commutative the square

$$\begin{array}{ccc} x' & \longrightarrow & t \\ {\scriptstyle\alpha}\big\downarrow & & \big\downarrow{\scriptstyle 1} \\ x & \longrightarrow & t \end{array}$$

This produces a map $x \longrightarrow t$ in the equivalence class of

$$\begin{array}{ccc} x' & \longrightarrow & t \\ {\scriptstyle\alpha}\big\downarrow & & \\ x & & \end{array}$$

proving the surjectivity of ϕ.

Now we have to prove ϕ injective. Suppose we are given a map $x \longrightarrow t$ in the kernel of ϕ; that is, the map becomes zero in $\mathcal{T}/\mathcal{S}$. By Lemma 2.1.26 the map $x \longrightarrow t$ must factor, in $\mathcal{T}$, as

$$x \longrightarrow s \longrightarrow t$$

with $s \in \mathcal{S}$. But t is $\mathcal{S}$–local, so the map $s \longrightarrow t$ vanishes. Hence so does the composite $x \longrightarrow s \longrightarrow t$. □

NOTATION 9.1.6. Suppose a Bousfield localisation functor exists for the pair $\mathcal{S} \subset \mathcal{T}$. Since the objects of $\mathcal{T}$ and $\mathcal{T}/\mathcal{S}$ are the same, we will freely confuse them. Given an object $t \in \mathcal{T}$, there is the unit of adjunction

$$\eta_t : t \longrightarrow GFt;$$

we remind the reader that it is the map in $\mathcal{T}(t, GFt)$ which corresponds to the identity $1 \in \mathcal{T}/\mathcal{S}\big(Ft, Ft\big)$ under the natural isomorphism

$$\mathcal{T}(t, GFt) \quad = \quad \mathcal{T}/\mathcal{S}\big(Ft, Ft\big).$$

Since t and Ft are really the same (F is the identity on objects), we will feel free to write

$$\eta_t : t \longrightarrow Gt$$

for the unit of adjunction.

LEMMA 9.1.7. *Suppose a Bousfield localisation functor exists for the pair $\mathcal{S} \subset \mathcal{T}$. Let t be an object of $\mathcal{T}$. Then the map $\eta_t : t \longrightarrow Gt$ is an isomorphism in $\mathcal{T}/\mathcal{S}$.*

Proof: First let us untangle precisely what this means. From the adjuction, we have a natural transformation, the unit of adjunction

$$\eta : 1 \longrightarrow GF.$$

For every $t \in \mathcal{T}$, this gives a map, in $\mathcal{T}$, $t \longrightarrow GFt$. We want to show that this map becomes an isomorphism in $\mathcal{T}/\mathcal{S}$. In other words, we wish to prove that

$$F\eta : F \longrightarrow FGF$$

is an isomorphism. We also have the counit

$$\varepsilon : FG \longrightarrow 1.$$

Very generally, we know that the composite

$$F \xrightarrow{F\eta} FGF \xrightarrow{\varepsilon F} F$$

is always the identity. That is, the map $F\eta$ always has a left inverse; the problem is to show that it is a two–sided inverse.

Let $x, t \in \mathcal{T}$ be arbitrary. By the adjunction,

$$\mathcal{T}/\mathcal{S}\big(Fx, Ft\big) \quad = \quad \mathcal{T}(x, GFt).$$

The natural map takes $f : Fx \longrightarrow Ft$ to the composite

$$x \xrightarrow{\eta_x} GFx \xrightarrow{Gf} GFt.$$

On the other hand, GFt is $\mathcal{S}$–local by Lemma 9.1.2, and by Lemma 9.1.5

$$\mathcal{T}/\mathcal{S}\big(Fx, FGFt\big) \quad = \quad \mathcal{T}(x, GFt).$$

More precisely, taking $g \in \mathcal{T}(x, GFt)$ to $Fg \in \mathcal{T}/\mathcal{S}\big(Fx, FGFt\big)$ induces the isomorphism. Thus

$$\mathcal{T}/\mathcal{S}\big(Fx, Ft\big) \quad = \quad \mathcal{T}/\mathcal{S}\big(Fx, FGFt\big),$$

and more precisely the isomorphism is induced by taking $f : Fx \longrightarrow Ft$ to

$$Fx \xrightarrow{F\eta_x} FGFx \xrightarrow{FGf} FGFt.$$

By the naturality of $\varepsilon : FG \longrightarrow 1$, we have a commutative square

$$\begin{array}{ccc} FGFx & \xrightarrow{FGf} & FGFt \\ {\scriptstyle \varepsilon_{Fx}}\Big\downarrow & & {\scriptstyle \varepsilon_{Ft}}\Big\downarrow \\ Fx & \xrightarrow{f} & Ft. \end{array}$$

It follows that the composite

$$Fx \xrightarrow{F\eta_x} FGFx \xrightarrow{FGf} FGFt \xrightarrow{\varepsilon_{Ft}} Ft$$

is equal to the composite

$$Fx \xrightarrow{F\eta_x} FGFx \xrightarrow{\varepsilon_{Fx}} Fx \xrightarrow{f} Ft$$

which is just the map $f : Fx \longrightarrow Ft$. In other words, composition with $\varepsilon_{Ft} : FGFt \longrightarrow Ft$ induces the inverse to our isomorphism

$$\mathcal{T}/\mathcal{S}\big(Fx, Ft\big) \quad = \quad \mathcal{T}/\mathcal{S}\big(Fx, FGFt\big),$$

and hence ε_{Ft} is an isomorphism. But $\varepsilon_{Ft} F\eta_t = 1$, and $F\eta_t$ must be the two–sided inverse of ε_{Ft}. In particular, $F\eta_t$ is an isomorphism. □

PROPOSITION 9.1.8. *Suppose $\mathcal{S}$ is a thick subcategory of the triangulated category $\mathcal{T}$, and suppose a Bousfield localisation functor exists for the pair $\mathcal{S} \subset \mathcal{T}$. Let t be an object of $\mathcal{T}$. In the triangle*

$$t_{\mathcal{S}} \longrightarrow t \xrightarrow{\eta_t} Gt \longrightarrow \Sigma t_{\mathcal{S}}$$

the object Gt is $\mathcal{S}$–local, while the object $t_{\mathcal{S}}$ lies in $\mathcal{S}$.

Proof: The fact that Gt is $\mathcal{S}$–local is Lemma 9.1.2. By Lemma 9.1.7 the map η_t is an isomorphism in $\mathcal{T}/\mathcal{S}$. By Proposition 2.1.35, it follows that $t_{\mathcal{S}}$, the third vertex of the triangle on the morphism η_t, is the direct summand of an object of $\mathcal{S}$. There exists an object $s \in \mathcal{T}$ so that $s \oplus t_{\mathcal{S}} \in \mathcal{S}$. But $\mathcal{S}$ is thick, and in particular contains all the direct summands of its objects. Hence $t_{\mathcal{S}} \in \mathcal{S}$. □

COROLLARY 9.1.9. *Suppose $\mathcal{S}$ is a thick subcategory of the triangulated category $\mathcal{T}$, and suppose a Bousfield localisation functor exists for the pair $\mathcal{S} \subset \mathcal{T}$. Let t be any $\mathcal{S}$-local object. Then the map $t \longrightarrow Gt$ is an isomorphism in $\mathcal{T}$ (and not just in $\mathcal{T}/\mathcal{S}$, as in Lemma 9.1.7).*

Proof: Let t be a $\mathcal{S}$–local object. By Proposition 9.1.8, in the triangle

$$t_{\mathcal{S}} \longrightarrow t \xrightarrow{\eta_t} Gt \longrightarrow \Sigma t_{\mathcal{S}}$$

the object $t_{\mathcal{S}}$ lies in $\mathcal{S}$. The map $t_{\mathcal{S}} \longrightarrow t$ must vanish, being a map from an object $t_{\mathcal{S}} \in \mathcal{S}$ to a $\mathcal{S}$–local object t. But then we must have $Gt \simeq t \oplus \Sigma t_{\mathcal{S}}$. By Lemma 9.1.2, the object Gt is also $\mathcal{S}$–local. The map

$$\Sigma t_{\mathcal{S}} \longrightarrow Gt \simeq t \oplus \Sigma t_{\mathcal{S}}$$

is a map from an object $\Sigma t_{\mathcal{S}} \in \mathcal{S}$ to a $\mathcal{S}$–local object Gt. Hence it must vanish. But it is the inclusion of a direct summand. Hence $t_{\mathcal{S}}$ is isomorphic to 0, and

$$t \xrightarrow{\eta_t} Gt$$

is an isomorphism. □

Next some definitions.

DEFINITION 9.1.10. *Let $\mathcal{T}$ be a triangulated category, $\mathcal{S}$ a class of objects in $\mathcal{T}$. Then the category ${}^{\perp}\mathcal{S}$ is defined to be the full subcategory of all $\mathcal{S}$-local objects. That is*

$${}^{\perp}\mathcal{S} \quad = \quad \{x \in \mathcal{T} \,|\, \forall s \in \mathcal{S},\, \mathcal{T}(s,x) = 0\}\,.$$

DEFINITION 9.1.11. *The dual of ${}^{\perp}\mathcal{S}$ will be denoted $\mathcal{S}^{\perp}$. Explicitly, let $\mathcal{S}$ be a class of objects in $\mathcal{T}$. We define $\mathcal{S}^{\perp}$ to be the full subcategory of all $\mathcal{S}$-colocal objects in $\mathcal{T}$. That is,*

$$\mathcal{S}^{\perp} \quad = \quad \{x \in \mathcal{T} \,|\, \forall s \in \mathcal{S},\, \mathcal{T}(x,s) = 0\}\,.$$

LEMMA 9.1.12. *Let $\mathcal{S}$ be any class of objects in the triangulated category $\mathcal{T}$. Then both ${}^{\perp}\mathcal{S}$ and $\mathcal{S}^{\perp}$ are thick subcategories of $\mathcal{T}$. If $\mathcal{T}$ satisfies [TR5], then $\mathcal{S}^{\perp}$ is localising; if $\mathcal{T}$ satisfies [TR5*], then ${}^{\perp}\mathcal{S}$ is colocalising.*

Proof: Obvious. □

THEOREM 9.1.13. *Let $\mathcal{T}$ be a triangulated category, $\mathcal{S}$ a thick subcategory. Then a Bousfield localisation functor exists for the pair $\mathcal{S} \subset \mathcal{T}$ if and only if, for every object $t \in \mathcal{T}$, there is a triangle in $\mathcal{T}$*

$$t_{\mathcal{S}} \longrightarrow t \longrightarrow {}_{\{^\perp\mathcal{S}\}}t \longrightarrow \Sigma t_{\mathcal{S}}$$

with $t_{\mathcal{S}} \in \mathcal{S}$ and ${}_{\{^\perp\mathcal{S}\}}t \in {}^\perp\mathcal{S}$.

Proof: If a Bousfield localisation functor exists for the pair $\mathcal{S} \subset \mathcal{T}$, then Proposition 9.1.8 establishes the existence of a triangle

$$t_{\mathcal{S}} \longrightarrow t \longrightarrow {}_{\{^\perp\mathcal{S}\}}t \longrightarrow \Sigma t_{\mathcal{S}}$$

as required. We are left with the reverse implication $\Longleftarrow$.

Suppose therefore that for every $t \in \mathcal{T}$ there is a triangle

$$t_{\mathcal{S}} \longrightarrow t \stackrel{\alpha}{\longrightarrow} {}_{\{^\perp\mathcal{S}\}}t \longrightarrow \Sigma t_{\mathcal{S}}$$

with $t_{\mathcal{S}} \in \mathcal{S}$ and ${}_{\{^\perp\mathcal{S}\}}t \in {}^\perp\mathcal{S}$. Since $t_{\mathcal{S}} \in \mathcal{S}$, the morphism α lies in $Mor_{\mathcal{S}}$ and becomes an isomorphism in $\mathcal{T}/\mathcal{S}$. Thus

$$\mathcal{T}/\mathcal{S}(x,t) \quad = \quad \mathcal{T}/\mathcal{S}\Big(x, {}_{\{^\perp\mathcal{S}\}}t\Big).$$

On the other hand ${}_{\{^\perp\mathcal{S}\}}t$ is in ${}^\perp\mathcal{S}$, that is it is $\mathcal{S}$–local. By Lemma 9.1.5

$$\mathcal{T}/\mathcal{S}\Big(x, {}_{\{^\perp\mathcal{S}\}}t\Big) \quad = \quad \mathcal{T}\Big(x, {}_{\{^\perp\mathcal{S}\}}t\Big).$$

Combining the two, we get

$$\mathcal{T}/\mathcal{S}\Big(x, t\Big) \quad = \quad \mathcal{T}\Big(x, {}_{\{^\perp\mathcal{S}\}}t\Big).$$

In other words, if we define Gt to be ${}_{\{^\perp\mathcal{S}\}}t$, we deduce a natural isomorphism

$$\mathcal{T}/\mathcal{S}(-,t) \quad = \quad \mathcal{T}(-,Gt).$$

This shows that $Gt = {}_{\{^\perp\mathcal{S}\}}t$ is unique up to canonical isomorphism, and that G extends to a functor, adjoint to F. □

COROLLARY 9.1.14. *Let $\mathcal{S}$ be a thick subcategory of the triangulated category $\mathcal{T}$, and suppose a Bousfield localisation functor exists for the pair $\mathcal{S} \subset \mathcal{T}$. Then a Bousfield localisation functor also exists for the pair $\{^\perp\mathcal{S}\}^{op} \subset \mathcal{T}^{op}$, and*

$$\{^\perp\mathcal{S}\}^\perp = \mathcal{S}.$$

Proof: Because a Bousfield localisation functor exists for the pair $\mathcal{S} \subset \mathcal{T}$, Theorem 9.1.13 says that for every $t \in \mathcal{T}$ there is a triangle

$$t_{\mathcal{S}} \longrightarrow t \stackrel{\alpha}{\longrightarrow} {}_{\{^\perp\mathcal{S}\}}t \longrightarrow \Sigma t_{\mathcal{S}},$$

with $t_{\mathcal{S}} \in \mathcal{S}$ and ${}_{\{^\perp\mathcal{S}\}}t \in {}^\perp\mathcal{S}$. But since $\mathcal{S}$ is clearly contained in $\left\{^\perp\mathcal{S}\right\}^\perp$, we may view $t_{\mathcal{S}}$ as lying in $\left\{^\perp\mathcal{S}\right\}^\perp$, and then the dual of Theorem 9.1.13 tells us that that there is a left adjoint (the dual of a right adjoint) to the natural map

$$U : \mathcal{T} \longrightarrow \mathcal{T}/{}^\perp\mathcal{S},$$

that is Bousfield localisation holds for the pair $\{^\perp\mathcal{S}\}^{op} \subset \mathcal{T}^{op}$. Call this adjoint

$$L : \mathcal{T}/{}^\perp\mathcal{S} \longrightarrow \mathcal{T}.$$

Now for every object $t \in \mathcal{T}$ we have a triangle in $\mathcal{T}$

$$LUt \longrightarrow t \longrightarrow c \longrightarrow \Sigma LUt$$

with $c \in {}^\perp\mathcal{S}$ and $LUt \in \left\{^\perp\mathcal{S}\right\}^\perp$. But above we identified this with the triangle

$$t_{\mathcal{S}} \longrightarrow t \stackrel{\alpha}{\longrightarrow} {}_{\{^\perp\mathcal{S}\}}t \longrightarrow \Sigma t_{\mathcal{S}}$$

and in particular we have that $LUt = t_{\mathcal{S}}$ lies in $\mathcal{S} \subset \left\{^\perp\mathcal{S}\right\}^\perp$, for every t. Suppose $t \in \left\{^\perp\mathcal{S}\right\}^\perp$, that is t is ${}^\perp\mathcal{S}$–colocal. By the dual of Corollary 9.1.9, the map $LUt \longrightarrow t$ is an isomorphism. But by the above LUt lies in $\mathcal{S}$. Hence $t \in \mathcal{S}$, and $\mathcal{S} = \left\{^\perp\mathcal{S}\right\}^\perp$. □

REMARK 9.1.15. What we have learned is that Bousfield localisations are very symmetric. Suppose in a triangulated category $\mathcal{T}$ we are given two thick subcategories $\mathcal{S}$ and $\mathcal{S}'$. Suppose they are perpendicular to each other, in the sense that $\mathcal{S} \subset \{\mathcal{S}'\}^\perp$ or equivalently $\mathcal{S}' \subset {}^\perp\mathcal{S}$. These statements are equivalent as both say that, for all $s \in \mathcal{S}$ and $s' \in \mathcal{S}'$, the group $\mathcal{T}(s, s')$ vanishes.

If for every object $t \in \mathcal{T}$ there is a triangle

$$t_{\mathcal{S}} \longrightarrow t \longrightarrow {}_{\mathcal{S}'}t \longrightarrow \Sigma t_{\mathcal{S}}$$

with $t_{\mathcal{S}} \in \mathcal{S}$ and ${}_{\mathcal{S}'}t \in \mathcal{S}'$, then both the natural projections

$$F : \mathcal{T} \longrightarrow \mathcal{T}/\mathcal{S} \qquad \text{and} \qquad U : \mathcal{T} \longrightarrow \mathcal{T}/\mathcal{S}'$$

have adjoints, F a right adjoint, U a left adjoint. Furthermore, using essentially the argument by which we proved $\left\{^\perp\mathcal{S}\right\}^\perp = \mathcal{S}$ in Corollary 9.1.14, one shows easily that the inclusions $\mathcal{S} \subset \{\mathcal{S}'\}^{\perp}$ and $\mathcal{S}' \subset {}^{\perp}\mathcal{S}$ are both equalities. That is,

$$\mathcal{S} = \{\mathcal{S}'\}^{\perp} \qquad \text{and} \qquad \mathcal{S}' = {}^{\perp}\mathcal{S}.$$

THEOREM 9.1.16. *Let $\mathcal{S}$ be a thick subcategory of the triangulated category $\mathcal{T}$, and suppose a Bousfield localisation functor exists for the pair $\mathcal{S} \subset \mathcal{T}$. Then the subcategory ${}^{\perp}\mathcal{S} \subset \mathcal{T}$ is equivalent to $\mathcal{T}/\mathcal{S}$. Precisely, the composite*

$$ {}^{\perp}\mathcal{S} \quad \subset \quad \mathcal{T} \xrightarrow{\ F\ } \mathcal{T}/\mathcal{S}$$

is an equivalence of categories.

Proof: The fact that the composite is fully faithful follows directly from Lemma 9.1.5. In Lemma 9.1.5 we proved that, for any $x \in \mathcal{T}$, $y \in {}^{\perp}\mathcal{S}$,

$$\mathcal{T}(x,y) \quad = \quad \mathcal{T}/\mathcal{S}\big(x,y\big).$$

Of course, if $x \in {}^{\perp}\mathcal{S} \subset \mathcal{T}$, the statement still holds, which establishes that morphisms between $\mathcal{S}$–local objects are the same in $\mathcal{T}$ as in $\mathcal{T}/\mathcal{S}$.

The fact that F is an essential equivalence is Lemma 9.1.2 coupled with Lemma 9.1.7. In Lemma 9.1.7 we proved that $\eta_t : t \longrightarrow Gt$ is always an isomorphism in $\mathcal{T}/\mathcal{S}$, and in Lemma 9.1.2 we proved that Gt is $\mathcal{S}$–local, that is $Gt \in {}^{\perp}\mathcal{S}$. Thus every object $t \in \mathcal{T}/\mathcal{S}$ is isomorphic to an object in ${}^{\perp}\mathcal{S}$. □

REMARK 9.1.17. Let $\mathcal{S}$ be a thick subcategory of the triangulated category $\mathcal{T}$, and suppose a Bousfield localisation functor exists for the pair $\mathcal{S} \subset \mathcal{T}$. It follows from Theorem 9.1.16 that the quotient $\mathcal{T}/\mathcal{S}$ is embedded as a full subcategory in $\mathcal{T}$. If $\mathcal{T}$ has small Hom–sets, so must $\mathcal{T}/\mathcal{S}$.

The functor $F : \mathcal{T} \longrightarrow \mathcal{T}/\mathcal{S}$, having a right adjoint, must respect coproducts in $\mathcal{T}$. If $\mathcal{T}$ satisfies [TR5], then coproducts exist in $\mathcal{T}$. Take any set of objects $\{X_\lambda, \lambda \in \Lambda\}$ in $\mathcal{S} \subset \mathcal{T}$. Their coproduct exists in $\mathcal{T}$, and since F respects coproducts,

$$F\left(\coprod_{\lambda\in\Lambda} X_\lambda\right) \quad = \quad \coprod_{\lambda\in\Lambda} F(X_\lambda).$$

On the other hand, as $X_\lambda \in \mathcal{S}$, we must have $F\big(X_\lambda\big) = 0$. The coproduct on the right, that is $\displaystyle\coprod_{\lambda\in\Lambda} F(X_\lambda)$, must vanish, forcing $F\left(\displaystyle\coprod_{\lambda\in\Lambda} X_\lambda\right)$ to also vanish. In other words, $\displaystyle\coprod_{\lambda\in\Lambda} X_\lambda$ is an object of $\mathcal{T}$ mapped to 0 under the functor $F : \mathcal{T} \longrightarrow \mathcal{T}/\mathcal{S}$. By Lemma 2.1.33, $\displaystyle\coprod_{\lambda\in\Lambda} X_\lambda$ must be a direct summand of an object in $\mathcal{S}$. But $\mathcal{S}$ is thick, hence closed under direct summands. The object $\coprod_{\lambda\in\Lambda} X_\lambda$ must lie in $\mathcal{S}$. The subcategory $\mathcal{S} \subset \mathcal{T}$ contains all coproducts of its objects. It is localising.

We have learned that if $\mathcal{T}$ is a triangulated category satisfying [TR5] with small Hom–sets, then for a thick subcategory $\mathcal{S} \subset \mathcal{T}$ to have a chance that a Bousfield localisation functor exist, the Hom–sets in $\mathcal{T}/\mathcal{S}$ must be small and $\mathcal{S} \subset \mathcal{T}$ must be localising. Example 8.4.5 tells us that a Bousfield localisation functor will exist if we also assume that the representability theorem holds for $\mathcal{T}$. The representability theorem is not necessary, but the other hypotheses are.

Dually, to have a chance for the existence of a right adjoint to the functor $F : \mathcal{T} \longrightarrow \mathcal{T}/\mathcal{S}$, one would have to have that $\mathcal{T}/\mathcal{S}$ has small Hom–sets, and if $\mathcal{T}$ satisfies [TR5*], $\mathcal{S}$ had better be colocalising as well.

In Example 8.4.5 we saw that a Bousfield localisation functor may be constructed for a pair $\mathcal{S} \subset \mathcal{T}$, as long as the representability theorem held for $\mathcal{T}$. But it turns out that, using the results of this Section, we can show that a Bousfield localisation functor exists for a pair $\mathcal{S} \subset \mathcal{T}$ even if the representability theorem holds for $\mathcal{S}$ instead. In the next two Propositions, let us state this carefully.

PROPOSITION 9.1.18. *Suppose $\mathcal{S}$ is a thick subcategory of the triangulated category $\mathcal{T}$. Suppose $\mathcal{T}$ has small* Hom*–sets. A Bousfield localisation functor exists for the pair $\mathcal{S} \subset \mathcal{T}$ if and only if the inclusion $\mathcal{S} \longrightarrow \mathcal{T}$ has a right adjoint.*

Proof: If a Bousfield localisation functor exists for $\mathcal{S} \subset \mathcal{T}$, then by Corollary 9.1.14, the natural projection $\mathcal{T} \longrightarrow \mathcal{T}/{}^\perp\mathcal{S}$ has a left adjoint, which embeds $\mathcal{T}/{}^\perp\mathcal{S}$ as $\mathcal{S} = \{{}^\perp\mathcal{S}\}^\perp \subset \mathcal{T}$. That is, the embedding $\mathcal{S} \subset \mathcal{T}$ has a right adjoint $\mathcal{T} \longrightarrow \mathcal{T}/{}^\perp\mathcal{S}$.

Now suppose the embedding $I : \mathcal{S} \longrightarrow \mathcal{T}$ has a right adjoint $J : \mathcal{T} \longrightarrow \mathcal{S}$. For any $t \in \mathcal{T}$, consider the unit of adjunction $IJt \longrightarrow t$. It may be completed to a triangle

$$IJt \longrightarrow t \longrightarrow z \longrightarrow \Sigma IJt.$$

Clearly, $IJt \in \mathcal{S}$. Because of the adjunction,

$$\mathcal{T}(Ix, t) \quad = \quad \mathcal{S}(x, Jt) \quad = \quad \mathcal{T}(Ix, IJt)$$

where the equality $\mathcal{S}(x, Jt) = \mathcal{T}(Ix, IJt)$ is because $\mathcal{S} \subset \mathcal{T}$ is a full subcategory; the functor I is fully faithful. From the long exact sequence for the triangle

$$IJt \longrightarrow t \longrightarrow z \longrightarrow \Sigma IJt$$

we learn that, for any $x \in \mathcal{S}$, $\mathcal{T}(Ix, z) = 0$. That is, $z \in {}^{\perp}\mathcal{S}$. We have produced a triangle

$$t_{\mathcal{S}} \longrightarrow t \longrightarrow {}_{\{^{\perp}\mathcal{S}\}}t \longrightarrow \Sigma t_{\mathcal{S}}$$

with $t_{\mathcal{S}} = IJt \in \mathcal{S}$ and ${}_{\{^{\perp}\mathcal{S}\}}t = z \in {}^{\perp}\mathcal{S}$. By Theorem 9.1.13 there is a Bousfield localisation for the pair $\mathcal{S} \subset \mathcal{T}$. □

PROPOSITION 9.1.19. *Let $\mathcal{T}$ be a triangulated category satisfying [TR5], and $\mathcal{S} \subset \mathcal{T}$ a localising subcategory. Suppose the representability theorem holds for $\mathcal{S}$. Assume also that $\mathcal{T}$ has small* Hom*–sets. Then a Bousfield localisation functor exists for the pair $\mathcal{S} \subset \mathcal{T}$.*

Proof: Because $\mathcal{S}$ is localising, the inclusion $\mathcal{S} \longrightarrow \mathcal{T}$ is triangulated and preserves coproducts. We are assuming that $\mathcal{S}$ satisfies the representability theorem. By Proposition 8.4.2 there is a right adjoint to the map $\mathcal{S} \longrightarrow \mathcal{T}$, and from Proposition 9.1.18 it now follows that there is a Bousfield localisation functor for the pair $\mathcal{S} \subset \mathcal{T}$. □

REMARK 9.1.20. Suppose $\mathcal{T}$ is a triangulated category with small Hom–sets satisfying [TR5], and $\mathcal{S} \subset \mathcal{T}$ is a localising subcategory for which the representability theorem holds. Then Proposition 9.1.19 tells us that a Bousfield localisation functor exists for the pair $\mathcal{S} \subset \mathcal{T}$. But in Theorem 9.1.16 the Verdier quotient $\mathcal{T}/\mathcal{S}$ is identified with ${}^{\perp}\mathcal{S} \subset \mathcal{T}$, and hence $\mathcal{T}/\mathcal{S}$ has small Hom–sets. In this sense, Proposition 9.1.19 is better than Example 8.4.5. In Example 8.4.5, where we used Brown representability for $\mathcal{T}$ to construct the right adjoint to $F : \mathcal{T} \longrightarrow \mathcal{T}/\mathcal{S}$, we had to assume that the Hom–sets in $\mathcal{T}/\mathcal{S}$ are small. In Proposition 9.1.19, which uses instead Brown representability for $\mathcal{S}$, the existence of the right adjoint to $F : \mathcal{T} \longrightarrow \mathcal{T}/\mathcal{S}$ comes without having to assume the Hom–sets in $\mathcal{T}/\mathcal{S}$ are small, and the fact that they must be small is one of the conclusions we may draw.

9.2. The six gluing functors

In Section 9.1 we studied the formal properties of a Bousfield localisation functor for a pair $\mathcal{S} \subset \mathcal{T}$. There is one special case of this that deserves mention, since it leads to an extensive theory. It is the case of the *six gluing functors*.

DEFINITION 9.2.1. *Let $\mathcal{T}$ be a triangulated category, and suppose $\mathcal{S}$ is a thick subcategory. If both a Bousfield localisation functor and a Bousfield colocalisation functor exist for the pair $\mathcal{S} \subset \mathcal{T}$, we say that we are in the situation of the* six gluing functors.

Because Bousield localisation and colocalisation functors exist for the pair $\mathcal{S} \subset \mathcal{T}$, the natural functor $\mathcal{T} \longrightarrow \mathcal{T}/\mathcal{S}$ has both a right and a left adjoint. This gives three functors.

But now because a Bousfield localisation functor exists for the pair $\mathcal{S} \subset \mathcal{T}$, Proposition 9.1.18 tells us that $\mathcal{S} \longrightarrow \mathcal{T}$ has a right adjoint. Dually, because Bousfield colocalisation functor exists for the pair $\mathcal{S} \subset \mathcal{T}$, the inclusion $\mathcal{S} \longrightarrow \mathcal{T}$ has a left adjoint. We have three more functors; the inclusion $\mathcal{S} \longrightarrow \mathcal{T}$ and its right and left adjoints.

It is customary to represent the six functors by a diagram

$$\mathcal{S} \;\substack{\longleftarrow \\ \longrightarrow \\ \longleftarrow}\; \mathcal{T} \;\substack{\longleftarrow \\ \longrightarrow \\ \longleftarrow}\; \mathcal{T}/\mathcal{S}$$

Let i_* be the inclusion functor $i_* : \mathcal{S} \longrightarrow \mathcal{T}$. Its left adjoint will be denoted $i^* : \mathcal{T} \longrightarrow \mathcal{S}$, its right adjoint $i^! : \mathcal{T} \longrightarrow \mathcal{S}$. Let the natural projection $\mathcal{T} \longrightarrow \mathcal{T}/\mathcal{S}$ be denoted j^*. Its left adjoint will be denoted $j_!$, its right adjoint j_*. The fact that the composite $j^* i_*$ vanishes gives us, by left and right adjunction, the identities

$$i^* j_! = 0, \qquad\qquad i^! j_* = 0.$$

Furthermore, for every object $t \in \mathcal{T}$ we have two distinguished triangles, one for the Bousfield localisation functor and one for the Bousfield colocalisation functor of the pair $\mathcal{S} \subset \mathcal{T}$. The existence of these triangles was proved in Proposition 9.1.8 (see also Theorem 9.1.13). Concretely, these triangles in $\mathcal{T}$ are

$$i_* i^! t \longrightarrow t \longrightarrow j_* j^* t \longrightarrow \Sigma i_* i^! t$$

$$j_! j^* t \longrightarrow t \longrightarrow i_* i^* t \longrightarrow \Sigma j_! j^* t.$$

All this should be very familiar to the readers who have seen it before. The six functors arise very naturally in algebraic geometry, and give "gluing data". We will see them again in the discussion of t-structures.

9.3. History of the results in Chapter 9

The results of Section 9.1 were undoubtedly all known to Bousfield; see his articles [**6**] and [**5**]. What I have tried to do here is give a clean exposition, pointing out the very formal nature of the argument. The theorems of Section 9.1 have little real content. They simply encapsulate the symmetry that is present whenever we have an adjoint to a Verdier quotient map. Because this does not seem to be written down anywhere, the author felt obliged to summarise the theory briefly.

The "six functors" have a long and venerable history in algebraic geometry. It was an old observation of Grothendieck that a sheaf on the étale site can be pieced together from its restrictions to an open set and its closed complement, using the units and counits of adjunction as in the six functors. The theory also plays an improtant rôle in the construction of perverse t–structures. See Beilinson, Bernstein and Deligne's [**1**].

APPENDIX A

Abelian categories

A.1. Locally presentable categories

One of the crucial properties of the category $\mathcal{E}x(\mathcal{S}^{op}, \mathcal{A}b)$ of Definition 6.1.3 is that it is *locally presentable.* The reader is referred to the book by Gabriel and Ulmer, [**16**], for a very extensive treatment of such categories. In this Section, we content ourselves with reminding the reader of the definitions and very elementary examples of locally presentable categories. Needless to say, nothing in this Section makes any claim to originality, and most of the facts are very easy.

DEFINITION A.1.1. *Let β be an infinite cardinal, $\mathcal{I}$ a category. We say that $\mathcal{I}$ is* β–filtered *if any subcategory of $\mathcal{I}$ of cardinality $< \beta$ is coned off in $\mathcal{I}$. That is, if $I \subset \mathcal{I}$ is a subcategory, and I has fewer than β morphisms, then there exists an object $c \in \mathcal{I}$, and for each $i \in I$ a morphism $i \longrightarrow c$, giving a natural transformation from the inclusion $I \subset \mathcal{I}$ to the collapsing functor, which sends every morphism in I to 1_c.*

DEFINITION A.1.2. *Let $\mathcal{A}$ be a category, closed under the formation of small colimits. Let $\mathcal{I}$ be a small category. A functor $\mathcal{I} \longrightarrow \mathcal{A}$ is said to be* β–filtered *if the category $\mathcal{I}$ is β–filtered. The colimit of a β–filtered functor $\mathcal{I} \longrightarrow \mathcal{A}$ is called a* β–filtered colimit.

LEMMA A.1.3. *Let the notation be as in Chapter 6. The small category $\mathcal{S}$ satisfies Hypothesis 6.1.1, the category $\mathcal{C}at(\mathcal{S}^{op}, \mathcal{A}b)$ is the category of all additive functors $\mathcal{S}^{op} \longrightarrow \mathcal{A}b$, while*

$$\mathcal{E}x(\mathcal{S}^{op}, \mathcal{A}b) \qquad \subset \qquad \mathcal{C}at(\mathcal{S}^{op}, \mathcal{A}b)$$

is the full subcategory of functors which take coproducts of fewer than α objects in $\mathcal{S}$ to products in $\mathcal{A}b$. Let $F : \mathcal{I} \longrightarrow \mathcal{E}x(\mathcal{S}^{op}, \mathcal{A}b)$ be an α–filtered functor. Then its colimits in $\mathcal{E}x(\mathcal{S}^{op}, \mathcal{A}b)$ and $\mathcal{C}at(\mathcal{S}^{op}, \mathcal{A}b)$ agree.

Proof: We need to prove that the colimit in $\mathcal{C}at(\mathcal{S}^{op}, \mathcal{A}b)$ is an object of the subcategory $\mathcal{E}x(\mathcal{S}^{op}, \mathcal{A}b)$. That is, we need to show that it respects coproducts of fewer than α objects. Let $\{x_\lambda, \lambda \in \Lambda\}$ be a set of fewer than

α objects in $\mathcal{S}$. There is in any case a natural map

$$\underrightarrow{\operatorname{colim}}[F]\left(\coprod_{\lambda\in\Lambda} x_\lambda\right) \xrightarrow{\ \phi\ } \prod_{\lambda\in\Lambda} \underrightarrow{\operatorname{colim}}[F](x_\lambda).$$

We need to show that this map is an isomorphism. Let us prove surjectivity first.

Pick an element $\rho \in \prod_{\lambda\in\Lambda} \underrightarrow{\operatorname{colim}}[F](x_\lambda)$. The element $\rho = \prod_{\lambda\in\Lambda}\rho_\lambda$ is, for each $\lambda \in \Lambda$, an element $\rho_\lambda \in \underrightarrow{\operatorname{colim}}[F](x_\lambda)$. Such an element ρ_λ may be lifted to $\big[F(i_\lambda)\big](x_\lambda)$, for some $i_\lambda \in \mathcal{I}$. This is because in the category $\mathcal{C}at\big(\mathcal{S}^{op}, \mathcal{A}b\big)$ the colimit is formed pointwise. We have, for each $\lambda \in \Lambda$, an object $i_\lambda \in \mathcal{I}$. The set Λ has cardinality $< \alpha$ while $\mathcal{I}$ is assumed α–filtered. There must be an object $i \in \mathcal{I}$ coning off the $i_\lambda \in \mathcal{I}$. Therefore our element $\rho \in \prod_{\lambda\in\Lambda} \underrightarrow{\operatorname{colim}}[F](x_\lambda)$ lies in the image of the map

$$\prod_{\lambda\in\Lambda} \big[F(i)\big](x_\lambda) \longrightarrow \prod_{\lambda\in\Lambda} \underrightarrow{\operatorname{colim}}[F](x_\lambda).$$

On the other hand, $F(i)$ lies in $\mathcal{E}x\big(\mathcal{S}^{op}, \mathcal{A}b\big)$, and hence the vertical map below is an isomrphism

$$\begin{array}{ccc}
\big[F(i)\big]\left(\coprod_{\lambda\in\Lambda} x_\lambda\right) & & \\
\downarrow & & \\
\prod_{\lambda\in\Lambda} \big[F(i)\big](x_\lambda) & \longrightarrow & \prod_{\lambda\in\Lambda} \underrightarrow{\operatorname{colim}}[F](x_\lambda)
\end{array}$$

We have therefore shown that ρ lies in the image of the composite above. On the other hand, the square

$$\begin{array}{ccc}
\big[F(i)\big]\left(\coprod_{\lambda\in\Lambda} x_\lambda\right) & \longrightarrow & \underrightarrow{\operatorname{colim}}[F]\left(\coprod_{\lambda\in\Lambda} x_\lambda\right) \\
\downarrow & & \downarrow{\scriptstyle \phi} \\
\prod_{\lambda\in\Lambda} \big[F(i)\big](x_\lambda) & \longrightarrow & \prod_{\lambda\in\Lambda} \underrightarrow{\operatorname{colim}}[F](x_\lambda)
\end{array}$$

commutes, showing that ρ lies in the image ϕ. This proves that ϕ is surjective.

Next we need the injectivity of ϕ. Suppose we have an element in the kernel of ϕ; that is, we have an element in $\rho \in \underrightarrow{\operatorname{colim}}[F]\left(\coprod_{\lambda\in\Lambda} x_\lambda\right)$ mapping to zero under ϕ. By the pointwise nature of colimits in $\mathcal{C}at\big(\mathcal{S}^{op}, \mathcal{A}b\big)$, there is an $i \in \mathcal{I}$ so that ρ lifts via

$$[F(i)]\left(\coprod_{\lambda\in\Lambda} x_\lambda\right) \longrightarrow \operatorname{\underrightarrow{colim}}[F]\left(\coprod_{\lambda\in\Lambda} x_\lambda\right);$$

there exists an $\tau \in [F(i)]\left(\coprod_{\lambda\in\Lambda} x_\lambda\right)$ mapping to ρ. But by hypothesis ρ is in the kernel of ϕ; therefore τ is in the kernel of the equal composites

$$\begin{array}{ccc}
[F(i)]\left(\coprod_{\lambda\in\Lambda} x_\lambda\right) & \longrightarrow & \operatorname{\underrightarrow{colim}}[F]\left(\coprod_{\lambda\in\Lambda} x_\lambda\right) \\
\downarrow & & \phi\downarrow \\
\prod_{\lambda\in\Lambda}[F(i)](x_\lambda) & \longrightarrow & \prod_{\lambda\in\Lambda}\operatorname{\underrightarrow{colim}}[F](x_\lambda)
\end{array}$$

and in particular its image maps to zero by

$$\prod_{\lambda\in\Lambda}[F(i)](x_\lambda) \longrightarrow \prod_{\lambda\in\Lambda}\operatorname{\underrightarrow{colim}}[F](x_\lambda).$$

For every $\lambda \in \Lambda$ there must exist a morphism $i \to i_\lambda$ in $\mathcal{I}$, so that the image of τ under the composite

$$\begin{array}{ccc}
[F(i)]\left(\coprod_{\lambda\in\Lambda} x_\lambda\right) & & \\
\downarrow & & \\
\prod_{\lambda\in\Lambda}[F(i)](x_\lambda) & \longrightarrow & \prod_{\lambda\in\Lambda}[F(i_\lambda)](x_\lambda)
\end{array}$$

already vanishes. But the subcategory of $\mathcal{I}$ containing all the morphisms $i \to i_\lambda$ has cardinality $< \alpha$, and $\mathcal{I}$ is assumed α–filtered. It follows that the subcategory may be coned off in $\mathcal{I}$. There exists an object $j \in \mathcal{I}$ coning it off, and the composite

$$\begin{array}{ccc}
[F(i)]\left(\coprod_{\lambda\in\Lambda} x_\lambda\right) & & \\
\downarrow & & \\
\prod_{\lambda\in\Lambda}[F(i)](x_\lambda) & \longrightarrow & \prod_{\lambda\in\Lambda}[F(j)](x_\lambda)
\end{array}$$

annihilates τ.

But the commutative square

$$\begin{array}{ccc} [F(i)]\left(\coprod_{\lambda\in\Lambda} x_\lambda\right) & \longrightarrow & [F(j)]\left(\coprod_{\lambda\in\Lambda} x_\lambda\right) \\ \downarrow & & \downarrow{\scriptstyle h} \\ \prod_{\lambda\in\Lambda} [F(i)](x_\lambda) & \longrightarrow & \prod_{\lambda\in\Lambda} [F(j)](x_\lambda) \end{array}$$

tells us that either composite annihilates τ. Since $F(j)$ lies in $\mathcal{E}x\big(\mathcal{S}^{op}, \mathcal{A}b\big)$, the map h in the square above is an isomorphism, and hence the image of τ dies already in $\big[F(j)\big]\big(\coprod_{\lambda\in\Lambda} x_\lambda\big)$. But that means that ρ, which is the image of τ by the composite

$$[F(i)]\left(\coprod_{\lambda\in\Lambda} x_\lambda\right) \longrightarrow [F(j)]\left(\coprod_{\lambda\in\Lambda} x_\lambda\right) \longrightarrow \underrightarrow{\operatorname{colim}}[F]\left(\coprod_{\lambda\in\Lambda} x_\lambda\right)$$

vanishes. The kernel of ϕ is trivial and ϕ is injective. □

Definition A.1.4. *Let $\mathcal{A}$ be a category, closed under the formation of small colimits. An object $a \in \mathcal{A}$ is called β–*presentable *if $\mathcal{A}(a, -)$ commutes with β–filtered colimits. That is, given any β–filtered functor $F : \mathcal{I} \longrightarrow \mathcal{A}$, the natural map*

$$\underrightarrow{\operatorname{colim}}\big[\mathcal{A}(a, Fi)\big] \longrightarrow \mathcal{A}\Big(a, \underrightarrow{\operatorname{colim}}\big[F(i)\big]\Big)$$

is an isomorphism.

Definition A.1.5. *Let $\mathcal{A}$ be a category, closed under the formation of small colimits. The category $\mathcal{A}$ is called* locally presentable *if every object $a \in \mathcal{A}$ is β–presentable for some infinite cardinal β. The cardinal β for which the object a is β–presentable in general depends on a; it should perhaps be denoted $\beta(a)$.*

Remark A.1.6. The careful reader will note that our definition of locally presentable categories is slightly different from Gabriel and Ulmer's. Unlike Gabriel and Ulmer's [**16**], we do not postulate that locally presentable categories must have a generator.

Lemma A.1.7. *The category $\mathcal{A}b$ of abelian groups is locally presentable.*

Proof: Let a be an abelian group, and choose any regular cardinal β greater than the cardinality of a. We assert that a is β–presentable. For suppose $F : \mathcal{I} \longrightarrow \mathcal{A}b$ is a β–filtered functor. There is a natural map

$$\underrightarrow{\operatorname{colim}}\big[\mathcal{A}(a, Fi)\big] \stackrel{\phi}{\longrightarrow} \mathcal{A}\Big(a, \underrightarrow{\operatorname{colim}}\big[F(i)\big]\Big)$$

and we need to show it an isomorphism. We begin by showing surjectivity.

Choose therefore an element in $\mathcal{A}\Big(a, \underrightarrow{\text{colim}}\big[F(i)\big]\Big)$, that is a map $\rho : a \longrightarrow \underrightarrow{\text{colim}}\big[F(i)\big]$. Recall that a is an abelian group of cardinality $< \beta$. For each element $\lambda \in a$ choose an $i_\lambda \in \mathcal{I}$ and an element $p_\lambda \in F(i_\lambda)$ so that the image of p_λ via

$$F(i_\lambda) \longrightarrow \underrightarrow{\text{colim}}\big[F(i)\big]$$

is the same as the image of $\lambda \in a$ via

$$a \xrightarrow{\ \rho\ } \underrightarrow{\text{colim}}\big[F(i)\big].$$

For each pair $(\lambda, \mu) \in a \times a$ we have that the image of $p_{\lambda-\mu} \in F(i_{\lambda-\mu})$ via

$$F(i_{\lambda-\mu}) \longrightarrow \underrightarrow{\text{colim}}\big[F(i)\big]$$

is the difference of the images of $p_\lambda \in F(i_\lambda)$ and $p_\mu \in F(i_\mu)$ via

$$F(i_\lambda) \longrightarrow \underrightarrow{\text{colim}}\big[F(i)\big]$$

$$F(i_\mu) \longrightarrow \underrightarrow{\text{colim}}\big[F(i)\big].$$

For each pair $(\lambda, \mu) \in a \times a$ we may choose an $i_{(\lambda,\mu)} \in \mathcal{I}$ with maps

$$\begin{array}{ccccc} & & i_{\lambda-\mu} & & \\ & & \downarrow & & \\ i_\lambda & \longrightarrow & i_{(\lambda,\mu)} & \longleftarrow & i_\mu \end{array}$$

so that the equality is realised already in $F\big(i_{(\lambda,\mu)}\big)$. Since the set $a \times a$ has cardinality $< \beta$, there exists an $i \in \mathcal{I}$ coning off all the objects and morphisms we have picked. It follows that the map

$$\rho : a \longrightarrow \underrightarrow{\text{colim}}\big[F\big]$$

factors through $F(i)$. Thus ρ lies in the image of

$$\mathcal{A}(a, Fi) \longrightarrow \mathcal{A}\Big(a, \underrightarrow{\text{colim}}\big[F(i)\big]\Big).$$

Since this map factors as

$$\mathcal{A}(a, Fi) \longrightarrow \underrightarrow{\text{colim}}\big[\mathcal{A}(a, Fi)\big] \xrightarrow{\ \phi\ } \mathcal{A}\Big(a, \underrightarrow{\text{colim}}\big[F(i)\big]\Big)$$

it follows that ρ lies in the image of ϕ. This proves ϕ surjective.

Next we need the injectivity of ϕ. Suppose we have an element of the kernel of ϕ. That is, we are given an element

$$\rho \quad \in \quad \underrightarrow{\text{colim}}\big[\mathcal{A}(a, Fi)\big]$$

which is annihilated by ϕ. First, ρ may be lifted via

$$\mathcal{A}(a, Fi) \longrightarrow \underrightarrow{\text{colim}}\big[\mathcal{A}(a, Fi)\big]$$

for some $i \in \mathcal{I}$. There exists $\tau : a \longrightarrow F(i)$ whose image is ρ. But the fact that ϕ annihilates ρ means that the composite

$$a \longrightarrow F(i) \longrightarrow \underrightarrow{\mathrm{colim}}[F]$$

vanishes. Pick $\lambda \in a$. The image of λ vanishes via

$$a \longrightarrow F(i) \longrightarrow \underrightarrow{\mathrm{colim}}[F],$$

and hence we may choose a morphism $i \to i_\lambda$ so that the composite

$$a \longrightarrow F(i) \longrightarrow F(i_\lambda)$$

annihilates λ. Choose such an $i \to i_\lambda$ for each $\lambda \in a$. The cardinality of a is less than β. Since $\mathcal{I}$ is β–filtered, the choices we have made may be coned off, and there exists a morphism $i \longrightarrow j$ in $\mathcal{I}$ so that every $\lambda \in a$ vanishes under the composite

$$a \longrightarrow F(i) \longrightarrow F(j).$$

But then the map

$$\mathcal{A}(a, Fi) \longrightarrow \mathcal{A}(a, Fj)$$

kills $\tau \in \mathcal{A}(a, Fi)$. Thus $\tau \in \mathcal{A}(a, Fi)$ must die under the longer composite

$$\mathcal{A}(a, Fi) \longrightarrow \mathcal{A}(a, Fj) \longrightarrow \underrightarrow{\mathrm{colim}}\big[\mathcal{A}(a, Fi)\big]$$

but here its image is ρ. This proves $\rho = 0$, and the kernel of ϕ is trivial. □

LEMMA A.1.8. *Let $\mathcal{S}$ be an essentially small additive category. Let $\mathcal{C}at\big(\mathcal{S}^{op}, \mathcal{A}b\big)$ be the category of all additive functors $\mathcal{S}^{op} \longrightarrow \mathcal{A}b$. Then $\mathcal{C}at\big(\mathcal{S}^{op}, \mathcal{A}b\big)$ is locally presentable.*

Proof: Colimits in $\mathcal{C}at\big(\mathcal{S}^{op}, \mathcal{A}b\big)$ are formed pointwise, and hence this reduces immediately to the case of Lemma A.1.7. □

PROPOSITION A.1.9. *With the notation as in Chapter 6, the category $\mathcal{E}x\big(\mathcal{S}^{op}, \mathcal{A}b\big)$ is locally presentable.*

Proof: Let a be an object of $\mathcal{E}x\big(\mathcal{S}^{op}, \mathcal{A}b\big)$. By Lemma A.1.8 there is an infinite cardinal β so that a is β–presentable as an object of the larger category $\mathcal{C}at\big(\mathcal{S}^{op}, \mathcal{A}b\big)$. That is, the functor $\mathrm{Hom}(a, -)$ commutes with β–filtered colimits, as long as the colimits are understood in the category $\mathcal{C}at\big(\mathcal{S}^{op}, \mathcal{A}b\big)$.

But by Lemma A.1.3, if $\beta > \alpha$, β–filtered colimits in $\mathcal{E}x\big(\mathcal{S}^{op}, \mathcal{A}b\big)$ agree with the same colimits in $\mathcal{C}at\big(\mathcal{S}^{op}, \mathcal{A}b\big)$. Replace β by $\beta + \alpha$; then $\mathrm{Hom}(a, -)$ commutes with β–filtered colimits, even in the category $\mathcal{E}x\big(\mathcal{S}^{op}, \mathcal{A}b\big)$. □

REMARK A.1.10. Everything in this Section, and much more about locally presentable categories, may be found in Gabriel and Ulmer's [**16**].

In Remark A.1.6, we noted that our definition of locally presentable categories differs slightly from Gabriel and Ulmer's. Unlike Gabriel and Ulmer's [**16**], we do not require that locally presentable categories have a generator. Of course, that categories $\mathcal{E}x(\mathcal{S}^{op}, \mathcal{A}b)$ in fact do, and hence they are locally presentable even in the more restrictive sense of Gabriel and Ulmer.

A.2. Formal properties of quotients

We begin by recalling the definition of a Serre subcategory of an abelian category.

DEFINITION A.2.1. *Let $\mathcal{A}$ be an abelian category. Let $\mathcal{B}$ be a full subcategory. The subcategory $\mathcal{B} \subset \mathcal{A}$ is called a* Serre subcategory *if*

A.2.1.1. *Every object of $\mathcal{A}$ isomorphic to an object in $\mathcal{B}$ is in $\mathcal{B}$.*

A.2.1.2. *Every $\mathcal{A}$–quotient object and every $\mathcal{A}$–subobject of an object in $\mathcal{B}$ lies in $\mathcal{B}$.*

A.2.1.3. *Every $\mathcal{A}$–extension of objects in $\mathcal{B}$ lies in $\mathcal{B}$.*

If $\mathcal{B} \subset \mathcal{A}$ is a Serre subcategory, it follows immediately from the definition that $\mathcal{B}$ is an abelian category and the inclusion in $\mathcal{A}$ is an exact functor. After all, if $f : x \longrightarrow y$ is a morphism in $\mathcal{B}$, then $\mathrm{Im}(f)$ is a subobject of y, $\mathrm{Coker}(f)$ ia a quotient object of y and $\mathrm{Ker}(f)$ is a subobject of x. By A.2.1.2 all three must lie in $\mathcal{B}$.

Given an abelian category $\mathcal{A}$ and a Serre subcategory $\mathcal{B} \subset \mathcal{A}$, one can form the quotient $\mathcal{A}/\mathcal{B}$ as follows

DEFINITION A.2.2. *The category $\mathcal{A}/\mathcal{B}$ is defined as follows:*

A.2.2.1. *The objects of $\mathcal{A}/\mathcal{B}$ are the same as the objects of $\mathcal{A}$.*

A.2.2.2. *The morphisms are given by*

$$\mathcal{A}/\mathcal{B}(x,y) \quad = \quad \operatorname*{colim}_{\substack{x' \subset x, \quad x/x' \in \mathcal{B} \\ y''=y/y', \quad y' \in \mathcal{B}}} \mathcal{A}(x', y'').$$

That is, a morphism $x \longrightarrow y$ in $\mathcal{A}/\mathcal{B}$ is an equivalence class of diagrams in $\mathcal{A}$

$$\begin{array}{ccccccccc} & & 0 & \longrightarrow & x' & \longrightarrow & x & \longrightarrow & x'' \longrightarrow 0 \\ & & & & \downarrow & & & & \\ 0 & \longrightarrow & y' & \longrightarrow & y & \longrightarrow & y'' & \longrightarrow & 0 \end{array}$$

where the rows are exact and $x'', y' \in \mathcal{B}$.

The natural functor $\mathcal{A} \longrightarrow \mathcal{A}/\mathcal{B}$ is the identity on objects, and sends a morphism to its equivalence class.

The following lemma may be found in [**15**], more precisely Corollaire 2, in Paragraphe 1 of Chapitre III; see page 368.

LEMMA A.2.3. *The category $\mathcal{A}/\mathcal{B}$ is an abelian category. The functor $F : \mathcal{A} \longrightarrow \mathcal{A}/\mathcal{B}$ is exact, and takes the objects of $\mathcal{B}$ to objects in $\mathcal{A}/\mathcal{B}$ isomorphic to zero. Furthermore, F is universal with this property. The subcategory $\mathcal{B} \subset \mathcal{A}$ is the full subcategory of all objects $b \in \mathcal{A}$ so that Fb is isomorphic to zero.*

REMARK A.2.4. Unlike in the case of triangulated categories, which we studied in Chapter 2, there is usually no set theoretic problem in forming $\mathcal{A}/\mathcal{B}$. If $\mathcal{A}$ is a well–powered abelian category with small *Hom*–sets, then the objects x and y have sets of subobjects and quotient objects, and hence the colimit in Definition A.2.2.2 is a colimit of sets indexed over a set; it is a set. Thus $\mathcal{A}/\mathcal{B}$ will have small *Hom*–sets.

As with triangulated categories, it is very interesting to study the situation in which the quotient map $\mathcal{A} \longrightarrow \mathcal{A}/\mathcal{B}$ has an adjoint, right or left. We will for now restrict attention to right adjoints; the case of a left adjoint is dual. The Serre subcategory $\mathcal{B} \subset \mathcal{A}$ is called *localizant* if the functor $\mathcal{A} \longrightarrow \mathcal{A}/\mathcal{B}$ has a right adjoint.

LEMMA A.2.5. *Let $F : \mathcal{A} \longrightarrow \mathcal{T}$ be an exact functor of abelian categories, $G : \mathcal{T} \longrightarrow \mathcal{A}$ its right adjoint. Let $\mathcal{B} \subset \mathcal{A}$ be the full subcategory of all objects $b \in \mathcal{A}$ so that Fb is isomorphic to zero. If an object $y \in \mathcal{A}$ is isomorphic to Gt for some $t \in \mathcal{T}$, then*

A.2.5.1. *For every object $b \in \mathcal{B}$, $\mathcal{A}(b, y) = 0$.*

A.2.5.2. *For every object $b \in \mathcal{B}$, $\mathrm{Ext}^1_{\mathcal{A}}(b, y) = 0$.*

Proof: Suppose $t \in \mathcal{T}$, $y = Gt$. Then by the adjunction, for all $x \in \mathcal{A}$

$$\mathcal{A}(x, y) \quad = \quad \mathcal{A}(x, Gt) \quad = \quad \mathcal{T}(Fx, t).$$

In particular, if $x = b \in \mathcal{B}$, then $Fb = 0$ and we deduce the isomorphisms $\mathcal{A}(b, y) = \mathcal{T}(Fb, t) = 0$.

Suppose z is an extension of $y = Gt$ by some object $b \in \mathcal{B}$. That is, we have an exact sequence in $\mathcal{A}$

$$0 \longrightarrow y \longrightarrow z \longrightarrow b \longrightarrow 0.$$

Applying F to it, we get an exact sequence in $\mathcal{T}$

$$0 \longrightarrow Fy \longrightarrow Fz \longrightarrow Fb \longrightarrow 0,$$

and since $Fb = 0$ we must have that $Fy \longrightarrow Fz$ is an isomorphism. But then

$$\mathcal{T}(Fz, t) \longrightarrow \mathcal{T}(Fy, t)$$

is also an isomorphism. Consider the commutative square below

$$\begin{array}{ccc} \mathcal{A}(z, Gt) & \longrightarrow & \mathcal{A}(y, Gt) \\ \wr\downarrow & & \wr\downarrow \\ \mathcal{T}(Fz, t) & \xrightarrow[\ \ ]{\sim} & \mathcal{T}(Fy, t). \end{array}$$

The vertical maps are isomorphisms by adjunction. The bottom row is an isomorphism by the above. Hence so is the top row. Recalling that $Gt = y$, we have that the identity $1 : y \longrightarrow y$ must be in the image of $\mathcal{A}(z, y)$; there is a map $z \longrightarrow y$ so that the composite $y \longrightarrow z \longrightarrow y$ is the identity. The short exact sequence

$$0 \longrightarrow y \longrightarrow z \longrightarrow b \longrightarrow 0$$

splits, and $\mathrm{Ext}^1_{\mathcal{A}}(b, y) = 0$. □

DEFINITION A.2.6. *Let $\mathcal{A}$ be an abelian category, $\mathcal{B} \subset \mathcal{A}$ a Serre subcategory. An object $y \in \mathcal{A}$ will be called* $\mathcal{B}$–local *if the following two conditions hold, as in Lemma A.2.5.*

A.2.5.1: *For every object $b \in \mathcal{B}$, $\mathcal{A}(b, y) = 0$.*
A.2.5.2: *For every object $b \in \mathcal{B}$, $\mathrm{Ext}^1_{\mathcal{A}}(b, y) = 0$.*

Lemma A.2.5 asserts that, if $F : \mathcal{A} \longrightarrow \mathcal{T}$ is exact and has a right adjoint G, then every $y = Gt$ is $\mathcal{B}$–local, where $\mathcal{B}$ is the full subcategory of all $b \in \mathcal{A}$ with $Fb \simeq 0$.

LEMMA A.2.7. *Let $\mathcal{A}$ be an abelian category, $\mathcal{B} \subset \mathcal{A}$ a Serre subcategory. If $y \in \mathcal{A}$ is a $\mathcal{B}$–local object, then for every $x \in \mathcal{A}$*

$$\mathcal{A}(x, y) \quad = \quad \mathcal{A}/\mathcal{B}(x, y).$$

Proof: Let y be a $\mathcal{B}$–local object in $\mathcal{A}$. There is a natural map

$$\mathcal{A}(x, y) \longrightarrow \mathcal{A}/\mathcal{B}(x, y);$$

we need to prove it an isomorphism. Before we start, let us make one helpful observation. Recall that a map in $\mathcal{A}/\mathcal{B}(x, y)$ is an equivalence class of diagrams in $\mathcal{A}$

$$\begin{array}{ccccccccc} & & & & 0 & \longrightarrow & x' & \longrightarrow & x & \longrightarrow & x'' & \longrightarrow & 0 \\ & & & & & & \downarrow & & & & & & \\ 0 & \longrightarrow & y' & \longrightarrow & y & \longrightarrow & y'' & \longrightarrow & 0 & & & & \end{array}$$

where the rows are exact and $x'', y' \in \mathcal{B}$. Since y is $\mathcal{B}$–local, by A.2.5.1 any map $b \longrightarrow y$, $b \in \mathcal{B}$ must vanish. But we are given $y' \in \mathcal{B}$, and hence the monomorphism $y' \longrightarrow y$ vanishes. Therefore $y' = 0$, and $y = y''$. In other words, if y is $\mathcal{B}$–local, morphisms $x \longrightarrow y$ in $\mathcal{A}/\mathcal{B}$ are equivalence classes of diagrams

$$\begin{array}{ccc} x' & \longrightarrow & y \\ \downarrow & & \\ x & & \end{array}$$

In the larger diagrams

$$\begin{array}{ccc} & & y \\ & & \downarrow \\ x' & \longrightarrow & y'' \\ \downarrow & & \\ x & & \end{array}$$

the map $y \longrightarrow y''$ must be an isomorphism.

Next we want to prove the map

$$\mathcal{A}(x,y) \longrightarrow \mathcal{A}/\mathcal{B}(x,y)$$

an isomorphism. Let us first prove surjectivity. Take any map in $\mathcal{A}/\mathcal{B}(x,y)$, that is a diagram

$$\begin{array}{ccc} x' & \longrightarrow & y \\ \downarrow & & \\ x & & \end{array}$$

as above. Now the map $x' \longrightarrow x$ is mono, so we may push out to get a bicartesian square

$$\begin{array}{ccc} x' & \longrightarrow & y \\ \downarrow & & \downarrow \\ x & \longrightarrow & z \end{array}$$

and this extends to a map of exact sequences

$$\begin{array}{ccccccccc} 0 & \longrightarrow & x' & \longrightarrow & x & \longrightarrow & x'' & \longrightarrow & 0 \\ & & \downarrow & & \downarrow & & \downarrow \wr & & \\ 0 & \longrightarrow & y & \longrightarrow & z & \longrightarrow & x'' & \longrightarrow & 0 \end{array}$$

But $x'' \in \mathcal{B}$, and by A.2.5.2, $\mathrm{Ext}^1_{\mathcal{A}}(x'',y) = 0$. In other words, the exact sequence

$$0 \longrightarrow y \longrightarrow z \longrightarrow x'' \longrightarrow 0$$

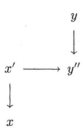

splits, and using any splitting we have a commutative diagram

$$\begin{array}{ccccc} x' & \longrightarrow & y & & \\ \downarrow & & \downarrow & & \\ x & \longrightarrow & z & \longrightarrow & y \end{array}$$

which gives us a map $x \longrightarrow y$ in $\mathcal{A}$, in the equivalence class of

$$\begin{array}{ccc} x' & \longrightarrow & y \\ \downarrow & & \\ x & & \end{array}$$

The map

$$\mathcal{A}(x,y) \longrightarrow \mathcal{A}/\mathcal{B}(x,y)$$

is surjective.

Next we want to prove it injective. Suppose we have a map $x \longrightarrow y$ in $\mathcal{A}$, which becomes zero in $\mathcal{A}/\mathcal{B}$. This means that in the equivalence class there is a diagram

$$\begin{array}{ccc} x' & \longrightarrow & y \\ \downarrow & & \\ x & & \end{array}$$

where $x' \longrightarrow y$ vanishes. More precisely, this means that for $x' \subset x$ with $x/x' \in \mathcal{B}$, the composite $x' \subset x \longrightarrow y$ vanishes. But then the exact sequence

$$0 \longrightarrow x' \longrightarrow x \longrightarrow x'' \longrightarrow 0$$

tells us that $x \longrightarrow y$ must factor through $x \longrightarrow x'' \longrightarrow y$. By hypothesis, $x/x' = x'' \in \mathcal{B}$ and y is $\mathcal{B}$–local; hence $x'' \longrightarrow y$ vanishes, and therefore so does the composite $x \longrightarrow x'' \longrightarrow y$. □

LEMMA A.2.8. *Let $\mathcal{A}$ be an abelian category, $\mathcal{B} \subset \mathcal{A}$ a localizant subcategory. Let $F : \mathcal{A} \longrightarrow \mathcal{A}/\mathcal{B}$ be the projection, G its right adjoint. The counit of adjunction $FG \Longrightarrow 1$ is an isomorphism.*

Proof: Let $x \in \mathcal{A}$, $y \in \mathcal{A}/\mathcal{B}$. Then by adjunction

$$\mathcal{A}/\mathcal{B}(Fx,y) \quad = \quad \mathcal{A}\big(x,Gy\big).$$

By Lemma A.2.5, Gy is $\mathcal{B}$–local. By Lemma A.2.7,

$$\mathcal{A}(x,Gy) \quad = \quad \mathcal{A}/\mathcal{B}\big(Fx,FGy\big).$$

Hence

$$\mathcal{A}/\mathcal{B}(Fx,y) \quad = \quad \mathcal{A}\big(x,Gy\big) \quad = \quad \mathcal{A}/\mathcal{B}\big(Fx,FGy\big).$$

In other words, the natural map

$$\varepsilon_y : FGy \longrightarrow y$$

induces, for every $x \in \mathcal{A}$, an isomorphism

$$\mathcal{A}/\mathcal{B}(Fx, FGy) \longrightarrow \mathcal{A}/\mathcal{B}(Fx, y).$$

Since every object of $\mathcal{A}/\mathcal{B}$ is of the form Fx, it follows that the map

$$\varepsilon_y : FGy \longrightarrow y$$

must be an isomorphism in $\mathcal{A}/\mathcal{B}$. □

LEMMA A.2.9. *Let $\mathcal{A}$ and $\mathcal{T}$ be any categories. Let $F : \mathcal{A} \longrightarrow \mathcal{T}$ be a functor, $G : \mathcal{T} \longrightarrow \mathcal{A}$ its right adjoint. The counit of adjunction*

$$\varepsilon : FG \longrightarrow 1$$

is an isomorphism if and only if G is fully faithful.

Proof: For any $x, y \in \mathcal{T}$, composition with the counit gives a map

$$\mathcal{T}(x, y) \xrightarrow{\mathcal{T}(\varepsilon_x, y)} \mathcal{T}(FGx, y),$$

and by adjunction

$$\mathcal{T}(FGx, y) \quad = \quad \mathcal{A}(Gx, Gy).$$

The composite is the natural map

$$\mathcal{T}(x, y) \longrightarrow \mathcal{A}(Gx, Gy);$$

it is an isomorphism if and only if

$$\mathcal{T}(x, y) \xrightarrow{\mathcal{T}(\varepsilon_x, y)} \mathcal{T}(FGx, y)$$

is an isomorphism. To say that

$$\mathcal{T}(x, y) \longrightarrow \mathcal{A}(Gx, Gy)$$

is an isomorphism for every x and y is to say that G is fully faithful, while

$$\mathcal{T}(x, y) \xrightarrow{\mathcal{T}(\varepsilon_x, y)} \mathcal{T}(FGx, y)$$

will be an isomorphism for every x and y if and only if

$$\varepsilon : FG \longrightarrow 1$$

is an isomorphism. □

PROPOSITION A.2.10. *Let $\mathcal{A}$ be an abelian category, $\mathcal{B} \subset \mathcal{A}$ a localizant subcategory. Let $F : \mathcal{A} \longrightarrow \mathcal{A}/\mathcal{B}$ be the projection to the quotient, $G : \mathcal{A}/\mathcal{B} \longrightarrow \mathcal{A}$ its right adjoint. The functor $G : \mathcal{A}/\mathcal{B} \longrightarrow \mathcal{A}$ is fully faithful, the counit of adjunction*

$$\varepsilon : FG \longrightarrow 1$$

is an isomorphism, and the objects isomorphic to $Gt, t \in \mathcal{A}/\mathcal{B}$ *are exactly the* $\mathcal{B}$*–local objects. More precisely, if* y *is a* $\mathcal{B}$*–local object of* $\mathcal{A}$*, then the unit of adjunction*

$$\eta_y : y \longrightarrow GFy$$

is an isomorphism.

Proof: By Lemma A.2.8, the counit of adjunction is an isomorphism

$$\varepsilon : FG \longrightarrow 1.$$

By Lemma A.2.9 this means that G is fully faithful. It only remains to show that if $y \in \mathcal{A}$ is $\mathcal{B}$–local, then the unit of adjunction

$$\eta_y : y \longrightarrow GFy$$

is an isomorphism.

In any case, we have that the composite

$$Fy \xrightarrow{F\eta_y} FGFy \xrightarrow{\varepsilon_{Fy}} Fy$$

is the identity, while the second map is an isomorphism by Lemma A.2.8. Therefore the first must be its two–sided inverse;

$$F\eta_y : Fy \longrightarrow FGFy$$

is an isomorphism.

Because y is $\mathcal{B}$–local, Lemma A.2.7 tells us that for any $x \in \mathcal{T}$,

$$\mathcal{A}(x, y) \quad = \quad \mathcal{A}/\mathcal{B}\big(Fx, Fy\big).$$

In particular, if we let $x = GFy$,

$$\mathcal{A}(GFy, y) \longrightarrow \mathcal{A}/\mathcal{B}\big(FGFy, Fy\big)$$

is an isomorphism. There exists a unique map

$$F^{-1}\varepsilon_{Fy} : GFy \longrightarrow y$$

lifting

$$\varepsilon_{Fy} : FGFy \longrightarrow Fy.$$

The fact that y is $\mathcal{B}$–local means also that the map

$$\mathcal{A}(y, y) \longrightarrow \mathcal{A}/\mathcal{B}\big(Fy, Fy\big)$$

is an isomorphism. The composite

$$y \xrightarrow{\eta_y} GFy \xrightarrow{F^{-1}\varepsilon_{Fy}} y$$

is a lifting of

$$Fy \xrightarrow{F\eta_y} FGFy \xrightarrow{\varepsilon_{Fy}} Fy,$$

and the latter is $1 : Fy \longrightarrow Fy$. We deduce that

$$y \xrightarrow{\eta_y} GFy \xrightarrow{F^{-1}\varepsilon_{Fy}} y$$

must be the identity. Finally, the object GFy is $\mathcal{B}$–local by Lemma A.2.5, and so by Lemma A.2.7 the map

$$\mathcal{A}(GFy, GFy) \longrightarrow \mathcal{A}/\mathcal{B}\big(FGFy, FGFy\big)$$

is an isomorphism. The composite

$$GFy \xrightarrow{F^{-1}\varepsilon_{Fy}} y \xrightarrow{\eta_y} GFy$$

maps to the identity under F, hence must be the identity. This establishes that η_y has a two–sided inverse, namely $F^{-1}\varepsilon_{Fy}$. □

Proposition A.2.10 can also be turned into a construction of the adjoint. We have

PROPOSITION A.2.11. *Let $\mathcal{A}$ be an abelian category, $\mathcal{B} \subset \mathcal{A}$ a Serre subcategory. Let $F : \mathcal{A} \longrightarrow \mathcal{A}/\mathcal{B}$ be the projection to the quotient. A right adjoint $G : \mathcal{A}/\mathcal{B} \longrightarrow \mathcal{A}$ will exist if and only if for every object $t \in \mathcal{A}$ there is a morphism $t \longrightarrow y$ in $\mathcal{A}$ such that*

A.2.11.1. *y is $\mathcal{B}$–local.*

A.2.11.2. *$Ft \longrightarrow Fy$ is an isomorphism in $\mathcal{A}/\mathcal{B}$.*

Proof: If there is a right adjoint $G : \mathcal{A}/\mathcal{B} \longrightarrow \mathcal{A}$, then consider the unit of adjunction $\eta_t : t \longrightarrow GFt$. By Lemma A.2.5 the object GFt is $\mathcal{B}$–local, while Lemma A.2.7 tells us that $\varepsilon_{Ft} : FGFt \longrightarrow Ft$ is an isomorphism. The composite

$$Ft \xrightarrow{F\eta_t} FGFt \xrightarrow{\varepsilon_{Ft}} Ft$$

is the identity, making $F\eta_t$ the two–sided inverse of the invertible ε_{Ft}.

Now we need to prove the converse. Suppose every object $t \in \mathcal{A}$ admits a map $t \longrightarrow y$ satisfying A.2.11.1 and A.2.11.2. We need to show that the functor $F : \mathcal{A} \longrightarrow \mathcal{A}/\mathcal{B}$ has a right adjoint G. Let t be any object of $\mathcal{A}/\mathcal{B}$, which is the same as an object of $\mathcal{A}$. We wish to show that the functor

$$x \quad \mapsto \quad \mathcal{A}/\mathcal{B}\big(Fx, Ft\big)$$

is representable, as a functor on $\mathcal{A}$. But we can choose a map $t \longrightarrow y$ satisfying A.2.11.1 and A.2.11.2. By A.2.11.2 the map $Ft \longrightarrow Fy$ is an isomorphism. Hence

$$\mathcal{A}/\mathcal{B}\big(Fx, Ft\big) \quad = \quad \mathcal{A}/\mathcal{B}\big(Fx, Fy\big).$$

But by A.2.11.1 the object $y \in \mathcal{A}$ is $\mathcal{B}$–local. Hence Lemma A.2.7 tells us that

$$\mathcal{A}/\mathcal{B}\big(Fx, Fy\big) \quad = \quad \mathcal{A}(x, y).$$

This proves the representability. □

PROPOSITION A.2.12. *Let $F : \mathcal{A} \longrightarrow \mathcal{T}$ be an exact functor of abelian categories, and $G : \mathcal{T} \longrightarrow \mathcal{A}$ its right adjoint. If the counit of adjunction*

$$\varepsilon : FG \longrightarrow 1$$

is an isomorphism, then $\mathcal{T} = \mathcal{A}/\mathcal{B}$ for a localizant subcategory $\mathcal{B} \subset \mathcal{A}$, and F is the projection. In fact, $\mathcal{B}$ is the full subcategory of all $b \in \mathcal{A}$ with Fb isomorphic to zero.

Proof: Define $\mathcal{B}$ to be the full subcategory of all $b \in \mathcal{A}$ with Fb isomorphic to zero. Observe that every object isomorphic to an object in $\mathcal{B}$ is in $\mathcal{B}$. Furthermore, if

$$0 \longrightarrow x' \longrightarrow x \longrightarrow x'' \longrightarrow 0$$

is an exact sequence in $\mathcal{A}$, the fact that the functor F is exact guarantees that

$$0 \longrightarrow Fx' \longrightarrow Fx \longrightarrow Fx'' \longrightarrow 0$$

is exact in $\mathcal{T}$. But then Fx is isomorphic to zero if and only if both Fx' and Fx'' are. In other words, $x \in \mathcal{B}$ if and only if $x', x'' \in \mathcal{B}$. The category $\mathcal{B}$ is a Serre subcategory of $\mathcal{A}$; see Definition A.2.1. One can form the universal quotient functor $\mathcal{A} \longrightarrow \mathcal{A}/\mathcal{B}$. The exact functor $F : \mathcal{A} \longrightarrow \mathcal{T}$ factors (uniquely) as a composite of exact functors

$$\mathcal{A} \longrightarrow \mathcal{A}/\mathcal{B} \longrightarrow \mathcal{T}.$$

We wish to show that the functor $\mathcal{A}/\mathcal{B} \longrightarrow \mathcal{T}$ is an equivalence.

For any object $t \in \mathcal{T}$, the object Gt is $\mathcal{B}$–local, by Lemma A.2.5. Let a be an object of $\mathcal{A}$. The object GFa is $\mathcal{B}$–local, and the unit of adjunction

$$\eta_a : a \longrightarrow GFa$$

satisfies the property that the composite

$$Fa \xrightarrow{F\eta_a} FGFa \xrightarrow{\varepsilon_{Fa}} Fa$$

is the identity. Since we are assuming that

$$FGFa \xrightarrow{\varepsilon_{Fa}} Fa$$

is an isomorphism, it follows that

$$Fa \xrightarrow{F\eta_a} FGFa$$

is its two–sided inverse, in particular is invertible in $\mathcal{T}$. But now consider the exact sequence

$$0 \longrightarrow k \longrightarrow a \xrightarrow{\eta_a} GFa \longrightarrow q \longrightarrow 0.$$

The functor F is exact; hence we get an exact sequence

$$0 \longrightarrow Fk \longrightarrow Fa \xrightarrow{F\eta_a} FGFa \longrightarrow Fq \longrightarrow 0.$$

Since $F\eta_a$ is an isomorphism, it follows that $Fk = 0 = Fq$. Thus $k, q \in \mathcal{B}$. But this means that η_a is an isomorphism already in $\mathcal{A}/\mathcal{B}$. In other words, for every object $a \in \mathcal{A}$ we have produced a morphism

$$\eta_a : a \longrightarrow GFa$$

so that GFa is $\mathcal{B}$–local, and η_a becomes an isomorphism in $\mathcal{A}/\mathcal{B}$. It follows from Proposition A.2.11 that the projection $F' : \mathcal{A} \longrightarrow \mathcal{A}/\mathcal{B}$ has a right adjoint $G' : \mathcal{A}/\mathcal{B} \longrightarrow \mathcal{A}$. By Lemma A.2.8 the counit of adjunction $\varepsilon' : F'G' \Longrightarrow 1$ is an isomorphism. Lemma A.2.9 now tells us that both G and G' are fully faithful. In other words, $G : \mathcal{T} \longrightarrow \mathcal{A}$ and $G' : \mathcal{A}/\mathcal{B} \longrightarrow \mathcal{A}$ are, up to equivalence, inclusions of full subcategories. Their left adjoints F and F' will agree if and only if G and G' agree; that is, if the images of G' and G

$$G'(\mathcal{A}/\mathcal{B}) \subset \mathcal{A} \qquad\qquad G(\mathcal{T}) \subset \mathcal{A}$$

are equivalent subcategories.

For every object $t \in \mathcal{T}$, Lemma A.2.5 tells us that Gt is $\mathcal{B}$–local; by Proposition A.2.10 this implies that Gt is isomorphic to an object of the form $G'(p)$, for some $p \in \mathcal{A}/\mathcal{B}$. Up to equivalence, $G(\mathcal{T}) \subset G'(\mathcal{A}/\mathcal{B})$. On the other hand, we showed above that the map

$$\eta_a : a \longrightarrow GFa$$

becomes an isomorphism in $\mathcal{A}/\mathcal{B}$. But $GFa \in G(\mathcal{T})$. This means that the composite

$$G(\mathcal{T}) \quad \subset \quad G'(\mathcal{A}/\mathcal{B}) \quad \subset \quad \mathcal{A} \longrightarrow \mathcal{A}/\mathcal{B}$$

is surjective. Since the composite

$$G'(\mathcal{A}/\mathcal{B}) \quad \subset \quad \mathcal{A} \xrightarrow{F'} \mathcal{A}/\mathcal{B}$$

is an equivalence by Lemma A.2.9, we deduce that $G'(\mathcal{A}/\mathcal{B}) = G(\mathcal{T})$. □

An additive functor $F : \mathcal{A} \longrightarrow \mathcal{T}$ has a right adjoint $G : \mathcal{T} \longrightarrow \mathcal{A}$ if and only if, for every $t \in \mathcal{T}$, the functor

$$\mathcal{T}\big(F(-), t\big) : \mathcal{A}^{op} \longrightarrow \mathcal{A}b$$

is representable; that is,

$$\mathcal{T}\big(F(x), t\big) \quad \simeq \quad \mathcal{A}\big(x, G(t)\big)$$

for some $G(t) \in \mathcal{A}$, and naturally in x. This much is completely standard. Now observe

LEMMA A.2.13. *Let $F : \mathcal{A} \longrightarrow \mathcal{T}$ be an additive functor of abelian categories. Suppose we are given a class I of objects in $\mathcal{T}$ so that*

A.2.13.1. *Every object $t \in \mathcal{T}$ admits a copresentation (an exact sequence)*

$$0 \longrightarrow t \longrightarrow i \longrightarrow j$$

with $i, j \in I$.

A.2.13.2. *For each $i \in I$, the functor*

$$\mathcal{T}\big(F(-), i\big) : \mathcal{A}^{op} \longrightarrow \mathcal{A}b$$

is representable.

Then the functor F has a right adjoint $G : \mathcal{T} \longrightarrow \mathcal{A}$.

Proof: For each $i \in I$, choose an object $G(i) \in \mathcal{A}$ so that

$$\mathcal{T}\big(F(x), i\big) \qquad \simeq \qquad \mathcal{A}\big(x, G(i)\big)$$

with the isomorphism natural in x. Let t be an object of $\mathcal{T}$. Choose a copresentation

$$0 \longrightarrow t \longrightarrow i \longrightarrow j$$

with $i, j \in I$. The map $i \longrightarrow j$ induces a natural transformation

$$\mathcal{T}\big(F(-), i\big) \longrightarrow \mathcal{T}\big(F(-), j\big)$$

and hence a natural transformation

$$\mathcal{A}\big(-, G(i)\big) \longrightarrow \mathcal{A}\big(-, G(j)\big).$$

By Yoneda's lemma, this is induced by a morphism $G(i) \longrightarrow G(j)$. Let T be the kernel in the exact sequence

$$0 \longrightarrow T \longrightarrow G(i) \longrightarrow G(j).$$

We deduce a commutative diagram with exact rows

$$\begin{array}{ccccccc}
0 & \longrightarrow & \mathcal{T}\big(F(-), t\big) & \longrightarrow & \mathcal{T}\big(F(-), i\big) & \longrightarrow & \mathcal{T}\big(F(-), j\big) \\
 & & & & \wr\big\downarrow & & \wr\big\downarrow \\
0 & \longrightarrow & \mathcal{A}\big(-, T\big) & \longrightarrow & \mathcal{A}\big(-, G(i)\big) & \longrightarrow & \mathcal{A}\big(-, G(j)\big)
\end{array}$$

The diagram allows us to identify the kernels; we get a natural isomorphism

$$\mathcal{T}\big(F(-), t\big) \qquad \simeq \qquad \mathcal{A}\big(-, T\big)$$

□

REMARK A.2.14. For example, let $F : \mathcal{A} \longrightarrow \mathcal{T}$ be an additive functor of abelian categories, and suppose the category $\mathcal{T}$ has enough injectives. Every object $t \in \mathcal{T}$ admits an injective copresentation; that is, an exact sequence

$$0 \longrightarrow t \longrightarrow i \longrightarrow j$$

with i and j injective. To prove the existence of the right adjoint $G : \mathcal{T} \longrightarrow \mathcal{A}$ it suffices to show that, for every injective object $i \in \mathcal{T}$, the functor $\mathcal{T}\big(F(-), i\big)$ is representable. What is more, if F is exact, then it is possible to check on the i's that the counit of adjunction is an isomorphism.

LEMMA A.2.15. *Let $F : \mathcal{A} \longrightarrow \mathcal{T}$ be an exact functor of abelian categories. Suppose a right adjoint $G : \mathcal{T} \longrightarrow \mathcal{A}$ exists. Suppose I is a class of objects in $\mathcal{T}$ satisfying A.2.13.1. We remind the reader:*

A.2.13.1: *Every object $t \in \mathcal{T}$ admits a copresentation*

$$0 \longrightarrow t \longrightarrow i \longrightarrow j$$

with $i, j \in I$.

Suppose that for every $i \in I$, the counit of adjunction

$$\varepsilon_i : FGi \longrightarrow i$$

is an isomorphism. Then for every $t \in \mathcal{T}$, the counit $\varepsilon_t : FGt \longrightarrow t$ is an isomorphism.

Proof: Let t be an object of $\mathcal{T}$, and choose a copresentation

$$0 \longrightarrow t \longrightarrow i \longrightarrow j$$

with $i, j \in I$. The functor G, being a right adjoint, is left exact. It takes the exact sequence above to an exact sequence

$$0 \longrightarrow Gt \longrightarrow Gi \longrightarrow Gj.$$

By hypothesis, the functor F is exact. Hence the sequence

$$0 \longrightarrow FGt \longrightarrow FGi \longrightarrow FGj$$

is also exact. Now the naturality of ε_t gives a commutative diagram

$$\begin{array}{ccccccc} 0 & \longrightarrow & FGt & \longrightarrow & FGi & \longrightarrow & FGj \\ & & {\scriptstyle \varepsilon_t}\big\downarrow & & \big\downarrow\wr & & \big\downarrow\wr \\ 0 & \longrightarrow & t & \longrightarrow & i & \longrightarrow & j \end{array}$$

from which we immediately deduce that the counit $\varepsilon_t : FGt \longrightarrow t$ is an isomorphism. $\square$

These lemmas give us a practical criterion to identify when a functor $F : \mathcal{A} \longrightarrow \mathcal{T}$ is the map to a quotient by a localizant subcategory.

PROPOSITION A.2.16. *Let $F : \mathcal{A} \longrightarrow \mathcal{T}$ be an exact functor of abelian categories. Suppose we are given a class I of objects in $\mathcal{T}$ so that*

A.2.13.1: *Every object $t \in \mathcal{T}$ admits a copresentation*

$$0 \longrightarrow t \longrightarrow i \longrightarrow j$$

with $i, j \in I$.

A.2.13.2: *For each $i \in I$, the functor*

$$\mathcal{T}\big(F(-), i\big) : \mathcal{A}^{op} \longrightarrow \mathcal{A}b$$

is representable.

By A.2.13.2, whenever $i \in I$ we have an isomorphism

$$\mathcal{T}\big(F(-), i\big) \quad = \quad \mathcal{T}\big(-, G(i)\big).$$

We have a map from a representable functor $\mathcal{T}\big(-, G(i)\big)$ to the functor $\mathcal{T}\big(F(-), i\big)$, and by Yoneda this correspond to an element of $\mathcal{T}\big(FGi, i\big)$, in other words a map

$$\varepsilon_i : FGi \longrightarrow i.$$

Suppose further that all the maps ε_i are isomorphisms. Then the projection $F : \mathcal{A} \longrightarrow \mathcal{T}$ is equivalent to $\mathcal{A} \longrightarrow \mathcal{A}/\mathcal{B}$, where $\mathcal{B} \subset \mathcal{A}$ is the full subcategory on which $F : \mathcal{A} \longrightarrow \mathcal{T}$ vanishes.

Proof: By Lemma A.2.13, conditions A.2.13.1 and A.2.13.2 already imply that the functor F has a right adjoint G. We are supposing further that for $i \in I$, the maps

$$\varepsilon_i : FGi \longrightarrow i$$

are isomorphisms. By Lemma A.2.15 the counit of adjunction ε is an isomorphism. But then Proposition A.2.12 tells us that $\mathcal{T} = \mathcal{A}/\mathcal{B}$. □

Until now, we have mostly presented results permitting us to prove that some given map $F : \mathcal{A} \longrightarrow \mathcal{T}$ can be identified as $\mathcal{A} \longrightarrow \mathcal{A}/\mathcal{B}$. That is, if it turns out that F has a right adjoint G and that $\varepsilon : FG \longrightarrow 1$ is an isomorphism, we know that $\mathcal{T} = \mathcal{A}/\mathcal{B}$. Proposition A.2.16 tells us how to check that the adjoint G exists, and that $\varepsilon : FG \longrightarrow 1$ is an isomorphism.

We might find ourselves in the situation where we are given $\mathcal{T} = \mathcal{A}/\mathcal{B}$. Can one give practical hints on how to produce the adjoint G? The next few lemmas should help.

LEMMA A.2.17. *Suppose $F : \mathcal{A} \longrightarrow \mathcal{A}/\mathcal{B}$ is the projection of an abelian category $\mathcal{A}$, to the quotient by a Serre subcategory $\mathcal{B}$. Suppose F has a right adjoint G. Then in every object $a \in \mathcal{A}$, there is a maximal subobject belonging to $\mathcal{B}$.*

Proof: Let k be the kernel of the unit $\eta_a : a \longrightarrow GFa$. I assert that k is maximal, among subobjects of a belonging to $\mathcal{B}$.

Because F is exact, Fk is the kernel of the isomorphism $F\eta_a : Fa \longrightarrow FGFa$. Hence $Fk \simeq 0$, that is $k \in \mathcal{B}$. Next we need to show that k is maximal with this property.

Let $b \subset a$ be a subobject of a, belonging to the category $\mathcal{B}$. We form the composite

$$b \longrightarrow a \xrightarrow{\eta_a} GFa.$$

It is a morphism from $b \in \mathcal{B}$ to the $\mathcal{B}$–local object GFa; by A.2.5.1, it must vanish. Hence

$$b \quad \subset \quad k \quad = \quad \mathrm{Ker}\{\eta_a : a \longrightarrow GFa\}.$$

Therefore k is maximal. □

NOTATION A.2.18. Suppose $F : \mathcal{A} \longrightarrow \mathcal{A}/\mathcal{B}$ is the projection of an abelian category $\mathcal{A}$, to the quotient by a Serre subcategory $\mathcal{B}$. Let a be an object of $\mathcal{A}$. If a has a maximal subobject belonging to $\mathcal{B}$, we will denote this maximal $\mathcal{B}$–subobject $a_m \subset a$. Note that if $b \subset a$ is another subobject belonging to $\mathcal{B}$, then the union $b \cup a_m$ also belongs to $\mathcal{B}$. By the maximality of a_m, we deduce

$$b \cup a_m \quad = \quad a_m,$$

that is $b \subset a_m$. In other words, a_m contains all other $\mathcal{B}$–subobjects of a, and is therefore unique.

LEMMA A.2.19. *Suppose $F : \mathcal{A} \longrightarrow \mathcal{A}/\mathcal{B}$ is the projection of an abelian category $\mathcal{A}$, to the quotient by a Serre subcategory $\mathcal{B}$. Suppose the category $\mathcal{A}$ has enough injectives. If every injective object $i \in \mathcal{A}$ contains a maximal $\mathcal{B}$–subobject $i_m \subset i$, then so does every object $a \in \mathcal{A}$.*

Proof: Let a be an object of $\mathcal{A}$. Since $\mathcal{A}$ has enough injectives, we may embed a in an injective i. By hypothesis, $i \in \mathcal{A}$ contains a maximal $\mathcal{B}$–subobject $i_m \subset i$. If $b \subset a \subset i$ is a subobject contained in $\mathcal{B}$, then by the argument of Notation A.2.18, $b \subset i_m$. But then

$$b \quad \subset \quad a \cap i_m \quad \subset \quad a,$$

and $a \cap i_m$ is a maximal $\mathcal{B}$–subobject of a. □

PROPOSITION A.2.20. *Suppose $F : \mathcal{A} \longrightarrow \mathcal{A}/\mathcal{B}$ is the projection of an abelian category $\mathcal{A}$, to the quotient by a Serre subcategory $\mathcal{B}$. Suppose the category $\mathcal{A}$ has enough injectives. There is a right adjoint $G : \mathcal{A}/\mathcal{B} \longrightarrow \mathcal{A}$ if and only if every injective object $i \in \mathcal{A}$ has a maximal $\mathcal{B}$–subobject $i_m \subset i$.*

Proof: The necessity is clear. Lemma A.2.17 established that if the right adjoint G exists, then every object $a \in \mathcal{A}$ (not just the injectives) has a maximal $\mathcal{B}$–subobject $a_m \subset a$.

We need to prove the sufficiency. Suppose therefore that every injective object $i \in \mathcal{A}$ has a maximal $\mathcal{B}$–subobject $i_m \subset i$. We must prove the existence of a right adjoint $G : \mathcal{A}/\mathcal{B} \longrightarrow \mathcal{A}$.

The objects of $\mathcal{A}$ and $\mathcal{A}/\mathcal{B}$ are the same. Pick an object $a \in \mathrm{Ob}(\mathcal{A}) = \mathrm{Ob}(\mathcal{A}/\mathcal{B})$. By Lemma A.2.19, it has a maximal $\mathcal{B}$–subobject $a_m \subset a$ (even though a need not be injective). We can find an exact sequence in $\mathcal{A}$

$$0 \longrightarrow a/a_m \longrightarrow i \longrightarrow j,$$

with i and j injective in $\mathcal{A}$. We can complete this to a commutative diagram with exact rows

$$\begin{array}{ccccccc} 0 & \longrightarrow & a/a_m & \longrightarrow & i & \longrightarrow & j \\ & & \downarrow & & \downarrow & & \downarrow \\ 0 & \longrightarrow & Ga & \longrightarrow & i/i_m & \longrightarrow & j/j_m \end{array}$$

and I assert that this definition of Ga works; that is,

$$\mathcal{A}/\mathcal{B}(Fx, a) \quad = \quad \mathcal{A}(x, Ga).$$

Now we must prove this fact.

First of all, we are given a map, in $\mathcal{A}$,

$$a \longrightarrow a/a_m \longrightarrow Ga.$$

It clearly is an isomorphism in $\mathcal{A}/\mathcal{B}$. The map $a \longrightarrow a/a_m$ is division by $a_m \in \mathcal{B}$, hence an isomorphism in $\mathcal{A}/\mathcal{B}$. And if we apply the exact functor F to the diagram

$$\begin{array}{ccccccc} 0 & \longrightarrow & a/a_m & \longrightarrow & i & \longrightarrow & j \\ & & \downarrow & & \downarrow & & \downarrow \\ 0 & \longrightarrow & Ga & \longrightarrow & i/i_m & \longrightarrow & j/j_m \end{array}$$

we get a diagram

$$\begin{array}{ccccccc} 0 & \longrightarrow & F\{a/a_m\} & \longrightarrow & Fi & \longrightarrow & Fj \\ & & \downarrow & & \downarrow \wr & & \downarrow \wr \\ 0 & \longrightarrow & FGa & \longrightarrow & F\{i/i_m\} & \longrightarrow & F\{j/j_m\} \end{array}$$

where the rows are exact, and two of the columns are clearly isomorphisms. Hence so is the third; the map $a/a_m \longrightarrow Ga$ is an isomorphism in $\mathcal{A}/\mathcal{B}$. Therefore the composite

$$a \longrightarrow a/a_m \longrightarrow Ga$$

is also an isomorphism in $\mathcal{A}/\mathcal{B}$.

This gives us a natural map

$$\mathcal{A}(x, Ga) \longrightarrow \mathcal{A}/\mathcal{B}(Fx, a).$$

We take a map in $\mathcal{A}(x, Ga)$, view it as a map in $\mathcal{A}/\mathcal{B}(Fx, FGa)$, and compose with the inverse of the isomorphism $Fa \longrightarrow FGa$. We need to prove that this natural map is an isomorphism $\mathcal{A}(x, Ga) \longrightarrow \mathcal{A}/\mathcal{B}(Fx, a)$.

First let us prove surjectivity. Suppose therefore that we are given a map in $\mathcal{A}/\mathcal{B}(Fx, a)$; we need to show that it is the image of something in $\mathcal{A}(x, Ga)$. But now recall that a map in $\mathcal{A}/\mathcal{B}(Fx, a)$ is an equivalence class of diagrams in $\mathcal{A}$

$$\begin{array}{ccccccccccc} & & & & 0 & \longrightarrow & x' & \longrightarrow & x & \longrightarrow & x'' & \longrightarrow & 0 \\ & & & & & & \downarrow & & & & & & \\ 0 & \longrightarrow & a' & \longrightarrow & a & \longrightarrow & a'' & \longrightarrow & 0 & & & & \end{array}$$

with a', x'' in $\mathcal{B}$. Because a' is a $\mathcal{B}$–subobject of a, it is contained in the maximal one, a_m. The above is therefore equivalent to a diagram

$$\begin{array}{ccccccccccccc} & & & & 0 & \longrightarrow & x' & \longrightarrow & x & \longrightarrow & x'' & \longrightarrow & 0 \\ & & & & & & \downarrow & & & & & & \\ 0 & \longrightarrow & a_m & \longrightarrow & a & \longrightarrow & a/a_m & \longrightarrow & 0 & & & & \end{array}$$

Now recall the exact sequence

$$0 \longrightarrow a/a_m \longrightarrow i \longrightarrow j$$

that we chose above. The composite $x' \longrightarrow a/a_m \longrightarrow i$ is a map from x' to an injective object i, and therefore factors through the inclusion $x' \longrightarrow x$. We deduce a commutative square with exact rows

$$\begin{array}{ccccccccc} 0 & \longrightarrow & x' & \longrightarrow & x & \longrightarrow & x'' & \longrightarrow & 0 \\ & & \downarrow & & \downarrow & & \downarrow & & \\ 0 & \longrightarrow & a/a_m & \longrightarrow & i & \longrightarrow & j & & \end{array}$$

which we can further extend to a diagram

$$\begin{array}{ccccccccc} 0 & \longrightarrow & x' & \longrightarrow & x & \longrightarrow & x'' & \longrightarrow & 0 \\ & & \downarrow & & \downarrow & & \downarrow & & \\ 0 & \longrightarrow & a/a_m & \longrightarrow & i & \longrightarrow & j & & \\ & & \downarrow & & \downarrow & & \downarrow & & \\ 0 & \longrightarrow & Ga & \longrightarrow & i/i_m & \longrightarrow & j/j_m & & \end{array}$$

Now $x'' \longrightarrow j$ is a map from an object $x'' \in \mathcal{B}$ to j. Its image is therefore a $\mathcal{B}$–subobject of j, hence contained in the maximal $j_m \subset j$. Hence the composite $x'' \longrightarrow j \longrightarrow j/j_m$ must vanish. From the commutative diagram, the map $x \longrightarrow i/i_m \longrightarrow j/j_m$ must also vanish, and hence $x \longrightarrow i/i_m$ must factor through $Ga \subset i/i_m$. This produces a map $x \longrightarrow Ga$, in the equivalence class of the composite of the map $x \longrightarrow a$ given by the diagram

$$\begin{array}{ccccccccc} & & & & 0 & \longrightarrow & x' & \longrightarrow & x & \longrightarrow & x'' & \longrightarrow & 0 \\ & & & & & & \downarrow & & & & & & \\ 0 & \longrightarrow & a' & \longrightarrow & a & \longrightarrow & a'' & \longrightarrow & 0 & & & & \end{array}$$

and the map

$$a \longrightarrow a'' \longrightarrow a/a_m \longrightarrow Ga.$$

In other words, we have shown the surjectivity of

$$\mathcal{A}(x, Ga) \longrightarrow \mathcal{A}/\mathcal{B}(Fx, a).$$

It only remains to prove the injectivity.

Let $f : x \longrightarrow Ga$ be a map whose image in $\mathcal{A}/\mathcal{B}(Fx, a)$ vanishes. That is, the composite of Ff with the isomorphism $FGa \longrightarrow Fa$ vanishes. Hence Ff must vanish. It follows that the image of $f : x \longrightarrow Ga$ is a $\mathcal{B}$–subobject of Ga. By consrtuction, Ga embeds in i/i_m. Therefore $\mathrm{Im}(f)$ is a $\mathcal{B}$–subobject of i/i_m. But then $\mathrm{Im}(f) = 0$, and so $f = 0$. This proves that the kernel of the map

$$\mathcal{A}(x, Ga) \longrightarrow \mathcal{A}/\mathcal{B}(Fx, a)$$

is trivial. □

After this extensive discussion of how to construct the adjoint, maybe we should briefly comment on the implications of its existence. If the functor $F : \mathcal{A} \longrightarrow \mathcal{A}/\mathcal{B}$ has a right adjoint G, then F must preserve coproducts. We note

LEMMA A.2.21. *Let $\mathcal{A}$ be an abelian category satisfying [AB4]; coproducts exist in $\mathcal{A}$, and coproducts of exact sequences are exact. Let $\mathcal{B}$ be a Serre subcategory. The functor $F : \mathcal{A} \longrightarrow \mathcal{A}/\mathcal{B}$ preserves coproducts if and only if $\mathcal{B}$ is closed under the formation of $\mathcal{A}$–coproducts of its objects.*

Proof: Suppose F preserves coproducts. Let $\{b_\lambda, \lambda \in \Lambda\}$ be a set of objects in $\mathcal{B}$. For each b_λ we have $F(b_\lambda) \simeq 0$. But F preserves coproducts. Hence

$$F\left(\coprod_{\lambda\in\Lambda} b_\lambda\right) = \coprod_{\lambda\in\Lambda} F(b_\lambda) \simeq 0.$$

Therefore $\coprod_{\lambda\in\Lambda} b_\lambda$ lies in the subcategory $\mathcal{B}$ on which the functor F vanishes.

Coversely, suppose the category $\mathcal{B}$ is closed under coproducts. We wish to show that the functor $F : \mathcal{A} \longrightarrow \mathcal{A}/\mathcal{B}$ preserves coproducts. Let $\{a_\lambda, \lambda \in \Lambda\}$ be a set of objects in $\mathcal{A}$. We can form the coproduct in $\mathcal{A}$. We wish to show that it also has the universal property of a coproduct in $\mathcal{A}/\mathcal{B}$.

Suppose therefore that, for every $\lambda \in \Lambda$, we are given a morphism $a_\lambda \longrightarrow y$ in $\mathcal{A}/\mathcal{B}$. That means an equivalence class of diagrams in $\mathcal{A}$

$$\begin{array}{ccccccccccccc}
 & & & & 0 & \longrightarrow & a'_\lambda & \longrightarrow & a_\lambda & \longrightarrow & a''_\lambda & \longrightarrow & 0 \\
 & & & & & & \downarrow & & & & & & \\
0 & \longrightarrow & y'_\lambda & \longrightarrow & y & \longrightarrow & y''_\lambda & \longrightarrow & 0 & & & &
\end{array}$$

where the rows are exact and $a''_\lambda, y'_\lambda \in \mathcal{B}$. By hypothesis, the coproduct of the objects y'_λ lies in $\mathcal{B}$. Therefore for each $\lambda \in \Lambda$ the diagram above is equivalent to

$$\begin{array}{ccccccccccc}
 & & & & 0 & \longrightarrow & a'_\lambda & \longrightarrow & a_\lambda & \longrightarrow & a''_\lambda \longrightarrow 0 \\
 & & & & & & \downarrow & & & & \\
\coprod_{\lambda \in \Lambda} y'_\lambda & \longrightarrow & y & \longrightarrow & y'' & \longrightarrow & 0 & & & &
\end{array}$$

and these assemble to

$$\begin{array}{ccccccccccc}
 & & & 0 & \longrightarrow & \coprod_{\lambda \in \Lambda} a'_\lambda & \longrightarrow & \coprod_{\lambda \in \Lambda} a_\lambda & \longrightarrow & \coprod_{\lambda \in \Lambda} a''_\lambda & \longrightarrow 0 \\
 & & & & & \downarrow & & & & & \\
\coprod_{\lambda \in \Lambda} y'_\lambda & \longrightarrow & y & \longrightarrow & & y'' & \longrightarrow & 0 & & &
\end{array}$$

which is a map $\coprod_{\lambda \in \Lambda} a_\lambda \longrightarrow y$ in $\mathcal{A}/\mathcal{B}$.

To give a map $\coprod_{\lambda \in \Lambda} a_\lambda \longrightarrow y$ in $\mathcal{A}/\mathcal{B}$ is to give a diagram

$$\begin{array}{ccccccccccccc}
 & & & & 0 & \longrightarrow & A' & \longrightarrow & \coprod_{\lambda \in \Lambda} a_\lambda & \longrightarrow & A'' & \longrightarrow & 0 \\
 & & & & & & \downarrow & & & & & & \\
0 & \longrightarrow & y' & \longrightarrow & y & \longrightarrow & y'' & \longrightarrow & 0 & & & &
\end{array}$$

where the rows are exact and $A'', y' \in \mathcal{B}$. For each $\lambda \in \Lambda$ we can extend to a larger diagram

$$\begin{array}{ccccccccccccc}
 & & & & 0 & \longrightarrow & a'_\lambda & \longrightarrow & a_\lambda & \longrightarrow & a''_\lambda & \longrightarrow & 0 \\
 & & & & & & \downarrow & & \downarrow & & \downarrow & & \\
 & & & & 0 & \longrightarrow & A' & \longrightarrow & \coprod_{\lambda \in \Lambda} a_\lambda & \longrightarrow & A'' & \longrightarrow & 0 \\
 & & & & & & \downarrow & & & & & & \\
0 & \longrightarrow & y' & \longrightarrow & y & \longrightarrow & y'' & \longrightarrow & 0 & & & &
\end{array}$$

where the rows are exact and $a''_\lambda, A'', y' \in \mathcal{B}$. We can for example take a'_λ to be the kernel of the composite

$$\begin{array}{ccc} a_\lambda & & \\ \downarrow & & \\ \coprod_{\lambda\in\Lambda} a_\lambda & \longrightarrow & A''. \end{array}$$

To say that in $\mathcal{A}/\mathcal{B}$ the composite

$$a_\lambda \longrightarrow \coprod_{\lambda\in\Lambda} a_\lambda \longrightarrow y$$

vanishes is to say that, in the diagram above, the image of $a'_\lambda \longrightarrow y''$ lies in $\mathcal{B}$. Replacing a'_λ by the kernel of $a'_\lambda \longrightarrow y''$, we may assume that in the diagram above, the composite

$$a'_\lambda \longrightarrow A' \longrightarrow y''$$

vanishes. But now the diagram

$$\begin{array}{ccccccccccc} & & & & 0 & \longrightarrow & \coprod_{\lambda\in\Lambda} a'_\lambda & \longrightarrow & \coprod_{\lambda\in\Lambda} a_\lambda & \longrightarrow & \coprod_{\lambda\in\Lambda} a''_\lambda \longrightarrow 0 \\ & & & & & & \downarrow & & \downarrow{\scriptstyle 1} & & \downarrow \\ & & & & 0 & \longrightarrow & A' & \longrightarrow & \coprod_{\lambda\in\Lambda} a_\lambda & \longrightarrow & A'' \longrightarrow 0 \\ & & & & & & \downarrow & & & & \\ 0 & \longrightarrow & y' & \longrightarrow & y & \longrightarrow & y'' & \longrightarrow & 0 & & \end{array}$$

immediately implies the vanishing, in $\mathcal{A}/\mathcal{B}$, of

$$\coprod_{\lambda\in\Lambda} a_\lambda \longrightarrow y.$$

□

A.3. Derived functors of limits

Let $\mathcal{A}$ be an abelian category satisfying [AB3*]; products exist in $\mathcal{A}$. Then of course all small limits exist in $\mathcal{A}$. Let $\mathcal{I}$ be a small category. The category $\mathcal{C}at(\mathcal{I}^{op}, \mathcal{A})$ is the category of all functors

$$\mathcal{I}^{op} \longrightarrow \mathcal{A}.$$

This is of course an abelian category. There is a functor

$$\varprojlim : \mathcal{C}at(\mathcal{I}^{op}, \mathcal{A}) \longrightarrow \mathcal{A},$$

taking an object $F \in \mathcal{C}at(\mathcal{I}^{op}, \mathcal{A})$ to its limit. The limit exists since we are assuming $\mathcal{A}$ satisfies [AB3*]. The functor $\varprojlim$ is trivially left exact. It is natural to wonder if it has right derived functors. There are two well–known sufficient conditions which guarantee the existence of right derived functors. The first is that the category $\mathcal{C}at(\mathcal{I}^{op}, \mathcal{A})$ have enough injectives. We will discuss this further in the next Section. The second is that the category $\mathcal{A}$ satisfy [AB4*]. In this Section we remind the reader how this goes. But we wish to treat a slightly more general case.

DEFINITION A.3.1. *Let $\mathcal{A}$ be an abelian category. Let α be an infinite cardinal. We say that $\mathcal{A}$ satisfies [AB3*(α)] if $\mathcal{A}$ is closed under products of fewer than α of its objects. We say that $\mathcal{A}$ satisfies [AB4*(α)] if it satisfies [AB3*(α)], and products of fewer than α exact sequences are exact.*

LEMMA A.3.2. *Let α be an infinite cardinal. Let $\mathcal{A}$ be an abelian category satisfying [AB4*(α)]. Suppose $\mathcal{I}$ is a small category of cardinality $< \alpha$. This means there are fewer than α morphisms in $\mathcal{I}$. Then the functor*

$$\varprojlim : \mathcal{C}at(\mathcal{I}^{op}, \mathcal{A}) \longrightarrow \mathcal{A}$$

has right derived functors, denoted $\varprojlim^n$. Furthermore, the $\varprojlim^n$ are functorial in exact functors preserving products. Precisely, suppose

A.3.2.1. *$\mathcal{A}$, $\mathcal{T}$ are abelian categories satisfying [AB4*(α)].*

A.3.2.2. *$\phi : \mathcal{A} \longrightarrow \mathcal{T}$ is an exact functor preserving products.*

Then, for any $F : \mathcal{I}^{op} \longrightarrow \mathcal{A}$,

$$\varprojlim{}^n\{\phi F\} \quad = \quad \phi \varprojlim{}^n F.$$

Proof: Recall that the nerve of the category $\mathcal{I}$ is defined as the simplicial set $\mathcal{N}.(\mathcal{I})$, where $\mathcal{N}_k(\mathcal{I})$ (the k–simplices) are sequences of k composable morphisms

$$i_0 \longrightarrow i_1 \longrightarrow i_2 \longrightarrow \cdots \longrightarrow i_k.$$

If the cardinality of $\mathcal{I}$ is $< \alpha$, so is the cardinality of each $\mathcal{N}_k(\mathcal{I})$.

Now $\mathcal{A}$ satisfies [AB3*(α)]. Given a functor $F : \mathcal{I}^{op} \longrightarrow \mathcal{A}$, we form a chain complex

$$N_0(F) \xrightarrow{\ \partial_0\ } N_1(F) \xrightarrow{\ \partial_1\ } N_2(F) \xrightarrow{\ \partial_2\ } \cdots$$

where

$$N_k(F) \quad = \prod_{\{i_0 \to i_1 \to \cdots \to i_k\} \in \mathcal{N}_k(\mathcal{I})} F(i_0).$$

The products exist, being products of fewer than α objects in $\mathcal{A}$. The differential $\partial_k : N_k(F) \longrightarrow N_{k+1}(F)$ is given by the usual alternating sum.

That is,

$$\partial_k = \sum_{j=0}^{k+1} \{-1\}^j \partial_k^j,$$

and ∂_k^j is the map induced by deleting the jth term in the sequence

$$i_0 \longrightarrow i_1 \longrightarrow i_2 \longrightarrow \cdots \longrightarrow i_k \longrightarrow i_{k+1}.$$

We define $T^n(F)$ to be the nth cohomology of the complex

$$N_0(F) \xrightarrow{\partial_0} N_1(F) \xrightarrow{\partial_1} N_2(F) \xrightarrow{\partial_2} \cdots$$

and I assert that, if $\mathcal{A}$ satisfies [AB4*(α)], then the collection of functors T^n are the right derived functor of $\underleftarrow{\lim}$.

Suppose we have a short exact sequence in $\mathcal{C}at(\mathfrak{I}^{op}, \mathcal{A})$, that is an exact sequence of functors $\mathfrak{I}^{op} \longrightarrow \mathcal{A}$

$$0 \longrightarrow F' \longrightarrow F \longrightarrow F'' \longrightarrow 0.$$

By [AB4*(α)], products of fewer than α exact sequences are exact in $\mathcal{A}$, and so the sequence

$$0 \longrightarrow N_k(F') \longrightarrow N_k(F) \longrightarrow N_k(F'') \longrightarrow 0$$

is exact. This being true for all k, we have a short exact sequence of chain complexes, hence a long exact sequence in cohomology. It follows that T is a δ–functor. It is also clear that T^n commute with products. That is,

$$T^n\left\{\prod_{\lambda\in\Lambda} F_\lambda\right\} = \prod_{\lambda\in\Lambda} T^n(F_\lambda).$$

To prove that the T^n give the derived functor of $\underleftarrow{\lim}$, it suffices to show two things.

A.3.2.3. *There is a natural isomorphism* $\underleftarrow{\lim} = T^0$.

A.3.2.4. *Every object* $F \in \mathcal{C}at(\mathfrak{I}^{op}, \mathcal{A})$ *can be embedded in an object* F', *with* $T^n(F') = 0$ *for all* $n > 0$.

The proof that T^0 is naturally isomorphic to $\underleftarrow{\lim}$ is obvious. It therefore remains to prove A.3.2.4. We must show that any object can be embedded in a T–acyclic.

For every $i \in \mathfrak{I}$ and every object $a \in \mathcal{A}$, we define a functor $F_a^i : \mathfrak{I} \longrightarrow \mathcal{A}$ by the rule:

$$F_a^i(j) = \prod_{\mathfrak{I}(i,j)} a.$$

That is, $F^i_a(j)$ is the product over all morphisms $i \longrightarrow j$ of a. The chain complex

$$N_0(F^i_a) \xrightarrow{\partial_0} N_1(F^i_a) \xrightarrow{\partial_1} N_2(F^i_a) \xrightarrow{\partial_2} \cdots$$

is obviously homotopy equivalent to the chain complex

$$a \longrightarrow 0 \longrightarrow 0 \longrightarrow \cdots$$

Hence the functor F^i_a is T–acyclic. Since the functor T commutes with products, the product of any F^i_a's is T–acyclic. But any functor G can be embedded in the product over all i of F^i_{Gi}.

This proves that, in an abelian category satisfying [AB4*(α)], the functors T^n compute $\varprojlim^n$. But $T^n(F)$ is defined as the cohomology of the chain complex

$$N_0(F) \xrightarrow{\partial_0} N_1(F) \xrightarrow{\partial_1} N_2(F) \xrightarrow{\partial_2} \cdots$$

and any functor $\phi : \mathcal{A} \longrightarrow \mathcal{T}$ preserving products will send the above chain complex to

$$N_0(\phi F) \xrightarrow{\partial_0} N_1(\phi F) \xrightarrow{\partial_1} N_2(\phi F) \xrightarrow{\partial_2} \cdots$$

If the functor ϕ is also exact, ϕ of the cohomology of a chain complex is the cohomology of ϕ of the chain complex. In our specific case above,

$$\varprojlim{}^n\{\phi F\} \quad = \quad \phi \varprojlim{}^n F.$$

□

REMARK A.3.3. We proved Lemma A.3.2 in the generality of any functor $F : \mathcal{I}^{op} \longrightarrow \mathcal{A}$. In practice, most of the time we will consider only very special $\mathcal{I}$'s. Our $\mathcal{I}$'s will be partially ordered sets, and most will be even more special than that. In the remainder of this section and in the next, we consider only partially ordered sets $\mathcal{I}$.

We will need Lemma A.3.2 in the generality stated, that is where $\mathcal{I}$ is a category, exactly once. It occurs in Appendix B. All we need there is the definition and construction. The other facts we prove, about derived functors of limits, play no rôle. For this reason, after Lemma A.3.2, the category $\mathcal{I}$ will be assumed a partially ordered set, usually even a rather special one.

We are particularly interested in limits of sequences. That is, we study functors $\mathcal{I}^{op} \longrightarrow \mathcal{A}$ for very special $\mathcal{I}$.

DEFINITION A.3.4. *Let γ be an ordinal. The partially ordered set $\mathcal{I} = \mathcal{I}(\gamma)$ will be the set of ordinals $< \gamma$.*

DEFINITION A.3.5. *Let $\mathcal{A}$ be an abelian category.* A sequence of length γ in $\mathcal{A}$ *is a functor $F : \mathcal{I}(\gamma)^{op} \longrightarrow \mathcal{A}$.*

REMARK A.3.6. If we fix γ, then we will write $\mathfrak{I}$ for $\mathfrak{I}(\gamma)$, and speak of sequences without mentioning their length. If ω is the smallest infinite ordinal, then a sequence of length ω is an inverse sequence in the usual sense

$$\cdots \longrightarrow a_2 \longrightarrow a_1 \longrightarrow a_0.$$

For sequences of length ω, the chain complex

$$N_0(F) \xrightarrow{\partial_0} N_1(F) \xrightarrow{\partial_1} N_2(F) \xrightarrow{\partial_2} \cdots$$

can be replaced by the much shorter complex

$$\prod_{i=0}^{\infty} a_i \xrightarrow{1-shift} \prod_{i=0}^{\infty} a_i.$$

By a proof virtually identical with that of Lemma A.3.2, one can show that the cohomology of this short complex gives the right derived functors of $\varprojlim$. In particular, for sequences of length ω, we have $\varprojlim^n = 0$ when $n > 1$.

The next useful idea is that of a Mittag–Leffler sequence. To make the concept more natural, we study first the analogy with sheaves on a space.

REMARK A.3.7. We can make $\mathfrak{I}$ into a topological space, by declaring a subset $U \subset \mathfrak{I}$ to be open if it satisfies the condition

$$\{i \in U \text{ and } j < i\} \Longrightarrow j \in U.$$

The complement of an open set U contains a minimal element β. For any ordinal β, define

$$U(\beta) \quad = \quad \{j \in \mathfrak{I} \,|\, j < \beta\}.$$

All the open sets must be of the form $U = U(\beta)$. To give a presheaf F on the topological space $\mathfrak{I}$ with values in the abelian category $\mathcal{A}$ is to give for each $U(\beta)$ an object $F[U(\beta)] \in \mathcal{A}$, with restriction maps. The stalk of the presheaf F at the point $i \in \mathfrak{I}$ is just $F[U(i+1)]$. The reason is that $U(i+1)$ is the smallest open set containing i. Assume the category $\mathcal{A}$ satisfies [AB3*]; limits exist. The presheaf will be a sheaf if, for every limit ordinal β,

$$F[U(\beta)] \quad = \quad \lim_{i<\beta} F[U(i+1)].$$

In other words, let F be a sheaf. Define $G(i) = F[U(i+1)]$, that is $G(i)$ is the stalk at i. Then $G : \mathfrak{I}^{op} \longrightarrow \mathcal{A}$ is a functor, and it completely determines the sheaf F. For a successor cardinal $i+1$, $F[U(i+1)] = G(i)$, while for a limit ordinal β

$$F[U(\beta)] \quad = \quad \lim_{i<\beta} G(i).$$

Let $\mathcal{A}$ be an abelian category satisfying [AB3*]. The category of sheaves on $\mathfrak{I}$ is therefore just the category $\mathcal{C}at(\mathfrak{I}^{op}, \mathcal{A})$ of functors $\mathfrak{I}^{op} \longrightarrow \mathcal{A}$. This isomorphism of categories is (of course) exact; a sequence of sheaves on $\mathfrak{I}$ is exact if and only if the sequence of stalks is exact at every point. And the correspondence between sheaves F on $\mathfrak{I}$ and functors $G : \mathfrak{I}^{op} \longrightarrow \mathcal{A}$ takes a sheaf to the functor taking i to the stalk at i.

Most importantly, the global section functor H^0 is just $\varprojlim$. It follow that the derived functors $\varprojlim^n$ are just the sheaf cohomology functors H^n.

We know that it is a good idea, when studying sheaf cohomology, to look at flabby sheaves. This leads to the next two definitions.

DEFINITION A.3.8. *Let $G : \mathfrak{I}^{op} \longrightarrow \mathcal{A}$ be a sequence in $\mathcal{A}$. We say that the sequence G is* flabby *if the corresponding sheaf F, as in Remark A.3.7, is a flabby sheaf. For a more concrete discussion of the condition this places on $G(i)$, see Remark A.3.12.*

LEMMA A.3.9. *Let $\mathcal{A}$ be an abelian category satisfying [AB4*]. If $G : \mathfrak{I}^{op} \longrightarrow \mathcal{A}$ is a flabby sequence, then for all $n \geq 1$, $\varprojlim^n G = 0$.*

Proof: By Remark A.3.7, the derived functors of limits of sequences agree with the sheaf cohomology of the corresponding sheaves. And flabby sheaves have vanishing sheaf cohomology. □

DEFINITION A.3.10. *Let $G : \mathfrak{I}^{op} \longrightarrow \mathcal{A}$ be a sequence in $\mathcal{A}$. We say that the sequence G is* Mittag–Leffler *if*

A.3.10.1. *For any pair of ordinals $i > j$ in $\mathfrak{I}$, the map $Gi \longrightarrow Gj$ is epi.*

A.3.10.2. *For any limit ordinal $j \in \mathfrak{I}$, the map*

$$Gj \longrightarrow \lim_{i<j} Gi$$

is epi.

REMARK A.3.11. More concisely, we could restate this as follows. A sequence $G : \mathfrak{I}^{op} \longrightarrow \mathcal{A}$ is Mittag–Leffler is and only if

A.3.11.1. *For any $j \leq i \in \mathfrak{I}$ the map*

$$Gi \longrightarrow \lim_{k<j} Gk.$$

is epi.

If $i = j$, A.3.11.1 reduces to A.3.10.2. To obtain A.3.10.1 as a consequence of A.3.11.1, let $i > j$ be a pair of ordinals as in A.3.10.1, and apply A.3.11.1 to the map

$$G_i \longrightarrow \lim_{k<j+1} Gk \quad = \quad Gj.$$

The fact that A.3.11.1 is a consequence of A.3.10.1 and A.3.10.2 is also easy; see Remark A.3.12.

REMARK A.3.12. The definition of Mitag–Leffler sequences is strongly reminiscent of flabbiness. It says that some of the restriction morphisms $F[U(i)] \longrightarrow F[U(j)]$ are epi. Flabbiness asserts, of course, that they all are.

Precisely, A.3.10.1 tells us that if both $i+1$ and $j+1$ are successor ordinals, then the map from $F[U(i+1)] = G(i)$ to $F[U(j+1)] = G(j)$ is epi. From A.3.10.2 we learn that if j is a limit ordinal, then the map from $F[U(j+1)] = G(j)$ to $F[U(j)] = \lim_{i<j} Gi$ is epi. By combining the two we have that if $i+1 > j$, $i+1$ is a successor ordinal and j a limit ordinal, then the composite

$$F[U(i+1)] \longrightarrow F[U(j+1)] \longrightarrow F[U(j)]$$

is epi. This amounts to A.3.11.1. It therefore formally follows that A.3.11.1 is a consequence of A.3.10.1 and A.3.10.2.

To have a flabby sheaf, we need all the restrictions to be epi. In particular, we must show that, if i is a limit ordinal and $j+1$ is a successor ordinal, then the restriction map $F[U(i)] \longrightarrow F[U(j+1)]$ is epi. The fourth case, where both ordinals are limit ordinals, is a consequence of the first three, since any such map can be factored

$$F[U(i)] \longrightarrow F[U(k+1)] \longrightarrow F[U(j)]$$

where $k+1$ is a successor ordinal.

LEMMA A.3.13. *Suppose $\mathcal{A} = \mathcal{A}b$, the category of abelian groups. Then any Mittag–Leffler sequence $G : \mathcal{I}^{op} \longrightarrow \mathcal{A}b$ corresponds, under the correspondence of Remark A.3.7, with a flabby sheaf.*

Proof: By Remark A.3.12 it suffices to show that, if i is a limit ordinal and $j+1 < i$ is a successor ordinal, then the restriction map $F[U(i)] \longrightarrow F[U(j+1)]$ is epi. In other words, the map

$$\lim_{k<i} G(k) \longrightarrow G(j)$$

must be shown onto. And this is true since any element of the abelian group $G(j)$ can be lifted inductively to a sequence of elements in $G(k), k > j$. That is, it can be lifted to the inverse limit. □

COROLLARY A.3.14. *If $G : \mathcal{I}^{op} \longrightarrow \mathcal{A}b$ is Mittag–Leffler and $n > 0$, then $\varprojlim^n G = 0$.*

Proof: By Lemma A.3.13, G corresponds to a flabby sheaf F. But Lemma A.3.9 then tells us that $\varprojlim^n G = 0$. □

LEMMA A.3.15. *Let α be an infinite cardinal. Suppose $\mathcal{A}$ is an abelian category satisfying [AB3*(α)]; products of fewer than α objects exist in $\mathcal{A}$. Suppose further that $\mathcal{A}$ has enough projectives. Then*

A.3.15.1. *The category $\mathcal{A}$ satisfies [AB4*(α)].*

A.3.15.2. *Let $\mathfrak{I} = \mathfrak{I}(\gamma)$ be the set of ordinals $< \gamma$, and assume $\gamma < \alpha$. For any Mittag–Leffler sequence $G : \mathfrak{I}^{op} \longrightarrow \mathcal{A}$ and any integer $n \geq 1$, $\varprojlim^n G = 0$.*

Proof: First we prove A.3.15.1. That is, we must show [AB4*(α)] holds in $\mathcal{A}$. Suppose Λ is a set of cardinality $< \alpha$, and for all $\lambda \in \Lambda$, we are given a short exact sequences in $\mathcal{A}$

$$0 \longrightarrow a'_\lambda \longrightarrow a_\lambda \longrightarrow a''_\lambda \longrightarrow 0.$$

Then the sequence

$$0 \longrightarrow \prod_{\lambda\in\Lambda} a'_\lambda \longrightarrow \prod_{\lambda\in\Lambda} a_\lambda \longrightarrow \prod_{\lambda\in\Lambda} a''_\lambda$$

is clearly exact, and we must show the map

$$\prod_{\lambda\in\Lambda} a_\lambda \longrightarrow \prod_{\lambda\in\Lambda} a''_\lambda$$

to be epi. Let q be the cokernel of the map. That is, we have an exact sequence

$$\prod_{\lambda\in\Lambda} a_\lambda \xrightarrow{\ a\ } \prod_{\lambda\in\Lambda} a''_\lambda \xrightarrow{\ b\ } q \longrightarrow 0$$

We are assuming $\mathcal{A}$ has enough projectives; there must be a epimorphism $p \longrightarrow q$ with p projective. But then we have that the surjection $p \longrightarrow q$ must factor through the epimorphism b above.

On the other hand, for each $\lambda \in \Lambda$ we have an surjective map

$$a_\lambda \longrightarrow a''_\lambda.$$

Since the object p is projective, this gives a surjective map

$$\mathcal{A}(p, a_\lambda) \longrightarrow \mathcal{A}(p, a''_\lambda).$$

Since in the category of abelian groups products are exact, we deduce the surjectivity of

$$\prod_{\lambda\in\Lambda} \mathcal{A}(p, a_\lambda) \longrightarrow \prod_{\lambda\in\Lambda} \mathcal{A}(p, a''_\lambda).$$

This map is canonically identified with

$$\mathcal{A}\left(p, \prod_{\lambda\in\Lambda} a_\lambda\right) \longrightarrow \mathcal{A}\left(p, \prod_{\lambda\in\Lambda} a''_\lambda\right).$$

We are given that the epi $p \longrightarrow q$ factors as

$$p \longrightarrow \prod_{\lambda\in\Lambda} a''_\lambda \longrightarrow q,$$

and the surjectivity of

$$\mathcal{A}\left(p, \prod_{\lambda\in\Lambda} a_\lambda\right) \longrightarrow \mathcal{A}\left(p, \prod_{\lambda\in\Lambda} a''_\lambda\right)$$

means it factors further as

$$p \longrightarrow \prod_{\lambda\in\Lambda} a_\lambda \longrightarrow \prod_{\lambda\in\Lambda} a''_\lambda \longrightarrow q.$$

But the composite

$$\prod_{\lambda\in\Lambda} a_\lambda \longrightarrow \prod_{\lambda\in\Lambda} a''_\lambda \longrightarrow q$$

vanishes, and hence $p \longrightarrow q$ must vanish. Since it was chosen to be an epimophism, this means $q \simeq 0$. The exact sequence

$$\prod_{\lambda\in\Lambda} a_\lambda \xrightarrow{\ a\ } \prod_{\lambda\in\Lambda} a''_\lambda \longrightarrow q \longrightarrow 0$$

now tells us that a is epi. This proves [AB4*(α)].

Let $\mathcal{I}$ be a partially ordered set of cardinality $< \alpha$. Now that we know the category $\mathcal{A}$ satisfies [AB4*(α)], Lemma A.3.2 tells us that the higher derived functors of

$$\underleftarrow{\lim} : \mathcal{C}at(\mathcal{I}^{op}, \mathcal{A}) \longrightarrow \mathcal{A}$$

exist in $\mathcal{A}$. Let p be any projective object of $\mathcal{A}$. The functor $\mathcal{A}(p, -)$ is a functor

$$\mathcal{A}(p, -) : \mathcal{A} \longrightarrow \mathcal{A}b$$

which is exact and takes products to products. By Lemma A.3.2 it also preserves $\underleftarrow{\lim}^n$.

We wish to apply this to the case of sequences. Let $\mathcal{I} = \mathcal{I}(\gamma)$ where $\gamma < \alpha$. For any functor $G : \mathcal{I}^{op} \longrightarrow \mathcal{A}$,

$$\mathcal{A}\big(p, \underleftarrow{\lim}^n G(-)\big) \quad = \quad \underleftarrow{\lim}^n \mathcal{A}\big(p, G(-)\big).$$

Consider now a Mittag–Leffler sequence $G : \mathcal{I}^{op} \longrightarrow \mathcal{A}$. Suppose p is a projective object in $\mathcal{A}$, as above. For $i > j$, the map $Gi \longrightarrow Gj$ are epi, and hence

$$\mathcal{A}(p, Gi) \longrightarrow \mathcal{A}(p, Gj)$$

must be epi. Furthermore, if j is a limit ordinal,

$$Gj \longrightarrow \lim_{i<j} Gi$$

is epi; hence

$$\mathcal{A}(p, Gj) \longrightarrow \lim_{i<j} \mathcal{A}(p, Gi)$$

is also epi. These statements combine to say that the sequence $\mathcal{A}\big(p, G(-)\big)$ is Mittag–Leffler in the category $\mathcal{A}b$.

By Corollary A.3.14, if $n \geq 1$ and $F : \mathcal{I}^{op} \longrightarrow \mathcal{A}b$ is Mittag–Leffler, then $\underleftarrow{\lim}^n F$ vanishes. Hence $\underleftarrow{\lim}^n \mathcal{A}\big(p, G(-)\big)$ must vanish. But we proved above

$$\mathcal{A}\big(p, \underleftarrow{\lim}^n G(-)\big) \quad = \quad \underleftarrow{\lim}^n \mathcal{A}\big(p, G(-)\big).$$

Hence $\mathcal{A}\big(p, \underleftarrow{\lim}^n G(-)\big)$ vanishes, whenever G is Mittag–Leffler and p projective.

Let $n \geq 1$, G a Mittag–Leffler sequence in $\mathcal{A}$. Because $\mathcal{A}$ has enough projectives, we can find a surjective map $p \longrightarrow \underleftarrow{\lim}^n G(-)$, with p projective. On the other hand, we have proved that the group $\mathcal{A}\big(p, \underleftarrow{\lim}^n G(-)\big)$ vanishes, for any projective p. Thus the map

$$p \longrightarrow \underleftarrow{\lim}^n G(-)$$

is an epimorphism which vanishes. It follows that $\underleftarrow{\lim}^n G(-) = 0$. The higher derived limits of Mittag–Leffler sequences vanish. □

Lemma A.3.15 suggests that we make the following definition.

DEFINITION A.3.16. *Let α be an infinite cardinal. An abelian category $\mathcal{A}$ is said to satisfy [AB4.5*(α)] if it satisfies [AB4*(α)], and for every Mittag–Leffler sequence*

$$F : \mathcal{I}(\gamma) \longrightarrow \mathcal{A}$$

of length $\gamma < \alpha$, $\lim^n F = 0$ for $n \geq 1$.

If $\mathcal{A}$ satisfies [AB4.5(α)] for every infinite cardinal α, we say it satisfies [AB4.5*]. If the dual satisfies [AB4.5*(α)], then $\mathcal{A}$ is said to satisfy [AB4.5(α)].*

REMARK A.3.17. In the language of Definition A.3.16, Lemma A.3.15 asserts that an abelian category $\mathcal{A}$ satisfying [AB3*(α)] and having enough projectives automatically satisfies [AB4.5*(α)]. We will see in Section A.5, more particularly Proposition A.5.12, that there are abelian categories which satisfy [AB4*(α)] but not [AB4.5*(α)].

A.4. Derived functors of limits via injectives

In Section A.3 we started with an abelian category $\mathcal{A}$ satisfying [AB4*], and showed the existence and functoriality of the derived functors of the limit. If $\mathcal{A}$ also had enough projectives, this treatment was very convenient for proving the vanishing of $\varprojlim^n$ of Mittag–Leffler sequences. It turns out that if $\mathcal{A}$ has enough injectives, one can also prove the existence of the derived functor of the limit. In this Section, we explore some of the consequences.

LEMMA A.4.1. *Let $\mathcal{A}$ be an abelian category. Let $\mathfrak{I}$ be a partially ordered set. As in the previous section, let $\mathfrak{Cat}(\mathfrak{I}^{op},\mathcal{A})$ be the category of functors $\mathfrak{I}^{op} \longrightarrow \mathcal{A}$. Given $i \in \mathfrak{I}$, there is a natural evaluation functor*

$$e : \mathfrak{Cat}(\mathfrak{I}^{op},\mathcal{A}) \longrightarrow \mathcal{A}$$

given by $e(F) = F(i)$. The functor e is exact and has a right adjoint.

Proof: The exactness of e is obvious. The right adjoint takes an object $a \in \mathcal{A}$ to the functor we have been denoting F_a^i, given by the formula

$$F_a^i(j) \quad = \quad \begin{cases} a & \text{if } j \geq i \\ 0 & \text{otherwise} \end{cases}$$

□

LEMMA A.4.2. *Let α be an infinite cardinal. Let $\mathcal{A}$ be an abelian category satisfying [AB3*(α)]. That means, products of fewer than α objects of $\mathcal{A}$ exist. Let $\mathfrak{I}$ be a partially ordered set. Then the abelian category $\mathfrak{Cat}(\mathfrak{I}^{op},\mathcal{A})$ satisfies [AB3*(α)].*

Proof: Let Λ be a set of cardinality $< \alpha$. Suppose for each $\lambda \in \Lambda$ we are given an object $F_\Lambda \in \mathfrak{Cat}(\mathfrak{I}^{op},\mathcal{A})$. The product is defined by the formula

$$\left\{\prod_{\lambda\in\Lambda} F_\Lambda\right\}(i) \quad = \quad \prod_{\lambda\in\Lambda} F_\Lambda(i).$$

The formula makes sense since the right hand side is a product of fewer than α objects of $\mathcal{A}$, and exists as $\mathcal{A}$ satisfies [AB3*(α)]. It is trivial that, with this definition, $\prod_{\lambda\in\Lambda} F_\Lambda$ is indeed the product in the category $\mathfrak{Cat}(\mathfrak{I}^{op},\mathcal{A})$.
□

LEMMA A.4.3. *Let α be an infinite cardinal. Let $\mathcal{A}$ be an abelian category satisfying [AB3*(α)]. Let $\mathfrak{I}$ be a partially ordered set of cardinality $< \alpha$. If the category $\mathcal{A}$ has enough injectives, then so does the category $\mathfrak{Cat}(\mathfrak{I}^{op},\mathcal{A})$.*

Proof: For $i \in \mathfrak{I}$, let $e : \mathfrak{Cat}(\mathfrak{I}^{op},\mathcal{A}) \longrightarrow \mathcal{A}$ be the evaluation functor of Lemma A.4.1. Let $f : \mathcal{A} \longrightarrow \mathfrak{Cat}(\mathfrak{I}^{op},\mathcal{A})$ be the right adjoint. Then f has an exact left adjoint, and must therefore take injective objects in $\mathcal{A}$ to injective objects in $\mathfrak{Cat}(\mathfrak{I}^{op},\mathcal{A})$. In other words, if a is an injective object of $\mathcal{A}$, then F_a^i is injective in $\mathfrak{Cat}(\mathfrak{I}^{op},\mathcal{A})$.

Take any object $F \in \mathcal{C}at(\mathfrak{I}^{op}, \mathcal{A})$. We know that the category $\mathcal{A}$ has enough injectives. For each $i \in \mathfrak{I}$ we may choose an embedding

$$Fi \longrightarrow a_i$$

with a_i an injective object of $\mathcal{A}$. But then F embeds into the product

$$F \longrightarrow \prod_{i \in \mathfrak{I}} F^i_{a_i}.$$

By Lemma A.4.2 the category $\mathcal{C}at(\mathfrak{I}^{op}, \mathcal{A})$ satisfies $[\text{AB3}^*(\alpha)]$. The product above is over the objects in $\mathfrak{I}$, which is assumed to have cardinality $< \alpha$; hence the product exists. Finally, being a product of injectives, it is injective. □

REMARK A.4.4. There is a refinement which is useful. Every object $F \in \mathcal{C}at(\mathfrak{I}^{op}, \mathcal{A})$ may be embedded in a product of $F^i_{a_i}$'s, a_i injective in $\mathcal{A}$. Even more precisely, suppose we are given a class I of injective objects of $\mathcal{A}$, so that every object $a \in \mathcal{A}$ admits an embedding $a \longrightarrow a_i$ with $a_i \in I$. Then any $F \in \mathcal{C}at(\mathfrak{I}^{op}, \mathcal{A})$ admits an embedding

$$F \longrightarrow \prod_{i \in \mathfrak{I}} F^i_{a_i}$$

with all the a_i in I.

COROLLARY A.4.5. *Let α be an infinite cardinal. Let $\mathcal{A}$ be an abelian category with enough injectives. Suppose $\mathcal{A}$ satisfies $[AB3^*(\alpha)]$. Let $\mathfrak{I}$ be a partially ordered set of cardinality $< \alpha$. Then the functor*

$$\varprojlim : \mathcal{C}at(\mathfrak{I}^{op}, \mathcal{A}) \longrightarrow \mathcal{A}$$

has right derived functors.

Proof: Because we are assuming that the cardinality of $\mathfrak{I}$ is $< \alpha$ and that products of fewer than α objects exist in $\mathcal{A}$, we can form limits over $\mathfrak{I}$; the functor

$$\varprojlim : \mathcal{C}at(\mathfrak{I}^{op}, \mathcal{A}) \longrightarrow \mathcal{A}$$

is well–defined, and is clearly left exact.

We are assuming that $\mathcal{A}$ has enough injectives. This being the case, Lemma A.4.3 tells us that the category $\mathcal{C}at(\mathfrak{I}^{op}, \mathcal{A})$ also has enough injectives. But then any left exact functor on $\mathcal{C}at(\mathfrak{I}^{op}, \mathcal{A})$ admits right derived functors. □

REMARK A.4.6. Everything so far was true for a general partially ordered set $\mathfrak{I}$. Now we wish to remind ourselves concretely what this all means for the special case where $\mathfrak{I} = \mathfrak{I}(\gamma)$ for some ordinal γ.

In Lemma A.4.3, we learned that the category $\mathcal{C}at(\mathcal{I}^{op}, \mathcal{A})$ has enough injectives. More concretely we learned, in the proof, that every object $F \in \mathcal{C}at(\mathcal{I}^{op}, \mathcal{A})$ admits an embedding into an injective object which is a product of F_a^i's. Remark A.4.4 tells us that, given a class I of injective objects of $\mathcal{A}$, so that every object $a \in \mathcal{A}$ admits an embedding $a \longrightarrow a_i$ with $a_i \in I$, then an object $F \in \mathcal{C}at(\mathcal{I}^{op}, \mathcal{A})$ can even be embedded into a product of F_a^i's with $a \in I$. When $\mathcal{I} = \mathcal{I}(\gamma)$, F_a^i is the sequence

$$\cdots \xrightarrow{\ 1\ } a \xrightarrow{\ 1\ } a \xrightarrow{\ 1\ } a \longrightarrow 0 \longrightarrow 0 \longrightarrow 0 \longrightarrow \cdots$$

When we take products of such sequences, we get a sequence $F : \mathcal{I}^{op} \longrightarrow \mathcal{A}$ which satisfies

   A.4.6.1. For any $i \in \mathcal{I}$, Fi is an injective object of $\mathcal{A}$.

   A.4.6.2. For any $i < j \in \mathcal{I}$, the map $Fj \longrightarrow Fi$ is a split epimorphism.

   A.4.6.3. For any limit ordinal j, the natural map

$$Fj \longrightarrow \varprojlim_{i<j} Fi$$

is a split epimorphism.

The sequences satisfying A.4.6.1, A.4.6.2 and A.4.6.3 are precisely the products of injectives F_a^i's. Therefore they are injective. It may be proved that every injective object in $\mathcal{C}at(\mathcal{I}^{op}, \mathcal{A})$ satisfies A.4.6.1, A.4.6.2 and A.4.6.3; this amounts to showing that a retract of a sequence satisfying the conditions also satisfies them. Sequences satisfying A.4.6.2 and A.4.6.3 are Mittag–Leffler; see Definition A.3.10. Note that A.4.6.2 is almost identical with A.3.10.1, and A.4.6.3 is almost identical with A.3.10.2; the difference is that for injective sequences (as opposed to Mittag–Leffler ones) one replaces epimorphisms by split epimorphisms.

There are a few useful facts that come very easily from the construction of injectives in $\mathcal{C}at(\mathcal{I}^{op}, \mathcal{A})$ as above. We begin with

LEMMA A.4.7. *Suppose $\mathcal{A}$ is an abelian category with enough injectives, satisfying $[AB3^*(\alpha)]$. Let β, γ be ordinals $< \alpha$. Suppose $f : \mathcal{I}(\beta) \longrightarrow \mathcal{I}(\gamma)$ is a map of partially ordered sets. Let*

$$F = \prod_{i<\gamma} F_{a_i}^i$$

be an injective object of the category $\mathcal{C}at(\mathcal{I}(\gamma)^{op}, \mathcal{A})$. Then the restriction of F to $\mathcal{I}(\beta)$, that is the composite functor

$$\mathcal{I}(\beta) \xrightarrow{\ f\ } \mathcal{I}(\gamma) \xrightarrow{\ F\ } \mathcal{A},$$

is an injective object in the category $\mathcal{C}at(\mathcal{I}(\beta)^{op}, \mathcal{A})$.

Proof: If F satisfies A.4.6.1, A.4.6.2 and A.4.6.3, then so does $F \circ f$. Thus if F can be written as a product of $F^i_{a_i}$ with a_i injective in $\mathcal{A}$, then so can $F \circ f$. $\square$

PROPOSITION A.4.8. *Suppose $\mathcal{A}$ is an abelian category with enough injectives, satisfying [AB3*(α)]. Let β, γ be ordinals $< \alpha$. Suppose $f : \mathfrak{I}(\beta) \longrightarrow \mathfrak{I}(\gamma)$ is a map of partially ordered sets. Suppose the map is cofinal: for every ordinal $i \in \mathfrak{I}(\gamma)$, there exists $j \in \mathfrak{I}(\beta)$ with $i \leq f(j)$. Suppose $F : \mathfrak{I}(\gamma) \longrightarrow \mathcal{A}$ is any functor. Then*

$$\varprojlim{}^n F \quad = \quad \varprojlim{}^n\{F \circ f\}.$$

That is, the derived functors of the limit of $F : \mathfrak{I}(\gamma) \longrightarrow \mathcal{A}$ agree with the derived functors of the limit of $Ff : \mathfrak{I}(\beta) \longrightarrow \mathcal{A}$.

Proof: Take an injective resolution of F in the category of functors $\mathfrak{I}(\gamma) \longrightarrow \mathcal{A}$, and choose the injectives to all be products of $F^i_{a_i}$, with a_i injective in $\mathcal{A}$. That is, we have an injective resolution

$$0 \longrightarrow F \longrightarrow G_0 \longrightarrow G_1 \longrightarrow G_2 \longrightarrow \cdots$$

Composing with $f : \mathfrak{I}(\beta) \longrightarrow \mathfrak{I}(\gamma)$, we obtain a resolution

$$0 \longrightarrow F \circ f \longrightarrow G_0 \circ f \longrightarrow G_1 \circ f \longrightarrow G_2 \circ f \longrightarrow \cdots$$

and the $G_i \circ f$ are injectives by Lemma A.4.7. The derived functors of the limit are obtained, in their respective categories, by applying the limit functor to the resolutions and computing homology. On the other hand, since the inclusion f is cofinal, we have an isomorphism of complexes

$$\begin{array}{ccccccc}
\varprojlim G_0 & \longrightarrow & \varprojlim G_1 & \longrightarrow & \varprojlim G_2 & \longrightarrow & \cdots \\
\Big\downarrow \wr & & \Big\downarrow \wr & & \Big\downarrow \wr & & \\
\varprojlim \{G_0 \circ f\} & \longrightarrow & \varprojlim \{G_1 \circ f\} & \longrightarrow & \varprojlim \{G_2 \circ f\} & \longrightarrow & \cdots
\end{array}$$

and hence the Lemma. $\square$

REMARK A.4.9. Proposition A.4.8 remains true for other cofinal inclusions of partially ordered sets. But since we do not use the result, I proved only the less general statement.

What does the standard construction of derived functors using injectives give, in the special case for Mittag–Leffler sequences? As above, given a sequence, to obtain $\varprojlim{}^n$ of it, we first resolve it by injectives in $\mathcal{C}at(\mathfrak{I}^{op}, \mathcal{A})$, then apply $\varprojlim$ to the resolution, then compute homology. Let us do this explicitly for Mittag–Leffler sequences in $\mathcal{C}at(\mathfrak{I}^{op}, \mathcal{A})$.

CONSTRUCTION A.4.10. Let α be an infinite cardinal. Let $\mathcal{A}$ be an abelian category satisfying $[AB3^*(\alpha)]$. Let $\mathcal{I} = \mathcal{I}(\gamma)$, with $\gamma < \alpha$. Suppose the category $\mathcal{A}$ has enough injectives.

Let F be an object of $\mathcal{C}at\big(\mathcal{I}^{op}, \mathcal{A}\big)$, that is a sequence in $\mathcal{A}$ of length $\gamma < \alpha$. Suppose F is Mittag–Leffler. We wish to compute $\varprojlim{}^n F$. Then we choose in $\mathcal{C}at\big(\mathcal{I}^{op}, \mathcal{A}\big)$ a resolution of F by injective objects. That is, a resolution

$$0 \longrightarrow F \longrightarrow i^0 \longrightarrow i^1 \longrightarrow i^2 \longrightarrow \cdots$$

For any ordinal $l < \gamma$, evaluating at l gives an exact sequence in $\mathcal{A}$

$$0 \longrightarrow F(l) \longrightarrow i^0(l) \longrightarrow i^1(l) \longrightarrow i^2(l) \longrightarrow \cdots$$

We are assuming that F is Mittag–Leffler. For any ordinal l, the map $F(l+1) \longrightarrow F(l)$ is epi. The map $i^k(l+1) \longrightarrow i^k(l)$ is not merely epi, but also a split epi. We therefore obtain a short exact sequence of resolutions

$$\begin{array}{ccccccccc}
 & & 0 & & 0 & & 0 & & \\
 & & \downarrow & & \downarrow & & \downarrow & & \\
0 & \longrightarrow & k_{l+1} & \longrightarrow & i^0_{l+1} & \longrightarrow & i^1_{l+1} & \longrightarrow & \cdots \\
 & & \downarrow & & \downarrow & & \downarrow & & \\
0 & \longrightarrow & F(l+1) & \longrightarrow & i^0(l+1) & \longrightarrow & i^1(l+1) & \longrightarrow & \cdots \\
 & & \downarrow & & \downarrow & & \downarrow & & \\
0 & \longrightarrow & F(l) & \longrightarrow & i^0(l) & \longrightarrow & i^1(l) & \longrightarrow & \cdots \\
 & & \downarrow & & \downarrow & & \downarrow & & \\
 & & 0 & & 0 & & 0 & &
\end{array}$$

where all but the first column are split exact.

If we assume that for all limit ordinals $l < \gamma$, the restriction of F to $\mathcal{I}(l) \subset \mathcal{I}(\gamma)$ has vanishing $\varprojlim{}^n$, then for any such l the sequence

$$0 \longrightarrow \lim_{m<l}\{F(m)\} \longrightarrow \lim_{m<l}\{i^0(m)\} \longrightarrow \lim_{m<l}\{i^1(m)\} \longrightarrow \cdots$$

is exact; the cohomology of the sequence is $\varprojlim{}^n$. The fact that

$$F(l) \longrightarrow \lim_{m<l}\{F(m)\} \qquad \text{and} \qquad i^k(l) \longrightarrow \lim_{m<l}\{i^k(m)\}$$

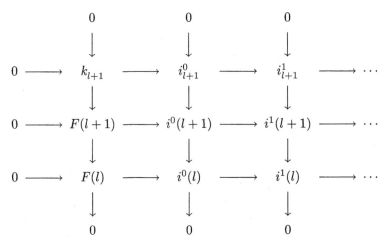

are, respectively, epi and split epi, allows us to obtain an exact sequence of acyclic chain complexes

$$\begin{array}{ccccccccc}
 & & 0 & & 0 & & 0 & & \\
 & & \downarrow & & \downarrow & & \downarrow & & \\
0 & \longrightarrow & k_l & \longrightarrow & i_l^0 & \longrightarrow & i_l^1 & \longrightarrow & \cdots \\
 & & \downarrow & & \downarrow & & \downarrow & & \\
0 & \longrightarrow & F(l) & \longrightarrow & i^0(l) & \longrightarrow & i^1(l) & \longrightarrow & \cdots \\
 & & \downarrow & & \downarrow & & \downarrow & & \\
0 & \longrightarrow & \lim\limits_{m<l}\{F(l)\} & \longrightarrow & \lim\limits_{m<l}\{i^0(l)\} & \longrightarrow & \lim\limits_{m<l}\{i^1(l)\} & \longrightarrow & \cdots \\
 & & \downarrow & & \downarrow & & \downarrow & & \\
 & & 0 & & 0 & & 0 & &
\end{array}$$

In other words, for every ordinal $l < \gamma$, successor or limit, we have constructed an exact sequence

$$0 \longrightarrow k_l \longrightarrow i_l^0 \longrightarrow i_l^1 \longrightarrow \cdots$$

with i_l^k all injective objects in $\mathcal{A}$. For any $l < \gamma$, there is an isomorphism

$$i^k(l) \quad = \quad \prod_{m<l} i_m^k.$$

The derived limit $\underleftarrow{\lim}^n F$ is the nth cohomology of the sequence

$$0 \longrightarrow \prod_{l<\gamma} i_l^0 \longrightarrow \prod_{l<\gamma} i_l^1 \longrightarrow \prod_{l<\gamma} i_l^2 \longrightarrow \cdots$$

Unfortunately, this is not the product of the sequences

$$0 \longrightarrow i_l^0 \longrightarrow i_l^1 \longrightarrow i_l^2 \longrightarrow \cdots$$

since the differential is twisted.

The reader can easily untwine the construction, to discover that if we resolve F by injective objects of the form $\prod_{i\in\mathcal{I}} F_{a_i}^i$, that is we give a resolution

$$0 \longrightarrow F \longrightarrow \prod_{i\in\mathcal{I}} F_{a_i^0}^i \longrightarrow \prod_{i\in\mathcal{I}} F_{a_i^1}^i \longrightarrow \cdots$$

then the terms in the exact resolution of k_l

$$0 \longrightarrow k_l \longrightarrow i_l^0 \longrightarrow i_l^1 \longrightarrow \cdots$$

are precisely the a_i's. Precisely, with the notation as above, $i_l^k = a_l^k$. Recall that, by Remark A.4.4, these may be chosen to lie in any class of injectives I large enough to admit embeddings of every object of $\mathcal{A}$.

We will see in Proposition A.5.12 that it is possible to have Mittag–Leffler sequences with non–vanishing $\lim^n$, even in categories satisfying [AB4*]; the fact that products of exact sequences are exact does not guarantee the exactness of twisted products.

A.5. A Mittag–Leffler sequence with non–vanishing $\lim^n$

We first define the category we wish to use for our construction.

DEFINITION A.5.1. *Let $\mathcal{S}$ be the category whose objects are complete, non–archimedean, normed abelian groups of cardinality $\leq 2^{\aleph_0}$, and whose morphisms are the contractions.*

REMARK A.5.2. Definition A.5.1 is quite a mouthful, so let us paraphrase it. An object of $\mathcal{S}$ is an abelian group A of cardinality $\leq 2^{\aleph_0}$, having a norm map. That is

A.5.2.1. For every $a \in A$, there is a number $\|a\| \in \mathbb{R}$. These numbers satisfy the inequality $\|a\| \geq 0$, with equality if and only if $a = 0$.

A.5.2.2. The norm is non–archimedean. It satisfies the inequality

$$\|a - b\| \leq \max(\|a\|, \|b\|).$$

A.5.2.3. The group A is complete with respect to the metric induced by the norm.

The morphisms in the category $\mathcal{S}$ are the contractions. They are homomorphisms of abelian groups $f : A \longrightarrow B$ satisfying

$$\|f(a)\| \leq \|a\|.$$

LEMMA A.5.3. *The category $\mathcal{S}$ contains coproducts of fewer than $\aleph_1$ of its objects.*

Proof: Suppose we are given fewer than $\aleph_1$ objects of $\mathcal{S}$, that is countably many objects $\{A_0, A_1, A_2, \cdots\}$. The A_i are all abelian groups of cardinality $\leq 2^{\aleph_0}$. Therefore the set theoretic product group

$$\prod_{i=0}^{\infty} A_i$$

has cardinality

$$\leq \left\{2^{\aleph_0}\right\}^{\aleph_0} \quad = \quad 2^{\aleph_0 \times \aleph_0} \quad = \quad 2^{\aleph_0}.$$

Define a norm map on $\prod_{i=0}^{\infty} A_i$ by the formula

$$\left\| \prod_{i=0}^{\infty} a_i \right\| \quad = \quad \sup_{i=0}^{\infty} \|a_i\|.$$

This norm takes its value in $\mathbb{R} \cup \{\infty\}$. The coproduct of the objects A_i in the category $\mathcal{S}$ is the subset of all elements of the set theoretic product, which are sequences whose norm tends to zero. That is,

$$\coprod_{i=0}^{\infty} A_i \quad \subset \quad \prod_{i=0}^{\infty} A_i,$$

and the condition for a sequence $\{a_0, a_1, a_2, \cdots\} \in \prod_{i=0}^{\infty} A_i$ to lie in the smaller $\coprod_{i=0}^{\infty} A_i$ is that

$$\lim_{i \to \infty} \|a_i\| = 0.$$

We give the subset

$$\coprod_{i=0}^{\infty} A_i \quad \subset \quad \prod_{i=0}^{\infty} A_i$$

the subspace norm; the reader can easily check that the subspace is complete. We need to establish that it satisfies the universal property of the coproduct.

Suppose for each $0 \leq i < \infty$ we have, in the category $\mathcal{S}$, a map $f_i : A_i \longrightarrow B$. That is, we have a contraction. Define

$$f : \coprod_{i=0}^{\infty} A_i \longrightarrow B$$

by the formula

$$f(a_0, a_1, a_2, \cdots) \quad = \quad \sum_{i=0}^{\infty} f_i(a_i).$$

This sum converges since as $i \to \infty$, we have first $\|a_i\| \to 0$, but as $\|f_i(a_i)\| \leq \|a_i\|$, we deduce $\|f_i(a_i)\| \longrightarrow 0$. Since the norm is non–archimedean,

$$\left\| \sum_{i=m}^{n} f_i(a_i) \right\| \quad \leq \quad \sup_{i=m}^{n} \|f_i(a_i)\| \quad \longrightarrow \quad 0$$

as $m, n \to \infty$. The partial sums form a cauchy sequence, which converges in the complete metric space B.

The fact that f is a contraction, and uniqueness of f given its composites with the inclusions

$$A_i \longrightarrow f : \coprod_{i=0}^{\infty} A_i$$

are both obvious. □

LEMMA A.5.4. *The category $\mathcal{S}$ is an additive category.*

Proof: Given two morphisms $f, g : A \longrightarrow B$ in $\mathcal{S}$, we form $f - g$ by the formula

$$\{f - g\}(a) = f(a) - g(a).$$

Since f and g are contractions, $\|f(a)\| \leq \|a\|$ and $\|g(a)\| \leq \|a\|$. This makes

$$\begin{aligned} \big\|\{f - g\}(a)\big\| &= \big\|f(a) - g(a)\big\| \\ &\leq \max\big(\|f(a)\|, \|g(a)\|\big) \\ &\leq \|a\|. \end{aligned}$$

Hence $f - g$ is a contraction, that is a morphism in $\mathcal{S}$.

This gives the *Hom*–sets $\mathcal{S}(A, B)$ the structure of abelian groups. Now observe that by, Lemma A.5.3, the category $\mathcal{S}$ contains countable coproducts of its objects, hence certainly finite coproducts. The reader can easily check that finite coproducts, as given in the proof of Lemma A.5.3, also satisfy the universal property of finite products. We conclude that the category $\mathcal{S}$ is additive. □

LEMMA A.5.5. *The category $\mathcal{S}$ contains kernels for all its morphisms.*

Proof: Let $f : A \longrightarrow B$ be a morphism in $\mathcal{S}$. The set theoretic kernel of f, given the subspace norm in A, is a closed subgroup of A and hence complete. It is the categorical kernel. □

LEMMA A.5.6. *Suppose we are given countably many morphisms in $\mathcal{S}$*

$$\{f_i : A_i \longrightarrow B_i \mid 0 \leq i < \infty\}.$$

The kernel of the coproduct map

$$\coprod_{i=0}^{\infty} f_i : \coprod_{i=0}^{\infty} A_i \longrightarrow \coprod_{i=0}^{\infty} B_i$$

is the coproduct of the kernels.

Proof: Both the kernel of the coproduct map and the coproduct of the kernels consist of sequences $\{a_0, a_1, a_2, \cdots\}$, with $a_i \in A_i$, so that $\|a_i\| \to 0$ and $f_i(a_i) = 0$. □

PROPOSITION A.5.7. *Let α be the cardinal $\aleph_1$. Then, for this regular cardinal α, the category $\mathcal{S}$ satisfies Hypothesis 6.1.1.*

Proof: The objects of $\mathcal{S}$ are abelian groups of cardinality $\leq 2^{\aleph_0}$, together with a norm. There is clearly only a set of these, up to isomorphism. Hence the category $\mathcal{S}$ is essentially small, and additive by Lemma A.5.4. This establishes 6.1.1.1.

Given fewer than $\alpha = \aleph_1$ objects of $\mathcal{S}$, their coproduct exists in $\mathcal{S}$ by Lemma A.5.3. This establishes 6.1.1.2.

The category $\mathcal{S}$ contains kernels, by Lemma A.5.5. Given a diagram

$$\begin{array}{ccc} & & x \\ & & \downarrow \\ x' & \longrightarrow & y \end{array}$$

the kernel of the map $x \oplus x' \longrightarrow y$ provides a genuine pullback, hence certainly a homotopy pullback. This establishes 6.1.1.3.

Finally, by Lemma A.5.6, the coproduct of $< \alpha$ (that is, countably many) kernels is the kernel. Hence the coproduct of $< \alpha$ homotopy pullback squares is a homotopy pullback square. That establishes 6.1.1.4. □

COROLLARY A.5.8. *The theory of Chapter 6 now applies. We deduce that $\mathcal{E}x(\mathcal{S}^{op}, \mathcal{A}b)$ is an abelian category (Lemma 6.1.4) satisfying [AB4*] (Lemma 6.3.1) and [AB4] (Lemma 6.3.2).* □

CONSTRUCTION A.5.9. Consider now the sequence of objects and morphisms in $\mathcal{S}$

$$\mathbb{Z}_p \xrightarrow{\ p\ } \mathbb{Z}_p \xrightarrow{\ p\ } \mathbb{Z}_p \xrightarrow{\ p\ } \cdots$$

where $\mathbb{Z}_p$ is the p–adic numbers with the usual topology, and the connecting maps are multiplication by p. The Yoneda functor

$$\mathcal{S} \longrightarrow \mathcal{E}x(\mathcal{S}^{op}, \mathcal{A}b)$$

takes this to a sequence in $\mathcal{E}x(\mathcal{S}^{op}, \mathcal{A}b)$. We remind the reader: the Yoneda functor takes an object $s \in \mathcal{S}$ to the representable functor $\mathcal{S}(-, s)$. In the rest of this Section, we will freely confuse the sequence in $\mathcal{S}$ with its image in $\mathcal{E}x(\mathcal{S}^{op}, \mathcal{A}b)$.

LEMMA A.5.10. *The sequence*

$$\mathbb{Z}_p \xrightarrow{\ p\ } \mathbb{Z}_p \xrightarrow{\ p\ } \mathbb{Z}_p \xrightarrow{\ p\ } \cdots$$

is a co–Mittag–Leffler sequence in $\mathcal{E}x(\mathcal{S}^{op}, \mathcal{A}b)$.

Proof: The kernel of $p : \mathbb{Z}_p \longrightarrow \mathbb{Z}_p$ is trivial, and hence the map

$$\mathcal{S}(-, \mathbb{Z}_p) \xrightarrow{\ p\ } \mathcal{S}(-, \mathbb{Z}_p)$$

is injective. □

LEMMA A.5.11. *The sequence in $\mathcal{E}x(\mathcal{S}^{op}, \mathcal{A}b)$*

$$\mathbb{Z}_p \xrightarrow{\ p\ } \mathbb{Z}_p \xrightarrow{\ p\ } \mathbb{Z}_p \xrightarrow{\ p\ } \cdots$$

has a vanishing colimit (and also a vanishing $\underrightarrow{\mathrm{colim}}^1$).

Proof: The colimit and $\underrightarrow{\text{colim}}^1$ are, respectively, the cokernel and kernel of the map

$$\coprod_{i=0}^{\infty} \mathcal{S}(-,\mathbb{Z}_p) \xrightarrow{1-shift} \coprod_{i=0}^{\infty} \mathcal{S}(-,\mathbb{Z}_p).$$

By Lemma 6.1.17, the natural map gives an isomorphism

$$\coprod_{i=0}^{\infty} \mathcal{S}(-,\mathbb{Z}_p) \longrightarrow \mathcal{S}\left(-,\coprod_{i=0}^{\infty} \mathbb{Z}_p\right);$$

in the commutative square below the vertical maps are isomorphisms

$$\begin{array}{ccc} \coprod_{i=0}^{\infty} \mathcal{S}(-,\mathbb{Z}_p) & \xrightarrow{1-shift} & \coprod_{i=0}^{\infty} \mathcal{S}(-,\mathbb{Z}_p) \\ \downarrow \wr & & \downarrow \wr \\ \mathcal{S}\left(-,\coprod_{i=0}^{\infty} \mathbb{Z}_p\right) & \xrightarrow{1-shift} & \mathcal{S}\left(-,\coprod_{i=0}^{\infty} \mathbb{Z}_p\right) \end{array}$$

It therefore suffices to show that the map

$$\coprod_{i=0}^{\infty} \mathbb{Z}_p \xrightarrow{1-shift} \coprod_{i=0}^{\infty} \mathbb{Z}_p.$$

is an isomorphism. But its inverse is given by

$$\{1-shift\}^{-1} \quad = \quad 1+\{shift\}+\{shift\}^2+\cdots$$

and the right hand side converges since the shift map is divisible by p. □

PROPOSITION A.5.12. *There exists in $\mathcal{E}x\big(\mathcal{S}^{op},\mathcal{A}b\big)$ a co–Mittag–Leffler sequence with a non–vanishing* $\underrightarrow{\text{colim}}^1$.

Proof: Consider the short exact sequence of sequences in $\mathcal{E}x\big(\mathcal{S}^{op},\mathcal{A}b\big)$

$$
\begin{array}{ccccccccc}
0 & & 0 & & 0 & & 0 & & \\
\downarrow & & \downarrow & & \downarrow & & \downarrow & & \\
\mathbb{Z}_p & \xrightarrow{1} & \mathbb{Z}_p & \xrightarrow{1} & \mathbb{Z}_p & \xrightarrow{1} & \mathbb{Z}_p & \xrightarrow{1} & \cdots \\
{\scriptstyle 1}\downarrow & & {\scriptstyle p}\downarrow & & {\scriptstyle p^2}\downarrow & & {\scriptstyle p^3}\downarrow & & \\
\mathbb{Z}_p & \xrightarrow{p} & \mathbb{Z}_p & \xrightarrow{p} & \mathbb{Z}_p & \xrightarrow{p} & \mathbb{Z}_p & \xrightarrow{p} & \cdots \\
\downarrow & & \downarrow & & \downarrow & & \downarrow & & \\
0 & \longrightarrow & Q_1 & \longrightarrow & Q_2 & \longrightarrow & Q_3 & \longrightarrow & \cdots \\
\downarrow & & \downarrow & & \downarrow & & \downarrow & & \\
0 & & 0 & & 0 & & 0 & &
\end{array}
$$

We deduce an exact sequence

$$
\begin{array}{ccc}
\underrightarrow{\mathrm{colim}}^1 Q_i & \longrightarrow & \underrightarrow{\mathrm{colim}}\{\mathbb{Z}_p \xrightarrow{1} \mathbb{Z}_p \xrightarrow{1} \cdots\} \\
& & \downarrow \\
& & \underrightarrow{\mathrm{colim}}\{\mathbb{Z}_p \xrightarrow{p} \mathbb{Z}_p \xrightarrow{p} \cdots\}
\end{array}
$$

In Lemma A.5.11 we computed that the term at the bottom vanishes. Hence our exact sequence becomes

$$
\underrightarrow{\mathrm{colim}}^1 Q_i \longrightarrow \mathbb{Z}_p \longrightarrow 0
$$

which establishes that the sequence $\{Q_i\}$ has a non–vanishing colim^1. But the sequence $\{Q_i\}$ is the quotient of the co–Mittag–Leffler sequence (see Lemma A.5.10)

$$
\mathbb{Z}_p \xrightarrow{p} \mathbb{Z}_p \xrightarrow{p} \mathbb{Z}_p \xrightarrow{p} \cdots
$$

by the constant term $\mathbb{Z}_p$, hence must be co–Mittag–Leffler. □

REMARK A.5.13. The category $\mathcal{E}x(\mathcal{S}^{op}, \mathcal{A}b)$ is therefore an example of an abelian category satisfying [AB4] but not [AB4.5]. See Definition A.3.16.

A.6. History of the results of Appendix A

The new results in the Appendix are the definitions and properties of abelian categories satisfying [AB4.5], and the counterexample of Section A.5, which shows that [AB4] does not imply [AB4.5]. More about this later.

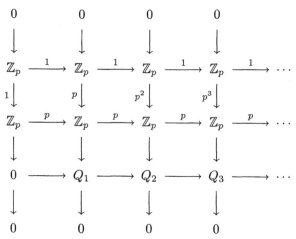

Section A.1 deals with locally presentable categories. Everything in the Section is discussed more thoroughly and in greater generality in the book [**16**] by Gabriel and Ulmer. We provide only what is needed here.

Section A.2 covers Gabriel's theory of quotient abelian categories. All the results may be found in Gabriel's thesis, [**15**]. For more recent treatments, the reader may wish to consult Golan's [**17**] or Stenström's [**33**].

Sections A.3 and A.4 deal with the derived functors of limits. Much of the material can be found in Roos' articles. The notable exception, as was mentioned above, is the treatment of Mittag–Leffler sequences. The reader is referred to Roos' articles [**29**], [**30**], [**31**] and [**32**]. The definitions and elementary properties of derived functors of the limit are given in [**29**]. The fact that Mittag–Leffler sequences have vanishing $\lim^1$ may also be found there. But as the reader will see by examining the four articles, especially the (unpublished) [**30**], Roos believed that the vanishing of $\lim^1$ of Mittag–Leffler sequences ought to be a formal consequence of [AB4*]. He conjectures, in fact, that Grothendieck categories satisfying [AB4*] should have enough projectives.

Section A.5 gives a counterexample to a "generalised" Roos conjecture. We produce a category $\mathcal{E}x\big(\mathcal{S}^{op}, \mathcal{A}b\big)$, whose dual satisfies [AB4*] but not [AB4.5*]. The dual therefore does not have enough projectives. This means we have a category satisfying [AB4*] without enough projectives. It is not a counterexample to Roos' conjecture, as he stated it, since it is not Grothendieck. Neither $\mathcal{E}x\big(\mathcal{S}^{op}, \mathcal{A}b\big)$ nor its dual satisfy [AB5].

The point of Sections A.3, A.4 and A.5 is to highlight the relation between the existence of projectives and [AB4.5*], that is the vanishing of $\lim^1$ for Mittag–Leffler sequences. Roos' articles, most especially [**30**], highlight the relation between the existence of projectives and [AB4*].

APPENDIX B

Homological functors into [AB5$^\alpha$] categories

B.1. A filtration

Let $\mathcal{T}$ be a trianagulated category satisfying [TR5]. Let α be a regular cardinal. If $\mathcal{T}$ is α–compactly generated, any object $X \in \mathcal{T}$ can be given a filtration. We wish to discuss this. First we remind the reader of an easy fact.

LEMMA B.1.1. *Let α be a regular cardinal. Let $\mathcal{T}$ be a triangulated category satisfying [TR5]. Put $\mathcal{S} = \mathcal{T}^\alpha$.*

Given any set $\{s_\lambda, \lambda \in \Lambda\}$ of objects of $\mathcal{S}$, and an object $y \in \mathcal{T}$, the natural map

$$\mathcal{T}\left(\coprod_{\lambda\in\Lambda} s_\lambda \, , \, y\right) \longrightarrow \mathcal{E}x(\mathcal{S}^{op}, \mathcal{A}b)\left\{\mathcal{T}\left(-, \coprod_{\lambda\in\Lambda} s_\lambda\right)\bigg|_{\mathcal{S}} \, , \, \mathcal{T}(-,y)|_{\mathcal{S}}\right\}$$

is an isomorphism.

Proof: (cf. proof of Proposition 6.5.3.) To give an element of the group

$$\mathcal{E}x(\mathcal{S}^{op}, \mathcal{A}b)\left\{\mathcal{T}\left(-, \coprod_{\lambda\in\Lambda} s_\lambda\right)\bigg|_{\mathcal{S}} \, , \, \mathcal{T}(-,y)|_{\mathcal{S}}\right\}$$

is to give a map in $\mathcal{E}x(\mathcal{S}^{op}, \mathcal{A}b)$

$$\mathcal{T}\left(-, \coprod_{\lambda\in\Lambda} s_\lambda\right)\bigg|_{\mathcal{S}} \longrightarrow \mathcal{T}(-,y)|_{\mathcal{S}}.$$

Now, the functor

$$\mathcal{T} \longrightarrow \mathcal{E}x(\mathcal{S}^{op}, \mathcal{A}b)$$

respects coproducts, and hence

$$\begin{aligned}\mathcal{T}\left(-, \coprod_{\lambda\in\Lambda} s_\lambda\right)\bigg|_{\mathcal{S}} &= \coprod_{\lambda\in\Lambda} \mathcal{T}(-, s_\lambda)|_{\mathcal{S}} \\ &= \coprod_{\lambda\in\Lambda} \mathcal{S}(-, s_\lambda).\end{aligned}$$

On the other hand, a map

$$\coprod_{\lambda\in\Lambda} \mathcal{S}(-,s_\lambda) \longrightarrow \mathcal{T}(-,y)|_{\mathcal{S}}$$

is, by the univesal property of the coproduct, a collections of maps

$$\mathcal{S}(-,s_\lambda) \longrightarrow \mathcal{T}(-,y)|_{\mathcal{S}},$$

that is, maps in $\mathcal{T}$

$$s_\lambda \longrightarrow y.$$

But these combine to a map

$$\coprod_{\lambda\in\Lambda} s_\lambda \longrightarrow y.$$

Summarising, maps

$$\mathcal{T}\left(-, \coprod_{\lambda\in\Lambda} s_\lambda\right)\Bigg|_{\mathcal{S}} \longrightarrow \mathcal{T}(-,y)|_{\mathcal{S}}$$

are in natural bijective correspondence with maps

$$\coprod_{\lambda\in\Lambda} s_\lambda \longrightarrow y.$$

□

One more reminder.

REMINDER B.1.2. Let $\mathcal{I}$ be a small category. The nerve of the category $\mathcal{I}$ is defined as the simplicial set $\mathcal{N}.(\mathcal{I})$, where $\mathcal{N}_k(\mathcal{I})$ (the k–simplices) are sequences of k composable morphisms

$$i_0 \longrightarrow i_1 \longrightarrow i_2 \longrightarrow \cdots \longrightarrow i_k.$$

Let $\mathcal{A}$ be an additive category, closed under coproducts. Given a functor $f : \mathcal{I} \longrightarrow \mathcal{A}$, we form a chain complex

$$\cdots \xrightarrow{\partial_2} N_2(f) \xrightarrow{\partial_1} N_1(f) \xrightarrow{\partial_0} N_0(f)$$

where

$$N_k(f) \quad = \coprod_{\{i_0\to i_1\to\cdots\to i_k\}\in\mathcal{N}_k(\mathcal{I})} f(i_0).$$

The differential $\partial_k : N_{k+1}(f) \longrightarrow N_k(f)$ is given by the usual alternating sum. That is,

$$\partial_k \quad = \quad \sum_{j=0}^{k+1}\{-1\}^j \partial_k^j,$$

and ∂_k^j is the map induced by deleting the jth term in the sequence

$$i_0 \longrightarrow i_1 \longrightarrow i_2 \longrightarrow \cdots \longrightarrow i_k \longrightarrow i_{k+1}.$$

We will denote the entire chain complex

$$\cdots \xrightarrow{\partial_2} N_2(f) \xrightarrow{\partial_1} N_1(f) \xrightarrow{\partial_0} N_0(f)$$

by the symbol $N(f)$. If $\mathcal{A}$ is an abelian category satisfying [AB4], then Lemma A.3.2 proved that the nth homology of the complex $N(f)$ is naturally $\underrightarrow{\operatorname{colim}}^n f$. But the chain complex makes sense in any additive category with coproducts.

LEMMA B.1.3. *Let α be a regular cardinal. Let $\mathcal{T}$ be an α–compactly generated triangulated category. Let X be an object of $\mathcal{T}$. There exists a (countable) sequence of objects of $\mathcal{T}$*

$$X_0 \longrightarrow X_1 \longrightarrow X_2 \longrightarrow \cdots$$

so that

B.1.3.1. $X = \underrightarrow{\operatorname{Hocolim}} X_i$.

B.1.3.2. *Put $Y_0 = X_0$. For any $i \geq 1$, let Y_i come from the triangle*

$$X_{i-1} \longrightarrow X_i \longrightarrow Y_i \longrightarrow \Sigma X_{i-1}.$$

B.1.3.3. *The composites*

$$Y_i \longrightarrow \Sigma X_{i-1} \longrightarrow \Sigma Y_{i-1}$$

give a chain complex

$$\cdots \longrightarrow \Sigma^{-2} Y_2 \longrightarrow \Sigma^{-1} Y_1 \longrightarrow Y_0.$$

The entire collection of data can be so chosen, so that there exists an α–filtered functor $f : \mathcal{I} \longrightarrow \mathcal{T}^\alpha$ and the sequence of Y's is nothing more nor less than $N(f)$, the realisation of the nerve of the functor f.

Proof: Let $\mathcal{S} = \mathcal{T}^\alpha$. We are assuming that $\mathcal{T}$ is α–compactly generated, hence $\mathcal{S}$ is essentially small. Observe first that $\mathcal{T}(-,X)|_\mathcal{S}$ is a homological object in $\mathcal{E}x\big(\mathcal{S}^{op}, \mathcal{A}b\big)$. By Lemma 7.2.4, it follows that $\mathcal{T}(-,X)|_\mathcal{S}$ is an α–filtered colimit of representable functors. There is an α–filtered category $\mathcal{I}$ and a functor $f : \mathcal{I} \longrightarrow \mathcal{S}$, so that $\mathcal{T}(-,X)|_\mathcal{S}$ is the colimit of f in $\mathcal{E}x\big(\mathcal{S}^{op}, \mathcal{A}b\big)$.

We can form, in $\mathcal{T}$, the chain complex $N(f)$. We have a chain complex

$$\cdots \xrightarrow{\partial_2} N_2(f) \xrightarrow{\partial_1} N_1(f) \xrightarrow{\partial_0} N_0(f).$$

Since the functor $\mathcal{T} \longrightarrow \mathcal{E}x\big(\mathcal{S}^{op}, \mathcal{A}b\big)$ respects coproducts, the image of this sequence in $\mathcal{E}x\big(\mathcal{S}^{op}, \mathcal{A}b\big)$ is the usual chain complex computing $\underrightarrow{\operatorname{colim}}^n$.

On the other hand, α–filtered colimits in $\mathcal{E}x\big(\mathcal{S}^{op}, \mathcal{A}b\big)$ agree with those in $\mathcal{C}at\big(\mathcal{S}^{op}, \mathcal{A}b\big)$, and are exact. For $n \geq 1$, $\underrightarrow{\operatorname{colim}}^n f = 0$. We deduce an exact sequence, in $\mathcal{E}x\big(\mathcal{S}^{op}, \mathcal{A}b\big)$,

$$\xrightarrow{\partial_1} \mathcal{T}\big(-, N_1(f)\big)|_{\mathcal{S}} \xrightarrow{\partial_0} \mathcal{T}\big(-, N_0(f)\big)|_{\mathcal{S}} \longrightarrow \mathcal{T}\big(-, X\big)|_{\mathcal{S}} \longrightarrow 0.$$

We define $\Sigma^{-i} Y_i = N_i(f)$. Now the problem is to prove the existence of the X_i, and that their homotopy colimit is X.

Note first that, in any case, we have that each Y_k is given by

$$\Sigma^{-k} Y_k = N_k(f) \quad = \coprod_{\{i_0 \to i_1 \to \cdots \to i_k\} \in \mathcal{N}_k(\mathcal{I})} f(i_0)$$

is a coproduct of objects in $\mathcal{S}$. Hence

B.1.3.4. *For any Y_i as above and any object $Z \in \mathcal{T}$, Lemma B.1.1 asserts that*

$$\mathcal{T}(Y_i, Z) \quad = \quad \mathcal{E}x\big(\mathcal{S}^{op}, \mathcal{A}b\big)\Big\{\mathcal{T}\big(-, Y_i\big)|_{\mathcal{S}}\ ,\ \mathcal{T}\big(-, Z\big)|_{\mathcal{S}}\Big\}.$$

We define $X_0 = Y_0$. We are given a map in $\mathcal{E}x\big(\mathcal{S}^{op}, \mathcal{A}b\big)$

$$\mathcal{T}\big(-, X_0\big)|_{\mathcal{S}} \quad = \quad \mathcal{T}\big(-, Y_0\big)|_{\mathcal{S}} \longrightarrow \mathcal{T}\big(-, X\big)|_{\mathcal{S}}.$$

By B.1.3.4, this comes from a (unique) map

$$X_0 \longrightarrow X.$$

We also have a map $\Sigma^{-1} Y_1 \longrightarrow Y_0 = X_0$, and the composite

$$\Sigma^{-1} Y_1 \longrightarrow X_0 \longrightarrow X$$

induces

$$\mathcal{T}\big(-, \Sigma^{-1} Y_1\big)|_{\mathcal{S}} \longrightarrow \mathcal{T}\big(-, X_0\big)|_{\mathcal{S}} \longrightarrow \mathcal{T}\big(-, X\big)|_{\mathcal{S}}$$

which vanishes. By B.1.3.4, it follows that the composite

$$\Sigma^{-1} Y_1 \longrightarrow X_0 \longrightarrow X$$

vanishes already in $\mathcal{T}$. If we form the triangle

$$\Sigma^{-1} Y_1 \longrightarrow X_0 \longrightarrow X_1 \longrightarrow Y_1,$$

we deduce that the map

$$X_0 \longrightarrow X$$

factors as

$$X_0 \longrightarrow X_1 \xrightarrow{g} X.$$

Now recall that the sequence

$$\begin{array}{ccccccc} \longrightarrow & \mathcal{T}\big(-,\Sigma^{-2}Y_2\big)\big|_{\mathcal{S}} & \longrightarrow & \mathcal{T}\big(-,\Sigma^{-1}Y_1\big)\big|_{\mathcal{S}} & \longrightarrow & \mathcal{T}(-,Y_0)|_{\mathcal{S}} \\ & & & & & \downarrow \\ & & & & & \mathcal{T}(-,X)|_{\mathcal{S}} \\ & & & & & \downarrow \\ & & & & & 0 \end{array}$$

is exact; it is, by construction, the sequence computing $\underrightarrow{\operatorname{colim}}^n$ of an α–filtered colimit in $\mathcal{E}x\big(\mathcal{S}^{op},\mathcal{A}b\big)$. In particular

$$\mathcal{T}\big(-,\Sigma^{-1}Y_1\big)\big|_{\mathcal{S}} \longrightarrow \mathcal{T}(-,X_0)|_{\mathcal{S}} \longrightarrow \mathcal{T}(-,X)|_{\mathcal{S}} \longrightarrow 0$$

is exact. On the other hand, from the triangle

$$\Sigma^{-1}Y_1 \longrightarrow X_0 \longrightarrow X_1 \longrightarrow Y_1$$

we deduce an exact sequence

$$\mathcal{T}\big(-,\Sigma^{-1}Y_1\big)\big|_{\mathcal{S}} \longrightarrow \mathcal{T}(-,X_0)|_{\mathcal{S}} \longrightarrow \mathcal{T}(-,X_1)|_{\mathcal{S}}.$$

We have a commutative diagram with exact rows

$$\begin{array}{ccccccc} \mathcal{T}\big(-,\Sigma^{-1}Y_1\big)\big|_{\mathcal{S}} & \longrightarrow & \mathcal{T}(-,X_0)|_{\mathcal{S}} & \longrightarrow & \mathcal{T}(-,X)|_{\mathcal{S}} & \longrightarrow & 0 \\ \| \downarrow \wr & & \| \downarrow \wr & & & & \\ \mathcal{T}\big(-,\Sigma^{-1}Y_1\big)\big|_{\mathcal{S}} & \longrightarrow & \mathcal{T}(-,X_0)|_{\mathcal{S}} & \longrightarrow & \mathcal{T}(-,X_1)|_{\mathcal{S}} & & \end{array}$$

which may be completed, uniquely, to

$$\begin{array}{ccccccc} \mathcal{T}\big(-,\Sigma^{-1}Y_1\big)\big|_{\mathcal{S}} & \longrightarrow & \mathcal{T}(-,X_0)|_{\mathcal{S}} & \longrightarrow & \mathcal{T}(-,X)|_{\mathcal{S}} & \longrightarrow & 0 \\ \| \downarrow \wr & & \| \downarrow \wr & & f \downarrow & & \\ \mathcal{T}\big(-,\Sigma^{-1}Y_1\big)\big|_{\mathcal{S}} & \longrightarrow & \mathcal{T}(-,X_0)|_{\mathcal{S}} & \longrightarrow & \mathcal{T}(-,X_1)|_{\mathcal{S}} & & \end{array}$$

The composite

$$\mathcal{T}(-,X_0)|_{\mathcal{S}} \longrightarrow \mathcal{T}(-,X)|_{\mathcal{S}} \xrightarrow{\ f\ } \mathcal{T}(-,X_1)|_{\mathcal{S}} \xrightarrow{\mathcal{T}(-,g)|_{\mathcal{S}}} \mathcal{T}(-,X)|_{\mathcal{S}}$$

is the natural surjection

$$\mathcal{T}(-,X_0)|_{\mathcal{S}} \longrightarrow \mathcal{T}(-,X)|_{\mathcal{S}}.$$

In other words, the two composites

$$\mathcal{T}(-,X_0)|_{\mathcal{S}} \longrightarrow \mathcal{T}(-,X)|_{\mathcal{S}} \quad \underset{1}{\overset{\{\mathcal{T}(-,g)|_{\mathcal{S}}\}\circ f}{\rightrightarrows}} \quad \mathcal{T}(-,X)|_{\mathcal{S}}$$

are equal, and we deduce that

$$\mathfrak{T}(-,X)|_{\mathcal{S}} \xrightarrow{\ f\ } \mathfrak{T}(-,X_1)|_{\mathcal{S}} \xrightarrow{\mathfrak{T}(-,g)|_{\mathcal{S}}} \mathfrak{T}(-,X)|_{\mathcal{S}}$$

must compose to the identity. Hence $\mathfrak{T}(-,X_1)|_{\mathcal{S}}$ contains $\mathfrak{T}(-,X)|_{\mathcal{S}}$ as a direct summand. The other direct summand has no choice. From the exact sequence

$$\begin{array}{llll} \mathfrak{T}\big(-,\Sigma^{-1}Y_1\big)\big|_{\mathcal{S}} & & & \\ \quad\downarrow & & & \\ \mathfrak{T}(-,X_0)|_{\mathcal{S}} \longrightarrow & \mathfrak{T}(-,X_1)|_{\mathcal{S}} \longrightarrow & \mathfrak{T}(-,Y_1)|_{\mathcal{S}} \longrightarrow & \mathfrak{T}(-,\Sigma X_0)|_{\mathcal{S}} \end{array}$$

it clearly has to be the kernel of

$$\mathfrak{T}(-,Y_1)|_{\mathcal{S}} \longrightarrow \mathfrak{T}(-,\Sigma X_0)|_{\mathcal{S}},$$

which is also the image of

$$\mathfrak{T}\big(-,\Sigma^{-1}Y_2\big)\big|_{\mathcal{S}} \longrightarrow \mathfrak{T}(-,Y_1)|_{\mathcal{S}}.$$

Suppose therefore, by induction, that up to some integer n, we have constructed a sequence

$$X_0 \longrightarrow X_1 \longrightarrow \cdots \longrightarrow X_{n-1} \longrightarrow X_n \longrightarrow X.$$

We already know that there is a map $\mathfrak{T}(-,X)|_{\mathcal{S}} \longrightarrow \mathfrak{T}(-,X_1)|_{\mathcal{S}}$, so that the composite

$$\mathfrak{T}(-,X)|_{\mathcal{S}} \longrightarrow \mathfrak{T}(-,X_1)|_{\mathcal{S}} \longrightarrow \mathfrak{T}(-,X)|_{\mathcal{S}}$$

is the identity. So $\mathfrak{T}(-,X)|_{\mathcal{S}}$ must be a direct summand of each $\mathfrak{T}(-,X_i)|_{\mathcal{S}}$, $1 \le i \le n$. Suppose further that

B.1.3.5. *For $i < n$, the other direct summand is the kernel of* $\mathfrak{T}(-,X_i)|_{\mathcal{S}} \longrightarrow \mathfrak{T}\big(-,X_{i+1}\big)\big|_{\mathcal{S}}$.

B.1.3.6. *For each $i \le n$, we have a triangle*

$$X_{i-1} \longrightarrow X_i \longrightarrow Y_i \longrightarrow \Sigma X_{i-1}$$

B.1.3.7. *The images of*

$$\begin{array}{rcl} \mathfrak{T}(-,X_i)|_{\mathcal{S}} & \longrightarrow & \mathfrak{T}(-,Y_i)|_{\mathcal{S}} \\ \mathfrak{T}(-,Y_i)|_{\mathcal{S}} & \longrightarrow & \mathfrak{T}\big(-,\Sigma X_{i-1}\big)\big|_{\mathcal{S}} \end{array}$$

agree, respectively, with

$$\begin{array}{l} Im\Big\{\mathfrak{T}\big(-,\Sigma^{-1}Y_{i+1}\big)\big|_{\mathcal{S}} \longrightarrow \mathfrak{T}(-,Y_i)|_{\mathcal{S}}\Big\} \\ Im\Big\{\mathfrak{T}(-,Y_i)|_{\mathcal{S}} \longrightarrow \mathfrak{T}\big(-,\Sigma Y_{i-1}\big)\big|_{\mathcal{S}}\Big\}. \end{array}$$

B.1.3.8. *For $i < n$, the isomorphisms of the image*

$$Im\Big\{\mathcal{T}\big(-,\Sigma^{-1}Y_{i+1}\big)\big|_{\mathcal{S}} \longrightarrow \mathcal{T}(-,Y_i)|_{\mathcal{S}}\Big\}$$

with

$$Im\Big\{\mathcal{T}\big(-,\Sigma^{-1}Y_{i+1}\big)\big|_{\mathcal{S}} \longrightarrow \mathcal{T}(-,X_i)|_{\mathcal{S}}\Big\}$$
$$Im\Big\{\mathcal{T}(-,X_i)|_{\mathcal{S}} \longrightarrow \mathcal{T}(-,Y_i)|_{\mathcal{S}}\Big\}$$

are compatible. That is, the composite

$$\mathcal{T}\big(-,\Sigma^{-1}Y_{i+1}\big)\big|_{\mathcal{S}} \longrightarrow \mathcal{T}(-,X_i)|_{\mathcal{S}} \longrightarrow \mathcal{T}(-,Y_i)|_{\mathcal{S}}$$

is the given map. After all, the sequence of the Y's is given. By B.1.3.4, this has to mean that the composite

$$\Sigma^{-1}Y_{i+1} \longrightarrow X_i \longrightarrow Y_i$$

is also correct in $\mathcal{T}$. Maps $Y \longrightarrow Z$ in $\mathcal{T}$ are in 1–to–1 correspondence with their images in $\mathcal{E}x\big(\mathcal{S}^{op},\mathcal{A}b\big)$.

We wish to show that the sequence of X's may be continued to $n+1$.

By the induction hypothesis B.1.3.6, we have a triangle

$$X_{n-1} \longrightarrow X_n \longrightarrow Y_n \longrightarrow \Sigma X_{n-1}.$$

Hence an exact sequence

$$\mathcal{T}\big(-,X_{n-1}\big)\big|_{\mathcal{S}} \longrightarrow \mathcal{T}(-,X_n)|_{\mathcal{S}} \longrightarrow \mathcal{T}(-,Y_n)|_{\mathcal{S}}.$$

By hypothesis B.1.3.5, the image of $\mathcal{T}\big(-,X_{n-1}\big)\big|_{\mathcal{S}} \longrightarrow \mathcal{T}(-,X_n)|_{\mathcal{S}}$ is the direct summand $\mathcal{T}(-,X)|_{\mathcal{S}} \subset \mathcal{T}\big(-,X_{n-1}\big)\big|_{\mathcal{S}}$. After all,

$$\mathcal{T}\big(-,X_{n-1}\big)\big|_{\mathcal{S}} \quad = \quad K \oplus \mathcal{T}(-,X)|_{\mathcal{S}},$$

where K is the kernel of the map. From the exact sequence, $\mathcal{T}(-,X_n)|_{\mathcal{S}}$ is a direct sum of $\mathcal{T}(-,X)|_{\mathcal{S}}$ and the image of

$$\mathcal{T}(-,X_n)|_{\mathcal{S}} \longrightarrow \mathcal{T}(-,Y_n)|_{\mathcal{S}}.$$

Again by induction hypothesis, this image is also the image of

$$\mathcal{T}\big(-,\Sigma^{-1}Y_{n+1}\big)\big|_{\mathcal{S}} \longrightarrow \mathcal{T}(-,Y_n)|_{\mathcal{S}}.$$

This image is a direct summand of $\mathcal{T}(-,X_n)|_{\mathcal{S}}$. It is the kernel of the map

$$\mathcal{T}(-,X_n)|_{\mathcal{S}} \longrightarrow \mathcal{T}(-,X)|_{\mathcal{S}}$$

induced by $X_n \longrightarrow X$. It follows that the map from $\mathcal{T}\big(-,\Sigma^{-1}Y_{n+1}\big)\big|_{\mathcal{S}}$ to this image also gives a map to the kernel above. We have an exact sequence

$$\mathcal{T}\big(-,\Sigma^{-1}Y_{n+1}\big)\big|_{\mathcal{S}} \longrightarrow \mathcal{T}(-,X_n)|_{\mathcal{S}} \longrightarrow \mathcal{T}(-,X)|_{\mathcal{S}},$$

and by construction, the composite

$$\mathcal{T}\big(-,\Sigma^{-1}Y_{n+1}\big)\big|_{\mathcal{S}} \longrightarrow \mathcal{T}(-,X_n)|_{\mathcal{S}} \longrightarrow \mathcal{T}(-,Y_n)|_{\mathcal{S}}$$

is right. Hence B.1.3.8 is proved for $n+1$, modulo the fact that we have yet to define a map $\Sigma^{-1}Y_{n+1} \longrightarrow X_n$ in $\mathcal{T}$ lifting the map constructed above in $\mathcal{E}x\big(\mathcal{S}^{op}, \mathcal{A}b\big)$.

But now B.1.3.4 applies. The map

$$\mathcal{T}\big(-, \Sigma^{-1}Y_{n+1}\big)\big|_{\mathcal{S}} \longrightarrow \mathcal{T}(-, X_n)|_{\mathcal{S}}$$

is a morphism from a $\mathcal{T}(-, Y)|_{\mathcal{S}}$, and must come from a morphism in $\mathcal{T}$

$$\Sigma^{-1}Y_{n+1} \longrightarrow X_n.$$

The vanishing of the composite

$$\mathcal{T}\big(-, \Sigma^{-1}Y_{n+1}\big)\big|_{\mathcal{S}} \longrightarrow \mathcal{T}(-, X_n)|_{\mathcal{S}} \longrightarrow \mathcal{T}(-, X)|_{\mathcal{S}}$$

implies the vanishing of

$$\Sigma^{-1}Y_{n+1} \longrightarrow X_n \longrightarrow X.$$

Form the triangle

$$\Sigma^{-1}Y_{n+1} \longrightarrow X_n \longrightarrow X_{n+1} \longrightarrow Y_{n+1}.$$

The vanishing of

$$\Sigma^{-1}Y_{n+1} \longrightarrow X_n \longrightarrow X$$

means that $X_n \longrightarrow X$ must factor as

$$X_n \longrightarrow X_{n+1} \longrightarrow X.$$

Finally, the map $\mathcal{T}\big(-, \Sigma^{-1}Y_{n+1}\big)\big|_{\mathcal{S}} \longrightarrow \mathcal{T}(-, X_n)|_{\mathcal{S}}$ was constructed so that its image is precisely the image of $\mathcal{T}\big(-, \Sigma^{-1}Y_{n+1}\big)\big|_{\mathcal{S}} \longrightarrow \mathcal{T}(-, Y_n)|_{\mathcal{S}}$. The triangle

$$\Sigma^{-1}Y_{n+1} \longrightarrow X_n \longrightarrow X_{n+1} \longrightarrow Y_{n+1}$$

gives an exact sequence

$$\mathcal{T}\big(-, X_{n+1}\big)\big|_{\mathcal{S}} \longrightarrow \mathcal{T}\big(-, Y_{n+1}\big)\big|_{\mathcal{S}} \longrightarrow \mathcal{T}(-, \Sigma X_n)|_{\mathcal{S}}.$$

This means that the image of $\mathcal{T}\big(-, X_{n+1}\big)\big|_{\mathcal{S}} \longrightarrow \mathcal{T}\big(-, Y_{n+1}\big)\big|_{\mathcal{S}}$ must be indentified with the kernel of $\mathcal{T}\big(-, Y_{n+1}\big)\big|_{\mathcal{S}} \longrightarrow \mathcal{T}(-, \Sigma Y_n)|_{\mathcal{S}}$, which is also the image of $\mathcal{T}\big(-, \Sigma^{-1}Y_{n+2}\big)\big|_{\mathcal{S}} \longrightarrow \mathcal{T}\big(-, Y_{n+1}\big)\big|_{\mathcal{S}}$.

Inductively, this allows us to construct the sequence

$$X_1 \longrightarrow X_2 \longrightarrow X_3 \longrightarrow \cdots$$

which maps to X. There is therefore a map from the homotopy colimit to X. Applying the homological functor

$$\mathcal{T} \longrightarrow \mathcal{E}x\big(\mathcal{S}^{op}, \mathcal{A}b\big)$$

to this sequence, we get

$$\mathcal{T}(-, X_1)|_{\mathcal{S}} \longrightarrow \mathcal{T}(-, X_2)|_{\mathcal{S}} \longrightarrow \mathcal{T}(-, X_3)|_{\mathcal{S}} \longrightarrow \cdots$$

and by construction, this is the direct sum of the two sequences

$$\begin{array}{ccccccc} \mathcal{T}(-,X)|_{\mathcal{S}} & \xrightarrow{1} & \mathcal{T}(-,X)|_{\mathcal{S}} & \xrightarrow{1} & \mathcal{T}(-,X)|_{\mathcal{S}} & \xrightarrow{1} & \cdots \\ K_1(-) & \xrightarrow{0} & K_2(-) & \xrightarrow{0} & K_3(-) & \xrightarrow{0} & \cdots \end{array}$$

It follows that the colimit of this sequence is $\mathcal{T}(-,X)|_{\mathcal{S}}$, while its $\underrightarrow{\mathrm{colim}}^1$ vanishes. We get an exact sequence

$$\begin{array}{ccccc} 0 \longrightarrow \coprod_{i=0}^{\infty} \mathcal{T}(-,X_i)|_{\mathcal{S}} & \xrightarrow{1-shift} & \coprod_{i=0}^{\infty} \mathcal{T}(-,X_i)|_{\mathcal{S}} & & \\ & & \downarrow & & \\ & & \mathcal{T}(-,X)|_{\mathcal{S}} & \longrightarrow & 0. \end{array}$$

Now we remind ourselves that the functor $\mathcal{T} \longrightarrow \mathcal{E}x\big(\mathcal{S}^{op}, \mathcal{A}b\big)$ respects coproducts; the above can be written as

$$\begin{array}{ccccc} 0 \longrightarrow \mathcal{T}\left(-,\coprod_{i=0}^{\infty} X_i\right)\Big|_{\mathcal{S}} & \xrightarrow{1-shift} & \mathcal{T}\left(-,\coprod_{i=0}^{\infty} X_i\right)\Big|_{\mathcal{S}} & & \\ & & \downarrow & & \\ & & \mathcal{T}(-,X)|_{\mathcal{S}} & \longrightarrow & 0 \end{array}$$

But the triangle

$$\coprod_{i=0}^{\infty} X_i \xrightarrow{1-shift} \coprod_{i=0}^{\infty} X_i \longrightarrow \underrightarrow{\mathrm{Hocolim}}\, X \longrightarrow \Sigma\left\{\coprod_{i=0}^{\infty} X_i\right\}$$

gives another exact sequence

$$\begin{array}{ccccc} \mathcal{T}\left(-,\coprod_{i=0}^{\infty} X_i\right)\Big|_{\mathcal{S}} & \xrightarrow{1-shift} & \mathcal{T}\left(-,\coprod_{i=0}^{\infty} X_i\right)\Big|_{\mathcal{S}} & \longrightarrow & \mathcal{T}\left(-, \underrightarrow{\mathrm{Hocolim}} X\right)\Big|_{\mathcal{S}} \\ & & & & \downarrow \\ & & & & \mathcal{T}\left(-,\Sigma\left\{\coprod_{i=0}^{\infty} X_i\right\}\right)\Big|_{\mathcal{S}} \end{array}$$

and the two must clearly agree; the map

$$\mathcal{T}\left(-, \underrightarrow{\mathrm{Hocolim}}\, X_i\right)\Big|_{\mathcal{S}} \longrightarrow \mathcal{T}(-,X)|_{\mathcal{S}}$$

is an isomorphism. Complete $\underrightarrow{\mathrm{Hocolim}}\, X_i \longrightarrow X$ to a triangle

$$\underrightarrow{\mathrm{Hocolim}}\, X_i \longrightarrow X \longrightarrow Z \longrightarrow \Sigma\left\{\underrightarrow{\mathrm{Hocolim}}\, X_i\right\}.$$

The homological functor to $\mathcal{E}x(\mathcal{S}^{op}, \mathcal{A}b)$ gives an exact sequence, from which we easily deduce that $\mathcal{T}(-,Z)|_{\mathcal{S}} = 0$. Since $\mathcal{S}$ generates, $Z = 0$ and X is isomorphic to $\underrightarrow{\text{Hocolim}}\, X_i$. □

B.2. Abelian categories satisfying [AB5$^\alpha$]

Let $\mathcal{T}$ be a triangulated category satisfying [TR5]. In this Section, we will be studying functors into triangulated categories satisfying weak versions of [AB5]. We first need to define the abelian categories we are dealing with.

DEFINITION B.2.1. *An abelian category $\mathcal{A}$ is said to satisfy [AB5$^\alpha$] if it satisfies [AB4], and α–filtered colimits in $\mathcal{A}$ are exact.*

LEMMA B.2.2. *Let α be a regular cardinal. Let $\mathcal{T}$ be an α–compactly generated triangulated category. Let X be an object in $\mathcal{T}$, and suppose we have a sequence*

$$X_0 \longrightarrow X_1 \longrightarrow X_2 \longrightarrow \cdots$$

as in Lemma B.1.3. We remind the reader:

B.1.3.1: $X \quad = \quad \underrightarrow{\text{Hocolim}}\, X_i$.

B.1.3.2: *Put $Y_0 = X_0$. For any $i \geq 1$, let Y_i come from the triangle*

$$X_{i-1} \longrightarrow X_i \longrightarrow Y_i \longrightarrow \Sigma X_{i-1}.$$

B.1.3.3: *The composites*

$$Y_i \longrightarrow \Sigma X_{i-1} \longrightarrow \Sigma Y_{i-1}$$

give a chain complex

$$\cdots \longrightarrow \Sigma^{-2} Y_2 \longrightarrow \Sigma^{-1} Y_1 \longrightarrow Y_0.$$

The entire collection of data can be so chosen, so that there exists an α–filtered functor $f : \mathcal{I} \longrightarrow \mathcal{T}^\alpha$ and the sequence of Y's is exactly $N(f)$, the realisation of the nerve of the functor f.

Let $H : \mathcal{T} \longrightarrow \mathcal{A}$ be a homological functor, preserving coproducts. Suppose the category $\mathcal{A}$ satisfies [AB5$^\alpha$]. Then the sequence

$$H\left(\Sigma^{-1} Y_1\right) \longrightarrow H\left(Y_0\right) \longrightarrow H(X) \longrightarrow 0$$

is exact in $\mathcal{A}$.

Proof: Let $H : \mathcal{T} \longrightarrow \mathcal{A}$ be a functor satisfying the hypothesis of the Lemma; that is, the category $\mathcal{A}$ satisfies [AB5$^\alpha$] and H is homological and respects coproducts. Observe first that for every integer $n \in \mathbb{Z}$, the functor $H\Sigma^n$ also satisfies the hypothesis; after all, $\Sigma : \mathcal{T} \longrightarrow \mathcal{T}$ respects coproducts, and up to sign also respects triangles. We may take the product over all $n \in \mathbb{Z}$ of these. We deduce a functor from $\mathcal{T}$ to the category of graded objects in $\mathcal{A}$, also satisfying the hypothesis of the Lemma. Replacing

$\mathcal{A}$ by the category of graded objects in $\mathcal{A}$, and H by the product of all $\{H\Sigma^n, n \in \mathbb{Z}\}$, we may assume that the abelian category $\mathcal{A}$ admits an automorphism Σ, and that we have a natural isomorphism

$$H\Sigma \simeq \Sigma H,$$

that is, H commutes with Σ. From now until the end of the proof, we assume this is the situation.

Recall that the chain complex

$$\cdots \longrightarrow \Sigma^{-2}Y_2 \longrightarrow \Sigma^{-1}Y_1 \longrightarrow Y_0$$

was chosen isomorphic to $N(f)$, for some α–filtered functor $f : \mathcal{I} \longrightarrow \mathcal{T}^\alpha$. The homological functor H respects coproducts. Hence

$$\cdots \longrightarrow H\left(\Sigma^{-2}Y_2\right) \longrightarrow H\left(\Sigma^{-1}Y_1\right) \longrightarrow H\left(Y_0\right)$$

is just $N(Hf)$. But by Lemma A.3.2, $N(Hf)$ is a chain complex computing $\underrightarrow{\mathrm{colim}}^n$, for the α–filtered functor $Hf : \mathcal{I} \longrightarrow \mathcal{A}$. Since $\mathcal{A}$ satisfies [AB5$^\alpha$], $\underrightarrow{\mathrm{colim}}^n$ vanishes if $n \geq 1$. It follows that the sequence

$$\longrightarrow H\left(\Sigma^{-2}Y_2\right) \longrightarrow H\left(\Sigma^{-1}Y_1\right) \longrightarrow H\left(Y_0\right) \longrightarrow Z \longrightarrow 0$$

is exact, where Z is defined to be the cokernel of

$$H\left(\Sigma^{-1}Y_1\right) \longrightarrow H\left(Y_0\right).$$

Recall that $X_0 = Y_0$, and we have, in $\mathcal{T}$, a triangle

$$X_0 \longrightarrow X_1 \longrightarrow Y_1 \longrightarrow \Sigma X_0.$$

This gives an exact sequence

$$H\left(\Sigma^{-1}Y_1\right) \longrightarrow H\left(X_0\right) \longrightarrow H\left(X_1\right).$$

We have a commutative diagram with exact rows

$$\begin{array}{ccccccc}
H\left(\Sigma^{-1}Y_1\right) & \longrightarrow & H\left(Y_0\right) & \longrightarrow & Z & \longrightarrow & 0 \\
\| \downarrow \wr & & \| \downarrow \wr & & & & \\
H\left(\Sigma^{-1}Y_1\right) & \longrightarrow & H\left(X_0\right) & \longrightarrow & H\left(X_1\right) & &
\end{array}$$

which means we can extend uniquely to

$$\begin{array}{ccccccc}
H\left(\Sigma^{-1}Y_1\right) & \longrightarrow & H\left(Y_0\right) & \longrightarrow & Z & \longrightarrow & 0 \\
\| \downarrow \wr & & \| \downarrow \wr & & \downarrow & & \\
H\left(\Sigma^{-1}Y_1\right) & \longrightarrow & H\left(X_0\right) & \longrightarrow & H\left(X_1\right) & &
\end{array}$$

This gives a monomorphism $Z \longrightarrow H\left(X_1\right)$, and the exact sequence

$$H\left(\Sigma^{-1}Y_1\right) \longrightarrow H\left(X_0\right) \longrightarrow H\left(X_1\right) \longrightarrow H\left(Y_1\right) \longrightarrow H\left(\Sigma X_0\right)$$

allows us to identify the cokernel of $Z \longrightarrow H\left(X_1\right)$ with the kernel of $H\left(Y_1\right) \longrightarrow H\left(\Sigma X_0\right)$.

Now we propose to prove, by induction on $n \geq 1$, that

B.2.2.1. *The composite*

$$Z \longrightarrow H(X_1) \longrightarrow H(X_n)$$

is a monomorphism, and its cokernel is the kernel of the morphism $H(Y_n) \longrightarrow H(\Sigma Y_{n-1})$. *What is more, the identification of the cokernel with a subobject of* $H(Y_n)$ *is from the natural map, the one induced by the triangle*

$$X_{n-1} \longrightarrow X_n \longrightarrow Y_n \longrightarrow \Sigma X_{n-1}.$$

B.2.2.2. *The composite*

$$Z \longrightarrow H(X_1) \longrightarrow H(X_n)$$

is a split monomorphism, and

$$H(X_n) = K_n \oplus Z$$

where K_n *is the kernel of*

$$H(X_n) \longrightarrow H(X_{n+1}).$$

Proof: The strategy of the proof will be to show that

$$\{\text{B.2.2.1 for } n \geq 1\} \implies \{\text{B.2.2.1 for } n+1\} \wedge \{\text{B.2.2.2 for } n\}.$$

Since we already know B.2.2.1 for $n = 1$, this suffices.

Suppose therefore that we know B.2.2.1 for n. The triangle

$$X_{n-1} \longrightarrow X_n \longrightarrow Y_n \longrightarrow \Sigma X_{n-1}$$

gives a map $X_n \longrightarrow Y_n$, and we have an exact sequence

$$0 \longrightarrow Z \longrightarrow H(X_n) \longrightarrow H(Y_n)$$

with the image of $H(X_n) \longrightarrow H(Y_n)$ being the kernel of $H(Y_n) \longrightarrow H(\Sigma Y_{n-1})$. This kernel is also the image of $H(\Sigma^{-1}Y_{n+1}) \longrightarrow H(Y_n)$. By hypothesis on the X's and Y's, the map $\Sigma^{-1}Y_{n+1} \longrightarrow Y_n$ factors as

$$\Sigma^{-1}Y_{n+1} \longrightarrow X_n \longrightarrow Y_n.$$

Hence, in the category $\mathcal{A}$, we have a factorisation

$$H(\Sigma^{-1}Y_{n+1}) \longrightarrow H(X_n) \longrightarrow H(Y_n).$$

Let $\Sigma^{-1}K_{n+1}$ be the kernel of $H(\Sigma^{-1}Y_{n+1}) \longrightarrow H(Y_n)$, K_n the image of $H(\Sigma^{-1}Y_{n+1}) \longrightarrow H(Y_n)$, which is also the kernel of $H(Y_n) \longrightarrow H(\Sigma Y_{n-1})$. We have a commutative diagram with exact rows

$$\begin{array}{ccccccccc}
0 & \longrightarrow & \Sigma^{-1}K_{n+1} & \longrightarrow & H(\Sigma^{-1}Y_{n+1}) & \longrightarrow & K_n & \longrightarrow & 0 \\
 & & \downarrow & & \downarrow & & \downarrow{\scriptstyle 1} & & \\
0 & \longrightarrow & Z & \longrightarrow & H(X_n) & \longrightarrow & K_n & \longrightarrow & 0
\end{array}$$

From the triangle

$$X_n \longrightarrow X_{n+1} \longrightarrow Y_{n+1} \longrightarrow \Sigma X_n$$

we have a diagram with exact rows and columns

$$\begin{array}{ccccccc}
 & & 0 & & 0 & & \\
 & & \downarrow & & \downarrow & & \\
H\left(\Sigma^{-1}X_{n+1}\right) & \longrightarrow & \Sigma^{-1}K_{n+1} & \longrightarrow & Z & \longrightarrow & H\left(X_{n+1}\right) \\
{\scriptstyle 1}\downarrow & & \downarrow & & \downarrow & & {\scriptstyle 1}\downarrow \\
H\left(\Sigma^{-1}X_{n+1}\right) & \longrightarrow & H\left(\Sigma^{-1}Y_{n+1}\right) & \longrightarrow & H\left(X_n\right) & \longrightarrow & H\left(X_{n+1}\right) \\
 & & \downarrow & & \downarrow & & \\
 & & K_n & \xrightarrow{1} & K_n & & \\
 & & \downarrow & & \downarrow & & \\
 & & 0 & & 0 & &
\end{array}$$

But the image of the composite

$$H\left(\Sigma^{-2}Y_{n+2}\right) \longrightarrow H\left(\Sigma^{-1}X_{n+1}\right) \longrightarrow H\left(\Sigma^{-1}Y_{n+1}\right)$$

is the kernel of

$$H\left(\Sigma^{-1}Y_{n+1}\right) \longrightarrow H\left(Y_n\right),$$

that is $\Sigma^{-1}K_{n+1} \subset H\left(\Sigma^{-1}Y_{n+1}\right)$. The commutative diagram

$$\begin{array}{ccccc}
H\left(\Sigma^{-2}Y_{n+2}\right) & \longrightarrow & H\left(\Sigma^{-1}X_{n+1}\right) & \longrightarrow & \Sigma^{-1}K_{n+1} \\
 & & {\scriptstyle 1}\downarrow & & \downarrow \\
 & & H\left(\Sigma^{-1}X_{n+1}\right) & \longrightarrow & H\left(\Sigma^{-1}Y_{n+1}\right)
\end{array}$$

shows that the composite

$$H\left(\Sigma^{-2}Y_{n+2}\right) \longrightarrow H\left(\Sigma^{-1}X_{n+1}\right) \longrightarrow \Sigma^{-1}K_{n+1}$$

must be surjective. The map $H\left(\Sigma^{-1}X_{n+1}\right) \longrightarrow \Sigma^{-1}K_{n+1}$ is onto, and hence the map $\Sigma^{-1}K_{n+1} \longrightarrow Z$ is zero. We deduce a short exact sequence

$$0 \longrightarrow Z \longrightarrow H\left(X_{n+1}\right) \longrightarrow K_{n+1} \longrightarrow 0.$$

This proves B.2.2.1 for $n+1$.

Finally, in the commutative square

$$\begin{array}{ccc}
\Sigma^{-1}K_{n+1} & \longrightarrow & Z \\
\downarrow & & \downarrow \\
H\left(\Sigma^{-1}Y_{n+1}\right) & \longrightarrow & H\left(X_n\right)
\end{array}$$

we now know that the map $\Sigma^{-1}K_{n+1} \longrightarrow Z$ vanishes. Hence so do the equal composites

$$\begin{array}{ccc} \Sigma^{-1}K_{n+1} & \longrightarrow & Z \\ & & \downarrow \\ & & H(X_n) \end{array} \qquad \begin{array}{ccc} \Sigma^{-1}K_{n+1} & & \\ \downarrow & & \\ H\left(\Sigma^{-1}Y_{n+1}\right) & \longrightarrow & H(X_n) \end{array}$$

It follows that the map $H\left(\Sigma^{-1}Y_{n+1}\right) \longrightarrow H(X_n)$ must factor through the cokernel of $\Sigma^{-1}K_{n+1} \longrightarrow H\left(\Sigma^{-1}Y_{n+1}\right)$, which is K_n. We deduce a map $K_n \longrightarrow H(X_n)$ splitting the exact sequence

$$0 \longrightarrow Z \longrightarrow H(X_n) \longrightarrow K_n \longrightarrow 0.$$

And the fact that

$$H\left(\Sigma^{-1}Y_{n+1}\right) \longrightarrow K_n \longrightarrow H(X_n) \longrightarrow H\left(X_{n+1}\right)$$

vanishes, while $H\left(\Sigma^{-1}Y_{n+1}\right) \longrightarrow K_n$ is surjective, tells us that

$$K_n \longrightarrow H(X_n) \longrightarrow H\left(X_{n+1}\right)$$

vanishes. Hence we have B.2.2.2 for n. By induction, we have proved B.2.2.1 and B.2.2.2 for all $n \geq 1$. □

Now, to end the proof of the Lemma, observe what we now know about the sequence

$$X_0 \longrightarrow X_1 \longrightarrow X_2 \longrightarrow \cdots$$

We know that, for $n \geq 1$, $H(X_n) = Z \oplus K_n$, and that this gives an isomorphism of the sequence

$$H(X_1) \longrightarrow H(X_2) \longrightarrow H(X_2) \longrightarrow \cdots$$

with the direct sum of the two sequences

$$\begin{array}{l} Z \xrightarrow{\ 1\ } Z \xrightarrow{\ 1\ } Z \xrightarrow{\ 1\ } \cdots \\ K_1 \xrightarrow{\ 0\ } K_2 \xrightarrow{\ 0\ } K_3 \xrightarrow{\ 0\ } \cdots \end{array}$$

Hence $\underrightarrow{\operatorname{colim}} H(X_n) = Z$, while $\underrightarrow{\operatorname{colim}}^1 H(X_n) = 0$. There is an exact sequence

$$0 \longrightarrow \coprod_{n=0}^{\infty} H(X_n) \xrightarrow{1 - shift} \coprod_{n=0}^{\infty} H(X_n) \longrightarrow Z \longrightarrow 0.$$

From the triangle

$$\coprod_{n=0}^{\infty} X_n \xrightarrow{1 - shift} \coprod_{n=0}^{\infty} X_n \longrightarrow X \longrightarrow \Sigma\left\{\coprod_{n=0}^{\infty} X_n\right\}$$

we deduce an exact sequence

$$\begin{array}{ccccc} H\left(\coprod_{n=0}^{\infty} X_n\right) & \xrightarrow{1-shift} & H\left(\coprod_{n=0}^{\infty} X_n\right) & \longrightarrow & H(X) \\ & & & & \downarrow \\ & & & & H\left(\Sigma\left\{\coprod_{n=0}^{\infty} X_n\right\}\right). \end{array}$$

Since H respects coproducts, the two exact sequences must agree, and we have $Z = H(X)$. □

Corollary B.2.3. *Let α be a regular cardinal. Assume $\mathcal{T}$ is an α–compactly generated triangulated category. Let $\mathcal{A}$ be an abelian category satisfying [AB5$^\alpha$]. Let X be an object of $\mathcal{T}$. Suppose $H : \mathcal{T} \longrightarrow \mathcal{A}$ is a homological functor respecting coproducts.*

Then there exists an object $Y \in \mathcal{T}$, with

$$Y = \coprod_{\lambda\in\Lambda} s_\lambda$$

where the objects s_λ lie in $\mathcal{S} = \mathcal{T}^\alpha$. There is a map $Y \longrightarrow X$ in $\mathcal{T}$, and

$$H(Y) \longrightarrow H(X)$$

is an epi in $\mathcal{A}$.

Proof: By Lemma B.1.3, there is a sequence

$$X_0 \longrightarrow X_1 \longrightarrow X_2 \longrightarrow \cdots$$

satisfying some special properties we will not repeat fully. Among these properties, we have also an object $\Sigma^{-1}Y_1$ and two maps

$$\Sigma^{-1}Y_1 \longrightarrow X_0 \longrightarrow X$$

which compose to zero. By Lemma B.2.2, these induce a short exact sequence in $\mathcal{A}$

$$H\left(\Sigma^{-1}Y_1\right) \longrightarrow H(X_0) \longrightarrow H(X) \longrightarrow 0.$$

But $X_0 = Y_0$ is a coproduct of objects in $\mathcal{S}$. In other words, the assertions of the Corollary are satisfies if $Y = Y_0 = X_0$, and the map $X_0 \longrightarrow X$ is as in Lemma B.1.3. □

Corollary B.2.4. *Let α be a regular cardinal. Assume $\mathcal{T}$ is an α–compactly generated triangulated category. Let $\mathcal{A}$ be an abelian category satisfying [AB5$^\alpha$]. Suppose $H : \mathcal{T} \longrightarrow \mathcal{A}$ is a homological functor respecting coproducts.*

Then for any α–phantom map $X \longrightarrow Z$ in $\mathcal{T}$, $H(X) \longrightarrow H(Z)$ vanishes in $\mathcal{A}$.

Proof: We should perhaps remind the reader; a map $f : X \longrightarrow Z$ is called *α–phantom* if the induced map

$$\mathcal{T}(-,X)|_{\mathcal{S}} \longrightarrow \mathcal{T}(-,Z)|_{\mathcal{S}}$$

vanishes in $\mathcal{E}x\big(\mathcal{S}^{op}, \mathcal{A}b\big)$. See Definition 6.5.7.

By Corollary B.2.3, we may choose Y a coproduct of objects in $\mathcal{S}$, and a map $g : Y \longrightarrow X$, so that $H(g) : H(Y) \longrightarrow H(X)$ is surjective in $\mathcal{A}$. But now the composite

$$\mathcal{T}(-,Y)|_{\mathcal{S}} \xrightarrow{\mathcal{T}(-,g)|_{\mathcal{S}}} \mathcal{T}(-,X)|_{\mathcal{S}} \xrightarrow{\mathcal{T}(-,f)|_{\mathcal{S}}} \mathcal{T}(-,Z)|_{\mathcal{S}}$$

vanishes, because $\mathcal{T}(-,f)|_{\mathcal{S}}$ does. On the other hand, Y is a coproduct of objects in $\mathcal{S}$, and by Lemma B.1.1, maps $Y \longrightarrow Z$ correspond 1–to–1 with maps $\mathcal{T}(-,Y)|_{\mathcal{S}} \longrightarrow \mathcal{T}(-,Z)|_{\mathcal{S}}$. Therefore the composite

$$Y \xrightarrow{\;g\;} X \xrightarrow{\;f\;} Z$$

vanishes in $\mathcal{T}$. But then so does

$$H(Y) \xrightarrow{H(g)} H(X) \xrightarrow{H(f)} H(Z).$$

By Corollary B.2.3, the map $H(g)$ is surjective; we conclude that $H(f)$ must vanish. □

THEOREM B.2.5. *Let α be a regular cardinal. Let $\mathcal{T}$ be an α–compactly generated triangulated category. Then the functor*

$$\mathcal{T} \longrightarrow \mathcal{E}x\Big(\{\mathcal{T}^\alpha\}^{op}, \mathcal{A}b\Big)$$

is universal among coproduct–preserving homological functors to [AB5$^\alpha$] abelian categories $\mathcal{A}$. Any coproduct–preserving homological functor $\mathcal{T} \longrightarrow \mathcal{A}$ factors uniquely, up to canonical equivalence, as

$$\mathcal{T} \longrightarrow \mathcal{E}x\Big(\{\mathcal{T}^\alpha\}^{op}, \mathcal{A}b\Big) \xrightarrow{\exists!} \mathcal{A}.$$

Furthermore, any natural transformation of coproduct–preserving homological functors $\mathcal{T} \longrightarrow \mathcal{A}$ factors through a natural transformation of the coproduct–preserving exact functors $\mathcal{E}x\Big(\{\mathcal{T}^\alpha\}^{op}, \mathcal{A}b\Big) \longrightarrow \mathcal{A}$.

Proof: By Theorem 5.1.18, any homological functor factors as

$$\mathcal{T} \longrightarrow A(\mathcal{T}) \xrightarrow{\exists!} \mathcal{A}.$$

Moreover, by Lemma 5.1.24, $\mathcal{T} \longrightarrow \mathcal{A}$ preserves coproducts if and only if $A(\mathcal{T}) \longrightarrow \mathcal{A}$ does. Now put $\mathcal{S} = \mathcal{T}^\alpha$. By Proposition 6.5.3, the natural functor

$$A(\mathcal{T}) \longrightarrow \mathcal{E}x\big(\mathcal{S}^{op}, \mathcal{A}b\big)$$

is a coproduct–preserving Gabriel quotient map. To show that a map $A(\mathcal{T}) \longrightarrow \mathcal{A}$ factors further as

$$A(\mathcal{T}) \longrightarrow \mathcal{E}x\big(\mathcal{S}^{op}, \mathcal{A}b\big) \longrightarrow \mathcal{A}$$

it is necessary and sufficient to prove that the map $A(\mathcal{T}) \longrightarrow \mathcal{A}$ annihilates any object in the kernel of $A(\mathcal{T}) \longrightarrow \mathcal{E}x\big(\mathcal{S}^{op}, \mathcal{A}b\big)$. But in Lemma 6.5.6 we computed the kernel of $A(\mathcal{T}) \longrightarrow \mathcal{E}x\big(\mathcal{S}^{op}, \mathcal{A}b\big)$. In the description $D(\mathcal{T}) = A(\mathcal{T})$, the objects mapping to zero are precisely the α–phantom maps.

Now let H be a coproduct–preserving functor into an [AB5$^\alpha$] abelian category $\mathcal{A}$. By virtue of being homological, it factors through $A(\mathcal{T}) \longrightarrow \mathcal{A}$. Corollary B.2.4 tells us that H takes α–phantom maps to zero. The map $A(\mathcal{T}) \longrightarrow \mathcal{A}$ kills the kernel of $A(\mathcal{T}) \longrightarrow \mathcal{E}x\big(\mathcal{S}^{op}, \mathcal{A}b\big)$. Thus H factors, uniquely, through the Gabriel quotient $\mathcal{E}x\big(\mathcal{S}^{op}, \mathcal{A}b\big)$.

The statement about natural transformations comes about because both Freyd's $A(\mathcal{T})$ and Gabriel quotients behave well with respect to natural transformations. □

B.3. History of the results in Appendix B

In the case $\alpha = \aleph_0$, and where $\mathcal{T}$ is the homotopy category of spectra, the result is in the recent work of Christensen and Strickland [**9**]. The argument they give is more down–to–earth and concrete. But it does not seem to generalise. It does not even generalise to other triangulated categories, but with α still equal $\aleph_0$. I would like to thank Christensen for emphasizing the relevance of the work.

APPENDIX C

Counterexamples concerning the abelian category $A(\mathcal{T})$

C.1. The submodules $p^i M$

Let R be a discrete valuation ring. That is, R is a commutative, regular, noetherian local ring of height 1. The (unique) maximal ideal of R is principal. Let p be a generator. That is, $Rp \subset R$ is the maximal ideal.

Let us remind the reader briefly of the injective R–modules. It is standard that, for any commutative ring R and any R–module M, the module M is injective as an R–module if and only if, for every non–zero ideal $I \subset R$,

$$\mathrm{Ext}^1(R/I, M) = 0.$$

In a discrete valuation ring R, every non–zero ideal $I \subset R$ is Rp^n for some $n \geq 0$. There is an exact sequence of R–modules

$$0 \longrightarrow R \xrightarrow{\ p^n\ } R \longrightarrow R/I \longrightarrow 0.$$

Hence there is an exact sequence of Ext–groups

$$\mathrm{Hom}(R, M) \xrightarrow{\ p^n\ } \mathrm{Hom}(R, M) \longrightarrow \mathrm{Ext}^1(R/I, M) \longrightarrow \mathrm{Ext}^1(R, M).$$

Now $\mathrm{Ext}^1(R, M) = 0$ since R is projective, while we have a natural isomorphism $\mathrm{Hom}(R, M) = M$. Hence the sequence becomes

$$M \xrightarrow{\ p^n\ } M \longrightarrow \mathrm{Ext}^1(R/I, M) \longrightarrow 0.$$

Thus M is injective if and only if for each $n \geq 0$, multiplication by p^n gives a surjective map $p^n : M \longrightarrow M$. Clearly, the case $n = 1$ implies the general case; M is injective if and only if $p : M \longrightarrow M$ is surjective.

In the rest of this Section, we choose and fix a discrete valuation ring R and a generator p of its maximal ideal.

Definition C.1.1. *Let M be an R–module. By transfinite induction, we define for every ordinal i, a submodule $p^i M \subset M$. The definition is:*

C.1.1.1. *For $i = 0$, $p^i M = p^0 M = M$.*

C.1.1.2. *For a successor ordinal $i+1$, we define*

$$p^{i+1}M \quad = \quad p\{p^i M\}.$$

That is, $p^{i+1}M$ is the set of all elements of $p^i M$ divisible (in $p^i M$) by p.

C.1.1.3. *For a limit ordinal i, we define*

$$p^i M \quad = \quad \bigcap_{j<i} p^j M.$$

LEMMA C.1.2. *Let M be an R–module.*

C.1.2.1. *For any pair of ordinals $j \leq i$, we have $p^j M \supset p^i M$.*

C.1.2.2. *If i is an ordinal such that $p^i M = p^{i+1}M$, then for all $j > i$ we have $p^j M = p^i M$.*

Proof: The proof is an easy transfinite induction. We remind the reader briefly how these go.

For C.1.2.1, fix the ordinal j and consider the set

$$I = \{i \text{ an ordinal } \mid i \geq j \text{ and } p^j M \supset p^i M\}$$

To prove C.1.2.1, it suffices to show that every $i \geq j$ lies in I. To do this, it suffices, by transfinite induction, to show that $j \in I$, that $i \in I \Longrightarrow i+1 \in I$, and that if $k \in I$ for all $j \leq k < i$ for some limit ordinal i, then $i \in I$.

Observe that $p^j M \supset p^j M$, and hence $j \in I$. Furthermore, if $i \in I$ then $p^j M \supset p^i M$, from which we deduce

$$p^j M \quad \supset \quad p^i M \quad \supset \quad p\{p^i M\} \quad = \quad p^{i+1}M;$$

that is, $i+1 \in I$. If i is any limit ordinal $> j$, then the set of ordinals $k < i$ contains j. Therefore

$$p^i M \quad = \quad \bigcap_{k<i} p^k M \quad \subset \quad p^j M$$

since an intersection is always contained in any one of the terms. From this we deduce $i \in I$. This establishes C.1.2.1.

Suppose next that i is an ordinal for which $p^i M = p^{i+1}M$. We wish to prove C.1.2.2; that is, we want to show that if $j > i$ then $p^i M = p^j M$. Let J be the set of ordinals

$$J = \{j \text{ an ordinal } \mid j \geq i \text{ and } p^j M = p^i M\}.$$

Once again, we use transfinite induction to show that all ordinals $j \geq i$ lie in J. Clearly, $i \in J$. Suppose $j \in J$. Then $p^j M = p^i M$. Hence

$$p^{j+1}M \quad = \quad p\{p^j M\} \quad = \quad p\{p^i M\} \quad = \quad p^{i+1}M \quad = \quad p^i M$$

This means $p^{j+1}M = p^i M$, and hence $j+1 \in J$.

Suppose j is a limit ordinal, $j > i$. Then

$$p^j M \quad = \quad \bigcap_{k<j} p^k M \quad = \quad \left\{ \bigcap_{k<i} p^k M \right\} \cap \left\{ \bigcap_{i \le k < j} p^k M \right\}.$$

If for all ordinals k with $i \le k < j$ we have $k \in J$, then for all $i \le k < j$, $p^k M = p^i M$. Hence

$$\left\{ \bigcap_{i \le k < j} p^k M \right\} \quad = \quad \left\{ \bigcap_{i \le k < j} p^i M \right\} \quad = \quad p^i M.$$

By C.1.2.1, for $k < i$ we have $p^k M \supset p^i M$. Hence

$$\left\{ \bigcap_{k<i} p^k M \right\} \qquad \supset \qquad p^i M.$$

Combining these, we have

$$\begin{aligned} p^j M \quad &= \quad \left\{ \bigcap_{k<i} p^k M \right\} \cap \left\{ \bigcap_{i \le k < j} p^k M \right\} \\ &= \quad p^i M. \end{aligned}$$

By transfinite induction, this establishes C.1.2.2. □

REMARK C.1.3. Let M be an R–module. If $p^i M = p^{i+1} M$, then

$$p^{i+1} M \quad = \quad p\{p^i M\} \quad = \quad p^i M$$

which means that the map $p : p^i M \longrightarrow p^i M$ is surjective. By the reminder at the beginning of the Section, this means that the module $p^i M$ is an injective R module. In C.1.2.2 we saw that this injective R–module $p^i M$ is also $p^j M$ for all $j \ge i$. If N is any injective submodule of M, then $p^i N = N$, and $N \subset p^i M$ for all i. Therefore $p^i M$ contains all injective submodules of M. If $p^i M = p^{i+1} M$ then $p^i M$ is injective, and it must be the largest injective submodule of M.

COROLLARY C.1.4. *Let M be an R–module. Suppose i is an ordinal with $p^i M \neq p^{i+1} M$. Then for any $j < i$, we must have $p^j M \neq p^{j+1} M$.*

Proof: Let $j < i$. If $p^j M = p^{j+1} M$, then C.1.2.2 implies that for all $k > j$, $p^k M = p^j M$. In particular,

$$p^i M \quad = \quad p^j M \quad = \quad p^{i+1} M.$$

Thus for $j < i$,

$$\{p^j M = p^{j+1} M\} \Longrightarrow \{p^i M = p^{i+1} M\}.$$

The contrapositive is that for $j < i$,

$$\{p^iM \neq p^{i+1}M\} \Longrightarrow \{p^jM \neq p^{j+1}M\},$$

which is the assertion of the Corollary. □

LEMMA C.1.5. *Let M be an R–module. Suppose i is an ordinal with $p^iM \neq p^{i+1}M$. Then the cardinality of M is at least the cardinality of i.*

Proof: Assume $p^iM \neq p^{i+1}M$. By Corollary C.1.4, we deduce that for every $j < i$, we must have $p^jM \neq p^{j+1}M$. By C.1.2.1, we know that $p^jM \supset p^{j+1}M$. Since they are not equal, we may choose, for each $j < i$, an element

$$m_j \in p^jM - p^{j+1}M.$$

If $j < k < i$, then $m_j \in p^jM - p^{j+1}M$, while $m_k \in p^kM \subset p^{j+1}M$. Since $m_j \notin p^{j+1}M$ while $m_k \in p^{j+1}M$, it follows that $m_j \neq m_k$. The elements m_j are all distinct. The set $\{m_j, j < i\}$ is a set of distinct elements of M, whose cardinality is the cardinality of i. □

REMARK C.1.6. In Remark C.1.3, we saw that if $p^iM = p^{i+1}M$ then p^iM is an injective R–module, and contains every injective R–submodule of M. In Lemma C.1.5, we saw that if M has cardinality $< \alpha$, then for the ordinal α we must have $p^{\alpha+1}M = p^{\alpha}M$. For every R–module, there exists some ordinal with $p^iM = p^{i+1}M$. Every module has a largest injective submodule.

Of course, the inclusion of the injective $p^iM \subset M$ must split. M is the direct sum of the injective module p^iM and a submodule isomorphic to M/p^iM. The module M/p^iM contains no injective submodule.

LEMMA C.1.7. *Let $\phi : M \longrightarrow N$ be a homomorphism of R–modules. Then for every ordinal i, $\phi(p^iM) \subset p^iN$.*

Proof: We prove this by induction on the ordinal i. If $i = 0$, this is obvious; the assertion is that

$$\phi(p^0M) \quad = \quad \phi(M) \quad \subset \quad N \quad = \quad p^0N.$$

Suppose we are given a ordinal i for which $\phi(p^iM) \subset p^iN$. Then

$$\begin{aligned} \phi(p^{i+1}M) &= \phi\Big(p\{p^iM\}\Big) \\ &= p\phi(p^iM) \\ &\subset p\{p^iN\} \\ &= p^{i+1}N, \end{aligned}$$

that is the ordinal $i+1$ satisfies $\phi(p^{i+1}M) \subset p^{i+1}N$. Finally, if i is a limit ordinal and for every $j < i$ we have $\phi(p^jM) \subset p^jN$, then

$$\begin{aligned}\phi(p^iM) &= \phi\left(\bigcap_{j<i} p^jM\right) \\ &\subset \bigcap_{j<i} \phi\left(p^jM\right) \\ &\subset \bigcap_{j<i} p^jN \\ &= p^iN.\end{aligned}$$

We conclude that for all i, $\phi(p^iM) \subset p^iN$. □

Next we apply the ideas above to deduce the following slightly technical lemma. The lemma states that certain homomorphisms of R–modules must have large images. This lemma will be applied in the next section.

LEMMA C.1.8. *Let α be an infinite cardinal. Let $\phi : M \longrightarrow N$ be a homomorphism of R–modules. Suppose further*

C.1.8.1. *There exists an element $m \in p^\alpha M$, with $\phi(m) \neq 0$.*

C.1.8.2. *N contains no non–zero injective R–submodule.*

Then the cardinality of the image of ϕ is at least α.

Proof: We are given $m \in p^\alpha M$. We have maps of R–modules

$$M \xrightarrow{\ \psi\ } \mathrm{Im}(\phi) \qquad \subset \qquad N.$$

By Lemma C.1.7 we have that $\psi(p^\alpha M) \subset p^\alpha \mathrm{Im}(\phi)$. But $m \in p^\alpha M$, and hence $\phi(m) \in p^\alpha\mathrm{Im}(\phi)$. On the other hand $\phi(m) \neq 0$, and hence $p^\alpha\mathrm{Im}(\phi) \neq 0$. By C.1.8.2, we are assuming that N contains no non–zero injective submodules. Therefore $p^\alpha\mathrm{Im}(\phi)$ cannot be an injective submodule of N; we cannot have $p^\alpha\mathrm{Im}(\phi) = p^{\alpha+1}\mathrm{Im}(\phi)$. But Lemma C.1.5 now tells us that the cardinality of $\mathrm{Im}(\phi)$ must be at least α. □

REMARK C.1.9. The next two sections are structured as follows. In Section C.2 we construct a module M and a non–zero element $m \in p^\alpha M$. Suppose for this module M with its chosen element m, we are given some map $\phi : M \longrightarrow M$ with $\phi(m) = m$. Then Lemma C.1.8 applies. We have $m \in p^\alpha M$, and $\phi(m) = m \neq 0$. If M contains no R–injective submodules, then the image of ϕ has cardinality $\geq \alpha$.

In Section C.3 we explain why this leads us to conclude that $A(\mathfrak{T})$ is not well–powered. The sections are logically independent; the reader may read them in either order.

C.2. A large R–module

CONSTRUCTION C.2.1. Let α be an infinite cardinal, and let β be an infinite cardinal $\beta > \alpha$. We define an R–module M by generators and relations. The generators are all symbols $[i_n, i_{n-1}, \cdots, i_1, i_0]$, where the $i_n < i_{n-1} < \cdots < i_1 < i_0 < \beta$ are any ordinals. The relations are

$$p[i_n, i_{n-1}, \cdots, i_1, i_0] \quad = \quad [i_{n-1}, \cdots, i_1, i_0]$$

whenever $n > 0$, and

$$p[i_0] = 0.$$

LEMMA C.2.2. *For the module M of construction C.2.1, we have that $p^i M$ is the submodule generated by $[i_n, i_{n-1}, \cdots, i_1, i_0]$, with $i_n \geq i$.*

Proof: The statement is clearly true for $p^0 M$. Suppose it is true for $p^i M$. That is, $p^i M$ is the submodule generated by $[i_n, i_{n-1}, \cdots, i_1, i_0]$, with $i_n \geq i$. But then the generators for $p^{i+1} M = p\{p^i M\}$ are

$$p[i_n, i_{n-1}, \cdots, i_1, i_0] = [i_{n-1}, \cdots, i_1, i_0]$$

with $i_{n-1} > i_n \geq i$, that is $i_{n-1} \geq i+1$.

Now let j be a limit ordinal, and assume that for all $i < j$, $p^i M$ is the submodule generated by $[i_n, i_{n-1}, \cdots, i_1, i_0]$ with $i_n \geq i$. Every element of M can be written as a linear combination of genrators $[i_n, i_{n-1}, \cdots, i_1, i_0]$, with the coefficents not divisible by p. For $i < j$, an element will lie in $p^i M$ if and only if, when expressed as such a linear combination, all the generators have $i_n \geq i$. An element lying in

$$p^j M \quad = \quad \bigcap_{i<j} p^i M$$

is therefore a linear combination of generators $[i_n, i_{n-1}, \cdots, i_1, i_0]$ with $i_n \geq j$. □

PROPOSITION C.2.3. *Let M be the R–module of Construction C.2.1. Suppose we have an endomorphism $\phi : M \longrightarrow M$ taking $[\alpha] \in M$ to itself. Then the image $Im(\phi)$ has cardinality at least α. By $[\alpha] \in M$ we mean of course the generator $[i_0]$, where $i_0 = \alpha$.*

Proof: The idea is to apply Lemma C.1.8, to the map $\phi : M \longrightarrow M$. By Lemma C.2.2, we have that $[\alpha] \in p^\alpha M$. Since $\phi([\alpha]) = [\alpha]$, we have $\phi([\alpha]) \neq 0$. Also, we computed in Lemma C.2.2, that $p^\beta M$ is generated by $[i_n, i_{n-1}, \cdots, i_1, i_0]$ with $i_n \geq \beta$. Since there are no such generators in M, it follows that $p^\beta M = 0$ and M contains no injective submodule. Lemma C.1.8 therefore applies, and the image of ϕ must have cardinality at least α. □

C.3. The category $A(\mathcal{S})$ is not well–powered

As in Section C.2, let us choose a discrete valuation ring R with maximal ideal Rp. Let $D(R)$ be the derived category of all chain complexes of any R–modules. Let R be the chain complex which is R in dimension 0, zero elsewhere. Let $A(D(R))$ be the abelian category associated to the triangulated category $D(R)$, as in Definition 5.1.3.

LEMMA C.3.1. *Let $\{R \to x_\lambda, \lambda \in \Lambda\}$ be a set of quotient objects of R in $A(D(R))$; see Proposition 5.2.6. There exists an infinite cardinal α, so that for any representation of any one of these quotient objects, that is for any $\{R \to y\}$ isomorphic as a quotient object in $A(D(R))$ to one of the $\{R \to x_\lambda\}$, there is an endomorphism $y \longrightarrow y$ making commutative the square*

$$\begin{array}{ccc} R & \longrightarrow & y \\ {\scriptstyle 1}\downarrow & & \downarrow{\scriptstyle \phi} \\ R & \longrightarrow & y \end{array}$$

and the image of $\phi : H^0(y) \longrightarrow H^0(y)$ is of cardinality $< \alpha$.

Proof: Let α be any cardinal greater than the sum of the cardinalities of all $\{H^0(x_\lambda), \lambda \in \Lambda\}$. Since we are assuming that $\{R \to y\}$ isomorphic as a quotient object in $A(D(R))$ to one of the $\{R \to x_\lambda\}$, there exists a $\lambda \in \Lambda$ and an isomorphism of the two objects $\{R \to y\}$ and $\{R \to x_\lambda\}$. By Proposition 5.2.6, this means that there exist commutative squares in $D(R)$

$$\begin{array}{ccc} R & \longrightarrow & y \\ {\scriptstyle 1}\downarrow & & \downarrow \\ R & \longrightarrow & x_\lambda \end{array} \qquad \begin{array}{ccc} R & \longrightarrow & x_\lambda \\ {\scriptstyle 1}\downarrow & & \downarrow \\ R & \longrightarrow & y \end{array}$$

Applying the functor H^0 to these commutative squares, we get a commutative square of R–modules

$$\begin{array}{ccc} R & \longrightarrow & H^0(y) \\ {\scriptstyle 1}\downarrow & & \downarrow{\scriptstyle \rho} \\ R & \longrightarrow & H^0(x_\lambda) \end{array} \qquad \begin{array}{ccc} R & \longrightarrow & H^0(x_\lambda) \\ {\scriptstyle 1}\downarrow & & \downarrow{\scriptstyle \sigma} \\ R & \longrightarrow & H^0(y) \end{array}$$

Combining the squares, we have a commutative square

$$\begin{array}{ccc} R & \longrightarrow & H^0(y) \\ {\scriptstyle 1}\downarrow & & \downarrow{\scriptstyle \sigma\rho} \\ R & \longrightarrow & H^0(y) \end{array}$$

where $\sigma\rho$ factors through $H^0(x_\lambda)$, whose cardinality is by hypothesis $< \alpha$. Thus the image of $\sigma\rho$ has cardinality $< \alpha$. □

PROPOSITION C.3.2. *Suppose R is a discrete valuation ring. Assume $\{R \to x_\lambda, \lambda \in \Lambda\}$ is some set of quotient objects of R in $A(D(R))$. Then there exists a quotient object $R \to M$ not isomorphic in $A(D(R))$ to any of the $\{R \to x_\lambda, \lambda \in \Lambda\}$.*

Proof: By Lemma C.3.1 we can, for our set of quotient objects $\{R \to x_\lambda, \lambda \in \Lambda\}$, choose a cardinal α satisfying the conclusions of Lemma C.3.1. Pick such an α. For this α, let M be the module of Construction C.2.1. Let $\theta : R \to M$ be the map

$$\theta(r) = r[\alpha].$$

This is of course an R–module homomorphism, which we may view as a map of complexes concentrated in degree 0, in the derived category. For any commutative diagram

$$\begin{array}{ccc} R & \xrightarrow{\theta} & M \\ {\scriptstyle 1}\downarrow & & \downarrow{\scriptstyle \phi} \\ R & \xrightarrow{\theta} & M \end{array}$$

we have that $\phi([\alpha]) = \phi\theta(1) = \theta(1) = [\alpha]$. From Proposition C.2.3 we deduce that the image of ϕ has cardinality at least α.

On the other hand, if $\{R \to M\}$ is isomorphic, as a quotient of R in $A(D(R))$, to one of $\{R \to x_\lambda, \lambda \in \Lambda\}$, then by Lemma C.3.1 there must exist a commutative square

$$\begin{array}{ccc} R & \xrightarrow{\theta} & H^0(M) = M \\ {\scriptstyle 1}\downarrow & & \downarrow{\scriptstyle \phi} \\ R & \xrightarrow{\theta} & H^0(M) = M \end{array}$$

with the image of ϕ of cardinality $< \alpha$. We deduce that $\{R \to M\}$ is not isomorphic, as a quotient of R in $A(D(R))$, to any of $\{R \to x_\lambda, \lambda \in \Lambda\}$. □

COROLLARY C.3.3. *Let R be a discrete valuation ring. The class of isomorphism classes of quotients of R in $A(D(R))$ is not a set. Given any set of quotients, there is a quotient not isomorphic to any of them.*

REMARK C.3.4. If R is the localisation of $\mathbb{Z}$ at a prime p, we have shown that R does not have a set of quotients in $A(D(R))$. But since $A(D(R))$ is a localisation, in the sense of Gabriel, of the category $A(D(\mathbb{Z}))$, it follows that R does not have a set of quotients in $A(D(\mathbb{Z}))$ either.

C.4. A category $\mathcal{E}x(\mathcal{S}^{op}, \mathcal{A}b)$ without a cogenerator

As throughout this Appendix, R is a discrete valuation ring with maximal ideal $Rp \subset R$. Let K be the quotient field of R.

LEMMA C.4.1. *Let α be an infinite cardinal. There exists a non-trivial extension (a non-split short exact sequence) of R–modules*

$$0 \longrightarrow M \longrightarrow M' \longrightarrow K/R \longrightarrow 0$$

so that, for any $\phi : M \longrightarrow N$, with N of cardinality $< \alpha$, the induced extension of K/R by N splits. That is, if we push out the exact sequence to get

$$\begin{array}{ccccccccc} 0 & \longrightarrow & M & \longrightarrow & M' & \longrightarrow & K/R & \longrightarrow & 0 \\ & & {\scriptstyle\phi}\big\downarrow & & \big\downarrow & & {\scriptstyle 1}\big\downarrow & & \\ 0 & \longrightarrow & N & \longrightarrow & N' & \longrightarrow & K/R & \longrightarrow & 0 \end{array}$$

then the sequence

$$0 \longrightarrow N \longrightarrow N' \longrightarrow K/R \longrightarrow 0$$

is split.

Proof: Choose an infinite cardinal $\beta > \alpha$, and let M be the R–module of Construction C.2.1. The facts that are relevant to us here are:

C.4.1.1. $p^\alpha M \neq 0$

C.4.1.2. $p^\beta M = 0$.

From Lemma C.1.2, more precisely by C.1.2.2, we deduce that $p^\alpha M \neq p^{\alpha+1} M$. Choose an element $x \in p^\alpha M - p^{\alpha+1} M$. It corresponds to a map $R \longrightarrow p^\alpha M \subset M$. We define the extension of K/R by M to be given by the pushout

$$\begin{array}{ccccccccc} 0 & \longrightarrow & R & \longrightarrow & K & \longrightarrow & K/R & \longrightarrow & 0 \\ & & \big\downarrow & & \big\downarrow & & {\scriptstyle 1}\big\downarrow & & \\ 0 & \longrightarrow & M & \longrightarrow & M' & \longrightarrow & K/R & \longrightarrow & 0. \end{array}$$

Now we need to prove two things. We must prove that the extension is non–trivial, and also that it becomes trivial after pushing out along any map $\phi : M \longrightarrow N$, if the cardinality of N is $< \alpha$.

Suppose first that $\phi : M \longrightarrow N$ is a map, and that the cardinality of N is $< \alpha$. We have a diagram of extensions

$$\begin{array}{ccccccccc}
0 & \longrightarrow & R & \longrightarrow & K & \longrightarrow & K/R & \longrightarrow & 0 \\
 & & \downarrow & & \downarrow & & \downarrow{\scriptstyle 1} & & \\
0 & \longrightarrow & M & \longrightarrow & M' & \longrightarrow & K/R & \longrightarrow & 0 \\
 & & {\scriptstyle \phi}\downarrow & & \downarrow & & \downarrow{\scriptstyle 1} & & \\
0 & \longrightarrow & N & \longrightarrow & N' & \longrightarrow & K/R & \longrightarrow & 0
\end{array}$$

But the map $R \longrightarrow M$ was chosen to factor through $p^\alpha M \subset M$, and by Lemma C.1.7, $\phi(p^\alpha M) \subset p^\alpha N$. Therefore the map $R \longrightarrow N$ factors through $p^\alpha N \subset N$. Now recall that the cardinality of N is assumed $< \alpha$, and hence Lemma C.1.5 asserts that

$$p^\alpha N = p^{\alpha+1} N.$$

In other words, $p^\alpha N$ is an injective R–module. It follows that any extension of K/R by $p^\alpha N$ splits; the extension given by the diagram

$$\begin{array}{ccccccccc}
0 & \longrightarrow & R & \longrightarrow & K & \longrightarrow & K/R & \longrightarrow & 0 \\
 & & \downarrow & & \downarrow & & \downarrow{\scriptstyle 1} & & \\
0 & \longrightarrow & p^\alpha N & \longrightarrow & \overline{N} & \longrightarrow & K/R & \longrightarrow & 0
\end{array}$$

is split. Hence so is the bottom row of the commutative diagram

$$\begin{array}{ccccccccc}
0 & \longrightarrow & R & \longrightarrow & K & \longrightarrow & K/R & \longrightarrow & 0 \\
 & & \downarrow & & \downarrow & & \downarrow{\scriptstyle 1} & & \\
0 & \longrightarrow & p^\alpha N & \longrightarrow & \overline{N} & \longrightarrow & K/R & \longrightarrow & 0 \\
 & & \downarrow & & \downarrow & & \downarrow{\scriptstyle 1} & & \\
0 & \longrightarrow & N & \longrightarrow & N' & \longrightarrow & K/R & \longrightarrow & 0.
\end{array}$$

This establishes that the extension of K/R by M becomes trivial after extending by maps $\phi : M \longrightarrow N$, with N of cardinality $< \alpha$.

To finish the proof, we must establish that the bottom row in the commutative diagram

$$\begin{array}{ccccccccc}
0 & \longrightarrow & R & \longrightarrow & K & \longrightarrow & K/R & \longrightarrow & 0 \\
 & & \downarrow & & \downarrow & & \downarrow{\scriptstyle 1} & & \\
0 & \longrightarrow & M & \longrightarrow & M' & \longrightarrow & K/R & \longrightarrow & 0
\end{array}$$

is not a split exact sequence. Equivalently, we must show that the map $R \longrightarrow M$ does not extend to a map $R \subset K \longrightarrow M$. The image of $1 \in R$

under the map $R \longrightarrow M$ is an element $x \in M$, more precisely

$$x \quad \in \quad p^{\alpha}M - p^{\alpha+1}M \quad \subset \quad M.$$

To extend to a map $K \longrightarrow M$ would be equivalent to finding, for all $n \geq 0$, the image of $p^{-n} \in K$. Let x_n be the image of p^{-n}. We must have

C.4.1.3. $x_0 = x$.

C.4.1.4. $px_{n+1} = x_n$.

We must show that there is no such sequence $\{x_n\}$.

Suppose there exists a sequence $\{x_n\}$ of elements of M, satisfying the conditions C.4.1.3 and C.4.1.4. Consider the set S of ordinals, given by

$$S \quad = \quad \{i \leq \alpha + 1 \mid \exists n \geq 0 \text{ with } x_n \notin p^i M\}.$$

The set S is non–empty; by our construction, $x = x_0 \notin p^{\alpha+1}M$, and so $\alpha + 1 \in S$. Because the set of ordinals $\leq \alpha + 1$ is well–ordered, there exists a minimal ordinal $k \in S$. Since k is minimal, for all $n \geq 0$ and all $j < k$, $x_n \in p^j M$. Thus for all n,

$$x_n \in \bigcap_{j<k} p^j M$$

and since at least one x_n does not lie in $p^k M$, we must have

$$p^k M \neq \bigcap_{j<k} p^j M.$$

This means that k cannot be a limit ordinal. For limit ordinals k, by definition of $p^k M$, we have

$$p^k M = \bigcap_{j<k} p^j M.$$

Hence k must be a successor ordinal. Put $k = i + 1$. Then for all ℓ, $x_\ell \in p^i M$. And there exists at least one n for which $x_n \notin p^{i+1}M$. Choose and fix n, so that $x_n \notin p^{i+1}M$.

But C.4.1.4 asserts that $px_{n+1} = x_n$. We know that for all ℓ, $x_\ell \in p^i M$; in particular $x_{n+1} \in p^i M$. Therefore

$$x_n = px_{n+1} \quad \in \quad p\{p^i M\} \quad = \quad p^{i+1}M,$$

and this is our contradiction; the sequence $\{x_n\}$ cannot exist. □

Before we apply this to the derived category of R, let us remind the reader of well–known facts.

LEMMA C.4.2. *Let R be a ring, of projective dimension ≤ 1 (e.g. a discrete valuation ring). Let X be an object in the derived category $D(R)$. Then*

$$X \quad = \quad \coprod_{n=-\infty}^{\infty} \Sigma^{-n} H^n(X) \quad = \quad \prod_{n=-\infty}^{\infty} \Sigma^{-n} H^n(X).$$

Proof: Choose a projective resolution for the R–module $H^n(X)$

$$0 \longrightarrow P_1 \longrightarrow P_0 \longrightarrow H^n(X) \longrightarrow 0.$$

Such a resolution exists since R is of projective dimension ≤ 1. We may also assume that P_0 and P_1 are both free. They are coproducts of the module R.

Since $H^n(X) = \text{Hom}(\Sigma^{-n}R, X)$, the map $P_0 \longrightarrow H^n(X)$ could be thought of as a map

$$P_0 \quad = \quad \coprod_{\lambda \in \Lambda} R \longrightarrow \text{Hom}(\Sigma^{-n}R, X) \quad = \quad H^n(X),$$

which may be viewed as the functor $\text{Hom}(\Sigma^{-n}R, -)$ applied to a morphism in $D(R)$

$$\Sigma^{-n} P_0 \quad = \quad \coprod_{\lambda \in \Lambda} \Sigma^{-n} R \longrightarrow X.$$

But now we have a composite

$$\Sigma^{-n} P_1 \longrightarrow \Sigma^{-n} P_0 \longrightarrow X.$$

Since P_1 is free, this is a map

$$\Sigma^{-n} P_1 \quad = \quad \coprod_{\mu \in M} \Sigma^{-n} R \longrightarrow X.$$

On the other hand, applying the functor $\text{Hom}(\Sigma^{-n}R, -)$, we get $P_1 \longrightarrow H^n(X)$, which vanishes by hypothesis. It follows that the composite

$$\Sigma^{-n} P_1 \longrightarrow \Sigma^{-n} P_0 \longrightarrow X$$

vanishes in $D(R)$. But the triangle

$$\Sigma^{-n} P_1 \longrightarrow \Sigma^{-n} P_0 \longrightarrow \Sigma^{-n} H^n(X) \longrightarrow \Sigma^{-n+1} P_1$$

asserts that the map $\Sigma^{-n} P_0 \longrightarrow X$ factors, in $D(R)$, as

$$\Sigma^{-n} P_0 \longrightarrow \Sigma^{-n} H^n(X) \longrightarrow X.$$

We have produced a map $\Sigma^{-n} H^n(X) \longrightarrow X$ which is an isomorphism in H^n. Producing such a map for every n, we have a morphism

$$\coprod_{n \in -\infty}^{\infty} \Sigma^{-n} H^n(X) \longrightarrow X,$$

which is an H^n–isomorphism for every n. Hence it is an isomorphism in $D(R)$. This establishes that X is a coproduct of suspensions of its cohomology modules.

But now consider the natural map

$$\coprod_{n\in -\infty}^{\infty} \Sigma^{-n} H^n(X) \longrightarrow \prod_{n\in -\infty}^{\infty} \Sigma^{-n} H^n(X).$$

It is also an H^n–isomorphism for every n, hence an isomorphism in $D(R)$. Thus X is also isomorphic to the product of suspensions of its cohomology modules. □

Let R be a ring of projective dimension ≤ 1. Let $\mathcal{T} = D(R)$ be its derived category.

By Lemma C.4.2, any object $x \in \mathcal{T}$ can be written as

$$x = \coprod_{n=-\infty}^{\infty} \Sigma^{-n} x_n = \prod_{n=-\infty}^{\infty} \Sigma^{-n} x_n,$$

where for each $n \in \mathbb{Z}$, x_n is just an R–module. If $f : x \longrightarrow y$ is a morphism in $\mathcal{T} = D(R)$, then we may write

$$x = \coprod_{n=-\infty}^{\infty} \Sigma^{-n} x_n \qquad\qquad y = \prod_{n=-\infty}^{\infty} \Sigma^{-n} y_n$$

and a map from a coproduct to a product is entirely determined by its components. In other words, to understand all possible maps in $\mathcal{T}$, it is enough to understand the maps

$$\Sigma^m x_m \longrightarrow \Sigma^n y_n$$

with x_m and y_n just ordinary R–modules. It is classical that these maps are in 1–to–1 correspodence with elements of $\mathrm{Ext}^{n-m}(x_m, y_n)$.

But what we really need is to understand the maps in $\mathcal{E}x\Big(\{\mathcal{T}^\alpha\}^{op}, \mathcal{A}b\Big)$. To this end, we prove the Lemma

LEMMA C.4.3. *Let R be a ring of projective dimension ≤ 1. Let $\mathcal{T} = D(R)$ be its derived category. Let α be a regular cardinal. Put $\mathcal{S} = \mathcal{T}^\alpha$. Let*

$$\pi : A(\mathcal{T}) \longrightarrow \mathcal{E}x\big(\mathcal{S}^{op}, \mathcal{A}b\big)$$

be the quotient map of Section 6.5. Let x and y be ordinary R–modules.

If $n \neq 1$, then

$$\mathcal{T}(x, \Sigma^n y) = \mathcal{E}x\big(\mathcal{S}^{op}, \mathcal{A}b\big)\Big\{\pi(x)\,,\, \pi(\Sigma^n y)\Big\}.$$

That is, the maps in $\mathcal{T}$ of the form $x \longrightarrow \Sigma^n y$ agree, via the natural map, with the maps in $\mathcal{E}x\big(\mathcal{S}^{op}, \mathcal{A}b\big)$ of the form $\pi(x) \longrightarrow \pi(\Sigma^n y)$.

Proof: We need to show that maps in $\mathcal{E}x(\mathcal{S}^{op}, \mathcal{A}b)$ of the form $\pi(x) \longrightarrow \pi(\Sigma^n y)$ correspond 1–to–1 with elements of $\mathrm{Ext}^n(x,y)$. Since the ring R has projective dimension ≤ 1, $\mathrm{Ext}^n(x,y) = 0$ unless $n \in \{0,1\}$. Thus we must show:

C.4.3.1. *If $n \neq 0, 1$, then all maps $\pi(x) \longrightarrow \pi(\Sigma^n y)$ vanish.*

C.4.3.2. *Maps $\pi(x) \longrightarrow \pi(y)$ correspond 1–to–1 with R–module homomorphisms $x \longrightarrow y$.*

The reader should note that the case $n = 1$ is specifically excluded. We do not know, nor care about, the maps $\pi(x) \longrightarrow \pi(\Sigma^1 y)$.

It helps to recall what the functor π is. From the discusion at the beginning of Section 6.5, (see also Lemma 6.5.2), we know the functor π very concretely. The functor π takes an object $x \in \mathcal{T} \subset A(\mathcal{T})$ to the functor

$$\mathcal{T}(-,x)|_{\mathcal{S}} : \mathcal{S}^{op} \longrightarrow \mathcal{A}b.$$

What we must establish is that, for any R–modules x and y,

C.4.3.3. *If $n \neq 0, 1$, then any natural transformation*

$$\mathcal{T}(-,x)|_{\mathcal{S}} \longrightarrow \mathcal{T}(-,\Sigma^n y)|_{\mathcal{S}}$$

vanishes.

C.4.3.4. *The natural transformations*

$$\mathcal{T}(-,x)|_{\mathcal{S}} \longrightarrow \mathcal{T}(-,y)|_{\mathcal{S}}$$

correspond 1–to–1 with R–module homomorphisms $x \longrightarrow y$.

Let us first treat C.4.3.4. The functor π takes an R–module homomorphism $f : x \longrightarrow y$ to a natural transformtion

$$\mathcal{T}(-,x)|_{\mathcal{S}} \xrightarrow{\mathcal{T}(-,f)|_{\mathcal{S}}} \mathcal{T}(-,y)|_{\mathcal{S}}.$$

But there is an inverse. We can evaluate a natural transformation

$$\mathcal{T}(-,x)|_{\mathcal{S}} \xrightarrow{\phi} \mathcal{T}(-,y)|_{\mathcal{S}}$$

at any object $s \in \mathcal{S}$. In particular, we wish to evaluate it on the free module R. We deduce a map of R–modules

$$x = \mathcal{T}(R,x) \longrightarrow \mathcal{T}(R,y) = y.$$

Call this map $\phi(R)$. Clearly, $\mathcal{T}(-,f)|_{\mathcal{S}}(R) = f$. What we need to show is that $\phi = \mathcal{T}(-,\phi(R))|_{\mathcal{S}}$. Replacing ϕ by $\phi - \mathcal{T}(-,\phi(R))|_{\mathcal{S}}$, we need to show

C.4.3.5. *Suppose $n \neq 1$, and suppose we are given a natural transformation*

$$\mathcal{T}(-,x)|_{\mathcal{S}} \xrightarrow{\phi} \mathcal{T}(-,\Sigma^n y)|_{\mathcal{S}}$$

which vanishes on the object R. Then $\phi = 0$.

Note that C.4.3.5 combines C.4.3.3 and C.4.3.4. We have just seen that, if $n = 0$, C.4.3.5 implies C.4.3.4. But if $n \notin \{0, 1\}$, then any map

$$\mathcal{T}(-,x)|_{\mathcal{S}} \xrightarrow{\phi} \mathcal{T}(-,\Sigma^n y)|_{\mathcal{S}}$$

vanishes on the object R, and hence C.4.3.5 implies the vanishing of all maps ϕ, that is C.4.3.3.

We wish to prove C.4.3.5, that is the vanishing of all maps

$$\mathcal{T}(-,x)|_{\mathcal{S}} \xrightarrow{\phi} \mathcal{T}(-,\Sigma^n y)|_{\mathcal{S}}$$

which vanish on R. Let ϕ be such a map. Of course, since ϕ vanishes on R, it also vanishes on any coproduct of R's, and on any direct summand of such coproducts. Therefore ϕ vanishes on any projective R–module.

Let s be any object of the category $\mathcal{S} = \mathcal{T}^\alpha$. We wish to show that ϕ vanishes on s. By Lemma C.4.2,

$$s = \coprod_{m=-\infty}^{\infty} \Sigma^{-m} s_m,$$

with s_m ordinary R–modules. It therefore suffices to prove that, for any $m \in \mathbb{Z}$, ϕ vanishes on $\Sigma^{-m} s_m$. And since $s \in \mathcal{S} = \mathcal{T}^\alpha$ and $\Sigma^{-m} s_m$ is a direct summand of s, we must have $\Sigma^{-m} s_m \in \mathcal{S}$. In other words, we are reduced to showing that, for any R–module s belonging to $\mathcal{S} = \mathcal{T}^\alpha$, and for any integer m, the map ϕ vanishes when evaluated on $\Sigma^{-m} s$.

The fact that s is an R–module and $s \in \mathcal{T}^\alpha$ means that s admits a resolution by projective R–modules

$$0 \longrightarrow P_1 \longrightarrow P_0 \longrightarrow s \longrightarrow 0$$

with P_0 and P_1 of rank $< \alpha$. This becomes a triangle in $\mathcal{S}$

$$\Sigma^{-1} s \longrightarrow P_1 \longrightarrow P_0 \longrightarrow s$$

Both the functor $\mathcal{T}(-,x)|_{\mathcal{S}}$ and the functor $\mathcal{T}(-,y)|_{\mathcal{S}}$ take triangles in $\mathcal{S}$ to long exact sequences, and the naturality of ϕ gives a map of long exact sequences

$$\begin{array}{ccccccc}
\mathcal{T}(s,x) & \longrightarrow & \mathcal{T}(P_0,x) & \longrightarrow & \mathcal{T}(P_1,x) & \longrightarrow & \mathcal{T}(\Sigma^{-1}s,x) \\
\downarrow & & \downarrow & & \downarrow & & \downarrow \\
\mathcal{T}(s,\Sigma^n y) & \longrightarrow & \mathcal{T}(P_0,\Sigma^n y) & \longrightarrow & \mathcal{T}(P_1,\Sigma^n y) & \longrightarrow & \mathcal{T}(\Sigma^{-1}s,\Sigma^n y)
\end{array}$$

Now note that in each row, all but four of the abelian groups are zero. Let us prove it for the top row; the case of the bottom row is parallel. The point is that $\mathcal{T}(\Sigma^{-m}P, x) = \mathrm{Ext}^m(P, x)$. And if P is projective, this vanishes unless $m = 0$. Thus the terms $\mathcal{T}((\Sigma^{-m}P_i, x)$, with $i = 0, 1$, must vanish unless $m = 0$. This gives two possible non–zero terms. And from the long exact sequence we learn that $\mathcal{T}(\Sigma^{-m}s, x)$ vanishes, unless $m \in \{0, 1\}$.

Now recall that by the hypothesis of C.4.3.5, the map ϕ vanishes on any projective object. The commutative diagram with exact rows

$$\begin{array}{ccccccc} \longrightarrow & \mathcal{T}(P_1, x) & \longrightarrow & \mathcal{T}(\Sigma^{-1}s, x) & \longrightarrow & 0 \\ & {\scriptstyle 0}\downarrow & & \downarrow & & \downarrow \\ \longrightarrow & \mathcal{T}(P_1, \Sigma^n y) & \longrightarrow & \mathcal{T}(\Sigma^{-1}s, \Sigma^n y) & \longrightarrow & \mathcal{T}(\Sigma^{-1}P_0, \Sigma^n y) \end{array}$$

gives, in particular, a commutative square

$$\begin{array}{ccc} \mathcal{T}(P_1, x) & \longrightarrow & \mathcal{T}(\Sigma^{-1}s, x) \\ {\scriptstyle 0}\downarrow & & \downarrow \\ \mathcal{T}(P_1, \Sigma^n y) & \longrightarrow & \mathcal{T}(\Sigma^{-1}s, \Sigma^n y) \end{array}$$

and the map

$$\mathcal{T}(P_1, x) \longrightarrow \mathcal{T}(P_1, \Sigma^n y)$$

vanishes since P_1 is projective. But then the composite

$$\begin{array}{ccc} \mathcal{T}(P_1, x) & \longrightarrow & \mathcal{T}(\Sigma^{-1}s, x) \\ & & \downarrow \\ & & \mathcal{T}(\Sigma^{-1}s, \Sigma^n y) \end{array}$$

must vanish, and the surjectivity of

$$\mathcal{T}(P_1, x) \longrightarrow \mathcal{T}(\Sigma^{-1}s, x)$$

implies the vanishing of

$$\mathcal{T}(\Sigma^{-1}s, x) \longrightarrow \mathcal{T}(\Sigma^{-1}s, \Sigma^n y).$$

This much was painless.

We also have a commutative diagram with exact rows

$$\begin{array}{ccccc} 0 & \longrightarrow & \mathcal{T}(s, x) & \longrightarrow & \mathcal{T}(P_0, x) \\ \downarrow & & \downarrow & & \downarrow \\ \mathcal{T}(\Sigma P_1, \Sigma^n y) & \longrightarrow & \mathcal{T}(s, \Sigma^n y) & \longrightarrow & \mathcal{T}(P_0, \Sigma^n y) \end{array}$$

Because we are assuming $n \neq 1$,

$$\mathcal{T}(\Sigma P_1, \Sigma^n y) \quad = \quad \mathcal{T}(P_1, \Sigma^{n-1} y) \quad = \quad 0;$$

this is because P_1 is projective, and $n - 1 \neq 0$. The commutative diagram with exact rows above becomes

$$\begin{array}{ccccc} 0 & \longrightarrow & \mathcal{T}(s, x) & \longrightarrow & \mathcal{T}(P_0, x) \\ & & \downarrow & & {\scriptstyle 0}\downarrow \\ 0 & \longrightarrow & \mathcal{T}(s, \Sigma^n y) & \longrightarrow & \mathcal{T}(P_0, \Sigma^n y) \end{array}$$

Since the map

$$\mathcal{T}(P_0, x) \longrightarrow \mathcal{T}(P_0, \Sigma^n y)$$

vanishes (because P_0 is projective, and by the hypothesis of C.4.3.5), it follows that the map also vanishes on the subobjects. That is, the map

$$\mathcal{T}(s, x) \longrightarrow \mathcal{T}(s, \Sigma^n y)$$

is zero. Note that this part of the argument depends on $n \neq 1$.

Summarising, we have shown that for any R–module $s \in \mathcal{S}$, the maps

$$\begin{array}{ccc} \mathcal{T}(s, x) & \longrightarrow & \mathcal{T}(s, \Sigma^n y) \\ \mathcal{T}(\Sigma^{-1}s, x) & \longrightarrow & \mathcal{T}(\Sigma^{-1}s, \Sigma^n y) \end{array}$$

both vanish. But for $m \notin \{0, 1\}$ the group $\mathcal{T}(\Sigma^{-m}s, x)$ is zero, and hence the map

$$\mathcal{T}(\Sigma^{-m}s, x) \longrightarrow \mathcal{T}(\Sigma^{-m}s, \Sigma^n y)$$

vanishes trivially. Thus the map vanishes for all m, completing the proof of the Lemma. □

PROPOSITION C.4.4. *Let R be a discrete valuation ring, K its quotient field. Put $\mathcal{T} = D(R)$, the derived category of R. Let α be regular cardinal $\geq \aleph_1$. Put $\mathcal{S} = \mathcal{T}^\alpha$. Then the category $\mathcal{E}x(\mathcal{S}^{op}, \mathcal{A}b)$ does not have a cogenerator.*

Proof: Observe first that K/R is an object in $\mathcal{T}^\alpha$. It has countably many generators, namely $\{p^{-n}, n \geq 1\}$, and there are countably many relations. In other words, there is an exact sequence of R–modules

$$0 \longrightarrow \coprod_{n=0}^{\infty} R \longrightarrow \coprod_{n=0}^{\infty} R \longrightarrow K/R \longrightarrow 0.$$

This gives a triangle

$$\coprod_{n=0}^{\infty} R \longrightarrow \coprod_{n=0}^{\infty} R \longrightarrow K/R \longrightarrow \Sigma\left\{\coprod_{n=0}^{\infty} R\right\}.$$

But $R \in \mathcal{T}^\alpha$, and hence so is the countable coproduct $\coprod_{n=0}^{\infty} R$. Two terms in the triangle above lie in $\mathcal{S} = \mathcal{T}^\alpha$, and hence so does the third, K/R.

Suppose $\mathcal{E}x(\mathcal{S}^{op}, \mathcal{A}b)$ had a cogenerator. There would be an object C, so that all other objects inject into products of C. Now recall Proposition 6.5.3; the category $\mathcal{E}x\Big(\{\mathcal{T}^\alpha\}^{op}, \mathcal{A}b\Big)$ is a Gabriel quotient of the category $A(\mathcal{T})$. There is an exact quotient map

$$\pi : A(\mathcal{T}) \longrightarrow \mathcal{E}x\Big(\{\mathcal{T}^\alpha\}^{op}, \mathcal{A}b\Big)$$

and it has a left adjoint L. But then $L(C)$ is an object of $A(\mathcal{T})$, and may be embedded in an injective object I. The functor π, being exact, takes this to a monomorphism

$$C = \pi L(C) \longrightarrow \pi(I).$$

If C is a cogenerator, any object may be embedded in a product of C's. But C embeds in $\pi(I)$, and hence any object may be embedded in a product of $\pi(I)$'s. That is, $\pi(I)$ must also be a cogenerator. We may therefore assume that our cogenerator is of the form $\pi(I)$, with I an injective object in $A(\mathcal{T})$. By Corollary 5.1.23, the injectivity of I means that $I \in \mathcal{T} \subset A(\mathcal{T})$. Choose and fix such an I.

Now let β be an infinite cardinal greater than the maximum cardinality of all the homology groups $H^n(I)$, with $I \in \mathcal{T} = D(R)$, and $\pi(I)$ a cogenerator of $\mathcal{E}x\Big(\{\mathcal{T}^\alpha\}^{op}, \mathcal{A}b\Big)$, as above. Let M be an R–module as in Lemma C.4.1, for the infinite cardinal β chosen above. There exists a non–trivial extension

$$0 \longrightarrow M \longrightarrow M' \longrightarrow K/R \longrightarrow 0$$

so that, for any $\phi : M \longrightarrow N$, with N of cardinality $< \beta$, the induced extension of K/R by N splits. Now M may be viewed as an object of $D(R) = \mathcal{T}$, and $\pi(M)$ becomes an object in $\mathcal{E}x\Big(\{\mathcal{T}^\alpha\}^{op}, \mathcal{A}b\Big)$. I assert that this object cannot possibly be embedded into a product of $\pi(I)$'s.

By Lemma C.4.2, the object $I \in \mathcal{T}$ admits a decomposition

$$I \quad = \quad \prod_{m=-\infty}^{\infty} \Sigma^m I_m,$$

with I_m all R–modules, and since $I_m = H^{-m}(I)$, then by the choice of β their cardinalities are all $< \beta$. Now the map π respects products, since it has a left adjoint L. Therefore

$$\pi(I) \quad = \quad \prod_{m=-\infty}^{\infty} \pi(\Sigma^m I_m).$$

But we are supposing that $\pi(I)$ is a cogenerator. There is therefore an embedding in $\mathcal{E}x\big(\mathcal{S}^{op}, \mathcal{A}b\big)$

$$\pi(M) \quad \subset \quad \prod_{\Lambda} \pi(I) \quad = \quad \prod_{\Lambda} \prod_{m=-\infty}^{\infty} \pi(\Sigma^m I_m).$$

For every integer m, this gives a map

$$\pi(M) \quad \subset \quad \prod_{\Lambda} \pi(I) \longrightarrow \prod_{\Lambda} \pi(\Sigma^m I_m).$$

Now Lemma C.4.3 tells us that unless $m \in \{0,1\}$, all such maps vanish. We conclude that the following is already an embedding, the other components being zero

$$\pi(M) \longrightarrow \left\{\prod_{\Lambda} \pi(I_0)\right\} \oplus \left\{\prod_{\Lambda} \pi(\Sigma^1 I_1)\right\}.$$

Now we wish to evaluate this natural transformation on the particular object $\Sigma^{-1}\{K/R\} \in \mathcal{S}$. In the interest of making the formulas below more legible, let us abbreviate

$$K/R = \ell, \qquad \text{that is} \qquad \Sigma^{-1}\{K/R\} = \Sigma^{-1}\ell.$$

We have a monomorphism of abelian groups

$$\mathcal{T}(\Sigma^{-1}\ell, M) \longrightarrow \left\{\prod_{\Lambda} \mathcal{T}(\Sigma^{-1}\ell, I_0)\right\} \oplus \left\{\prod_{\Lambda} \mathcal{T}(\Sigma^{-1}\ell, \Sigma^1 I_1)\right\}.$$

But $\mathcal{T}(\Sigma^{-1}\ell, \Sigma^1 I_1) = \operatorname{Ext}^2(\ell, I_1) = 0$. Hence we have that the map

$$\mathcal{T}(\Sigma^{-1}\ell, M) \longrightarrow \prod_{\Lambda} \mathcal{T}(\Sigma^{-1}\ell, I_0)$$

must be a monomorphism. But this map is nothing other that the natural transformation

$$\pi(M) \longrightarrow \prod_{\Lambda} \pi(I_0),$$

evaluated on the object $\Sigma^{-1}\ell \in \mathcal{S}$. By Lemma C.4.3, we know that the natural transformation above is induced by a map of modules $M \longrightarrow \prod_{\Lambda} I_0$. But our extension

$$0 \longrightarrow M \longrightarrow M' \longrightarrow K/R = \ell \longrightarrow 0$$

gives a non–zero map $\Sigma^{-1}\ell \longrightarrow M$, and for any map $M \longrightarrow I_0$, the extension $\Sigma^{-1}\ell \longrightarrow M \longrightarrow I_0$ must vanish, as the cardinality of I_0 is $< \beta$. In other words, we have found a class in $\mathcal{T}(\Sigma^{-1}\ell, M)$, which maps to zero under the natural map to $\mathcal{T}(\Sigma^{-1}\ell, \prod_{\Lambda} I_0)$. This contradicts the injectivity. $\square$

C.5. History of the results of Appendix C

The results of Sections C.1, C.2 and C.3 are certainly not new. As far as the author knows, the earliest version appeared in Freyd's article [**14**]. There were also accounts due to Grandis and to Morava, but those never appeared in print.

The author learned the construction of the large module M from Boardman. Boardman uses it in his unpublished paper, on conditionally convergent spectral sequences.

The application of these constructions, in Section C.4, is entirely new. Since injectives in the category $\mathcal{E}x\big(\mathcal{S}^{op}, \mathcal{A}b\big)$ are new to this book, it is new that one can use these large modules to show that in general, $\mathcal{E}x\big(\mathcal{S}^{op}, \mathcal{A}b\big)$ does not have enough injectives.

APPENDIX D

Where $\mathcal{T}$ is the homotopy category of spectra

D.1. Localisation with respect to homology

In this Appendix, $\mathcal{T}$ will be the homotopy category of spectra. For the reader unfamiar with spectra, Summary D.1.1 lists the properties we will need in the present section. The properties used in Section D.2 are more difficult to summarise briefly. What we need there is basically some familiarity with the results of [**24**].

SUMMARY D.1.1. The homotopy category of spectra is a triangulated category $\mathcal{T}$, closed under small coproducts. That is, it satisfies [TR5]. It has a smash product

$$\wedge : \mathcal{T} \times \mathcal{T} \longrightarrow \mathcal{T}$$

and an object $S^0 \in \mathcal{T}$, satisfying the following additional properties:

D.1.1.1. *The object $S^0 \in \mathcal{T}$ is $\aleph_0$-compact. That is, any map*

$$S^0 \longrightarrow \coprod_{\lambda \in \Lambda} t_\lambda$$

factors through a finite coproduct.

D.1.1.2. *Put $S^n = \Sigma^n S^0$. Let t be an object of $\mathcal{T}$. If, for every $n \in \mathbb{Z}$, $\mathcal{T}(S^n, t) = 0$, then $t \simeq 0$.*

D.1.1.3. *For any $n \in \mathbb{Z}$, the group $\mathcal{T}(S^0, S^n)$ has cardinality $\leq \aleph_0$.*

D.1.1.4. *The smash product takes triangles to triangles. That is, there are natural isomorphisms*

$$\{\Sigma a\} \wedge b \qquad \simeq \qquad \Sigma\{a \wedge b\} \qquad \simeq \qquad a \wedge \{\Sigma b\},$$

and given an object $t \in \mathcal{T}$ and a triangle

$$x \longrightarrow y \longrightarrow z \longrightarrow \Sigma x,$$

then both of the following are triangles

$$x \wedge t \longrightarrow y \wedge t \longrightarrow z \wedge t \longrightarrow \{\Sigma x\} \wedge t \quad = \quad \Sigma\{x \wedge t\},$$

$$t \wedge x \longrightarrow t \wedge y \longrightarrow t \wedge z \longrightarrow t \wedge \{\Sigma x\} \quad = \quad \Sigma\{t \wedge x\}.$$

D.1.1.5. *The smash product respects coproducts. That is,*

$$x \wedge \left\{ \coprod_{\lambda \in \Lambda} t_\lambda \right\} = \coprod_{\lambda \in \Lambda} \{x \wedge t_\lambda\},$$

$$\left\{ \coprod_{\lambda \in \Lambda} t_\lambda \right\} \wedge x = \coprod_{\lambda \in \Lambda} \{t_\lambda \wedge x\}.$$

D.1.1.6. *S^0 is a two-sided unit for the smash product; that is*

$$x \wedge S^0 = x = S^0 \wedge x.$$

D.1.1.7. *Any map*

$$S^0 \longrightarrow x \wedge y$$

factors as

$$S^0 \longrightarrow x' \wedge y' \xrightarrow{f \wedge g} x \wedge y$$

where x' and y' are objects in $\mathcal{T}^{\aleph_0}$.

Lemma D.1.2. *Let T be the set of suspensions of S^0; that is,*

$$T = \{S^n \mid n \in \mathbb{Z}\}.$$

Then $T \subset \mathcal{T}^{\aleph_0}$, and $T \cup \{0\}$ is α–perfect for all infinite α.

Proof: By D.1.1.1, the objects of $S^n \in T$ are all $\aleph_0$–small; see Definition 4.1.1. The set $T \cup \{0\}$ is $\aleph_0$–compact, since any class of objects containing 0 is. See Example 3.3.16. Thus $T \cup \{0\}$ is an $\aleph_0$–perfect class of objects in $\mathcal{T}^{(\aleph_0)}$, hence contained in the maximal one $\mathcal{T}^{\aleph_0}$.

Now $T \cup \{0\}$ is an $\aleph_0$–perfect class of objects in $\mathcal{T}^{(\aleph_0)}$. Lemma 4.2.1 applies, and we deduce that $T \cup \{0\}$ is α–perfect for all infinite α. □

Lemma D.1.3. *As in Lemma D.1.2, let T be the set of suspensions of S^0. Then*

D.1.3.1. *The category $\mathcal{T}$ satisfies the representability theorem.*

D.1.3.2. $\mathcal{T} = \langle T \rangle$.

Proof: By D.1.1.2, the set T generates the triangulated category $\mathcal{T}$; see Definition 8.1.1. In Lemma D.1.3, we saw that $T \cup \{0\}$ is α–perfect for all infinite α, in particular $\aleph_1$–perfect. By Definition 8.1.2, T is therefore an $\aleph_1$–perfect generating set. Theorem 8.3.3 applies, and we conclude

D.1.3.1: The category $\mathcal{T}$ satisfies the representability theorem.
D.1.3.2: $\mathcal{T} = \langle T \rangle$.

□

PROPOSITION D.1.4. *As in Lemma D.1.2, let T be the set of suspensions of S^0. Then for any regular cardinal α,*

$$\mathcal{T}^\alpha = \langle T \rangle^\alpha.$$

That is, the subcategory of α–compact objects in $\mathcal{T}$ agrees with the α–localising subcategory generated by T.

Proof: From Lemma D.1.3, we know that for our set T, $\mathcal{T} = \langle T \rangle$. From Lemma D.1.2, we also have that $T \subset \mathcal{T}^{\aleph_0}$. Therefore Lemma 4.4.5 applies. We deduce that, for all regular $\alpha \geq \aleph_0$,

$$\mathcal{T}^\alpha = \langle T \rangle^\alpha.$$

□

REMARK D.1.5. Since T is a set, Proposition 3.2.5 guarantees that, for every regular α, $\mathcal{T}^\alpha$ is essentially small. It is the smallest triangulated category containing all the spheres, and closed under triangles, as well as coproducts of fewer than α of its objects. This is usually referred to as the "category of spectra with fewer than α cells". An object in $\mathcal{T}^\alpha$ can be constructed out of fewer than α spheres, by attaching.

LEMMA D.1.6. *Let α be a regular cardinal. If x and y are objects in $\mathcal{T}^\alpha$, then so is $x \wedge y$.*

Proof: Fix an object $y \in \mathcal{T}^\alpha$. Define a full subcategory $\mathcal{S} \subset \mathcal{T}$ by

$$\mathrm{Ob}(\mathcal{S}) = \{x \in \mathcal{T} \mid x \wedge y \in \mathcal{T}^\alpha\}.$$

We will prove that $\mathcal{S}$ is an α–localising subcategory containing T, and hence must contain $\langle T \rangle^\alpha = \mathcal{T}^\alpha$.

First observe that $\mathcal{S}$ contains $T = \{S^n \mid n \in \mathbb{Z}\}$. For if $n \in \mathbb{Z}$,

$$\begin{aligned} S^n \wedge y &= \{\Sigma^n S^0\} \wedge y \\ &= \Sigma^n \{S^0 \wedge y\} \\ &= \Sigma^n y. \end{aligned}$$

We are given $y \in \mathcal{T}^\alpha$, therefore $S^n \wedge y = \Sigma^n y \in \mathcal{T}^\alpha$. Hence $S^n \in \mathcal{S}$.

Next note that $x \in \mathcal{S}$ if and only if $\Sigma x \in \mathcal{S}$. This is because

$$\{\Sigma x\} \wedge y \simeq \Sigma\{x \wedge y\},$$

and so $x \wedge y \in \mathcal{T}^\alpha$ if and only if $\{\Sigma x\} \wedge y = \Sigma\{x \wedge y\} \in \mathcal{T}^\alpha$.

Now suppose we have a triangle in $\mathcal{T}$

$$x \longrightarrow x' \longrightarrow x'' \longrightarrow \Sigma x,$$

and suppose $x, x' \in \mathcal{S}$. By D.1.1.4, the following is a triangle in $\mathcal{T}$

$$x \wedge y \longrightarrow x' \wedge y \longrightarrow x'' \wedge y \longrightarrow \{\Sigma x\} \wedge y.$$

Since $x, x' \in \mathcal{S}$, we know that $x \wedge y, x' \wedge y \in \mathcal{T}^\alpha$. But $\mathcal{T}^\alpha$ is triangulated, and we deduce that $x'' \wedge y \in \mathcal{T}^\alpha$, that is $x'' \in \mathcal{S}$. The category $\mathcal{S}$ is triangulated.

Let $\{x_\lambda, \lambda \in \Lambda\}$ is a family of fewer than α objects of $\mathcal{S}$. That is, for each $\lambda \in \Lambda$, $x_\lambda \wedge y \in \mathcal{T}^\alpha$. By D.1.1.5,

$$\left\{\coprod_{\lambda\in\Lambda} x_\lambda\right\} \wedge y \quad = \quad \coprod_{\lambda\in\Lambda}\{x_\lambda \wedge y\}.$$

But $\coprod_{\lambda\in\Lambda}\{x_\lambda \wedge y\}$ is a coproduct of fewer than α objects in $\mathcal{T}^\alpha$. By Lemma 4.2.5, the category $\mathcal{T}^\alpha$ is α–localising: it is closed under coproducts of fewer than α of its objects. Hence

$$\left\{\coprod_{\lambda\in\Lambda} x_\lambda\right\} \wedge y \quad = \quad \coprod_{\lambda\in\Lambda}\{x_\lambda \wedge y\} \quad \in \quad \mathcal{T}^\alpha,$$

and $\coprod_{\lambda\in\Lambda} x_\lambda \in \mathcal{S}$. The category $\mathcal{S}$ is closed under coproducts of fewer than α of its objects.

Since we are not assuming $\alpha > \aleph_0$, to verify that $\mathcal{S}$ is α–localising, we must also check that it is thick. Suppose $x \oplus x' \in \mathcal{S}$. That is,

$$\{x \oplus x'\} \wedge y \quad = \quad \{x \wedge y\} \oplus \{x' \wedge y\} \quad \in \quad \mathcal{T}^\alpha.$$

Since $\mathcal{T}^\alpha$ is thick, we have $x \wedge y \in \mathcal{T}^\alpha$, that is $x \in \mathcal{S}$. The category $\mathcal{S}$ contains all the direct summands of its objects, that is $\mathcal{S}$ is thick.

Thus $\mathcal{S}$ is an α–localising subcategory, containing T. Therefore $\mathcal{S}$ contains the smallest such, $\langle T\rangle^\alpha$. But by Proposition D.1.4, $\langle T\rangle^\alpha = \mathcal{T}^\alpha$. That is, $\mathcal{T}^\alpha \subset \mathcal{S}$. In other words, for our fixed (but arbitrary) $y \in \mathcal{T}^\alpha$, and any $x \in \mathcal{T}^\alpha$, we have $x \wedge y \in \mathcal{T}^\alpha$. □

LEMMA D.1.7. *Let α be a regular cardinal, $\alpha > \aleph_0$. Let x be an object of $\mathcal{T}^\alpha$. Then for any $n \in \mathbb{Z}$, the cardinality of $\mathcal{T}(S^n, x)$ is $< \alpha$.*

Proof: Define a full subcategory $\mathcal{S} \subset \mathcal{T}$ by

$$\mathrm{Ob}(\mathcal{S}) \quad = \quad \{x \in \mathcal{T} \mid \forall n \in \mathbb{Z}, \#\mathcal{T}(S^n, x) < \alpha\}.$$

Once again, we will prove that $\mathcal{S}$ is an α–localising subcategory containing T, and hence must contain $\langle T\rangle^\alpha = \mathcal{T}^\alpha$.

The fact that $\mathcal{S}$ contains T is just D.1.1.3.

Next, we have $x \in \mathcal{S}$ if and only if $\Sigma x \in \mathcal{S}$, since

$$\mathcal{T}(S^n, x) \quad = \quad \mathcal{T}(S^{n+1}, \Sigma x).$$

Given a triangle in $\mathcal{T}$

$$x \longrightarrow y \longrightarrow z \longrightarrow \Sigma x,$$

the exact sequence

$$\mathcal{T}(S^n, y) \longrightarrow \mathcal{T}(S^n, z) \longrightarrow \mathcal{T}(S^n, \Sigma x)$$

tells us that the cardinality of $\mathcal{T}(S^n, z)$ is bounded above by the product

$$\{\#\mathcal{T}(S^n, y)\}\{\#\mathcal{T}(S^n, \Sigma x)\}.$$

If x and y lie in $\mathcal{S} \subset \mathcal{T}$, this bound is $\beta\gamma = \max(\beta, \gamma)$, with $\beta < \alpha$, $\gamma < \alpha$. Thus z must also lie in $\mathcal{S}$. Since $\mathcal{S}$ is closed under triangles and suspensions, it is a triangulated subcategory of $\mathcal{T}$.

Now suppose that $\{x_\lambda, \lambda \in \Lambda\}$ is a family of fewer than α objects of $\mathcal{S}$. Pick $n \in \mathbb{Z}$. By D.1.1.1, any map

$$S^n \longrightarrow \coprod_{\lambda \in \Lambda} x_\lambda$$

factors through a finite coproduct. There is a finite set $\Lambda' \subset \Lambda$, and a factorisation

$$S^n \longrightarrow \coprod_{\lambda \in \Lambda'} x_\lambda \quad \subset \quad \coprod_{\lambda \in \Lambda} x_\lambda.$$

Since the cardinality of Λ is $< \alpha$, there are fewer than α finite subsets $\Lambda' \subset \Lambda$. For each Λ', there are fewer than α maps

$$S^n \longrightarrow \coprod_{\lambda \in \Lambda'} x_\lambda.$$

Therefore the collection of maps

$$S^n \longrightarrow \coprod_{\lambda \in \Lambda} x_\lambda$$

is a union of fewer than α sets, all of cardinality $< \alpha$. Since α is regular, we deduce

$$\#\mathcal{T}\left(S^n, \coprod_{\lambda \in \Lambda} x_\lambda\right) \quad < \quad \alpha.$$

Since this is true for any $n \in \mathbb{Z}$, we deduce that $\coprod_{\lambda \in \Lambda} x_\lambda$ is in $\mathcal{S}$. The subcategory $\mathcal{S} \subset \mathcal{T}$ is closed under coproducts of fewer than α of its objects. We are assuming $\alpha > \aleph_0$; the fact that $\mathcal{S}$ is triangulated and closed under countable coproducts, guarantees that $\mathcal{S}$ must be thick.

Thus $\mathcal{S}$ is an α–localising subcategory, containing T. Therefore $\mathcal{S}$ contains the smallest such, $\langle T \rangle^\alpha$. By Proposition D.1.4, $\langle T \rangle^\alpha = \mathcal{T}^\alpha$. That is, $\mathcal{T}^\alpha \subset \mathcal{S}$. In other words, for every object $x \in \mathcal{T}^\alpha$ and any $n \in \mathbb{Z}$,

$$\#\mathcal{T}(S^n, x) \quad < \quad \alpha.$$

□

DEFINITION D.1.8. *Let E be an object of $\mathcal{T}$. The category of E–*acyclic spectra, *denoted $\mathcal{T}_E$, is defined to be the full subcategory of $\mathcal{T}$ whose objects are given by*

$$\mathrm{Ob}(\mathcal{T}_E) \quad = \quad \{x \in \mathcal{T} \mid x \wedge E = 0\}.$$

LEMMA D.1.9. *Let E be an object of $\mathcal{T}$. The category $\mathcal{T}_E$ of E–acyclic spectra is a localising subactegory of $\mathcal{T}$.*

Proof: Observe that $x \in \mathcal{T}_E$ if and only if $\Sigma x \in \mathcal{T}_E$. After all,

$$\{\Sigma x\} \wedge E \quad = \quad \Sigma\{x \wedge E\},$$

and $x \wedge E = 0$ if and only if $\{\Sigma x\} \wedge E = \Sigma\{x \wedge E\} = 0$.

Suppose we have a triangle in $\mathcal{T}$

$$x \longrightarrow x' \longrightarrow x'' \longrightarrow \Sigma x,$$

and suppose $x, x' \in \mathcal{T}_E$. By D.1.1.4, the following is a triangle in $\mathcal{T}$

$$x \wedge E \longrightarrow x' \wedge E \longrightarrow x'' \wedge E \longrightarrow \{\Sigma x\} \wedge E.$$

By hypothesis, $x \wedge E = 0 = x' \wedge E$. From the triangle we have $x'' \wedge E = 0$, that is $x'' \in \mathcal{T}_E$. The subcategory $\mathcal{T}_E$ is triangulated.

Let $\{x_\lambda, \lambda \in \Lambda\}$ be a set objects of $\mathcal{T}_E$. That is, for each $\lambda \in \Lambda$, $x_\lambda \wedge E = 0$. By D.1.1.5,

$$\left\{\coprod_{\lambda\in\Lambda} x_\lambda\right\} \wedge E \quad = \quad \coprod_{\lambda\in\Lambda}\{x_\lambda \wedge E\} \quad = \quad 0.$$

Therefore $\coprod_{\lambda\in\Lambda} x_\lambda$ is an object of $\mathcal{T}_E$, and $\mathcal{T}_E$ is localising. □

Now we come to the key lemma.

LEMMA D.1.10. *As above, $\mathcal{T}$ is the homotopy category of spectra. Let α be a regular cardinal, $\alpha > \aleph_0$. Let E be an object of $\mathcal{T}^\alpha$. Let R be a set of representatives for all isomorphism classes of objects in $\mathcal{T}^\alpha \cap \mathcal{T}_E$. [By Remark D.1.5, $\mathcal{T}^\alpha$ is essentially small, so we may always choose a* set *R of representatives for any class of objects in $\mathcal{T}^\alpha$.] Then any map $t \longrightarrow x$, with $t \in \mathcal{T}^\alpha$ and $x \in \mathcal{T}_E$, factors as*

$$t \longrightarrow r \longrightarrow x$$

with $r \in R$.

Proof: Suppose we are given a map $t \longrightarrow x$, with $t \in \mathcal{T}^\alpha$ and $x \in \mathcal{T}_E$. We want to factor it. The strategy of the proof is first to show that any map $t \longrightarrow x$ as above admits a factorisation

$$t \longrightarrow \bar{t} \longrightarrow x,$$

where $\bar{t} \in \mathcal{T}^\alpha$, and

$$\mathcal{T}(S^n, t \wedge E) \longrightarrow \mathcal{T}(S^n, \bar{t} \wedge E)$$

vanishes. Then we let r be the homotopy colimit of the sequence

$$t \longrightarrow \bar{t} \longrightarrow \bar{\bar{t}} \longrightarrow \cdots$$

and show that this r works.

We begin by constructing, for any map $t \longrightarrow x$ with $t \in \mathcal{T}^\alpha$ and $x \in \mathcal{T}_E$, a factorisation $t \longrightarrow \bar{t} \longrightarrow x$. Complete $t \longrightarrow x$ to a triangle

$$k \longrightarrow t \longrightarrow x \longrightarrow \Sigma k.$$

By D.1.1.4, smashing with E gives a triangle

$$k \wedge E \longrightarrow t \wedge E \longrightarrow x \wedge E \longrightarrow \{\Sigma k\} \wedge E.$$

Since $x \in \mathcal{T}_E$, we have $x \wedge E = 0$. Therefore the natural map is an isomorphism

$$k \wedge E \longrightarrow t \wedge E.$$

Both t and E lie in $\mathcal{T}^\alpha$. By Lemma D.1.6, $t \wedge E \in \mathcal{T}^\alpha$. By Lemma D.1.7, for any $n \in \mathbb{Z}$, the cardinality of the group

$$\mathcal{T}(S^n, k \wedge E) \quad \simeq \quad \mathcal{T}(S^n, t \wedge E)$$

is $< \alpha$. Let

$$\Lambda = \bigcup_{n=-\infty}^{\infty} \mathcal{T}(S^n, t \wedge E).$$

The set Λ is a union of countably many sets, each of cardinality $< \alpha$. Since $\alpha > \aleph_0$ and α is regular, $\#\Lambda < \alpha$.

An element of the set Λ is a map $S^n \longrightarrow k \wedge E \simeq t \wedge E$. From D.1.1.7, we know that it is possible to choose $k' \longrightarrow k$ and $E' \longrightarrow E$, with $k', E' \in \mathcal{T}^{\aleph_0}$, so that $S^n \longrightarrow k \wedge E$ will factor as

$$S^n \longrightarrow k' \wedge E' \xrightarrow{f \wedge g} k \wedge E.$$

For each $\lambda \in \Lambda$, choose such a factorisation. That is, pick an object $k' = k_\lambda \in \mathcal{T}^{\aleph_0}$, and a factorisation

$$S^n \longrightarrow k_\lambda \wedge E \xrightarrow{f_\lambda \wedge 1} k \wedge E.$$

We have a map

$$\coprod_{\lambda \in \Lambda} k_\lambda \longrightarrow k \longrightarrow t.$$

We define the map $t \longrightarrow \bar{t}$ from the triangle

$$\coprod_{\lambda \in \Lambda} k_\lambda \longrightarrow t \longrightarrow \bar{t} \longrightarrow \Sigma \left\{ \coprod_{\lambda \in \Lambda} k_\lambda \right\}.$$

Now $t \in \mathcal{T}^\alpha$, and $\coprod_{\lambda \in \Lambda} k_\lambda$ is the coproduct of fewer than α objects, each in $\mathcal{T}^{\aleph_0} \subset \mathcal{T}^\alpha$. The coproduct lies in $\mathcal{T}^\alpha$, and from the triangle, we deduce

$\bar{t} \in \mathcal{T}^\alpha$. Because the composite $k \longrightarrow t \longrightarrow x$ vanishes, so does the longer composite

$$\coprod_{\lambda\in\Lambda} k_\lambda \longrightarrow k \longrightarrow t \longrightarrow x.$$

From the triangle

$$\coprod_{\lambda\in\Lambda} k_\lambda \longrightarrow t \longrightarrow \bar{t},$$

we deduce that the map $t \longrightarrow x$ must factor as $t \longrightarrow \bar{t} \longrightarrow x$. Also, from the triangle

$$\coprod_{\lambda\in\Lambda} k_\lambda \longrightarrow t \longrightarrow \bar{t}$$

we obtain, by smashing with E, a triangle

$$\left\{\coprod_{\lambda\in\Lambda} k_\lambda\right\} \wedge E \longrightarrow t \wedge E \longrightarrow \bar{t} \wedge E.$$

But by construction, any morphism $S^n \longrightarrow t \wedge E$ factored through $k_\lambda \wedge E$ for some $\lambda \in \Lambda$. In other words, $S^n \longrightarrow t \wedge E$ factors as

$$S^n \longrightarrow \left\{\coprod_{\lambda\in\Lambda} k_\lambda\right\} \wedge E \longrightarrow t \wedge E,$$

and we deduce that, for every $S^n \longrightarrow t \wedge E$, the composite

$$S^n \longrightarrow t \wedge E \longrightarrow \bar{t} \wedge E$$

vanishes.

Now we iterate the process. We define a sequence

$$t_0 \longrightarrow t_1 \longrightarrow t_2 \longrightarrow \cdots$$

in $\mathcal{T}^\alpha$, together with a map from the sequence to x. Define $t_0 \longrightarrow x$ to be $t \longrightarrow x$. Suppose we have defined, for each $i \leq n$, $t_i \longrightarrow x$ and maps $t_{i-1} \longrightarrow t_i$, with all the morphisms compatible. Then $t_n \longrightarrow t_{n+1} \longrightarrow x$ is defined to be

$$t_n \longrightarrow \overline{t_n} \longrightarrow x$$

as above. Put $r = \underrightarrow{\text{Hocolim}}\, t_n$. Clearly, the map $t \longrightarrow x$ factors as

$$t = t_0 \longrightarrow \underrightarrow{\text{Hocolim}}\, t_n \longrightarrow x$$

To complete the proof of the Lemma, we need to show that $\underrightarrow{\text{Hocolim}}\, t_n \in \mathcal{T}^\alpha \cap \mathcal{T}_E$. In other words, up to replacing r by an isomorph, we may choose $r \in R$.

We have a triangle in $\mathcal{T}$

$$\coprod_{i=0}^{\infty} t_i \xrightarrow{1-shift} \coprod_{i=0}^{\infty} t_i \longrightarrow \underrightarrow{\mathrm{Hocolim}}\, t_i \longrightarrow \Sigma\left\{\coprod_{i=0}^{\infty} t_i\right\}.$$

By construction, $t_i \in \mathcal{T}^\alpha$. Since $\alpha > \aleph_0$ and $\mathcal{T}^\alpha$ is α–localising, the coproduct of countably many objects in $\mathcal{T}^\alpha$ is in $\mathcal{T}^\alpha$. Thus in the triangle, the two coproducts lie in $\mathcal{T}^\alpha$. Since $\mathcal{T}^\alpha$ is triangulated, $\underrightarrow{\mathrm{Hocolim}}\, t_i \in \mathcal{T}^\alpha$.

Now we may smash the above triangle with E, and by D.1.1.4, we get a triangle

$$\left\{\coprod_{i=0}^{\infty} t_i\right\} \wedge E \xrightarrow{1-shift} \left\{\coprod_{i=0}^{\infty} t_i\right\} \wedge E \longrightarrow \left\{\underrightarrow{\mathrm{Hocolim}}\, t_i\right\} \wedge E.$$

By D.1.1.5, the smash product commutes with coproducts. The triangle above naturally identifies with

$$\coprod_{i=0}^{\infty}\{t_i \wedge E\} \xrightarrow{1-shift} \coprod_{i=0}^{\infty}\{t_i \wedge E\} \longrightarrow \left\{\underrightarrow{\mathrm{Hocolim}}\, t_i\right\} \wedge E.$$

But the third vertex in the triangle

$$\coprod_{i=0}^{\infty}\{t_i \wedge E\} \xrightarrow{1-shift} \coprod_{i=0}^{\infty}\{t_i \wedge E\} \longrightarrow \underrightarrow{\mathrm{Hocolim}}\,\{t_i \wedge E\}$$

is by definition $\underrightarrow{\mathrm{Hocolim}}\,\{t_i \wedge E\}$; we deduce a (non–canonical) isomorphism

$$\left\{\underrightarrow{\mathrm{Hocolim}}\, t_i\right\} \wedge E \quad = \quad \underrightarrow{\mathrm{Hocolim}}\,\{t_i \wedge E\}.$$

We will prove that $\underrightarrow{\mathrm{Hocolim}}\,\{t_i \wedge E\}$ vanishes. From the isomorphism above, we deduce that $\left\{\underrightarrow{\mathrm{Hocolim}}\, t_i\right\} \wedge E$ vanishes, that is

$$\left\{\underrightarrow{\mathrm{Hocolim}}\, t_i\right\} \quad \in \quad \mathcal{T}_E.$$

Now put $\mathcal{S} = \mathcal{T}^{\aleph_0}$, and let $\pi : \mathcal{T} \longrightarrow \mathcal{E}x\big(\mathcal{S}^{op}, \mathcal{A}b\big)$ be the usual homological functor, commuting with coproducts. It takes the triangle

$$\coprod_{i=0}^{\infty}\{t_i \wedge E\} \xrightarrow{1-shift} \coprod_{i=0}^{\infty}\{t_i \wedge E\} \longrightarrow \underrightarrow{\mathrm{Hocolim}}\,\{t_i \wedge E\}$$

to the long exact sequence

$$\rightarrow \coprod_{i=0}^{\infty}\pi\{t_i \wedge E\} \xrightarrow{1-shift} \coprod_{i=0}^{\infty}\pi\{t_i \wedge E\} \longrightarrow \pi\,\underrightarrow{\mathrm{Hocolim}}\,\{t_i \wedge E\} \rightarrow$$

Now note that $\mathcal{S} = \mathcal{T}^{\aleph_0}$, and $\mathcal{E}x\big(\mathcal{S}^{op}, \mathcal{A}b\big)$ satisfies [AB5]. The map $1 - shift$ is therefore injective, with cokernel $\underrightarrow{\mathrm{colim}}\,\pi\{t_i \wedge E\}$. From the long exact sequence,

$$\pi \underrightarrow{\text{Hocolim}} \{t_i \wedge E\} \quad = \quad \underrightarrow{\text{colim}}\, \pi\{t_i \wedge E\}.$$

But this is an equality of two objects in $\mathcal{E}x(\mathcal{S}^{op}, \mathcal{A}b)$, that is two functors $\mathcal{S} \longrightarrow \mathcal{A}b$. We can evaluate these at $S^n \in \mathcal{S} = \mathcal{T}^{\aleph_0}$. Recalling that $\pi t = \mathcal{T}(-,t)|_{\mathcal{S}}$, and so $\pi t(s) = \mathcal{T}(s,t)$, we have

$$\mathcal{T}(S^n, \underrightarrow{\text{Hocolim}} \{t_i \wedge E\}) \quad = \quad \underrightarrow{\text{colim}}\, \mathcal{T}(S^n, \{t_i \wedge E\}).$$

Now the map $t_i \longrightarrow t_{i+1}$ was constructed so that the induced map

$$\mathcal{T}(S^n, \{t_i \wedge E\}) \longrightarrow \mathcal{T}(S^n, \{t_{i+1} \wedge E\})$$

vanishes. Therefore $\underrightarrow{\text{colim}}\, \mathcal{T}(S^n, \{t_i \wedge E\}) = 0$. But then

$$\mathcal{T}(S^n, \underrightarrow{\text{Hocolim}} \{t_i \wedge E\}) = 0.$$

That is, $\underrightarrow{\text{Hocolim}} \{t_i \wedge E\}$ is an object in $\mathcal{T}$, so that every map

$$S^n \longrightarrow \underrightarrow{\text{Hocolim}} \{t_i \wedge E\}$$

vanishes. By D.1.1.2, $\underrightarrow{\text{Hocolim}} \{t_i \wedge E\} = 0$. □

LEMMA D.1.11. *Let α be a regular cardinal, $\alpha > \aleph_0$. Let E be an object of $\mathcal{T}^\alpha$. Let R be a set of representatives for all isomorphism classes of objects in $\mathcal{T}^\alpha \cap \mathcal{T}_E$. Then R is an $\aleph_1$–perfect generating set for $\mathcal{T}_E$.*

Proof: Let us first prove that R generates. It is a set of objects closed under suspension (up to isomorphism). If x is a non–zero object of $\mathcal{T}_E$, then by D.1.1.2, there exists a non–zero map in $\mathcal{T}$

$$S^n \longrightarrow x.$$

But $S^n \in \mathcal{T}^{\aleph_0} \subset \mathcal{T}^\alpha$, and $x \in \mathcal{T}_E$. Lemma D.1.10 now tells us that we may factor the above as

$$S^n \longrightarrow r \longrightarrow x,$$

with $r \in R$. Thus there is a non–zero map $r \longrightarrow x$, in $\mathcal{T}_E$.

Next we want to prove R to be $\aleph_1$–perfect. Note that $\mathcal{T}^\alpha \cap \mathcal{T}_E$ is the intersection of two triangulated subcategories of $\mathcal{T}$, hence is triangulated. And R is a set of representatives for the isomorphism classes. By Lemma 3.3.5, to show that R is $\aleph_1$–perfect, it suffices to prove that every element is $\aleph_1$–good. Suppose therefore that we are given a countable set of objects $\{x_i \in \mathcal{T}_E \mid 0 \le i < \infty\}$. Let r be an object of R, and suppose we have a map

$$r \longrightarrow \coprod_{i=0}^{\infty} x_i.$$

Now $r \in \mathcal{T}^\alpha$, and the objects of $\mathcal{T}^\alpha$ form an β–perfect class for every infinite cardinal β, in particular an $\aleph_1$–perfect class. The above map therefore factorises as

$$r \longrightarrow \coprod_{i=0}^{\infty} t_i \xrightarrow{\coprod_{i=0}^{\infty} f_i} \coprod_{i=0}^{\infty} x_i,$$

with $t_i \in \mathcal{T}^\alpha$. But for each i, the map $f_i : t_i \longrightarrow x_i$ is a morphism from $t_i \in \mathcal{T}^\alpha$ to $x_i \in \mathcal{T}_E$. By Lemma D.1.10, the map must factor as

$$t_i \xrightarrow{g_i} r_i \xrightarrow{h_i} x_i$$

with $r_i \in R$. The map $r \longrightarrow \coprod_{i=0}^{\infty} x_i$ therefore factors as

$$r \longrightarrow \coprod_{i=0}^{\infty} t_i \xrightarrow{\coprod_{i=0}^{\infty} g_i} \coprod_{i=0}^{\infty} r_i \xrightarrow{\coprod_{i=0}^{\infty} h_i} \coprod_{i=0}^{\infty} x_i,$$

and we conclude that r is $\aleph_1$–good, and since this is true for all $r \in R$, R is $\aleph_1$–perfect. □

THEOREM D.1.12. *As above, $\mathcal{T}$ is the homotopy category of spectra. Let α be a regular cardinal, $\alpha > \aleph_0$. Let E be an object of $\mathcal{T}^\alpha$. Let R be a set of representatives for all isomorphism classes of objects in $\mathcal{T}^\alpha \cap \mathcal{T}_E$. Then the following hold*

D.1.12.1. *$\mathcal{T}_E$ satisfies the representability theorem, and $\mathcal{T}_E = \langle R \rangle$.*

D.1.12.2. *There is a Bousfield localisation functor for the inclusion $\mathcal{T}_E \subset \mathcal{T}$.*

D.1.12.3. *For any regular cardinal $\beta \geq \alpha$, we have*

$$\{\mathcal{T}_E\}^\beta = \mathcal{T}^\beta \cap \mathcal{T}_E = \langle R \rangle^\beta,$$

and the category $\mathcal{T}_E$ is β–compactly generated.

D.1.12.4. *For any regular cardinal $\beta \geq \alpha$,*

$$\{\mathcal{T}/\mathcal{T}_E\}^\beta = \mathcal{T}^\beta / \{\mathcal{T}_E\}^\beta,$$

and the category $\mathcal{T}/\mathcal{T}_E$ is β–compactly generated.

D.1.12.5. *The representability theorem holds for $\mathcal{T}/\mathcal{T}_E$.*

Proof: In Lemma D.1.9, we learned that the subcategory $\mathcal{T}_E \subset \mathcal{T}$ is localising. In particular, it is a triangulated category satisfying [TR5]. In Lemma D.1.11, we learned that the set of objects $R \subset \mathcal{T}_E$ is an $\aleph_1$–perfect generating set. Theorem 8.3.3 therefore applies; we conclude that

$\mathcal{T}_E$ satisfies the representability theorem, and that $\mathcal{T}_E = \langle R \rangle$. That is, we conclude D.1.12.1.

Now observe that $\mathcal{T}_E$ is a localising subcategory of $\mathcal{T}$, and the representability theorem holds for $\mathcal{T}_E$. The category $\mathcal{T}$ has small Hom–sets, and we can therefore use Proposition 9.1.19, to conclude that the inclusion $\mathcal{T}_E \subset \mathcal{T}$ has a right adjoint, and a Bousfield localisation functor exists for the pair $\mathcal{T}_E \subset \mathcal{T}$. This proves D.1.12.2.

In D.1.12.1, we saw that $\mathcal{T}_E = \langle R \rangle$, where $R \subset \mathcal{T}^\alpha$. Therefore Theorem 4.4.9 now applies. If β is any regular cardinal $> \alpha$, we conclude that

$$\{\mathcal{T}_E\}^\beta = \mathcal{T}^\beta \cap \mathcal{T}_E = \langle R \rangle^\beta.$$

Since R is a set, the category $\{\mathcal{T}_E\}^\beta = \langle R \rangle^\beta$ is essentially small. Since $\{\mathcal{T}_E\}^\beta$ contains R, $\{\mathcal{T}_E\}^\beta$ generates $\mathcal{T}_E$. Therefore $\mathcal{T}_E$ is β–compactly generated. This establishes D.1.12.3.

But Theorem 4.4.9 goes on to tell us more. It asserts that, for all regular $\beta \geq \alpha$, the natural map

$$\mathcal{T}^\beta / \{\mathcal{T}_E\}^\beta \longrightarrow \{\mathcal{T}/\mathcal{T}_E\}^\beta$$

is an equivalence of categories. It follows that the category $\{\mathcal{T}/\mathcal{T}_E\}^\beta$ is essentially small; expressing it as $\mathcal{T}^\beta/\{\mathcal{T}_E\}^\beta$, we see that it is the quotient of two essentially small categories. Now consider the quotient functor

$$F : \mathcal{T} \longrightarrow \mathcal{T}/\mathcal{T}_E.$$

We have that $F^{-1}\{\mathcal{T}/\mathcal{T}_E\}^\beta \supset \mathcal{T}^\beta$. For any regular $\gamma > \beta$,

$$F^{-1}\Big\langle \{\mathcal{T}/\mathcal{T}_E\}^\beta \Big\rangle^\gamma$$

is a γ–localising subcategory of $\mathcal{T}$, containing T. Hence it contains $\langle T \rangle^\gamma$. Taking the union over all γ, we have

$$\mathcal{T} \quad \subset \quad \bigcup_\gamma F^{-1}\Big\langle \{\mathcal{T}/\mathcal{T}_E\}^\beta \Big\rangle^\gamma$$

and hence

$$\bigcup_\gamma \Big\langle \{\mathcal{T}/\mathcal{T}_E\}^\beta \Big\rangle^\gamma \quad = \quad \mathcal{T}/\mathcal{T}_E.$$

But then $\{\mathcal{T}/\mathcal{T}_E\}^\beta$ generates $\mathcal{T}/\mathcal{T}_E$; the two notions of generation coincide. See Proposition 8.4.1. This establishes that $\mathcal{T}/\mathcal{T}_E$ is β–compactly generated. Thus we have proved D.1.12.4.

Finally, the category $\mathcal{T}/\mathcal{T}_E$ is well–generated by D.1.12.4. Hence the representability theorem holds for $\mathcal{T}/\mathcal{T}_E$, by Theorem 8.3.3. This establishes D.1.12.5. $\square$

REMARK D.1.13. From D.1.12.2, we learn that a Bousfield localisation functor exists for the pair $\mathcal{T}_E \subset \mathcal{T}$. The category ${}^\perp\mathcal{T}_E$ is usually called the category of E–local spectra, for the homology theory E. From Theorem 9.1.16, we know that there is a natural equivalence

$$ {}^\perp\mathcal{T}_E \quad = \quad \mathcal{T}/\mathcal{T}_E. $$

In other words, in D.1.12.4 and D.1.12.5, we have proved that ${}^\perp\mathcal{T}_E$ is β–compactly generated, and satisfies the representability theorem.

The fact that a Bousfield localisation exists for the pair $\mathcal{T}_E \subset \mathcal{T}$ was first proved by Bousfield. What is new here is that we prove, for $\beta \geq \alpha$, that both ${}^\perp\mathcal{T}_E$ and $\mathcal{T}_E$ are β–compactly generated triangulated categories. Of course, since the concept did not exist before this book, this is new. But the concrete consequence is that both ${}^\perp\mathcal{T}_E$ and $\mathcal{T}_E$ satisfy the representability theorem. This is completely new.

The category $\mathcal{T}$ is $\aleph_0$–compactly generated, and if $\mathcal{S} = \mathcal{T}^{\aleph_0}$, then $\mathcal{E}x(\mathcal{S}^{op}, \mathcal{A}b)$ has enough injectives. Theorem 8.6.1 applies, and we learn that the representability theorem holds for the dual of $\mathcal{T}$.

It is natural to wonder whether the representability theorem holds for the dual of ${}^\perp\mathcal{T}_E$ and $\mathcal{T}_E$. For $\mathcal{T}_E$ the answer is yes. Let us quickly prove this.

LEMMA D.1.14. *Let E be an object of $\mathcal{T}$. The representability theorem holds for the dual of the category $\mathcal{T}_E$.*

Proof: There must be a regular cardinal α so that $E \in \langle T \rangle^\alpha = \mathcal{T}^\alpha$. Theorem D.1.12, more precisely D.1.12.2, asserts that a Bousfield localisation functor exists for the pair $\mathcal{T}_E \subset \mathcal{T}$. Let $J : \mathcal{T} \longrightarrow \mathcal{T}_E$ be right adjoint to the inclusion $I : \mathcal{T}_E \subset \mathcal{T}$, as in Proposition 9.1.18. Being a right adjoint of a triangulated functor, J is triangulated, and takes products to products.

Now let $H : \mathcal{T}_E \longrightarrow \mathcal{A}b$ be a homological functor taking products to products. Then $HJ : \mathcal{T} \longrightarrow \mathcal{A}b$ is a homological functor taking products to products. By Theorem 8.6.1, the representability theorem holds in the dual of $\mathcal{T}$. Therefore,

$$ HJ(-) \quad = \quad \mathcal{T}(h, -). $$

Now, for any $x \in {}^\perp\mathcal{T}_E$, we have $J(x) = 0$, and hence $\mathcal{T}(h,x) = HJ(x) = 0$. Thus $h \in \{{}^\perp\mathcal{T}_E\}^\perp = \mathcal{T}_E$, and we conclude that the functor H is representable in $\mathcal{T}_E$. $\Box$

REMARK D.1.15. It is very natural to ask, whether the duals of the categories ${}^\perp\mathcal{T}_E$ also satisfy the representability theorem. And the answer is that I do not know. There is no simple trick, allowing us to reduce the problem to a question about $\mathcal{T}$. And since the categories ${}^\perp\mathcal{T}_E$ are β–compactly generated only for large β, we cannot be sure that, if $\mathcal{S} = \mathcal{T}^\beta$, there will be enough injectives in $\mathcal{E}x\big(\mathcal{S}^{op}, \mathcal{A}b\big)$. In other words, Theorem 8.6.1 does not apply.

We should maybe stress the nature of our estimate for β. If β is a regular cardinal, and $E \in \mathcal{T}^\beta$, then the categories $\mathcal{T}_E$ and ${}^\perp\mathcal{T}_E$ are both β–compactly generated. This estimate is not best possible, but it illustrates that the bound depends on E.

D.2. The lack of injectives

In Section C.4, we saw that if $D(R)$ is the derived category of a discrete valuation ring R, if α is any regular cardinal $> \aleph_0$, and if $\mathcal{S} = D(R)^\alpha$, then the category $\mathcal{E}x\big(\mathcal{S}^{op}, \mathcal{A}b\big)$ does not have enough injectives. Even worse, it has no cogenerator.

In this Section, we want to transfer all of this to the homotopy category of spectra. For the remainder of this Section, $\mathcal{T}$ is the homotopy category of spectra. Let p be a prime number $p \neq 2$, and let $R = \mathbb{Z}_{(p)}$, the localisation of $\mathbb{Z}$ where all primes other than p are inverted. The ring R is a discrete valuation ring. Now we wish to quote some results from [**24**].

SUMMARY D.2.1. The proofs of the following facts may be found in [**24**].

D.2.1.1. *There is a functor $F : D(R) \longrightarrow \mathcal{T}$, taking a p–local abelian group to its Moore space. This functor is triangulated, and preserves coproducts.*

D.2.1.2. *Note that the category $D(R)$ satisfies the representability theorem. By Theorem 8.4.4 of the present book, the functor $F : D(R) \longrightarrow \mathcal{T}$ has a right adjoint $P : \mathcal{T} \longrightarrow D(R)$. In* [**24**], *we essentially compute the restriction of this right adjoint, to the subcategory $\mathcal{T}_p \subset \mathcal{T}$, of p–local spectra. The functor $P : \mathcal{T}_p \longrightarrow D(R)$ is denoted Π in* [**24**], *but here the letter π is reserved for the projection $\pi : A(\mathcal{T}) \longrightarrow \mathcal{E}x\big(\mathcal{S}^{op}, \mathcal{A}b\big)$.*

D.2.1.3. *On the category of p–local spectra, the functor P takes a spectrum to a direct sum of suspensions of its stable homotopy groups. Thus, on the subcategory $\mathcal{T}_p \subset \mathcal{T}$, the functor P commutes with coproducts. The functor F commutes with all coproducts by D.2.1.1, and takes any $x \in D(R)$ to $Fx \in \mathcal{T}_p$. It follows that the functor $PF : D(R) \longrightarrow D(R)$ commutes with coproducts.*

LEMMA D.2.2. *Let α be a regular cardinal, $\alpha > \aleph_0$. The functor F above, which takes an abelian group A to its Moore spectrum, satisfies the hypothesis*

$$F\big\{D(R)^\alpha\big\} \subset \mathcal{T}^\alpha.$$

Proof: The explicit description of F makes this obvious. The point is that the Moore space of an abelian group with fewer than α elements is a space with fewer than α cells. It is necessary here to take $\alpha > \aleph_0$, since an $\aleph_0$–compact object of $D(R)$ is a coproduct of suspensions of finitely generated $\mathbb{Z}_{(p)}$–modules, which need not be finitely generated abelian groups. The Moore spectrum on them is made up of finitely many p–local spheres, but countably many ordinary spheres. If we consider F as a functor to p–local spectra, then it is true that

$$F\left\{D(R)^{\aleph_0}\right\} \quad \subset \quad \left\{\mathcal{T}_p\right\}^{\aleph_0}.$$

It is for this reason, that the computation of the adjoint P is much easier on $\mathcal{T}_p \subset \mathcal{T}$. □

LEMMA D.2.3. *Let $F : D(R) \longrightarrow \mathcal{T}$ and $P : \mathcal{T} \longrightarrow D(R)$ be the pair of adjoint functors in D.2.1.2. For every object $x \in D(R)$, the unit of adjunction induces a map*

$$\eta_x : x \longrightarrow PFx.$$

This map is a split monomorphism in the triangulated category $D(R)$.

Proof: By Lemma C.4.2, the object $x \in D(R)$ is isomorphic to a coproduct of suspensions of R–modules;

$$x \quad \simeq \quad \coprod_{n=-\infty}^{\infty} \Sigma^n x_n.$$

Since PF respects coproducts by D.2.1.3, the unit of adjunction

$$\eta_x : x \longrightarrow PFx$$

can be identified as the coproduct map

$$\coprod_{n=-\infty}^{\infty} \Sigma^n x_n \xrightarrow{\coprod_{n=-\infty}^{\infty} \eta_{\Sigma^n x_n}} \coprod_{n=-\infty}^{\infty} PF\Sigma^n x_n.$$

It suffices to show that, for each $n \in \mathbb{Z}$, the map

$$\eta_{\Sigma^n x_n} : \Sigma^n x_n \longrightarrow PF\Sigma^n x_n$$

is a split monomorphism. Desuspending, we may assume $n = 0$. Thus we are reduced to proving that

$$\eta_x : x \longrightarrow PFx$$

is a split monomorphism, where x is an R–module, viewed as a complex in $D(R)$, concentrated in degree 0.

But now we know both F and P quite explicitly. Fx is the Moore spectrum on the abelian group x, and PFx is the stable homotopy of this

Moore spectrum. The map $x = H^0(x) \longrightarrow H^0(PFx)$ is an isomorphism, since the zeroth stable homotopy group of the Moore spectrum on a group x is just x. Now Lemma C.4.2 tells us that $H^0(PFx)$ is a direct summand of PFx. □

LEMMA D.2.4. *Let $\mathcal{T}$ be the category of spectra, $D(R)$ the category of p–local abelian groups, as in Summary D.2.1. Let α be a regular cardinal, $\alpha > \aleph_0$. Put*

$$\begin{aligned} G(\mathcal{T}) &= \mathcal{E}x\Big(\{\mathcal{T}^\alpha\}^{op}, \mathcal{A}b\Big), \\ G(R) &= \mathcal{E}x\big(\{D(R)^\alpha\}^{op}, \mathcal{A}b\big). \end{aligned}$$

We want to extend the adjoint functors F, P to $G(\mathcal{T})$ and $G(R)$. We will prove

D.2.4.1. *There is a functor $\overline{F} : G(R) \longrightarrow G(\mathcal{T})$, with a right adjoint $\overline{P} : G(\mathcal{T}) \longrightarrow G(R)$.*

D.2.4.2. *Let x be an object of $G(R)$. The unit of adjunction*

$$\eta_x : x \longrightarrow \overline{PF}x$$

is a monomorphism in the abelian category $G(R)$.

Proof: We are given a pair of adjoint functors

$$F : D(R) \longrightarrow \mathcal{T}, \qquad\qquad P : \mathcal{T} \longrightarrow D(R).$$

By Lemma 5.3.8, these extend to give a pair of adjoint functors on Freyd's universal abelian category $A(\mathcal{T})$. That is, we have a pair of exact functors of abelian categories

$$A(F) : A(D(R)) \longrightarrow A(\mathcal{T}), \qquad\qquad A(P) : A(\mathcal{T}) \longrightarrow A(D(R)),$$

and $A(F)$ is left adjoint to $A(P)$. Given any object $x \in A(D(R)$, we may embed it by a map $f : x \longrightarrow i$, with i an injective object of $A(D(R)$. Corollary 5.1.23 asserts that injective objects $i \in A(D(R)$ lie in the subcategory $D(R) \subset A(D(R)$. By the naturality of the unit of adjunction η, there is a commutative square

$$\begin{array}{ccc} x & \xrightarrow{\quad f \quad} & i \\ {\scriptstyle \eta_x}\downarrow & & \downarrow{\scriptstyle \eta_i} \\ A(P)A(F)x & \xrightarrow{A(P)A(F)f} & A(P)A(F)i \end{array}$$

Lemma D.2.3 tells us that, for objects $i \in D(R)$, the map $\eta_i : i \longrightarrow A(P)A(F)i = PFi$ is a split monomorphism. The composite

$$\begin{array}{ccc} x & \stackrel{f}{\longrightarrow} & i \\ & & \Big\downarrow{\scriptstyle \eta_i} \\ & & A(P)A(F)i \end{array}$$

is theorefore the composite of two monomorphisms, and is mono. It is equal to the composite

$$\begin{array}{ccc} x & & \\ {\scriptstyle \eta_x}\Big\downarrow & & \\ A(P)A(F)x & \xrightarrow{A(P)A(F)f} & A(P)A(F)i, \end{array}$$

and we deduce that the map $\eta_x : x \longrightarrow A(P)A(F)x$ must be mono, for every $x \in A(D(R))$.

Now we have to pass from $A(\mathcal{T})$ and $A(D(R))$ to the quotient categories

$$\begin{aligned} G(\mathcal{T}) &= \mathcal{E}x\Big(\{\mathcal{T}^\alpha\}^{op}, \mathcal{A}b\Big), \\ G(R) &= \mathcal{E}x\big(\{D(R)^\alpha\}^{op}, \mathcal{A}b\big). \end{aligned}$$

Recall that, by Proposition 6.5.3, for every triangulated category $\mathcal{T}$, there is a functor

$$\pi : A(\mathcal{T}) = D(\mathcal{T}) \longrightarrow \mathcal{E}x\Big(\{\mathcal{T}^\alpha\}^{op}, \mathcal{A}b\Big),$$

which is a Gabriel quoient map. In Lemma 6.5.6 and Remark 6.5.8, we even identified the kernel of π; viewed as a subcategory of $D(\mathcal{T})$, it is the α–phantom maps. We have diagrams

$$\begin{array}{ccc} A(D(R)) & \xrightarrow{A(F)} & A(\mathcal{T}) \\ {\scriptstyle \pi}\Big\downarrow & & \Big\downarrow{\scriptstyle \pi} \\ G(R) & & G(\mathcal{T}) \end{array} \qquad \text{and} \qquad \begin{array}{ccc} A(\mathcal{T}) & \xrightarrow{A(P)} & A(D(R)) \\ {\scriptstyle \pi}\Big\downarrow & & \Big\downarrow{\scriptstyle \pi} \\ G(\mathcal{T}) & & G(R) \end{array}$$

To prove that these diagrams can be completed, up to canonical equivalence, to

$$\begin{array}{ccc} A(D(R)) & \xrightarrow{A(F)} & A(\mathcal{T}) \\ {\scriptstyle \pi}\Big\downarrow & & \Big\downarrow{\scriptstyle \pi} \\ G(R) & \xrightarrow{\overline{F}} & G(\mathcal{T}) \end{array} \qquad \text{and} \qquad \begin{array}{ccc} A(\mathcal{T}) & \xrightarrow{A(P)} & A(D(R)) \\ {\scriptstyle \pi}\Big\downarrow & & \Big\downarrow{\scriptstyle \pi} \\ G(\mathcal{T}) & \xrightarrow{\overline{G}} & G(R) \end{array}$$

amounts to showing that $F : D(R) \longrightarrow \mathcal{T}$ and $G : \mathcal{T} \longrightarrow D(R)$ take α–phantom maps to α–phantom maps.

For F, one reasons as follows. The functor $F : D(R) \longrightarrow \mathcal{T}$ respects coproducts. By Proposition 6.2.6, so does the functor

$$\pi : \mathcal{T} \longrightarrow G(\mathcal{T}) \quad = \quad \mathcal{E}x\Big(\{\mathcal{T}^\alpha\}^{op}, \mathcal{A}b\Big).$$

Hence the composite

$$\begin{array}{ccc} D(R) & \xrightarrow{F} & \mathcal{T} \\ & & \downarrow{\scriptstyle \pi} \\ & & G(\mathcal{T}) \end{array}$$

is a homological functor respecting coproducts. But then the abelian category $G(\mathcal{T}) = \mathcal{E}x\Big(\{\mathcal{T}^\alpha\}^{op}, \mathcal{A}b\Big)$ is an abelian category satisfying [AB5$^{(\alpha)}$], and from Corollary B.2.4, the composite πF above must take α–phantom maps in $D(R)$ to zero in $G(\mathcal{T})$. So F takes α–phantom maps in $D(R)$ to the kernel of $\pi : D(\mathcal{T}) \longrightarrow G(\mathcal{T})$, that is to α–phantom maps in $\mathcal{T}$.

Now we want to show that G also takes α–phantom maps to α–phantom maps. Let $f : x \longrightarrow y$ be an α–phantom map in $\mathcal{T}$. Let s be an object of $D(R)^\alpha$. We are assuming $\alpha > \aleph_0$, and so by Lemma D.2.2,

$$F\big\{D(R)^\alpha\big\} \subset \mathcal{T}^\alpha.$$

Therefore, $Fs \in \mathcal{T}^\alpha$, and since $f : x \longrightarrow y$ is an α–phantom map, all composites

$$Fs \longrightarrow x \xrightarrow{f} y$$

must vanish. But under the adjunction, this corresponds to

$$s \longrightarrow Gx \xrightarrow{Gf} Gy.$$

Since all such composites vanish, for any object $s \in D(R)^\alpha$, it follows that the map $Gf : Gx \longrightarrow Gy$ is α–phantom.

We therefore have commutative squares

$$\begin{array}{ccc} A(D(R)) & \xrightarrow{A(F)} & A(\mathcal{T}) \\ {\scriptstyle \pi}\downarrow & & \downarrow{\scriptstyle \pi} \\ G(R) & \xrightarrow{\overline{F}} & G(\mathcal{T}) \end{array} \qquad \text{and} \qquad \begin{array}{ccc} A(\mathcal{T}) & \xrightarrow{A(P)} & A(D(R)) \\ {\scriptstyle \pi}\downarrow & & \downarrow{\scriptstyle \pi} \\ G(\mathcal{T}) & \xrightarrow{\overline{G}} & G(R) \end{array}$$

and by the universality of the factorisation through the Gabriel quotient, natural transformations descend. In particular, we have natural transformations

$$\overline{\eta} : 1 \longrightarrow \overline{GF} \qquad \text{and} \qquad \overline{\varepsilon} : \overline{FG} \longrightarrow 1,$$

and the composites

$$\overline{F} \xrightarrow{\overline{F}\overline{\eta}} \overline{FGF} \xrightarrow{\overline{\varepsilon}\overline{F}} \overline{F}$$

$$\overline{G} \xrightarrow{\overline{\eta}\overline{G}} \overline{GFG} \xrightarrow{\overline{G}\overline{\varepsilon}} \overline{G}$$

are both identities. In fact, the natural transformations are given by the formulas

$$\overline{\eta} = \pi\eta, \qquad \overline{\varepsilon} = \pi\varepsilon.$$

It follows formally that $\overline{F}$ is left adjoint to $\overline{G}$.

Finally, we know that for every $x \in A(D(R))$, the unit of adjunction

$$\eta_x : x \longrightarrow A(P)A(F)x$$

is mono. But the functor $\pi : A(D(R)) \longrightarrow G(R)$ is exact, and hence $\overline{\eta}_x = \pi\eta_x$ is also mono. □

PROPOSITION D.2.5. *Let α be a regular cardinal, $\alpha > \aleph_0$. Let $\mathfrak{T}$ be the homotopy category of spectra. Then the category $G(\mathfrak{T}) = \mathcal{E}x\Big(\{\mathfrak{T}^\alpha\}^{op}, \mathcal{A}b\Big)$ has no cogenerator.*

Proof: Let the notation be as in Lemma D.2.4. Suppose k is a cogenerator of the category $G(\mathfrak{T})$. I assert that $\overline{P}k$ must be a cogenerator of $G(R)$. We know, from Section C.4, that $G(R)$ has no cogenerator. Hence our contradiction.

It remains therefore to prove our assertion, that if k is a cogenerator of the category $G(\mathfrak{T})$, then $\overline{P}k$ must be a cogenerator of $G(R)$. Let x be an object of $G(R)$. Then $\overline{F}x$ is an object of $G(\mathfrak{T})$, and since k is a cogenerator, there is an embedding

$$\overline{F}x \longrightarrow \prod_{\Lambda} k.$$

Since $\overline{P} : G(\mathfrak{T}) \longrightarrow G(R)$ is exact and respects products, we have a monomorphism

$$\overline{PF}x \longrightarrow \prod_{\Lambda} \overline{P}k.$$

But we know that the unit of adjunction $x \longrightarrow \overline{PF}x$ is a monomorphism. Hence so is the composite

$$x \longrightarrow \overline{PF}x \longrightarrow \prod_{\Lambda} \overline{P}k;$$

we deduce a monomorphism from x to a product of $\overline{P}k$'s. Since x is arbitrary, $\overline{P}k$ is a cogenerator. □

D.3. History of the results in Appendix D

The existence of Bousfield localisation with respect to homology was proved by Bousfield in [**6**]. Up through Lemma D.1.10, the treatment given here is, at least in spirit, lifted directly from Bousfield. There are minor changes in detail, mostly because we do not assume the reader has any familiarity with spectra. But starting with our Lemma D.1.11, our argument is entirely different. Bousfield does not appeal to a representability theorem. He directly constructs the localisation functor, as a colimit of a long sequence. One problem with the construction, is that it depends on lifting to models, and does not generalise well to triangulated categories other that spectra.

Margolis [**21**] gave an argument based on a representability theorem, but he used the representability theorem for $\mathfrak{T}$. To apply it, one must show that the category $\mathfrak{T}/\mathfrak{T}_E$ has small Hom–sets. See [**21**] for details.

What we do here, is appeal to the representability theorem for $\mathfrak{T}_E$. Since the fact that such a theorem holds is new to this book, it is inevitable that our proof is, at this point, quite different from the older arguments. And along the way, we also prove the representability theorem for $\mathfrak{T}_E$, $\mathfrak{T}_E^{op}$ and ${}^{\perp}\mathfrak{T}_E$.

The results of Section D.2, about the absence of injective objects, are completely new to this book. The reader is encouraged to read Section 8.5, in particular Proposition 8.5.18 and Lemma 8.5.20, to see what consequences one can draw, about α–phantom maps. All these facts, about α–phantom maps between spectra, are very new. The theory developed in this book has new and surprising applications, even to very old problems.

APPENDIX E

Examples of non–perfectly–generated categories

E.1. If $\mathcal{T}$ is $\aleph_0$–compactly generated, $\mathcal{T}^{op}$ is not even well–generated

Assume $\mathcal{T}$ is a well–generated triangulated category, as in Remark 8.4.3. We remind the reader what this means. First, $\mathcal{T}$ must satisfy [TR5]. Choose α a large enough regular cardinal, and let $\mathcal{S} = \mathcal{T}^\alpha$. By Remark 8.4.3, $\mathcal{S}$ is essentially small and, since α is chosen big, $\mathcal{S}$ generates $\mathcal{T}$. Recall also what it means to generate $\mathcal{T}$.

REMINDER E.1.1. By Proposition 8.4.1, $\mathcal{S} = \mathcal{T}^\alpha \subset \mathcal{T}$ generates $\mathcal{T}$ if the following equivalent conditions hold:

E.1.1.1. *If x is an object of $\mathcal{T}$ and, for all objects $s \in \mathcal{S}$,*

$$\mathrm{Hom}(s, x) = 0,$$

then $x = 0$.

E.1.1.2.

$$\mathcal{T} = \langle \mathcal{S} \rangle.$$

We remind the reader that E.1.1.1 can be slightly rewritten as

E.1.1.3. *Suppose x is an object in $\mathcal{T}$. If*

$$\mathcal{T}(-, x)|_{\mathcal{S}} = 0 \qquad \in \qquad \mathcal{E}x\big(\mathcal{S}^{op}, \mathcal{A}b\big)$$

then x is isomorphic to zero. In other words, if $x \in \mathcal{T}$ maps to zero under the functor

$$\mathcal{T} \longrightarrow \mathcal{E}x\big(\mathcal{S}^{op}, \mathcal{A}b\big),$$

then $x = 0$.

Suppose $\mathcal{T}$ is α–compactly generated. Suppose further that the category $\mathcal{E}x\big(\mathcal{S}^{op}, \mathcal{A}b\big)$ has an injective cogenerator. If $\alpha = \aleph_0$, the existence of an injective cogenerator is automatic, but not for larger α. By Remark 8.5.22, this injective cogenerator is $\mathcal{T}(-,\mathbb{BC})|_{\mathcal{S}}$, for some object $\mathbb{BC} \in \mathcal{T}$. Furthermore, for any object $x \in \mathcal{T}$, by Proposition 8.5.2, there is a natural isomorphism

$$\mathcal{T}(x,\mathbb{BC}) \longrightarrow \mathcal{E}x\big(\mathcal{S}^{op},\mathcal{A}b\big)\Big[\mathcal{T}(-,x)|_{\mathcal{S}},\mathcal{T}(-,\mathbb{BC})|_{\mathcal{S}}\Big].$$

Now we are ready to state the main result of this Section.

PROPOSITION E.1.2. *Let $\mathcal{T}$ be an α–compactly generated triangulated category, and let $\mathcal{S} = \mathcal{T}^{\alpha}$. Suppose there exists an injective cogenerator in $\mathcal{E}x\big(\mathcal{S}^{op},\mathcal{A}b\big)$, and let $\mathbb{BC} \in \mathcal{T}$ be chosen so that $\mathcal{T}(-,\mathbb{BC})|_{\mathcal{S}}$ is an injective cogenerator. Then for any infinite cardinal β, the object $\mathbb{BC}$ is* not β–small *in the category $\mathcal{T}^{op}$.*

Proof: Choose a non–zero object $x \in \mathcal{T}$, and choose a regular cardinal $\gamma > \max(\alpha,\beta)$. Let $\mathcal{I}(\gamma)$ be the set of ordinals $< \gamma$. Here, we do not care about the order; we consider just the set. Let P be the set of all subsets of $\mathcal{I}(\gamma)$ of cardinality $< \gamma$. The set P is ordered by inclusion, and since γ is regular and $\alpha < \gamma$, the union of $< \alpha$ subsets of $\mathcal{I}(\gamma)$ of cardinality $< \gamma$ has cardinality $< \gamma$. That is, P is α–filtered.

We define a functor $F : P \longrightarrow \mathcal{E}x\big(\mathcal{S}^{op},\mathcal{A}b\big)$ by the formula

$$F(p) \quad = \quad \prod_{i\in p} \mathcal{T}(-,x)|_{\mathcal{S}}.$$

Here, $p \in P$ is a subset of $\mathcal{I}(\gamma)$, and $F(p)$ is the product of $\mathcal{T}(-,x)|_{\mathcal{S}}$ over the index set p. For each $p \in P$, we have a monomorphism

$$F(p) \longrightarrow \prod_{i\in\mathcal{I}(\gamma)} \mathcal{T}(-,x)|_{\mathcal{S}}.$$

Therefore there is a map

$$\operatorname*{colim}_{p\in P} F(p) \longrightarrow \prod_{i\in\mathcal{I}(\gamma)} \mathcal{T}(-,x)|_{\mathcal{S}}.$$

Because P is α–filtered, Lemma A.1.3 says that the colimit is the same, whether taken in $\mathcal{E}x\big(\mathcal{S}^{op},\mathcal{A}b\big)$ or $\mathcal{C}at\big(\mathcal{S}^{op},\mathcal{A}b\big)$. The map above clearly is not surjective. After all, $\mathcal{S}$ generates $\mathcal{T}$, so there is a non–zero element f in $\mathcal{T}(s,x)$, for some $s \in \mathcal{S}$. Now

$$\prod_{i\in\mathcal{I}(\gamma)} f$$

is not in the image of

$$\operatorname*{colim}_{p\in P} F(p) \longrightarrow \prod_{i\in\mathcal{I}(\gamma)} \mathcal{T}(-,x)|_{\mathcal{S}}.$$

This is clear, since the colimit may be formed in $\mathcal{C}at\big(\mathcal{S}^{op},\mathcal{A}b\big)$, where it is constructed pointwise.

We deduce an exact sequence

$$\operatorname*{colim}_{p\in P} F(p) \longrightarrow \prod_{i\in\mathcal{I}(\gamma)} \mathcal{T}(-,x)|_{\mathcal{S}} \longrightarrow Q \longrightarrow 0,$$

with $Q \in \mathcal{E}x\big(\mathcal{S}^{op},\mathcal{A}b\big)$ non–zero. Mapping into the injective cogenerator $\mathcal{T}(-,\mathbb{BC})|_{\mathcal{S}}$, we deduce that the kernel of the map

$$\begin{array}{c}\mathcal{E}x\big(\mathcal{S}^{op},\mathcal{A}b\big)\left[\prod_{i\in\mathcal{I}(\gamma)} \mathcal{T}(-,x)|_{\mathcal{S}}\ ,\ \mathcal{T}(-,\mathbb{BC})|_{\mathcal{S}}\right]\\ \big\downarrow \\ \mathcal{E}x\big(\mathcal{S}^{op},\mathcal{A}b\big)\left[\operatorname*{colim}_{p\in P} F(p)\ ,\ \mathcal{T}(-,\mathbb{BC})|_{\mathcal{S}}\right]\end{array}$$

is precisely

$$\mathcal{E}x\big(\mathcal{S}^{op},\mathcal{A}b\big)\Big[Q\ ,\ \mathcal{T}(-,\mathbb{BC})|_{\mathcal{S}}\Big].$$

Since Q is non–zero and $\mathcal{T}(-,\mathbb{BC})|_{\mathcal{S}}$ an injective cogenerator, the group

$$\mathcal{E}x\big(\mathcal{S}^{op},\mathcal{A}b\big)\Big[Q\ ,\ \mathcal{T}(-,\mathbb{BC})|_{\mathcal{S}}\Big]$$

must be non–zero. There is a non–zero kernel to the map

$$\begin{array}{c}\mathcal{E}x\big(\mathcal{S}^{op},\mathcal{A}b\big)\left[\prod_{i\in\mathcal{I}(\gamma)} \mathcal{T}(-,x)|_{\mathcal{S}}\ ,\ \mathcal{T}(-,\mathbb{BC})|_{\mathcal{S}}\right]\\ \big\downarrow \\ \mathcal{E}x\big(\mathcal{S}^{op},\mathcal{A}b\big)\left[\operatorname*{colim}_{p\in P} F(p)\ ,\ \mathcal{T}(-,\mathbb{BC})|_{\mathcal{S}}\right].\end{array}$$

Now by the universal property of colimits,

$$\begin{aligned}\mathcal{E}x\big(\mathcal{S}^{op},\mathcal{A}b\big)\left[\operatorname*{colim}_{p\in P} F(p)\ ,\ \mathcal{T}(-,\mathbb{BC})|_{\mathcal{S}}\right] \quad &= \\ &\lim_{p\in P}\Big\{\mathcal{E}x\big(\mathcal{S}^{op},\mathcal{A}b\big)\Big[F(p)\ ,\ \mathcal{T}(-,\mathbb{BC})|_{\mathcal{S}}\Big]\Big\}.\end{aligned}$$

Also,

$$F(p) \quad = \quad \prod_{i\in p} \mathcal{T}(-,x)|_{\mathcal{S}}.$$

On the other hand, the functor

$$\mathcal{T} \longrightarrow \mathcal{E}x\big(\mathcal{S}^{op},\mathcal{A}b\big)$$

respects products by Lemma 6.2.4. That is, for any index set Λ,

$$\mathcal{T}\left(-,\prod_{i\in\Lambda}x\right)\Bigg|_{\mathcal{S}} = \prod_{i\in\Lambda}\mathcal{T}(-,x)|_{\mathcal{S}}.$$

The map

$$\begin{array}{c}\mathcal{E}x\big(\mathcal{S}^{op},\mathcal{A}b\big)\left[\prod_{i\in\mathfrak{I}(\gamma)}\mathcal{T}(-,x)|_{\mathcal{S}}\ ,\ \mathcal{T}(-,\mathbb{BC})|_{\mathcal{S}}\right]\\ \Big\downarrow \\ \mathcal{E}x\big(\mathcal{S}^{op},\mathcal{A}b\big)\left[\operatorname*{colim}_{p\in P}F(p)\ ,\ \mathcal{T}(-,\mathbb{BC})|_{\mathcal{S}}\right]\end{array}$$

may therefore be identified with

$$\begin{array}{c}\mathcal{E}x\big(\mathcal{S}^{op},\mathcal{A}b\big)\left[\mathcal{T}\left(-,\prod_{i\in\mathfrak{I}(\gamma)}x\right)\Bigg|_{\mathcal{S}}\ ,\ \mathcal{T}(-,\mathbb{BC})|_{\mathcal{S}}\right]\\ \Big\downarrow \\ \lim\limits_{p\in P}\left\{\mathcal{E}x\big(\mathcal{S}^{op},\mathcal{A}b\big)\left[\mathcal{T}\left(-,\prod_{i\in p}x\right)\Bigg|_{\mathcal{S}}\ ,\ \mathcal{T}(-,\mathbb{BC})|_{\mathcal{S}}\right]\right\}.\end{array}$$

Now recall Proposition 8.5.2. The natural isomorphism

$$\mathcal{T}(x,\mathbb{BC}) \longrightarrow \mathcal{E}x\big(\mathcal{S}^{op},\mathcal{A}b\big)\Big[\mathcal{T}(-,x)|_{\mathcal{S}}\ ,\ \mathcal{T}(-,\mathbb{BC})|_{\mathcal{S}}\Big].$$

allows us to identify

$$\begin{array}{ccc}\mathcal{E}x\big(\mathcal{S}^{op},\mathcal{A}b\big)\left[\mathcal{T}\left(-,\prod_{i\in\mathfrak{I}(\gamma)}x\right)\Bigg|_{\mathcal{S}}\ ,\ \mathcal{T}(-,\mathbb{BC})|_{\mathcal{S}}\right] & = & \mathcal{T}\left(\prod_{i\in\mathfrak{I}(\gamma)}x\ ,\ \mathbb{BC}\right)\\ \Big\downarrow & & \Big\downarrow \\ \lim\limits_{p\in P}\left\{\mathcal{E}x\big(\mathcal{S}^{op},\mathcal{A}b\big)\left[\mathcal{T}\left(-,\prod_{i\in p}x\right)\Bigg|_{\mathcal{S}}\ ,\ \mathcal{T}(-,\mathbb{BC})|_{\mathcal{S}}\right]\right\} & = & \lim\limits_{p\in P}\mathcal{T}\left(\prod_{i\in p}x\ ,\ \mathbb{BC}\right)\end{array}$$

In other words, from the discussion above we conclude that the map

$$\mathcal{T}\left(\prod_{i\in\mathfrak{I}(\gamma)}x\ ,\ \mathbb{BC}\right) \longrightarrow \lim_{p\in P}\mathcal{T}\left(\prod_{i\in p}x\ ,\ \mathbb{BC}\right)$$

has a non–trivial kernel. There is a non–zero map

$$\prod_{i\in\mathfrak{I}(\gamma)}x \longrightarrow \mathbb{BC}$$

so that, for every $p \subset \mathcal{I}(\gamma)$ of cardinality $< \gamma$, the composite

$$\prod_{i\in p} x \longrightarrow \prod_{i\in\mathcal{I}(\gamma)} x \longrightarrow \mathbb{BC}$$

vanishes. But if $\mathbb{BC}$ were β–small as an object of $\mathcal{T}^{op}$, then the map

$$\prod_{i\in\mathcal{I}(\gamma)} x \longrightarrow \mathbb{BC}$$

must factor as

$$\prod_{i\in\mathcal{I}(\gamma)} x \longrightarrow \prod_{i\in\Lambda} x \stackrel{f}{\longrightarrow} \mathbb{BC}$$

where the cardinality of Λ is $< \beta < \gamma$. Thus Λ belongs to the set P. It follows that the composite

$$\prod_{i\in\Lambda} x \longrightarrow \prod_{i\in\mathcal{I}(\gamma)} x \longrightarrow \prod_{i\in\Lambda} x \stackrel{f}{\longrightarrow} \mathbb{BC}$$

must vanish. On the other hand

$$\prod_{i\in\Lambda} x \longrightarrow \prod_{i\in\mathcal{I}(\gamma)} x \longrightarrow \prod_{i\in\Lambda} x$$

is the identity, and hence f must vanish, and hence so must

$$\prod_{i\in\mathcal{I}(\gamma)} x \longrightarrow \prod_{i\in\Lambda} x \stackrel{f}{\longrightarrow} \mathbb{BC}.$$

Since the map does not vanish, $\mathbb{BC}$ cannot be β–small. □

COROLLARY E.1.3. *Let $\mathcal{T}$ be an α–compactly generated triangulated category, and suppose $\mathcal{E}x\Big(\{\mathcal{T}^\alpha\}^{op}, \mathcal{A}b\Big)$ has enough injectives. Then $\mathcal{T}^{op}$ is not well–generated.*

Proof: Suppose $\mathcal{T}^{op}$ is well–generated. By Proposition 8.4.2, or more precisely 8.4.2.3,

$$\mathcal{T}^{op} \quad = \quad \bigcup_{\beta} \{\mathcal{T}^{op}\}^{\beta}.$$

This means that for some infinite β, the object $\mathbb{BC}$ must lie in

$$\mathbb{BC} \qquad \in \qquad \{\mathcal{T}^{op}\}^{\beta} \qquad \subset \qquad \{\mathcal{T}^{op}\}^{(\beta)}.$$

In other words, $\mathbb{BC}$ would be β–small, contradicting Proposition E.1.2. □

E.2. An example of a non $\aleph_1$–perfectly generated $\mathcal{T}$

Corollary E.1.3 showed us that, if $\mathcal{T}$ is $\aleph_0$–compactly generated, $\mathcal{T}^{op}$ cannot even be well–generated. We can ask if $\mathcal{T}^{op}$ can have an α–perfect generating set. This at least seems weaker than well–generation. A well–generated triangulated category $\mathcal{T}$ has an α–perfect generating set for every α; pick $\mathcal{T}^\beta$, where β is large enough so that $\mathcal{T}^\beta$ generates.

Let $\mathcal{T}$ be $\aleph_0$–compactly generated. I suspect that $\mathcal{T}^{op}$ is never $\aleph_1$–perfectly generated. But I have never carefully gone through a general argument. Instead, we will show the following, easy special case.

LEMMA E.2.1. *Let $\mathcal{T} = D(\mathbb{Q})$ be the derived category of the category of vector spaces over $\mathbb{Q}$. Then $\mathcal{T}^{op}$ is not $\aleph_1$–perfectly generated.*

Proof: Suppose T is an $\aleph_1$–perfect generating set for $\mathcal{T}^{op}$. Then T is closed under suspension, by Definition 8.1.1. The collection $\overline{T}$ of all retracts (direct summands) of objects of T is a set, since any object in T has only a set of idempotent endomorphisms. By Lemma 3.3.3, $\overline{T}$ is still $\aleph_1$–perfect. Since it contains T and is closed under suspension, $\overline{T}$ is also an $\aleph_1$–perfect generating set for $\mathcal{T}^{op}$. Replacing T by $\overline{T}$, we may assume T is closed under suspension and direct summands.

Since T generates, it must contain a non–zero object; call it x. An object of the derived category of $\mathbb{Q}$ is a product of suspensions of vector spaces. That is,

$$x \quad = \quad \prod_{i=-\infty}^{\infty} \Sigma^i V_i,$$

where each V_i is a vector space over $\mathbb{Q}$. Since $x \neq 0$, for some i we must have $V_i \neq 0$. But then $\Sigma^i V_i$ is a direct summand of x, and hence lies in T. Since T is closed under Σ and Σ^{-1}, $\Sigma^0 V_i = V_i$ must also belong to T. But V_i is a non–zero vector space over $\mathbb{Q}$, hence must contain $\mathbb{Q}$ as a retract. It follows that $\mathbb{Q} \in T$.

For each object of T we define its *size* to be the cardinality of the direct sum of its cohomology. Since T is a set, the size of the objects of T is bounded. There is a regular cardinal $\beta > \aleph_0$, so that every object $x \in T$ is isomorphic to

$$x \quad = \quad \prod_{i=-\infty}^{\infty} \Sigma^i V_i,$$

where each V_i has cardinality $< \beta$.

We are supposing T is $\aleph_1$–perfect. Take any vector space W. Since $\mathbb{Q}$ is in T, any map

$$\prod_{i=1}^{\infty} W \longrightarrow \mathbb{Q}$$

must factor as

$$\prod_{i=1}^{\infty} W \xrightarrow{\prod_{i=1}^{\infty} f_i} \prod_{i=1}^{\infty} x_i \longrightarrow \mathbb{Q}$$

with $x_i \in T$. Applying the functor H^0, we deduce a factorisation

$$\prod_{i=1}^{\infty} W \xrightarrow{\prod_{i=1}^{\infty} f_i} \prod_{i=1}^{\infty} H^0(x_i) \longrightarrow \mathbb{Q}.$$

That is, for some vector spaces $V_i = H^0(x_i)$ of cardinality $< \beta$, we have a factorisation

$$\prod_{i=1}^{\infty} W \xrightarrow{\prod_{i=1}^{\infty} f_i} \prod_{i=1}^{\infty} V_i \longrightarrow \mathbb{Q}.$$

Replacing each V_i by $V = \coprod_{i=1}^{\infty} V_i$, we may assume that all the V_i's are the same, and the cardinality is the sum of $\aleph_0$ cardinals, all $< \beta$. Since β is regular and $\beta > \aleph_0$, this sum is $< \beta$.

In other words, we are given a cardinal β, which is determined by the generating set T. We want to show that T cannot be $\aleph_1$–perfect. Given β, we want to choose a cardinal γ and a vector space W of cardinality γ, so that not all maps

$$\prod_{i=1}^{\infty} W \longrightarrow \mathbb{Q}$$

can factor as

$$\prod_{i=1}^{\infty} W \xrightarrow{\prod_{i=1}^{\infty} f_i} \prod_{i=1}^{\infty} V \longrightarrow \mathbb{Q}$$

with V of cardinality β. Of course, we can increase β. Replace β by 2^{β}. It will suffice to show that there is a cardinal γ and a vector space W of cardinality γ, so that not all maps

$$\prod_{i=1}^{\infty} W \longrightarrow \mathbb{Q}$$

can factor as

$$\prod_{i=1}^{\infty} W \xrightarrow{\prod_{i=1}^{\infty} f_i} \prod_{i=1}^{\infty} V \longrightarrow \mathbb{Q}$$

with V of cardinality 2^β.

We propose to estimate the number of maps

$$\prod_{i=1}^{\infty} W \longrightarrow \mathbb{Q}$$

and the number of maps

$$\prod_{i=1}^{\infty} W \xrightarrow{\prod_{i=1}^{\infty} f_i} \prod_{i=1}^{\infty} V \longrightarrow \mathbb{Q}.$$

For the purpose of the estimate, we will assume γ is very large, to be chosen later.

E.2.1.1. **Upper bound for the number of maps**

$$\prod_{i=1}^{\infty} W \xrightarrow{\prod_{i=1}^{\infty} f_i} \prod_{i=1}^{\infty} V \longrightarrow \mathbb{Q}.$$

Let us estimate the number of maps of sets with this factorisation; clearly, the linear maps are fewer. For each

$$f_i : W \longrightarrow V$$

there are $\{2^\beta\}^\gamma$ choices. This is the number of maps from a set W of cardinality γ to a set V of cardinality 2^β. But we are assuming γ large, in particular $\gamma > \beta$. Hence

$$\{2^\beta\}^\gamma = 2^{\beta \cdot \gamma} = 2^\gamma.$$

The number of maps

$$\prod_{i=1}^{\infty} W \xrightarrow{\prod_{i=1}^{\infty} f_i} \prod_{i=1}^{\infty} V$$

is the number of countable sequences of f_i, that is

$$\{2^\gamma\}^{\aleph_0} = 2^{\gamma \cdot \aleph_0} = 2^\gamma.$$

The number of maps

$$\prod_{i=1}^{\infty} V \longrightarrow \mathbb{Q}$$

is independent of γ. The set $\prod_{i=1}^{\infty} V$ has cardinality

$$\{2^{\beta}\}^{\aleph_0} = 2^{\beta \cdot \aleph_0} = 2^{\beta}.$$

The number of maps

$$\prod_{i=1}^{\infty} V \longrightarrow \mathbb{Q}$$

is therefore

$$\{\aleph_0\}^{\{2^{\beta}\}} = 2^{2^{\beta}}.$$

The number of pairs of maps

$$\prod_{i=1}^{\infty} W \xrightarrow{\prod_{i=1}^{\infty} f_i} \prod_{i=1}^{\infty} V \qquad\qquad \prod_{i=1}^{\infty} V \longrightarrow \mathbb{Q}$$

is therefore bounded by

$$2^{\gamma} 2^{2^{\beta}}.$$

If γ is large, then this is 2^{γ}. Some pairs of maps will, of course, give the same composite. But we have that, for γ large, the number of composites

$$\prod_{i=1}^{\infty} W \xrightarrow{\prod_{i=1}^{\infty} f_i} \prod_{i=1}^{\infty} V \longrightarrow \mathbb{Q}$$

is bounded above by 2^{γ}.

E.2.1.2. **Lower bound for the number of maps**

$$\prod_{i=1}^{\infty} W \longrightarrow \mathbb{Q}.$$

The number of elements in the set $\prod_{i=1}^{\infty} W$ is $\gamma^{\aleph_0}$. But this is also the dimension of $\prod_{i=1}^{\infty} W$, as a vector space over $\mathbb{Q}$. This means we may choose a basis of cardinality $\gamma^{\aleph_0}$. Let $1 \in \mathbb{Q}$ be a basis for the 1–dimensional vector space $\mathbb{Q}$. I do not want to count all the linear maps

$$\prod_{i=1}^{\infty} W \longrightarrow \mathbb{Q}.$$

Instead, let us only count the ones with a matrix of zeros and ones. In other words, every basis element in $\prod_{i=1}^{\infty} W$ goes either to $0 \in \mathbb{Q}$ or $1 \in \mathbb{Q}$. The number of such linear maps, all of which are distinct, is the number of ways to divide the basis into two sets, the elements mapping to 1 and the elements mapping to 0. There are $2^{\gamma^{\aleph_0}}$ choices. Since this is only some of the linear maps, we have that the set of all linear maps is of cardinality bounded below by $2^{\gamma^{\aleph_0}}$.

To complete the proof, we need to show that for a suitable choice of a very large γ,

$$2^{\gamma^{\aleph_0}} > 2^{\gamma}.$$

This will mean that there are more maps

$$\prod_{i=1}^{\infty} W \longrightarrow \mathbb{Q}$$

than maps that admit factorisations

$$\prod_{i=1}^{\infty} W \xrightarrow{\prod_{i=1}^{\infty} f_i} \prod_{i=1}^{\infty} V \longrightarrow \mathbb{Q}.$$

Now we proceed to show this. More precisely, we will show

E.2.1.3. *Let β be any cardinal. There exists a $\gamma > \beta$ with*

$$2^{\gamma^{\aleph_0}} > 2^{\gamma}.$$

The author would like to thank Shelah for pointing out this argument.

Define a sequence of cardinals $c_i, i \in \mathbb{N}$ by

E.2.1.3.1. $c_0 = \beta$.

E.2.1.3.2. $c_{i+1} = 2^{c_i}$.

Now let $\gamma = \sum_{i=0}^{\infty} c_i$. We get the estimate that

$$\begin{aligned} \gamma^{\aleph_0} &\geq \prod_{i=1}^{\infty} c_i \\ &= \prod_{i=1}^{\infty} 2^{c_{i-1}} \qquad \text{since } c_i = 2^{c_{i-1}} \\ &= 2^{\sum_{i=0}^{\infty} c_i} \\ &= 2^{\gamma}. \end{aligned}$$

Therefore

$$2^{\gamma^{\aleph_0}} \geq 2^{2^{\gamma}}.$$

But we know that $2^\alpha > \alpha$, and if $\alpha = 2^\gamma$ this yields $2^{2^\gamma} > 2^\gamma$. Combining with the last inequality, we have

$$2^{\gamma^{\aleph_0}} \geq 2^{2^\gamma} > 2^\gamma.$$

□

E.3. For $\mathcal{T} = K(\mathbb{Z})$, neither $\mathcal{T}$ nor $\mathcal{T}^{op}$ is well–generated.

Recall that if $\mathbb{Z}$ is the ring of integers, the triangulated category $K(\mathbb{Z})$ is defined as follows. The objects are chain complexes of abelian groups. The morphisms are the homotopy equivalence classes of chain maps. The derived category $D(\mathbb{Z})$ is the Verdier quotient of $K(\mathbb{Z})$, where we divide by the subcategory of acyclic complexes.

It is well–known that $D(\mathbb{Z})$ is $\aleph_0$–compactly generated. In fact, the set

$$T = \{\Sigma^n\mathbb{Z} \mid n \in \mathbb{Z}\}$$

is a generating set, and is contained in $\{D(\mathbb{Z})\}^{\aleph_0} = \{D(\mathbb{Z})\}^{(\aleph_0)}$. In Corollary E.1.3, we saw that the category $D(\mathbb{Z})^{op}$ cannot be well–generated. Now we will deduce

LEMMA E.3.1. *The category $K(\mathbb{Z})^{op}$ is not well–generated.*

Proof: We assume $K(\mathbb{Z})^{op}$ is well–generated, and deduce a contradiction. Consider the abelian group $\mathbb{Q}/\mathbb{Z}$. It is an injective cogenerator in the category of abelian groups. For any object $x \in K(\mathbb{Z})$, we have

$$K(\mathbb{Z})\Big[x, \Sigma^n\{\mathbb{Q}/\mathbb{Z}\}\Big] \quad = \quad \mathcal{A}b\Big[H^{-n}(x), \mathbb{Q}/\mathbb{Z}\Big].$$

The right hand side vanishes only if $H^{-n}(x) = 0$. In other words,

$$K(\mathbb{Z})\Big[x, \Sigma^n\{\mathbb{Q}/\mathbb{Z}\}\Big] \quad = \quad 0$$

for all $n \in \mathbb{Z}$ if and only if x is acyclic. Let R be the set of all suspensions of $\mathbb{Q}/\mathbb{Z}$. That is,

$$R \quad = \quad \{\Sigma^n\{\mathbb{Q}/\mathbb{Z}\} \mid n \in \mathbb{Z}\}.$$

In the notation of Definition 9.1.11, $R^\perp$ is the class of acyclic complexes in $K(\mathbb{Z})$.

We are assuming $\mathcal{S} = K(\mathbb{Z})^{op}$ is well–generated. By Proposition 8.4.2,

$$\mathcal{S} \quad = \quad \bigcup_\beta \mathcal{S}^\beta.$$

Since R is a set of objects in $\mathcal{T}$, it must be contained in some $\mathcal{S}^\beta$. Let $\mathcal{R}$ be the category generated by R. That is,

$$\mathcal{R} = \langle R \rangle.$$

We now find ourselves in the situation of Thomason's localisation theorem 4.4.9. Put $\mathcal{T} = \mathcal{S}/\mathcal{R}$. Then for any $\gamma \geq \beta$,

$$\mathcal{R}^\gamma = \langle R \rangle^\gamma,$$

and

$$\mathcal{T}^\gamma = \mathcal{S}^\gamma/\mathcal{R}^\gamma.$$

In particular, $\mathcal{T}$ has small Hom–sets. We are assuming the category $\mathcal{S} = K(\mathbb{Z})^{op}$ is well–generated. By Theorem 8.3.3, $\mathcal{S}$ satisfies the representability theorem. The hypotheses of Example 8.4.5 are satisfied, and we conclude that the map

$$\mathcal{S} \longrightarrow \mathcal{S}/\mathcal{R} = \mathcal{T}$$

has a right adjoint. A Bousfield localisation functor exists for the pair $\mathcal{R} = \langle R \rangle \subset \mathcal{S}$.

By Corollary 9.1.14, we deduce that $\mathcal{R} = \mathcal{S}/\mathcal{R}^\perp$. On the other hand, $\mathcal{R}^\perp = \langle R \rangle^\perp = R^\perp$, and we computed that this is the category of acyclic complexes. This identifies $\mathcal{R}^{op}$ as the quotient of $\mathcal{S}^{op} = K(\mathbb{Z})$ by the subcategory of acyclics; that is,

$$\mathcal{R}^{op} = D(\mathbb{Z}).$$

But $\mathcal{R} = \langle R \rangle$, with $R \subset \mathcal{R}^\beta$. This makes $\mathcal{R} = D(\mathbb{Z})^{op}$ a well–generated triangulated category, contradicting Corollary E.1.3. $\square$

Now we want to prove

LEMMA E.3.2. *The category $K(\mathbb{Z})$ does not have a generating set.*

Proof: Once again, we suppose the category $K(\mathbb{Z})$ has a generating set, and prove a contradiction. Choose a generating set S. The objects are chain complexes of abelian groups. Since there is only a set of them, we may choose a regular cardinal $\alpha > \aleph_0$ exceeding the maximum size of these abelian groups. The objects of S are chain complexes of abelian groups, whose cardinality is $< \alpha$.

By Remark C.3.4, the category $\mathcal{A}(D(\mathbb{Z}))$ is not well–powered. Let $\mathcal{S} = D(\mathbb{Z})^\alpha$. Since the category $\mathcal{E}x\big(\mathcal{S}^{op}, \mathcal{A}b\big)$ is well–powered, the quotient map

$$\mathcal{A}(D(\mathbb{Z})) \longrightarrow \mathcal{E}x\big(\mathcal{S}^{op}, \mathcal{A}b\big)$$

cannot be an equivalence. By Proposition 6.5.3, the map is a Gabriel quotient map. Since it is not an equivalence, there must be objects in $\mathcal{A}(D(\mathbb{Z}))$ which map to zero. These objects can be identified with the α–phantom maps. See Lemma 6.5.6, Definition 6.5.7 and Remark 6.5.8. There is a non–zero object in $\mathcal{A}(D(\mathbb{Z}))$, that is a non–zero map in $D(\mathbb{Z})$

$$x \longrightarrow y,$$

so that the induced map

$$\mathcal{T}(-,x)|_{\mathcal{S}} \longrightarrow \mathcal{T}(-,y)|_{\mathcal{S}}$$

vanishes.

In the derived category $D(\mathbb{Z})$, any object x can be written as

$$x \quad = \quad \coprod_{i=-\infty}^{\infty} \Sigma^{-i}H^i(x) \quad = \quad \prod_{i=-\infty}^{\infty} \Sigma^{-i}H^i(x).$$

We are given a map $x \longrightarrow y$, that is a map

$$\coprod_{i=-\infty}^{\infty} \Sigma^{-i}H^i(x) \longrightarrow \prod_{i=-\infty}^{\infty} \Sigma^{-i}H^i(y).$$

A map from a coproduct to a product is given, by the universal properties of both coproduct and product, by a matrix of maps

$$\Sigma^{-i}H^i(x) \longrightarrow \Sigma^{-j}H^j(y).$$

Since the map $x \longrightarrow y$ is not zero, at least one of the components is not zero. Since the functor

$$A(D(\mathbb{Z})) \longrightarrow \mathcal{E}x\big(\mathcal{S}^{op}, \mathcal{A}b\big)$$

respects coproducts and products, it annihilates all the components. In other words, we deduce that there are two abelian groups C and A, a non–zero map

$$\Sigma^{-i}C \longrightarrow \Sigma^{-j}A,$$

so that

$$\mathcal{T}\big(-,\Sigma^{-i}C\big)\big|_{\mathcal{S}} \longrightarrow \mathcal{T}\big(-,\Sigma^{-j}A\big)\big|_{\mathcal{S}}$$

vanishes.

By suspending, we may assume $i = 0$. Since the map

$$C \longrightarrow \Sigma^{-j}A$$

is non–zero and $\mathbb{Z}$ is of projective dimension 1, we have $j = 0$ or -1. The case $j = 0$ is eliminated since there are no α–phantom maps of abelian groups $C \longrightarrow A$. To say that the map is phantom would assert that, for any abelian group G of cardinality $< \alpha$, the composite $G \longrightarrow C \longrightarrow A$ vanishes. Since $\alpha > \aleph_0$, we may choose $G = \mathbb{Z}$. But to say that

$$\mathbb{Z} \longrightarrow C \longrightarrow A$$

vanishes for every $\mathbb{Z} \longrightarrow C$ is to say that the map $C \longrightarrow A$ takes every element of C to $0 \in A$.

We deduce that there exists a non–zero α–phantom map

$$C \longrightarrow \Sigma A.$$

That is, the map is non–zero, but for any object $x \in D(\mathbb{Z})^\alpha$, the composite

$$x \longrightarrow C \longrightarrow \Sigma A$$

vanishes. In particular, if x is an abelian group of cardinality $< \alpha$, the composite vanishes.

Of course, a morphism in $D(\mathbb{Z})$

$$C \longrightarrow \Sigma A$$

is an extension of C by A; it corresponds to an exact sequence

$$0 \longrightarrow A \longrightarrow B \longrightarrow C \longrightarrow 0.$$

To say that the composite

$$x \longrightarrow C \longrightarrow \Sigma A$$

vanishes, is equivalent to asserting that when we pull back the extension via $x \longrightarrow C$, we get

$$\begin{array}{ccccccccc} 0 & \longrightarrow & A & \longrightarrow & B' & \longrightarrow & x & \longrightarrow & 0 \\ & & \downarrow & & \downarrow & & \downarrow & & \\ 0 & \longrightarrow & A & \longrightarrow & B & \longrightarrow & C & \longrightarrow & 0 \end{array}$$

and the extension

$$0 \longrightarrow A \longrightarrow B' \longrightarrow x \longrightarrow 0$$

is split. And this is true for all abelian groups x of cardinality $< \alpha$.

Now let s be any object of the generating set S for $K(\mathbb{Z})$. Write s as a complex

$$\longrightarrow x_0 \longrightarrow x_1 \longrightarrow x_2 \longrightarrow$$

By the choice of α, each of the groups x_i has cardinality $< \alpha$. Suppose we are given a map of chain complexes

$$\begin{array}{ccccccccccc} \longrightarrow & x_{-1} & \longrightarrow & x_0 & \longrightarrow & x_1 & \longrightarrow & x_2 & \longrightarrow & x_3 & \longrightarrow \\ & \downarrow & & \downarrow & & \downarrow & & \downarrow & & \downarrow & \\ \longrightarrow & 0 & \longrightarrow & A & \longrightarrow & B & \longrightarrow & C & \longrightarrow & 0 & \longrightarrow \end{array}$$

The map factors as

$$\begin{array}{ccccccc} \longrightarrow & x_0 & \longrightarrow & x_1 & \longrightarrow & x_2 & \longrightarrow \\ & \downarrow & & \downarrow & & \downarrow & \\ \longrightarrow & A & \longrightarrow & B' & \longrightarrow & x_2 & \longrightarrow \\ & \downarrow & & \downarrow & & \downarrow & \\ \longrightarrow & A & \longrightarrow & B & \longrightarrow & C & \longrightarrow \end{array}$$

where the diagram

$$\begin{array}{ccccccccc} 0 & \longrightarrow & A & \longrightarrow & B' & \longrightarrow & x_2 & \longrightarrow & 0 \\ & & \downarrow & & \downarrow & & \downarrow & & \\ 0 & \longrightarrow & A & \longrightarrow & B & \longrightarrow & C & \longrightarrow & 0 \end{array}$$

is obtained by pulling back the extension

$$0 \longrightarrow A \longrightarrow B \longrightarrow C \longrightarrow 0$$

along the map $x_2 \longrightarrow C$. Since the cardinality of x_2 is $< \alpha$, the above tells us that the sequence

$$0 \longrightarrow A \longrightarrow B' \longrightarrow x_2 \longrightarrow 0$$

is split exact. It is the zero object in $K(\mathbb{Z})$. The composite

$$\begin{array}{cccccccc} \longrightarrow & x_0 & \longrightarrow & x_1 & \longrightarrow & x_2 & \longrightarrow \\ & \downarrow & & \downarrow & & \downarrow & \\ \longrightarrow & A & \longrightarrow & B' & \longrightarrow & x_2 & \longrightarrow \\ & \downarrow & & \downarrow & & \downarrow & \\ \longrightarrow & A & \longrightarrow & B & \longrightarrow & C & \longrightarrow \end{array}$$

is therefore the zero map in $K(\mathbb{Z})$. We deduce that, for any object $s \in S$, any map

$$\begin{array}{ccccccccccc} \longrightarrow & x_{-1} & \longrightarrow & x_0 & \longrightarrow & x_1 & \longrightarrow & x_2 & \longrightarrow & x_3 & \longrightarrow \\ & \downarrow & & \downarrow & & \downarrow & & \downarrow & & \downarrow & \\ \longrightarrow & 0 & \longrightarrow & A & \longrightarrow & B & \longrightarrow & C & \longrightarrow & 0 & \longrightarrow \end{array}$$

must vanish. On the other hand,

$$\longrightarrow 0 \longrightarrow A \longrightarrow B \longrightarrow C \longrightarrow 0 \longrightarrow$$

is not the zero object in $K(\mathbb{Z})$, since the extension of C by A is not split. This contradicts the hypothesis that S generates. $\square$

SUMMARY E.3.3. In Lemma E.3.1, we proved that $K(\mathbb{Z})^{op}$ is not well–generated. In Lemma E.3.2, we saw that $K(\mathbb{Z})$ does not even have a generating set. Most definitely, it cannot be well–generated. Therefore $K(\mathbb{Z})$ is an example of a triangulated category satisfying [TR5] and [TR5*], so that neither $K(\mathbb{Z})^{op}$ nor $K(\mathbb{Z})$ is well–generated.

E.4. History of the results in Appendix E

The counterexamples of Appendix E are, to the best of the author's knowledge, all new. The only result in the literature that comes to mind is Boardman's [**2**]. Boardman proves that the category $\mathcal{T}$ of spectra is not self–dual. Since the category of spectra is $\aleph_0$–compactly generated, by Corollary E.1.3 the dual cannot be, and we are also able to deduce that the category $\mathcal{T}$ cannot be equivalent to its dual $\mathcal{T}^{op}$. In this sense, Corollary E.1.3 can be viewed as a generalisation of Boardman's theorem. We certainly prove that $\mathcal{T}$ and $\mathcal{T}^{op}$ are not equivalent. But we prove more; we in fact prove that $\mathcal{T}$ is not equivalent to $\mathcal{S}^{op}$, for any well–generated $\mathcal{S}$.

Bibliography

1. Alexander A. Beilinson, Joseph Bernstein, and Pierre Deligne, *Analyse et topologie sur les éspaces singuliers*, Astérisque, vol. 100, Soc. Math. France, 1982 (French).
2. J. M. Boardman, *Stable homotopy is not self-dual*, Proceedings Amer. Math. Soc. **26** (1970), 369–370.
3. Marcel Bökstedt and Amnon Neeman, *Homotopy limits in triangulated categories*, Compositio Math. **86** (1993), 209–234.
4. A.K. Bousfield, *The localization of spaces with respect to homology*, Topology **14** (1975), 133–150.
5. ———, *The boolean algebra of spectra*, Comm. Math. Helv. **54** (1979), 368–377.
6. ———, *The localization of spectra with respect to homology*, Topology **18** (1979), 257–281.
7. E. H. Brown, *Cohomology theories*, Annals of Math. **75** (1962), 467–484.
8. E. H. Brown and M. Comenetz, *Pontrjagin duality for generalized homology and cohomology*, Amer. J. Math. **98** (1976), 1–27.
9. J. Daniel Christensen and Neil P. Strickland, *Phantom maps and homology theories*, Topology **37** (1998), 339–364.
10. Pierre Deligne, *Cohomology à support propre en construction du foncteur* $f^!$, Residues and Duality, Lecture Notes in Mathematics, vol. 20, Springer–Verlag, 1966, pp. 404–421.
11. Jens Franke, *On the Brown representability theorem for triangulated categories*, (2000?), to appear in Topology.
12. Peter Freyd, *Abelian categories*, Harper, Heidelberg, 1966.
13. ———, *Stable homotopy*, Proc. Conf. Categorical Algebra, Springer–Verlag, 1966, pp. 121–172.
14. ———, *Homotopy is not concrete*, The Steenrod algebra and its applications, Lec. Notes in Math., vol. 168, Springer–Verlag, 1970, pp. 25–34.
15. Peter Gabriel, *Des catégories abéliennes*, Bull. Soc. Math. France **90** (1962), 323–448.
16. Peter Gabriel and Friedrich Ulmer, *Lokal präsentierbare Kategorien*, Lecture Notes in Mathematics, vol. 221, Springer–Verlag, 1971.
17. Jonathan S. Golan, *Torsion theories*, Pitman monographs and surveys in pure and appl. math., vol. 29, Longman Scientific & Technical, 1986.
18. Alexandre Grothendieck, *Sur quelques points d'algèbre homologique*, Tôhoku Math. J. **9** (1957), 119–221.
19. Robin Hartshorne, *Residues and duality*, Lecture Notes in Mathematics, vol. 20, Springer–Verlag, 1966.
20. Uwe Jannsen, *Continuous étale cohomology*, Math. Annalen **107** (1988), 207–245.
21. H. R. Margolis, *Spectra and the Steenrod algebra*, Elsevier Science Publishers B. V., 1983.

22. Amnon Neeman, *Some new axioms for triangulated categories*, J. Algebra **139** (1991), 221–255.
23. ———, *The connection between the K–theory localisation theorem of Thomason, Trobaugh and Yao, and the smashing subcategories of Bousfield and Ravenel*, Ann. Sci. École Normale Supérieure **25** (1992), 547–566.
24. ———, *Stable homotopy as a triangulated functor*, Inventiones Mathematicae **109** (1992), 17–40.
25. ———, *Brown representability for the dual*, Inventiones Mathematicae **133** (1998), 97–105.
26. Brian Parshall and Leonard L. Scott, *Derived categories, quasi-hereditary algebras, and algebraic groups*, Carlton U. Math. Notes **3** (1988), 1–104.
27. Dieter Puppe, *On the structure of stable homotopy theory*, Colloquium on algebraic topology, Aarhus Universitet Matematisk Institut, 1962, pp. 65–71.
28. Jeremy Rickard, *Derived categories and stable equivalence*, J. Pure and Appl. Algebra **61** (1989), 303–317.
29. Jan-Erik Roos, *Sur les foncteurs dérivés de* $\varprojlim$. *Applications*, Comptes Rendus Acad. Sci. Paris Ser. AB **252** (1961), 3702–3704 (French).
30. ———, *Derived functors of infinite products and projective objects in abelian categories*, (1962), unpublished.
31. ———, *Sur les foncteurs dérivés des produits infinis dans les catégories de Grothendieck. Exemples et contre–exemples*, Comptes Rend. Acad. Sci. Paris Ser. AB **263** (1966), 895–898 (French).
32. ———, *Sur la condition AB6 et ses variantes dans les catégories abéliennes*, Comptes Rend. Acad. Sci. Paris Ser. AB **264** (1967), 991–994 (French).
33. Bo Stenström, *Rings of quotients*, Grundlehren der math. Wissensch., vol. 217, Springer–Verlag, 1975.
34. Robert W. Thomason and Thomas F. Trobaugh, *Higher algebraic K–theory of schemes and of derived categories*, The Grothendieck Festschrift (a collection of papers to honor Grothendieck's 60'th birthday), vol. 3, Birkhäuser, 1990, pp. 247–435.
35. Jean-Louis Verdier, *Catégories dérivées, état 0*, SGA 4.5, Lec. Notes in Math., vol. 569, Springer–Verlag, 1977, pp. 262–308 (French).
36. ———, *Des catégories dérivées des catégories abeliennes*, Asterisque, vol. 239, Société Mathématique de France, 1996 (French).
37. Charles A. Weibel, *An introduction to homological algebra*, Cambridge Studies in Advanced Mathematics, vol. 38, Cambridge University Press, 1994.

Index